Springer Optimization and Its Applications

Volume 158

Aims and Scope

Optimization has continued to expand in all directions at an astonishing rate. New algorithmic and theoretical techniques are continually developing and the diffusion into other disciplines is proceeding at a rapid pace, with a spot light on machine learning, artificial intelligence, and quantum computing. Our knowledge of all aspects of the field has grown even more profound. At the same time, one of the most striking trends in optimization is the constantly increasing emphasis on the interdisciplinary nature of the field. Optimization has been a basic tool in areas not limited to applied mathematics, engineering, medicine, economics, computer science, operations research, and other sciences.

The series **Springer Optimization and Its Applications (SOIA)** aims to publish state-of-the-art expository works (monographs, contributed volumes, textbooks, handbooks) that focus on theory, methods, and applications of optimization. Topics covered include, but are not limited to, nonlinear optimization, combinatorial optimization, continuous optimization, stochastic optimization, Bayesian optimization, optimal control, discrete optimization, multi-objective optimization, and more. New to the series portfolio include Works at the intersection of optimization and machine learning, artificial intelligence, and quantum computing.

Volumes from this series are indexed by Web of Science, zbMATH, Mathematical Reviews, and SCOPUS.

More information about this series at http://www.springer.com/series/7393

Neculai Andrei

Nonlinear Conjugate Gradient Methods for Unconstrained Optimization

Springer

Neculai Andrei
Center for Advanced Modeling
and Optimization
Academy of Romanian Scientists
Bucharest, Romania

ISSN 1931-6828 ISSN 1931-6836 (electronic)
Springer Optimization and Its Applications
ISBN 978-3-030-42949-2 ISBN 978-3-030-42950-8 (eBook)
https://doi.org/10.1007/978-3-030-42950-8

Mathematics Subject Classification (2010): 49M37, 65K05, 90C30, 90C06, 90C90

This Springer imprint is published by the registered company Springer Nature Switzerland AG
The registered company address is: Gewerbestrasse 11, 6330 Cham, Switzerland

Preface

This book is on conjugate gradient methods for unconstrained optimization. The concept of conjugacy was introduced by Magnus Hestenes and Garrett Birkhoff in 1936 in the context of the variational theory. The history of conjugate gradient methods, surveyed by Golub and O'Leary (1989), began with the research studies of Cornelius Lanczos, Magnus Hestenes, George Forsythe, Theodore Motzkin, Barkley Rosser, and others at the Institute for Numerical Analysis as well as with the independent research of Eduard Steifel at Eidgenössische Technische Hochschule, Zürich. The first presentation of conjugate direction algorithms seems to be that of Fox, Huskey, and Wilkinson (1948), who considered them as direct methods, and of Forsythe, Hestenes, and Rosser (1951), Hestenes and Stiefel (1952), and Rosser (1953). The landmark paper published by Hestenes and Stiefel in 1952 presented both the method of the linear conjugate gradient and the conjugate direction methods, including conjugate Gram–Schmidt processes for solving symmetric, positive definite linear algebraic systems. A closely related algorithm was proposed by Lanczos (1952), who worked on algorithms for determining the eigenvalues of a matrix (Lanczos, 1950). His iterative algorithm yielded the similarity transformation of a matrix into the tridiagonal form which the eigenvalues can be well approximated. Hestenes, who worked on iterative methods for solving linear systems (Hestenes, 1951, 1955), was also interested in the Gram–Schmidt process for finding conjugate diameters of an ellipsoid. He was interested in developing a general theory of quadratic forms in Hilbert space (Hestenes, 1956a, 1956b). Initially, the linear conjugate gradient algorithm was called the Hestenes–Stiefel–Lanczos method (Golub & O'Leary, 1989).

The initial numerical experience with conjugate gradient algorithms was not very encouraging. Although widely used in the 1960s, their application to ill-conditioned problems gave rather poor results. At that time, preconditioning techniques were not well understood. They were developed in the 1970s together with methods intended for large sparse linear systems; these methods were prompted by the paper of Reid (1971), who reinforced them by showing their potential as iterative methods for sparse linear systems. Although Hestenes and Stiefel stated their algorithm for sets of linear systems of equations with positive definite matrices, from the beginning it was viewed as an optimization technique for minimizing quadratic functions. In the 1960s, conjugate gradient and conjugate direction methods were extended to the optimization of nonquadratic functions. The first algorithm for nonconvex problems was proposed by Feder (1962), who suggested using conjugate gradient algorithms for solving some problems in optics. The algorithms and the convergence study of several versions of conjugate gradient algorithms for nonquadratic functions were discussed by Fletcher and Reeves (1964), Polak and Ribière (1969), and Polyak (1969).

It is interesting to see that the work of Davidon (1959) on variable metric algorithms was followed by that of Fletcher and Powell (1963). Other variants of these methods were established by Broyden (1970), Fletcher (1970), Goldfarb (1970), and Shanno (1970), who established one of the most effective techniques for minimizing nonquadratic functions—the BFGS method. The main idea behind variable metric methods is the construction of a sequence of matrices to approximate the Hessian matrix (or its inverse) by applying a sequence of rank-one (or rank-two) update formulae. Details on the BFGS method can be found in the landmark papers of Dennis and Moré (1974, 1977). When applied to a quadratic function and under an exact evaluation to the Hessian, these methods give a solution in a finite number of iterates, and they are exactly conjugate gradient methods. Variable metric approximations to the Hessian matrix are dense matrices, and therefore, they are not suitable for large-scale problems, i.e., problems with many variables. However, the work of Nocedal (1980) on limited-memory quasi-Newton methods which use a variable metric updating procedure but within a prespecified memory storage enlarged the applicability of quasi-Newton methods. At the same time, the introduction of the inexact (truncated) Newton method by Dembo, Eisenstat, and Steihaug (1982) and its development by Nash (1985), and by Schlick and Fogelson (1992a, 1992b) gave the possibility of solving large-scale unconstrained optimization problems. The idea behind the inexact Newton method was that far away from a local minimum, it is not necessary to spend too much time computing an accurate Newton search vector. It is better to approximate the solution of the Newton system for the search direction computation. The limited-memory quasi-Newton and the truncated Newton are reliable methods, able to solve large-scale unconstrained optimization problems. However, as it is to be seen, there is a close connection between the conjugate gradient and the quasi-Newton methods. Actually, conjugate gradient methods are precisely the BFGS quasi-Newton method, where the approximation to the inverse Hessian of the minimizing function is restarted as the identity matrix at every iteration. The developments of the conjugate gradient methods subject both to the search direction and to the stepsize computation yielded algorithms and the corresponding reliable software with better numerical performances than the limited-memory quasi-Newton or inexact Newton methods.

The book is structured into 12 chapters. Chapter 1 has an introductory character by presenting the optimality conditions for unconstrained optimization and a thorough description and the properties of the main methods for unconstrained optimization (steepest descent, Newton, quasi-Newton, modifications of the BFGS method, quasi-Newton methods with diagonal updating of the Hessian, limited-memory quasi-Newton methods, truncated Newton, conjugate gradient, and trust-region methods). *It is common knowledge that the final test of a theory is its capacity to solve the problems which originated it.* Therefore, in this chapter a collection of 80 unconstrained optimization test problems with different structures and complexities, as well as five large-scale applications from the MINPACK-2 collection for testing the numerical performances of the algorithms described in this book, is presented. Some problems from this collection are quadratic, and some others are highly nonlinear. For some problems, the Hessian has a block-diagonal structure, for others it has a banded structure with small bandwidth. There are problems with sparse or dense Hessian. In Chapter 2, the linear conjugate gradient algorithm is detailed. The general convergence results for conjugate gradient methods are assembled in Chapter 3. The purpose is to put together the main convergence results both for conjugate gradient methods with standard Wolfe line search and for conjugate gradient methods with strong Wolfe line search. Since the search direction depends on a parameter, the conditions on this parameter which ensure the convergence of the algorithm are detailed. The global convergence results of conjugate gradient algorithms presented in this chapter follow from the conditions given by Zoutendijk and by Nocedal under classical assumptions. The remaining chapters are dedicated to the nonlinear conjugate gradient methods for unconstrained optimization, insisting both on the theoretical aspects of their convergence and on their numerical performances for solving large-scale problems and applications.

Plenty of nonlinear conjugate gradient methods are known. The difference among them is twofold: the way in which the search direction is updated and the procedure for the stepsize computation along this direction. The main requirement of the search direction of the conjugate gradient methods is to satisfy the descent or the sufficient descent condition. The stepsize is computed by using the Wolfe line search conditions or some variants of them. In a broad sense, the conjugate gradient algorithms may be classified as standard, hybrid, modifications of the standard conjugate gradient algorithms, memoryless BFGS preconditioned, three-term conjugate gradient algorithms, and others.

The most important standard conjugate gradient methods discussed in Chapter 4 are: Hestenes–Stiefel, Fletcher–Reeves, Polak–Ribière–Polyak, conjugate descent of Fletcher, Liu–Storey, and Dai–Yuan. If the minimizing function is strongly convex quadratic and the line search is exact, then, in theory, all choices for the search direction in standard conjugate gradient algorithms are equivalent. However, for nonquadratic functions, each choice of the search direction leads to standard conjugate gradient algorithms with very different performances. An important ingredient in conjugate gradient algorithms is the acceleration, discussed in Chapter 5.

Hybrid conjugate gradient algorithms presented in Chapter 6 try to combine the standard conjugate gradient methods in order to exploit the attractive features of each one. To obtain hybrid conjugate gradient algorithms, the standard schemes may be combined in two different ways. The first combination is based on the projection concept. The idea of these methods is to consider a pair of standard conjugate gradient methods and use one of them when a criterion is satisfied. As soon as the criterion has been violated, then the other standard conjugate gradient from the pair is used. The second class of the hybrid conjugate gradient methods is based on the convex combination of the standard methods. This idea of these methods is to choose a pair of standard methods and to combine them in a convex way, where the parameter in the convex combination is computed by using the conjugacy condition or the Newton search direction. In general, the hybrid methods based on the convex combination of the standard schemes outperform the hybrid methods based on the projection concept. The hybrid methods are more efficient and more robust than the standard ones.

An important class of conjugate gradient algorithms discussed in Chapter 7 is obtained by modifying the standard algorithms. Any standard conjugate gradient algorithm may be modified in such a way that the corresponding search direction is descent, and the numerical performances are improved. In this area of research, only some modifications of the Hestenes–Stifel standard conjugate gradient algorithm are presented. Today's best-performing conjugate gradient algorithms are the modifications of the Hestenes–Stiefel conjugate gradient algorithm: CG-DESCENT of Hager and Zhang (2005) and DESCON of Andrei (2013c). CG-DESCENT is a conjugate gradient algorithm with guaranteed descent. In fact, CG-DESCENT can be viewed as an adaptive version of the Dai and Liao conjugate gradient algorithm with a special value for its parameter. The search direction of CG-DESCENT is related to the memoryless quasi-Newton direction of Perry–Shanno. DESCON is a conjugate gradient algorithm with guaranteed descent and conjugacy conditions and with a modified Wolfe line search. Mainly, it is a modification of the Hestenes–Stiefel conjugate gradient algorithm. In CG-DESCENT, the stepsize is computed by using the standard Wolfe line search or an approximate Wolfe line search introduced by Hager and Zhang (2005, 2006a, 2006b), which is responsible for the high performances of the algorithm. In DESCON, the stepsize is computed by using the modified Wolfe line search introduced by Andrei (2013c), in which the parameter in the curvature condition of the Wolfe line search is adaptively modified at every iteration. Besides, DESCON is equipped with an acceleration scheme which improves its performances.

The first connection between the conjugate gradient algorithms and the quasi-Newton ones was presented by Perry (1976), who expressed the Hestenes–Stiefel search direction as a matrix multiplying the negative gradient. Later on, Shanno (1978a) showed that the conjugate gradient methods are exactly the BFGS quasi-Newton methods, where the approximation to the inverse Hessian is restarted as the identity matrix at every iteration. In other words, conjugate gradient methods are memoryless quasi-Newton methods. This was the starting point of a very prolific research area of memoryless quasi-Newton conjugate gradient methods, which is discussed in Chapter 8. The point was how the second-order information of the minimizing function should be introduced in the formula for updating the search direction. Using this idea to include the curvature of the minimizing function in the search direction computation, Shanno (1983) elaborated CONMIN as the first conjugate gradient algorithm memoryless BFGS preconditioned. Later on, by using a combination of the scaled memoryless BFGS method and the preconditioning, Andrei (2007a, 2007b, 2007c, 2008a) elaborated SCALCG as a double-quasi-Newton update scheme. Dai and Kou (2013) elaborated the CGOPT algorithm as a family of conjugate gradient methods based on the self-scaling memoryless BFGS method in which the search direction is computed in a one-dimensional manifold. The search direction in CGOPT is chosen to be closest to the Perry–Shanno direction. The stepsize in CGOPT is computed by using an improved Wolfe line search introduced by Dai and Kou (2013). CGOPT with improved Wolfe line search and a special restart condition is one of the best conjugate gradient algorithms. New conjugate gradient algorithms based on the self-scaling memoryless BFGS updating using the determinant or the trace of the iteration matrix or the measure function of Byrd and Nocedal are presented in this chapter.

Beale (1972) and Nazareth (1977) introduced the three-term conjugate gradient methods, presented, and analyzed in Chapter 9. The convergence rate of the conjugate gradient method may be improved from linear to n-step quadratic if the method is restarted with the negative gradient direction at every n iterations. One such restart technique was proposed by Beale (1972). In his restarting procedure, the restart direction is a combination of the negative gradient and the previous search direction which includes the second-order derivative information achieved by searching along the previous direction. Thus, a three-term conjugate gradient was obtained. In order to achieve finite convergence for an arbitrary initial search direction, Nazareth (1977) proposed a conjugate gradient method in which the search direction has three terms. Plenty of three-term conjugate gradient algorithms are known. This chapter presents only the three-term conjugate gradient with descent and conjugacy conditions, the three-term conjugate gradient method with subspace minimization, and the three-term conjugate gradient method with minimization of one-parameter quadratic model of the minimizing function. The three-term conjugate gradient concept is an interesting innovation. However, the numerical performances of these algorithms are modest.

Preconditioning of the conjugate gradient algorithms is presented in Chapter 10. This is a technique for accelerating the convergence of algorithms. In fact, preconditioning was used in the previous chapters as well, but it is here where the proper preconditioning by a change of variables which improves the eigenvalues distribution of the iteration matrix is emphasized.

Some other conjugate gradient methods, like those based on clustering the eigenvalues of the iteration matrix or on minimizing the condition number of this matrix, including the methods with guaranteed descent and conjugacy conditions are presented in Chapter 11. Clustering the eigenvalues of the iteration matrix and minimizing its condition number are two important approaches to basically pursue similar ideas for improving the performances of the corresponding conjugate gradient algorithms. However, the approximations of the Hessian used in these algorithms play a crucial role in capturing the curvature of the minimizing function. The methods with clustering the eigenvalues or minimizing the condition number of the iteration matrix are very close to those based on memoryless BFGS preconditioned, the best ones in this class, but they are strongly dependent on the approximation of the Hessian used in the search direction definition. The methods in which both the sufficient descent and the conjugacy conditions are satisfied do not perform very well. Apart from these two conditions, some additional ingredients are necessary for them to perform better. This chapter also focuses on some combinations between the conjugate gradient algorithm satisfying the sufficient descent and the conjugacy conditions and the limited-memory BFGS algorithms. Finally, the limited-memory L-BFGS preconditioned conjugate gradient algorithm (L-CG-DESCENT) of Hager and Zhang (2013) and the subspace minimization conjugate gradient algorithms based on cubic regularization (Zhao, Liu, & Liu, 2019) are discussed.

The last chapter details some discussions and conclusions on the conjugate gradient methods presented in this book, insisting on the performances of the algorithms for solving large-scale applications from MINPACK-2 collection (Averick, Carter, Moré, & Xue, 1992) up to 250,000 variables.

Optimization algorithms, particularly the conjugate gradient ones, involve some advanced mathematical concepts used in defining them and in proving their convergence and complexity. Therefore, Appendix A contains some key elements from: linear algebra, real analysis, functional analysis, and convexity. The readers are recommended to go through this appendix first. Appendix B presents the algebraic expression of 80 unconstrained optimization problems, included in the UOP collection, used for testing the performances of the algorithms described in this book.

The reader will find a well-organized book, written at an accessible level and presenting in a rigorous and friendly manner the recent theoretical developments of conjugate gradient methods for unconstrained optimization, computational results, and performances of algorithms for solving a large class of unconstrained optimization problems with different structures and complexities as well as performances and behavior of algorithms for solving large-scale unconstrained optimization engineering applications. A great deal of attention has been given to the computational performances and numerical results of these algorithms and comparisons for solving unconstrained optimization problems and large-scale applications. Plenty of Dolan and Moré (2002) performance profiles which illustrate the behavior of the algorithms have been given. Basically, the main purpose of the book has been to establish the computational power of the most known conjugate gradient algorithms for solving large-scale and complex unconstrained optimization problems.

The book is an invitation for researchers working in the unconstrained optimization area to understand, learn, and develop new conjugate gradient algorithms with better properties. It is of great interests to all those interested in developing and using new advanced techniques for solving unconstrained optimization complex problems. Mathematical programming researchers, theoreticians, and practitioners in operations research, practitioners in engineering and industry researchers as well as graduate students in mathematics, Ph.D., and master students in mathematical programming will find plenty of information and practical aspects for solving large-scale unconstrained optimization problems and applications by conjugate gradient methods.

I am grateful to the **Alexander von Humboldt Foundation** for its appreciation and generous financial support during the 2+ years at different universities in Germany. My thanks also go to Elizabeth Loew and to all the staff of Springer, for their encouragement, competent, and superb assistance with the preparation of this book. Finally, my deepest thanks go to my wife, Mihaela, for her constant understanding and support along the years.

Tohăniţa / Bran Resort, Neculai Andrei
Bucharest, Romania
January 2020

Contents

List of Figures

List of Tables

List of Algorithms

Chapter 1
Introduction: Overview of Unconstrained Optimization

Unconstrained optimization consists of minimizing a function which depends on a number of real variables without any restrictions on the values of these variables. When the number of variables is large, this problem becomes quite challenging. The most important gradient methods for solving unconstrained optimization problems are described in this chapter. These methods are iterative. They start with an initial guess of the variables and generate a sequence of improved estimates until they terminate with a set of values for variables. For checking that this set of values of variables is indeed the solution of the problem, the optimality conditions should be used. If the optimality conditions are not satisfied, they may be used to improve the current estimate of the solution. The algorithms described in this book make use of the values of the minimizing function, of the first and possibly of the second derivatives of this function. The following unconstrained optimization methods are mainly described: steepest descent, Newton, quasi-Newton, limited-memory quasi-Newton, truncated Newton, conjugate gradient and trust-region.

1.1 The Problem

In this book, the following unconstrained optimization problem

$$\min_{x \in \mathbb{R}^n} f(x) \tag{1.1}$$

is considered, where $f : \mathbb{R}^n \to \mathbb{R}$ is a real-valued function f of n variables, smooth enough on $\mathbb{R}^n$. The interest is in finding a *local minimizer* of this function, that is a point x^*, so that

$$f(x^*) \leq f(x) \text{ for all } x \text{ near } x^*. \tag{1.2}$$

N. Andrei, *Nonlinear Conjugate Gradient Methods for Unconstrained Optimization*, Springer Optimization and Its Applications 158,
https://doi.org/10.1007/978-3-030-42950-8_1

If $f(x^*) < f(x)$ for all x near x^*, then x^* is called a *strict local minimizer* of function f. Often, f is referred to as the *objective function,* while $f(x^*)$ as the *minimum* or the *minimum value*.

The local minimization problem is different from the *global minimization problem*, where a global minimizer, i.e., a point x^* so that

$$f(x^*) \le f(x) \text{ for all } x \in \mathbb{R}^n \tag{1.3}$$

is sought. This book deals with only the local minimization problems.

The function f in (1.1) may have any algebraic expression and we suppose that it is twice continuously differentiable on $\mathbb{R}^n$. Denote $\nabla f(x)$ as the gradient of f and $\nabla^2 f(x)$ its Hessian.

For solving (1.1), plenty of methods are known see: Luenberger (1973), (1984), Gill, Murray, and Wright (1981), Bazaraa, Sherali, and Shetty (1993), Bertsekas (1999), Nocedal and Wright (2006), Sun and Yuan (2006), Bartholomew-Biggs (2008), Andrei (1999), (2009e), (2015b). In general, for solving (1.1) the unconstrained optimization methods implement one of the following two strategies: the *line search* and the *trust-region*. Both these strategies are used for solving (1.1).

In the *line search strategy*, the corresponding algorithm chooses a direction d_k and searches along this direction from the current iterate x_k for a new iterate with a lower function value. Specifically, starting with an initial point x_0, the iterations are generated as:

$$x_{k+1} = x_k + \alpha_k d_k, \quad k = 0, 1, \ldots, \tag{1.4}$$

where $d_k \in \mathbb{R}^n$ is the *search direction* along which the values of function f are reduced and $\alpha_k \in \mathbb{R}$ is the *stepsize* determined by a line search procedure. The main requirement is that the search direction d_k, at iteration k should be a *descent direction.* In Section 1.3, it is proved that the algebraic characterization of descent directions is that

$$d_k^T g_k < 0, \tag{1.5}$$

which is a very important criterion concerning the effectiveness of an algorithm. In (1.5), $g_k = \nabla f(x_k)$ is the gradient of f in point x_k. In order to guarantee the global convergence, sometimes it is required that the search direction d_k satisfy the *sufficient descent* condition

$$g_k^T d_k \le -c\|g_k\|^2, \tag{1.6}$$

where c is a positive constant.

In the *trust-region strategy*, the idea is to use the information gathered about the minimizing function f to construct a model function m_k whose behavior near the current point x_k is similar to that of the actual objective function f. In other words, the step p is determined by approximately solving the following subproblem

$$\min_p m_k(x_k + p), \tag{1.7}$$

where the point $x_k + p$ lies inside the trust region. If the step p does not produce a sufficient reduction of the function values, then it follows that the trust-region is too large. In this case, the trust-region is shrinked and the model m_k in (1.7) is re-solved. Usually, the trust-region is a ball defined by $\|p\|_2 \leq \Delta$, where the scalar Δ is known as the *trust-region radius*. Of course, elliptical and box-shaped trust regions may be used.

Usually, the model m_k in (1.7) is defined as a quadratic approximation of the minimizing function f:

$$m_k(x_k + p) = f(x_k) + p^T \nabla f(x_k) + \frac{1}{2} p^T B_k p, \tag{1.8}$$

where B_k is either the Hessian $\nabla^2 f(x_k)$ or an approximation to it. Observe that each time when the size of the trust-region, i.e., the trust-region radius, is reduced after a failure of the current iterate, then the step from x_k to the new point will be shorter and usually points in a different direction from the previous point.

As a comparison, the line search and trust-region differ in the order in which they choose the *search direction* and the *stepsize* to move to the next iterate. Line search starts with a direction d_k and then determine an appropriate distance along this direction, namely the stepsize α_k. In trust-region, firstly the maximum distance is chosen, that is the trust-region radius Δ_k, and then a direction and a step p_k that determine the best improvement of the function values subject to this distance constraint is determined. If this step is not satisfactory, then the distance measure Δ_k is reduced and the process is repeated.

For the search direction computation, there is a large variety of methods. Some of the most important will be discussed in this chapter. For the moment, let us discuss the main procedures for stepsize determination in the frame of line search strategy for unconstrained optimization. After that an overview of the unconstrained optimization methods will be presented.

1.2 Line Search

Suppose that the minimizing function f is enough smooth on $\mathbb{R}^n$. Concerning the stepsize α_k which have to be used in (1.4), the greatest reduction of the function values is achieved when the *exact line search* is used, in which

$$\alpha_k = \arg\min_{\alpha \geq 0} f(x_k + \alpha d_k). \tag{1.9}$$

In other words, the exact line search determines a stepsize α_k as solution of the equation

$$\nabla f(x_k + \alpha_k d_k)^T d_k = 0. \tag{1.10}$$

However, being impractical, the exact line search is rarely used in optimization algorithms. Instead, an *inexact line search* is often used. Plenty of inexact line search methods were proposed: Goldstein (1965), Armijo (1966), Wolfe (1969, 1971), Powell (1976a), Lemaréchal (1981), Shanno (1983), Dennis and Schnabel (1983), Al-Baali and Fletcher (1984), Hager (1989), Moré and Thuente (1990), Lukšan (1992), Potra and Shi (1995), Hager and Zhang (2005), Gu and Mo (2008), Ou and Liu (2017), and many others. The challenges in finding a good stepsize α_k by inexact line search are both in avoiding that the stepsize is too long or too short. Therefore, the inexact line search methods concentrate on: a good initial selection of stepsize, criteria that assures that α_k are neither too long nor too short and construction of a sequence of updates that satisfies the above requirements.

Generally, the inexact line search procedures are based on quadratic or cubic polynomial interpolations of the values of the one dimensional function $\varphi_k(\alpha) = f(x_k + \alpha d_k)$, $\alpha \geq 0$. For minimizing the polynomial approximation of $\varphi_k(\alpha)$, the inexact line search procedures generate a sequence of stepsizes until one of these values of the stepsize satisfies some stopping conditions.

Backtracking—Armijo line search

One of the very simple and efficient line search procedure is particularly the *backtracking line search* (Ortega & Rheinboldt, 1970). This procedure considers the following scalars: $0 < \gamma < 1$, $0 < \beta < 1$ and $s_k = -g_k^T d_k / \|g_k\|^2$ and takes the following steps based on the Armijo's rule:

Algorithm 1.1 Backtracking-Armijo line search

1.	Consider the descent direction d_k for f at x_k. Set $\alpha = s_k$
2.	While $f(x_k + \alpha d_k) > f(x_k) + \gamma \alpha g_k^T d_k$, set $\alpha = \alpha\beta$
3.	Set $\alpha_k = \alpha$ ♦

Observe that this line search requires that the achieved reduction in f be at least a fixed fraction γ of the reduction promised by the first-order Taylor approximation of f at x_k. Typically, $\gamma = 0.0001$ and $\beta = 0.8$, meaning that a small portion of the decrease predicted by the linear approximation of f at the current point is accepted. Observe that, when $d_k = -g_k$, then $s_k = 1$.

Theorem 1.1 (*Termination of backtracking Armijo*) *Let f be continuously differentiable with gradient $g(x)$ Lipschitz continuous with constant $L > 0$, i.e., $\|g(x) - g(y)\| \le L\|x - y\|$, for any x, y from the level set $S = \{x : f(x) \le f(x_0)\}$. Let d_k be a descent direction at x_k, i.e., $g_k^T d_k < 0$. Then for fixed $\gamma \in (0, 1)$:*

1. *The Armijo condition $f(x_k + \alpha d_k) \le f(x_k) + \gamma \alpha g_k^T d_k$ is satisfied for all $\alpha \in [0, \alpha_k^{\max}]$, where*

$$\alpha_k^{\max} = \frac{2(\gamma - 1)g_k^T d_k}{L\|d_k\|_2^2},$$

2. *For fixed $\tau \in (0, 1)$ the stepsize generated by the backtracking-Armijo line search terminates with*

$$\alpha_k \ge \min\left\{ \alpha_k^0, \frac{2\tau(\gamma - 1)g_k^T d_k}{L\|d_k\|_2^2} \right\},$$

where α_k^0 is the initial stepsize at iteration k. ♦

Observe that in practice the Lipschitz constant L is unknown. Therefore, $\alpha_k^{\max}$ and α_k cannot simply be computed via the explicit formulae given by the Theorem 1.1.

Goldstein line search
One inexact line search is given by Goldstein (1965), where α_k is determined to satisfy the conditions:

$$\delta_1 \alpha_k g_k^T d_k \le f(x_k + \alpha_k d_k) - f(x_k) \le \delta_2 \alpha_k g_k^T d_k, \tag{1.11}$$

where $0 < \delta_2 < 1/2 < \delta_1 < 1$.

Wolfe line search
The most used line search conditions for the stepsize determination are the so called *standard Wolfe line search conditions* (Wolfe, 1969, 1971):

$$f(x_k + \alpha_k d_k) \le f(x_k) + \rho \alpha_k d_k^T g_k, \tag{1.12}$$

$$\nabla f(x_k + \alpha_k d_k)^T d_k \ge \sigma d_k^T g_k, \tag{1.13}$$

where $0 < \rho < \sigma < 1$. The first condition (1.12), called the *Armijo condition*, ensures a *sufficient reduction* of the objective function value, while the second condition (1.13), called the *curvature condition*, ensures unacceptable short stepsizes. It is worth mentioning that a stepsize computed by the Wolfe line search conditions (1.12) and (1.13) may not be sufficiently close to a minimizer of $\varphi_k(\alpha)$. In these situations, the *strong Wolfe line search conditions* may be used, which consist of (1.12), and, instead of (1.13), the following strengthened version

$$\left|\nabla f(x_k + \alpha_k d_k)^T d_k\right| \leq -\sigma d_k^T g_k \tag{1.14}$$

is used. From (1.14), we see that if $\sigma \to 0$, then the stepsize which satisfies (1.12) and (1.14) tends to be the optimal stepsize. Observe that if a stepsize α_k satisfies the strong Wolfe line search, then it satisfies the standard Wolfe conditions.

Proposition 1.1 *Suppose that the function f is continuously differentiable. Let d_k be a descent direction at point x_k and assume that f is bounded from below along the ray $\{x_k + \alpha d_k : \alpha > 0\}$. Then, if $0 < \rho < \sigma < 1$, there exists an interval of stepsizes α satisfying the Wolfe conditions* and *the strong Wolfe conditions.*

Proof Since $\varphi_k(\alpha) = f(x_k + \alpha d_k)$ is bounded from below for all $\alpha > 0$, the line $l(\alpha) = f(x_k) + \alpha\rho\nabla f(x_k)^T d_k$ must intersect the graph of φ at least once. Let $\alpha' > 0$ be the smallest intersection value of α, i.e.,

$$f(x_k + \alpha' d_k) = f(x_k) + \alpha'\rho\nabla f(x_k)^T d_k < f(x_k) + \rho\nabla f(x_k)^T d_k. \tag{1.15}$$

Hence, a sufficient decrease holds for all $0 < \alpha < \alpha'$.
Now, by the mean value theorem, there exists $\alpha'' \in (0, \alpha')$ so that

$$f(x_k + \alpha' d_k) - f(x_k) = \alpha' \ \nabla f(x_k + \alpha'' d_k)^T d_k. \tag{1.16}$$

Since $\rho < \sigma$ and $\nabla f(x_k)^T d_k < 0$, from (1.15) and (1.16) we get

$$\nabla f(x_k + \alpha'' d_k)^T d_k = \rho\nabla f(x_k)^T d_k > \sigma\nabla f(x_k)^T d_k. \tag{1.17}$$

Therefore, α'' satisfies the Wolfe line search conditions (1.12) and (1.13) and the inequalities are strict. By smoothness assumption on f, there is an interval around α'' for which the Wolfe conditions hold. Since $\nabla f(x_k + \alpha'' d_k)^T d_k < 0$, it follows that the strong Wolfe line search conditions (1.12) and (1.14) hold in the same interval. ♦

Proposition 1.2 *Suppose that d_k is a descent direction and ∇f satisfies the Lipschitz condition*

$$\|\nabla f(x) - \nabla f(x_k)\| \leq L\|x - x_k\|$$

for all x on the line segment connecting x_k and x_{k+1}, where L is a constant. If the line search satisfies the Goldstein conditions, then

$$\alpha_k \geq \frac{1 - \delta_1}{L} \frac{\left|g_k^T d_k\right|}{\|d_k\|^2}. \tag{1.18}$$

If the line search satisfies the standard Wolfe conditions, then

$$\alpha_k \geq \frac{1-\sigma}{L} \frac{\left|g_k^T d_k\right|}{\|d_k\|^2}. \tag{1.19}$$

Proof If the Goldstein conditions hold, then by (1.11) and the mean value theorem we have

$$\begin{aligned}\delta_1 \alpha_k g_k^T d_k &\leq f(x_k + \alpha_k d_k) - f(x_k) \\ &= \alpha_k \nabla f(x_k + \xi d_k)^T d_k \\ &\leq \alpha_k g_k^T d_k + L\alpha_k^2 \|d_k\|^2,\end{aligned}$$

where $\xi \in [0, \alpha_k]$. From the above inequality, we get (1.18).

Subtracting $g_k^T d_k$ from both sides of (1.13) and using the Lipschitz condition, it follows that

$$(\sigma - 1) g_k^T d_k \leq (g_{k+1} - g_k)^T d_k \leq \alpha_k L \|d_k\|^2.$$

But d_k is a descent direction and $\sigma < 1$, therefore (1.19) follows from the above inequality. ♦

A detailed presentation and a safeguarded Fortran implementation of the Wolfe line search (1.12) and (1.13) with cubic interpolation is given in Chapter 5.

Generalized Wolfe line search

In the *generalized Wolfe line search*, the absolute value in (1.14) is replaced by a pair of inequalities:

$$\sigma_1 d_k^T g_k \leq d_k^T g_{k+1} \leq -\sigma_2 d_k^T g_k, \tag{1.20}$$

where $0 < \rho < \sigma_1 < 1$ and $\sigma_2 \geq 0$. The particular case in which $\sigma_1 = \sigma_2 = \sigma$ corresponds to the strong Wolfe line search.

Hager-Zhang line search

Hager and Zhang (2005) introduced the *approximate Wolfe line search*

$$\sigma d_k^T g_k \leq d_k^T g_{k+1} \leq (2\rho - 1) d_k^T g_k, \tag{1.21}$$

where $0 < \rho < 1/2$ and $\rho < \sigma < 1$. Observe that the approximate Wolfe line search (1.21) has the same form as the generalized Wolfe line search (1.20), but with a special choice for σ_2. The first inequality in (1.21) is the same as (1.13). When f is quadratic, the second inequality in (1.21) is equivalent to (1.12).

In general, when $\varphi_k(\alpha) = f(x_k + \alpha d_k)$ is replaced by a quadratic interpolating $q(.)$ that matches $\varphi_k(\alpha)$ at $\alpha = 0$ and $\varphi_k'(\alpha)$ at $\alpha = 0$ and $\alpha = \alpha_k$, (1.12) reduces to the second inequality in (1.21). Observe that the decay condition (1.12) is a component of the generalized Wolfe line search, while in the approximate Wolfe line search the decay condition is approximately enforced through the second inequality in (1.21). As shown by Hager and Zhang (2005), the first Wolfe condition (1.12) limits the accuracy of a conjugate gradient method to the order of the square root of the machine precision, while with the approximate Wolfe line search, we can achieve accuracy to the order of the machine precision.

The approximate Wolfe line search is based on the derivative of $\varphi_k(\alpha)$. This can be achieved by using a quadratic approximation of φ_k. The quadratic interpolating polynomial q that matches $\varphi_k(\alpha)$ at $\alpha = 0$ and $\varphi'(\alpha)$ at $\alpha = 0$ and $\alpha = \alpha_k$ (which is unknown) is given by

$$q(\alpha) = \varphi_k(0) + \varphi_k'(0)\alpha + \frac{\varphi_k'(\alpha_k) - \varphi_k'(0)}{2\alpha_k}\alpha^2.$$

Observe that the first Wolfe condition (1.12) can be written as $\varphi_k(\alpha_k) \leq \varphi_k(0) + \rho\alpha_k\varphi_k'(0)$. Now, if φ_k is replaced by q in the first Wolfe condition, we get $q(\alpha_k) \leq q(0) + \rho q'(\alpha_k)$, which is rewritten as

$$\frac{\varphi_k'(\alpha_k) - \varphi_k'(0)}{2}\alpha_k + \varphi_k'(0)\alpha_k \leq \rho\alpha_k\varphi_k'(0),$$

and can be restated as

$$\varphi_k'(\alpha_k) \leq (2\rho - 1)\varphi_k'(0), \tag{1.22}$$

where $\rho < \min\{0.5, \sigma\}$, which is exactly the second inequality in (1.21).

In terms of function $\varphi_k(.)$, the approximate line search aims at finding the stepsize α_k which satisfies the Wolfe conditions:

$$\varphi_k(\alpha) \leq \varphi_k(0) + \rho\varphi_k'(0)\alpha, \text{ and } \varphi_k'(\alpha) \geq \sigma\varphi_k'(0), \tag{1.23}$$

which are called *LS1 conditions*, or the conditions (1.22) together with

$$\varphi_k(\alpha) \leq \varphi_k(0) + \varepsilon_k, \text{ and } \varepsilon_k = \varepsilon|f(x_k)|, \tag{1.24}$$

where ε is a small positive parameter ($\varepsilon = 10^{-6}$), which are called *LS2 conditions*. ε_k is an estimate for the error in the value of f at iteration k. With these, the approximate Wolfe line search algorithm is as follows:

Algorithm 1.2 Hager and Zhang line search

1.	Choose an initial interval $[a_0, b_0]$ and set $k = 0$
2.	If either LS1 or LS2 conditions are satisfied at a_k, stop
3.	Define a new interval $[a, b]$ by using the *secant*^{2} procedure: $[a, b] = secant^2\ (a_k, b_k)$
4.	If $b - a > \gamma(b_k - a_k)$, then $c = (a + b)/2$ and use the *update* procedure: $[a, b] = update(a, b, c)$, where $\gamma \in (0, 1)$. ($\gamma = 0.66$)
5.	Set $[a_k, b_k] = [a, b]$ and $k = k + 1$ and go to step 2 ♦

The update procedure changes the current bracketing interval $[a, b]$ into a new one $[\bar{a}, \bar{b}]$ by using an additional point which is either obtained by a bisection step or a secant step. The input data in the procedure update are the points a, b, c. The parameter in the procedure update is $\theta \in (0, 1)$ ($\theta = 0.5$). The output data are $\bar{a}, \bar{b}$.

The *update procedure*

1.	If $c \notin (a, b)$, then set $\bar{a} = a$, $\bar{b} = b$ and return
2.	If $\varphi'_k(c) \geq 0$, then set $\bar{a} = a$, $\bar{b} = c$ and return
3.	If $\varphi'_k(c) < 0$ and $\varphi_k(c) \leq \varphi_k(0) + \varepsilon_k$, then set $\bar{a} = c$, $\bar{b} = b$ and return
4.	If $\varphi'_k(c) < 0$ and $\varphi_k(c) > \varphi_k(0) + \varepsilon_k$, then set $\hat{a} = a$, $\hat{b} = c$ and perform the following steps: (a) Set $d = (1 - \theta)\hat{a} + \theta\hat{b}$. If $\varphi'_k(d) \geq 0$, set $\bar{b} = d$, $\bar{a} = \hat{a}$ and return, (b) If $\varphi'_k(d) < 0$ and $\varphi_k(d) \leq \varphi_k(0) + \varepsilon_k$, then set $\hat{a} = d$ and go to step (a), (c) If $\varphi'_k(d) < 0$ and $\varphi_k(d) > \varphi_k(0) + \varepsilon_k$, then set $\hat{b} = d$ and go to step (a) ♦

The update procedure finds the interval $[\bar{a}, \bar{b}]$ so that

$$\varphi_k(\bar{a}) < \varphi_k(0) + \varepsilon_k, \quad \varphi'_k(\bar{a}) < 0 \quad \text{and} \quad \varphi'_k(\bar{b}) \geq 0. \tag{1.25}$$

Eventually, a nested sequence of intervals $[a_k, b_k]$ is determined, which converges to the point that satisfies either LS1 (1.23) or LS2 (1.22) and (1.24) conditions.

The secant procedure updates the interval by secant steps. If c is obtained from a secant step based on the function values at a and b, then we write

$$c = \text{secant } (a, b) = \frac{a\varphi'_k(b) - b\varphi'_k(a)}{\varphi'_k(b) - \varphi'_k(a)}.$$

Since we do not know whether φ' is a convex or a concave function, then a pair of secant steps is generated by a procedure denoted secant^2, defined as follows. The input data are the points a and b. The outputs are $\bar{a}$ and $\bar{b}$ which define the interval $[\bar{a}, \bar{b}]$.

Procedure secant^2

1.	Set $c = \text{sec ant}\,(a, b)$ and $[A, B] = \text{update}(a, b, c)$
2.	If $c = B$, then $\bar{c} = \text{secant}(b, B)$
3.	If $c = A$, then $\bar{c} = \text{secant}(a, A)$
4.	If $c = A$ or $c = B$, then $[\bar{a}, \bar{b}] = \text{update } (A, B, \bar{c})$. Otherwise, $[\bar{a}, \bar{b}] = [A, B]$ ♦

The Hager and Zhang line search procedure finds the stepsize α_k satisfying either LS1 or LS2 in a finite number of operations, as it is stated in the following theorem proved by Hager and Zhang (2005).

Theorem 1.2 *Suppose that $\varphi_k(\alpha)$ is continuously differentiable on an interval $[a_0, b_0]$, where* (1.25) *holds. If $\rho \in (0, 1/2)$, then the Hager and Zhang line search procedure terminates at a point satisfying either LS1 or LS2 conditions.* ♦

Under some additional assumptions, the convergence analysis of the secant^2 procedure was given by Hager and Zhang (2005), proving that the interval width generated by it is tending to zero, with the root convergence order $1 + \sqrt{2}$. This line search procedure is implemented in CG-DESCENT, one of the most advanced conjugate gradient algorithms, which is presented in Chapter 7.

Dai and Kou line search
In practical computations, the first Wolfe condition (1.12) may never be satisfied because of the numerical errors, even for tinny values of ρ. In order to avoid the numerical drawback of the Wolfe line search, Hager and Zhang (2005) introduced a combination of the original Wolfe conditions and the approximate Wolfe conditions (1.21). Their line search is working well in numerical computations, but in theory it cannot guarantee the global convergence of the algorithm. Therefore, in order to overcome this deficiency of the approximate Wolfe line search, Dai and Kou (2013) introduced the so called *improved Wolfe line search*: "given a constant parameter $\varepsilon > 0$, a positive sequence $\{\eta_k\}$ satisfying $\sum_{k\geq 1}\eta_k < \infty$ as well as the parameters ρ and σ satisfying $0 < \rho < \sigma < 1$, Dai and Kou (2013) proposed the following modified Wolfe condition:

$$f(x_k + \alpha d_k) \leq f(x_k) + \min\left\{\varepsilon\left|g_k^T d_k\right|,\ \rho\alpha g_k^T d_k + \eta_k\right\}." \tag{1.26}$$

The line search satisfying (1.26) and (1.13) is called the improved Wolfe line search. If f is continuously differentiable and bounded from below, the gradient g is Lipschitz continuous and d_k is a descent direction (i.e., $g_k^T d_k < 0$), then there must exist a suitable stepsize satisfying (1.13) and (1.26), since they are weaker than the standard Wolfe conditions.

Nonmonotone line search Grippo, Lampariello, and Lucidi
The nonmonotone line search for Newton's methods was introduced by Grippo, Lampariello, and Lucidi (1986). In this method the stepsize α_k satisfies the following condition:

$$f(x_k + \alpha_k d_k) \leq \max_{0\leq j\leq m(k)} f(x_{k-j}) + \rho\alpha_k g_k^T d_k, \tag{1.27}$$

where $\rho \in (0,1)$, $m(0) = 0$, $0 \leq m(k) \leq \min\{m(k-1)+1, M\}$ and M is a prespecified nonnegative integer. Theoretical analysis and numerical experiments showed the efficiency and robustness of this line search for solving unconstrained optimization problems in the context of the Newton method. The r-linear convergence for the nonmonotone line search (1.27), when the objective function f is strongly convex, was proved by Dai (2002b).

Although these nonmonotone techniques based on (1.27) work well in many cases, there are some drawbacks. First, a good function value generated in any iteration is essentially discarded due to the max in (1.27). Second, in some cases, the numerical performance is very dependent on the choice of M see Raydan (1997). Furthermore, it has been pointed out by Dai (2002b) that although an iterative method is generating r-linearly convergent iterations for a strongly convex function, the iterates may not satisfy the condition (1.27) for k sufficiently large, for any fixed bound M on the memory.

Nonmonotone line search Zhang and Hager
Zhang and Hager (2004) proposed another nonmonotone line search technique by replacing the maximum function values in (1.27) with an average of function values. Suppose that d_k is a descent direction. Their line search determines a stepsize α_k as follows.

Algorithm 1.3 Zhang and Hager nonmonotone line search

1.	Choose a starting guess x_0 and the parameters: $0 \le \eta_{\min} \le \eta_{\max} \le 1$, $0 < \rho < \sigma < 1 < \beta$ and $\mu > 0$. Set $C_0 = f(x_0)$, $Q_0 = 1$ and $k = 0$
2.	If $\|\nabla f(x_k)\|$ is sufficiently small, then stop
3.	Line search update: Set $x_{k+1} = x_k + \alpha_k d_k$, where α_k satisfies either the nonmonotone Wolfe conditions: $$f(x_k + \alpha_k d_k) \le C_k + \rho \alpha_k g_k^T d_k, \quad (1.28)$$ $$\nabla f(x_k + \alpha_k d_k)^T d_k \ge \sigma d_k^T g_k, \quad (1.29)$$ or the nonmonotone Armijo conditions: $\alpha_k = \bar{\alpha}_k \beta^{h_k}$, where $\bar{\alpha}_k > 0$ is the trial step and h_k is the largest integer such that (1.28) holds and $\alpha_k \le \mu$
4.	Choose $\eta_k \in [\eta_{\min}, \eta_{\max}]$ and set: $$Q_{k+1} = \eta_k Q_k + 1, \quad (1.30)$$ $$C_{k+1} = \frac{\eta_k Q_k C_k + f(x_{k+1})}{Q_{k+1}} \quad (1.31)$$
5.	Set $k = k + 1$ and go to strp 2 ♦

Observe that C_{k+1} is a convex combination of C_k and $f(x_{k+1})$. Since $C_0 = f(x_0)$, it follows that C_k is a convex combination of the function values $f(x_0), f(x_1), \ldots, f(x_k)$. Parameter η_k control the degree of nonmonotonicity. If $\eta_k = 0$ for all k, then this nonmonotone line search reduces to the monotone Wolfe or Armijo line search. If $\eta_k = 1$ for all k, then $C_k = A_k$, where

$$A_k = \frac{1}{k+1} \sum_{i=0}^{n} f(x_i).$$

Theorem 1.3 *If $g_k^T d_k \le 0$ for each k, then for the iterates generated by the nonmonotone line search Zhang and Hager algorithm, we have $f(x_k) \le C_k \le A_k$ for each k. Moreover, if $g_k^T d_k < 0$ and $f(x)$ is bounded from below, then there exists α_k satisfying either Wolfe or Armijo conditions of the line search update.* ♦

Zhang and Hager (2004) proved the convergence of their algorithm.

Theorem 1.4 *Suppose that f is bounded from below and there exist the positive constants c_1 and c_2 such that $g_k^T d_k \le -c_1 \|g_k\|^2$ and $\|d_k\| \le c_2 \|g_k\|$ for all sufficiently large k. Then, under the Wolfe line search if ∇f is Lipschitz continuous, then the iterates x_k generated by the nonmonotone line search Zhang and Hager algorithm have the property that* $\liminf_{k \to \infty} \|\nabla f(x_k)\| = 0$. *Morover, if $\eta_{\max} < 1$, then* $\lim_{k \to \infty} \nabla f(x_k) = 0$. ♦

The numerical results reported by Zhang and Hager (2004) showed that this nonmonotone line search is superior to the nonmonotone technique (1.27).

Nonmonotone line search Gu and Mo
Recently, a modified version of the nonmonotone line search (1.27) has been proposed by Gu and Mo (2008). In this method, the current nonmonotone term is a convex combination of the previous nonmonotone term and the current value of the objective function, instead of an average of the successive objective function values introduced by Zhang and Hager (2004), i.e., the stepsize α_k is computed to satisfy the following line search condition:

$$f(x_k + \alpha_k d_k) \leq D_k + \rho \alpha_k g_k^T d_k, \tag{1.32}$$

where

$$\begin{cases} D_0 = f(x_0), & k = 0, \\ D_k = \theta_k D_{k-1} + (1 - \theta_k) f(x_k), & k \geq 1, \end{cases} \tag{1.33}$$

with $0 \leq \theta_k \leq \theta_{\max} < 1$ and $\rho \in (0, 1)$. Theoretical and numerical results, reported by Gu and Mo (2008), in the frame of the trust-region method, showed the efficiency of this nonmonotone line search scheme.

Nonmonotone line search Huang, Wan and Chen
Recently, Huang, Wan, and Chen (2014) proposed a new nonmonotone line search as an improved version of the nonmonotone line search technique proposed by Zhang and Hager. Their algorithm implementing the nonmonotone Armijo condition has the same properties as the nonmonotone line search algorithm of Zhang and Hager, as well as some other properties that certify its convergence in very mild conditions. Suppose that at x_k the search direction is d_k. The nonmonotone line search proposed by Huang, Wan, and Chen is as follows:

Algorithm 1.4 Huang-Wan-Chen nonmonotone line search

1.	Choose $0 \leq \eta_{\min} \leq \eta_{\max} < 1 < \beta$, $\delta_{\max} < 1$, $0 < \delta_{\min} < (1 - \eta_{\max})\delta_{\max}$, $\varepsilon > 0$ small enough and $\mu > 0$
2.	If $\|g_k\| \leq \varepsilon$, then the algorithm stop
3.	Choose $\eta_k \in [\eta_{\min}, \eta_{\max}]$. Compute Q_{k+1} and C_{k+1} by (1.30) and (1.31) respectively. Choose $\delta_{\min} \leq \delta_k \leq \delta_{\max}/Q_{k+1}$. Let $\alpha_k = \bar{\alpha}_k \beta^{h_k} \leq \mu$ be a stepsize satisfying $$C_{k+1} = \frac{\eta_k Q_k C_k + f(x_k + \alpha_k d_k)}{Q_{k+1}} \leq C_k + \delta_k \alpha_k g_k^T d_k, \tag{1.34}$$ where h_k is the largest integer such that (1.34) holds and Q_k, C_k, Q_{k+1}, and C_{k+1} are computed as in the nonmonotone line search of Zhang and Hager
4.	Set $x_{k+1} = x_k + \alpha_k d_k$. Set $k = k + 1$ and go to step 2 ♦

If the minimizing function f is continuously differentiable and if $g_k^T d_k \leq 0$ for each k, then there exists a trial step $\bar{\alpha}_k$ such that (1.34) holds. The convergence of this nonmonotone line search is obtained in the same conditions as in Theorem 1.4. The r-linear convergence is proved for strongly convex functions.

Nonmonotone line search Ou and Liu
Based on (1.32) a new modified nonmonotone memory gradient algorithm for unconstrained optimization was elaborated by Ou and Liu (2017). Given $\rho_1 \in (0,1)$, $\rho_2 > 0$ and $\beta \in (0,1)$ set $s_k = -(g_k^T d_k)/\|d_k\|^2$ and compute the stepsize $\alpha_k = \max\{s_k, s_k\beta, s_k\beta^2, \ldots\}$ satisfying the line search condition:

$$f(x_k + \alpha_k d_k) \leq D_k + \rho_1 \alpha_k g_k^T d_k - \rho_2 \alpha_k^2 \|d_k\|^2, \tag{1.35}$$

where D_k is defined by (1.33) and d_k is a descent direction, i.e., $g_k^T d_k < 0$. Observe that if $\rho_2 = 0$ and $s_k \equiv s$ for all k, then the nonmonotone line search (1.35) reduces to the nonmonotone line search (1.32). The algorithm corresponding to this nonmonotone line search presented by Ou and Liu is as follows.

Algorithm 1.5 Ou and Liu nonmonotone line search

1.	Consider a starting guess x_0 and select the parameters: $\varepsilon \geq 0$, $0 < \tau < 1$, $\rho_1 \in (0,1)$, $\rho_2 > 0$, $\beta \in (0,1)$ and an integer $m > 0$. Set $k = 0$
2.	If $\|g_k\| \leq \varepsilon$, then stop
3.	Compute the direction d_k by the following recursive formula: $$d_k = \begin{cases} -g_k, & \text{if } k \leq m, \\ -\lambda_k g_k - \sum_{i=1}^m \lambda_{k-i} d_{k-i} & \text{if } k \geq m+1, \end{cases} \quad (1.36)$$ where $$\lambda_{k-i} = \frac{\tau}{m} \frac{\|g_k\|^2}{\|g_k\|^2 + \lvert g_k^T d_{k-i} \rvert}, \quad i = 1, \ldots, m,$$ $$\lambda_k = 1 - \sum_{i=1}^m \lambda_{k-i}$$
4.	Using the above procedure, determine the stepsize α_k satisfying (1.35) and set $x_{k+1} = x_k + \alpha_k d_k$
5.	Set $k = k + 1$ and go to step 2 ♦

The algorithm has the following interesting properties. For any $k \geq 0$, it follows that $g_k^T d_k \leq -(1-\tau)\|g_k\|^2$. For any $k \geq m$, it follows that $\|d_k\| \leq \max_{1 \leq i \leq m} \{\|g_k\|, \|d_{k-i}\|\}$. Moreover, for any $k \geq 0$, $\|d_k\| \leq \max_{0 \leq j \leq k} \{\|g_j\|\}$.

Theorem 1.5 *If the objective function is bounded from below on the level set $S = \{x : f(x) \leq f(x_0)\}$ and the gradient $\nabla f(x)$ is Lipschitz continuous on an open convex set that contains S, then the algorithm of Ou and Liu terminates in a finite number of iterates. Moreover, if the algorithm generates an infinite sequence $\{x_k\}$, then $\lim_{k \to +\infty} \|g_k\| = 0$.* ♦

Numerical results, presented by Ou and Liu (2017), showed that this method is suitable for solving large-scale unconstrained optimization problems and is more stable than other similar methods.

A special nonmonotone line search is the Barzilai and Borwein (1988) method. In this method, the next approximation to the minimum is computed as $x_{k+1} = x_k - D_k g_k$, $k = 0, 1, \ldots$, where $D_k = \alpha_k I$, I being the identity matrix. The stepsize α_k is computed as solution of the problem $\min_{\alpha_k} \|s_k - D_k y_k\|$, or as solution of $\min_{\alpha_k} \|D_k^{-1} s_k - y_k\|$. In the first case $\alpha_k = (s_k^T y_k)/\|y_k\|^2$ and in the second one $\alpha_k = \|s_k\|^2/(s_k^T y_k)$, where $s_k = x_{k+1} - x_k$ and $y_k = g_{k+1} - g_k$. Barzilai and Borwein proved that their algorithm is superlinearly convergent. Many researcher studied the Barzilai and Borwein algorithm including: Raydan (1997), Grippo and Sciandrone (2002), Dai, Hager, Schittkowski, and Zhang (2006), Dai and Liao (2002), Narushima, Wakamatsu, Yabe, (2008), Liu and Liu (2019).

Nonmonotone line search methods have been investigated by many authors, for example, see Dai (2002b) and the references therein. Observe that all these nonmonotone line searchs concentrate on modifying the first Wolfe condition (1.12). Also, the approximate Wolfe line search (1.21) of Hager and Zhang and the improved Wolfe line search (1.26) and (1.13) of Dai and Kou modify the first Wolfe condition, responsible for a sufficient reduction of the objective function value. No numerical comparisons among these nonmonotone line searches have been given.

As for stopping the iterative scheme (1.4), one of the most popular criteria is $\|g_k\| \leq \varepsilon$, where ε is a small positive constant and $\|.\|$ is the Euclidian or l_∞ norm.
In the following, the optimality conditions for unconstrained optimization are presented and then the most important algorithms for the search direction d_k in (1.4) are shortly discussed.

1.3 Optimality Conditions for Unconstrained Optimization

In this section, we are interested in giving conditions under which a solution for the problem (1.1) exists. The purpose is to discuss the main concepts and the fundamental results in unconstrained optimization known as optimality conditions. Both necessary and sufficient conditions for optimality are presented. Plenty of very good books showing these conditions are known: Bertsekas (1999), Nocedal and Wright (2006), Sun and Yuan (2006), Chachuat (2007), Andrei (2017c), etc. To formulate the optimality conditions, it is necessary to introduce some concepts which characterize an improving direction along which the values of the function f decrease (see Appendix A).

Definition 1.1 (Descent Direction). Suppose that $f : \mathbb{R}^n \to \mathbb{R}$ is continuous at x^*. A vector $d \in \mathbb{R}^n$ is a descent direction for f at x^* if there exists $\delta > 0$ so that $f(x^* + \lambda d) < f(x^*)$ for any $\lambda \in (0, \delta)$. The cone of descent directions at x^*, denoted by $C_{dd}(x^*)$ is given by:

$$C_{dd}(x^*) = \{d : \text{there exists } \delta > 0 \text{ such that } f(x^* + \lambda d) < f(x^*), \text{ for any } \lambda \in (0, \delta)\}.$$

Assume that f is a differentiable function. To get an algebraic characterization for a descent direction for f at x^* let us define the set

$$C_0(x^*) = \{d : \nabla f(x^*)^T d < 0\}.$$

The following result shows that every $d \in C_0(x^*)$ is a descent direction at x^*.

Proposition 1.3 *(Algebraic Characterization of a Descent Direction). Suppose that $f : \mathbb{R}^n \to \mathbb{R}$ is differentiable at x^*. If there exists a vector d so that $\nabla f(x^*)^T d < 0$, then d is a descent direction for f at x^*, i.e., $C_0(x^*) \subseteq C_{dd}(x^*)$.*

Proof Since f is differentiable at x^*, it follows that

$$f(x^* + \lambda d) = f(x^*) + \lambda \nabla f(x^*)^T d + \lambda \|d\| o(\lambda d),$$

where $\lim_{\lambda \to 0} o(\lambda d) = 0$. Therefore,

$$\frac{f(x^* + \lambda d) - f(x^*)}{\lambda} = \nabla f(x^*)^T d + \|d\| o(\lambda d).$$

Since $\nabla f(x^*)^T d < 0$ and $\lim_{\lambda \to 0} o(\lambda d) = 0$, it follows that there exists a $\delta > 0$ so that $\nabla f(x^*)^T d + \|d\| o(\lambda d) < 0$ for all $\lambda \in (0, \delta)$. ♦

Theorem 1.6 *(First-Order Necessary Conditions for a Local Minimum). Suppose that $f : \mathbb{R}^n \to \mathbb{R}$ is differentiable at x^*. If x^* is a local minimum, then $\nabla f(x^*) = 0$.*

Proof Suppose that $\nabla f(x^*) \neq 0$. If we consider $d = -\nabla f(x^*)$, then $\nabla f(x^*)^T d = -\|\nabla f(x^*)\|^2 < 0$. By Proposition 1.3 there exists a $\delta > 0$ so that for any $\lambda \in (0, \delta)$, $f(x^* + \lambda d) < f(x^*)$. But this is in contradiction with the assumption that x^* is a local minimum for f. ♦

Observe that the above necessary condition represents a system of n algebraic nonlinear equations. All the points x^* which solve the system $\nabla f(x) = 0$ are called *stationary points*. Clearly, the stationary points need not all be local minima. They could very well be local maxima or even saddle points. In order to characterize a local minimum, we need more restrictive necessary conditions involving the Hessian matrix of the function f.

Theorem 1.7 *(Second-Order Necessary Conditions for a Local Minimum). Suppose that $f : \mathbb{R}^n \to \mathbb{R}$ is twice differentiable at point x^*. If x^* is a local minimum, then $\nabla f(x^*) = 0$ and $\nabla^2 f(x^*)$ is positive semidefinite.*

Proof Consider an arbitrary direction d. Then, using the differentiability of f at x^* we get

$$f(x^* + \lambda d) = f(x^*) + \lambda \nabla f(x^*)^T d + \frac{1}{2} \lambda^2 d^T \nabla^2 f(x^*) d + \lambda^2 \|d\|^2 o(\lambda d),$$

where $\lim_{\lambda \to 0} o(\lambda d) = 0$. Since x^* is a local minimum, $\nabla f(x^*) = 0$. Therefore,

$$\frac{f(x^* + \lambda d) - f(x^*)}{\lambda^2} = \frac{1}{2} d^T \nabla^2 f(x^*) d + \|d\|^2 o(\lambda d).$$

Since x^* is a local minimum, for λ sufficiently small, $f(x^* + \lambda d) \geq f(x^*)$. For $\lambda \to 0$ it follows from the above equality that $d^T \nabla^2 f(x^*) d \geq 0$. Since d is an arbitrary direction, it follows that $\nabla^2 f(x^*)$ is positive semidefinite. ♦

In the above theorems, we have presented the *necessary* conditions for a point x^* to be a local minimum, i.e., these conditions *must be satisfied* at every local minimum solution. However, a point satisfying these necessary conditions need not be a local minimum. In the following theorems, the *sufficient* conditions for a global minimum are given, provided that the objective function is *convex* on $\mathbb{R}^n$.

The following theorem can be proved. It shows that the convexity is crucial in global nonlinear optimization.

Theorem 1.8 *(First-Order Sufficient Conditions for a Strict Local Minimum). Suppose that* $f : \mathbb{R}^n \to \mathbb{R}$ *is differentiable at* x^* *and convex on* $\mathbb{R}^n$. *If* $\nabla f(x^*) = 0$, *then* x^* *is a global minimum of f on* $\mathbb{R}^n$.

Proof Since f is convex on $\mathbb{R}^n$ and differentiable at x^* then from the property of convex functions given by the Proposition A4.3 it follows that for any $x \in \mathbb{R}^n$ $f(x) \geq f(x^*) + \nabla f(x^*)^T (x - x^*)$. But x^* is a stationary point, i.e., $f(x) \geq f(x^*)$ for any $x \in \mathbb{R}^n$. ♦

The following theorem gives the second-order sufficient conditions characterizing a local minimum point for those functions which are strictly convex in a neighborhood of the minimum point.

Theorem 1.9 (*Second-Order Sufficient Conditions for a Strict Local Minimum*). *Suppose that* $f : \mathbb{R}^n \to \mathbb{R}$ *is twice differentiable at point* x^*. *If* $\nabla f(x^*) = 0$ *and* $\nabla^2 f(x^*)$ *is positive definite, then* x^* *is a local minimum of f.*

Proof Since f is twice differentiable, for any $d \in \mathbb{R}^n$, we can write:

$$f(x^* + d) = f(x^*) + \nabla f(x^*)^T d + \frac{1}{2} d^T \nabla^2 f(x^*) d + \|d\|^2 o(d),$$

where $\lim_{d \to 0} o(d) = 0$. Let λ be the smallest eigenvalue of $\nabla^2 f(x^*)$. Since $\nabla^2 f(x^*)$ is positive definite, it follows that $\lambda > 0$ and $d^T \nabla^2 f(x^*) d \geq \lambda \|d\|^2$. Therefore, since $\nabla f(x^*) = 0$, we can write:

$$f(x^* + d) - f(x^*) \geq \left[\frac{\lambda}{2} + o(d) \right] \|d\|^2.$$

Since $\lim_{d \to 0} o(d) = 0$, then there exists a $\eta > 0$ so that $|o(d)| < \lambda/4$ for any $d \in B(0, \eta)$, where $B(0, \eta)$ is the open ball of radius η centered at 0. Hence

$$f(x^* + d) - f(x^*) \geq \frac{\lambda}{4}\|d\|^2 > 0$$

for any $d \in B(0, \eta)\backslash\{0\}$, i.e., x^* is a strict local minimum of function f. ♦

If we assume f to be twice continuously differentiable, we observe that, since $\nabla^2 f(x^*)$ is positive definite, then $\nabla^2 f(x^*)$ is positive definite in a small neighborhood of x^* and therefore f is strictly convex in a small neighborhood of x^*. Hence, x^* is a strict local minimum, it is the unique global minimum over a small neighborhood of x^*.

1.4 Overview of Unconstrained Optimization Methods

In this section, let us present some of the most important unconstrained optimization methods based on the gradient computation, insisting on their definition, their advantages and disadvantages, as well as on their convergence properties. The main difference among these methods is the procedure for the search direction d_k computation. For stepsize α_k computation, the most used procedure is that of Wolfe (standard). The following methods are discussed: the steepest descent, Newton, quasi-Newton, limited-memory quasi-Newton, truncated Newton, conjugate gradient, trust-region, and p-regularized methods.

1.4.1 Steepest Descent Method

The fundamental method for the unconstrained optimization is the steepest descent. This is the simplest method, designed by Cauchy (1847), in which the search direction is selected as:

$$d_k = -g_k. \tag{1.37}$$

At the current point x_k, the direction of the negative gradient is the best direction of search for a minimum of f. However, as soon as we move in this direction, it ceases to be the best one and continues to deteriorate until it becomes orthogonal to $-g_k$, That is, the method begins to take small steps without making significant progress to minimum. This is its major drawback, the steps it takes are too long, i.e., there are some other points z_k on the line segment connecting x_k and x_{k+1}, where $-\nabla f(z_k)$ provides a better new search direction than $-\nabla f(x_{k+1})$. The steepest descent method is globally convergent under a large variety of inexact line search procedures. However, its convergence is only linear and it is badly affected by ill-conditioning (Akaike, 1959). The convergence rate of this method is strongly dependent on the distribution of the eigenvalues of the Hessian of the minimizing function.

Theorem 1.10 *Suppose that f is twice continuously differentiable. If the Hessian $\nabla^2 f(x^*)$ of function f is positive definite and has the smallest eigenvalue $\lambda_1 > 0$ and the largest eigenvalue $\lambda_n > 0$, then the sequence of objective values $\{f(x_k)\}$ generated by the steepest descent algorithm converges to $f(x^*)$ linearly with a convergence ratio no greater than*

$$\left(\frac{\lambda_n - \lambda_1}{\lambda_n + \lambda_1}\right)^2 = \left(\frac{\kappa - 1}{\kappa + 1}\right)^2, \tag{1.38}$$

i.e.,

$$f(x_{k+1}) \leq \left(\frac{\kappa - 1}{\kappa + 1}\right)^2 f(x_k), \tag{1.39}$$

where $\kappa = \lambda_n/\lambda_1$ is the condition number of the Hessian. ♦

This is one of the best estimation we can obtain for steepest decent in certain conditions. For strongly convex functions for which the gradient is Lipschitz continuous, Nemirovsky and Yudin (1983) define the global estimate of the rate of convergence of an iterative method as $f(x_{k+1}) - f(x^*) \leq ch(x_1 - x^*, m, L, k)$, where $h(.)$ is a function, c is a constant, m is a lower bound on the smallest eigenvalue of the Hessian $\nabla^2 f(x)$, L is the Lipschitz constant, and k is the iteration number. The faster the rate at which h converges to 0 as $k \to \infty$, the more efficient the algorithm.

The advantages of the steepest descent method are as follows. It is globally convergent to local minimizer from any starting point x_0. Many other optimization methods switch to steepest descent when they do not make sufficient progress. On the other hand, it has the following disadvantages. It is not scale invariant, i.e., changing the scalar product on $\mathbb{R}^n$ will change the notion of gradient. Besides, usually it is very (very) slow, i.e., its convergence is linear. Numerically, it is often not convergent at all. An acceleration of the steepest descent method with backtracking was given by Andrei (2006a) and discussed by Babaie-Kafaki and Rezaee (2018).

1.4.2 Newton Method

The Newton method is based on the quadratic approximation of the function f and on the exact minimization of this quadratic approximation. Thus, near the current point x_k, the function f is approximated by the truncated Taylor series

$$f(x) \cong f(x_k) + \nabla f(x_k)^T (x - x_k) + \frac{1}{2}(x - x_k)^T \nabla^2 f(x_k)(x - x_k), \tag{1.40}$$

known as the *local quadratic model* of f around x_k. Minimizing the right-hand side of (1.40), the search direction of the Newton method is computed as

$$d_k = -\nabla^2 f(x_k)^{-1} g_k, \tag{1.41}$$

Therefore, the Newton method is defined as:

$$x_{k+1} = x_k - \alpha_k \nabla^2 f(x_k)^{-1} g_k, \; k = 0, 1, \ldots, \tag{1.42}$$

where α_k is the stepsize. For the Newton method (1.42), we see that d_k is a descent direction if and only if $\nabla^2 f(x_k)$ is a positive definite matrix. If the starting point x_0 is close to x^*, then the sequence $\{x_k\}$ generated by the Newton method converges to x^* with a quadratic rate. More exactly:

Theorem 1.11 (*Local convergence of the Newton method*) *Let the function f be twice continuously differentiable on $\mathbb{R}^n$ and its Hessian $\nabla^2 f(x)$ be uniformly Lipschitz continuous on $\mathbb{R}^n$. Let iterates x_k be generated by the Newton method* (1.42) *with backtracking-Armijo line search using $\alpha_k^0 = 1$ and $\gamma < 1/2$. If the sequence $\{x_k\}$ has an accumulation point x^* where $\nabla^2 f(x^*)$ is positive definite, then:*

1. $\alpha_k = 1$ *for all k large enough,*
2. $\lim_{k\to\infty} x_k = x^*$,
3. *The sequence $\{x_k\}$ converges q-quadratically to x^*, that is, there exists a constant $K > 0$ such that*
   $\lim\limits_{k\to\infty} \frac{\|x_{k+1}-x^*\|}{\|x_k - x^*\|^2} \leq K.$ ♦

The machinery that makes Theorem 1.11 work is that once the sequence $\{x_k\}$ generated by the Newton method enters a certain domain of attraction of x^*, then it cannot escape from this domain and immediately the quadratic convergence to x^* starts. The main drawback of this method consists of computing and saving the Hessian matrix, which is an $n \times n$ matrix. Clearly, the Newton method is not suitable for solving large-scale problems. Besides, far away from the solution, the Hessian matrix may not be a positive definite matrix and therefore the search direction (1.41) may not be a descent one. Some modifications of the Newton method are discussed in this chapter, others are presented in (Sun & Yuan, 2006; Nocedal & Wright, 2006; Andrei, 2009e; Luenberger & Ye, 2016).

The following theorem shows the evolution of the error of the Newton method along the iterations, as well as the main characteristics of the method (Kelley, 1995, 1999).

Theorem 1.12 *Consider $e_k = x_k - x^*$ as the error at iteration k. Let $\nabla^2 f(x_k)$ be invertible and $\Delta_k \in \mathbb{R}^{n\times n}$ so that $\left\|\nabla^2 f(x_k)^{-1}\Delta_k\right\| < 1$. If for the problem* (1.1) *the Newton step*

$$x_{k+1} = x_k - \nabla^2 f(x_k)^{-1}\nabla f(x_k) \tag{1.43}$$

is applied by using $(\nabla^2 f(x_k) + \Delta_k)$ and $(\nabla f(x_k) + \delta_k)$ instead of $\nabla^2 f(x_k)$ and $\nabla f(x_k)$ respectively, then for Δ_k sufficiently small in norm, $\delta_k > 0$ and x_k sufficiently close to x^*.

$$\|e_{k+1}\| \leq K\left(\|e_k\|^2 + \|\Delta_k\|\|e_k\| + \|\delta_k\|\right), \tag{1.44}$$

for some positive constant K. ♦

The interpretation of (1.44) is as follows. Observe that in the norm of the error e_{k+1}, given by (1.44), the inaccuracy evaluation of the Hessian, given by $\|\Delta_k\|$, is multiplied by the norm of the previous error. On the other hand, the inaccuracy evaluation of the gradient, given by $\|\delta_k\|$, is not multiplied by the previous error and has a direct influence on $\|e_{k+1}\|$. In other words, in the norm of the error, the inaccuracy in the Hessian has a smaller influence than the inaccuracy of the gradient. Therefore, in this context, from (1.44) the following remarks may be emphasized:

1. If both Δ_k and δ_k are zero, then the quadratic convergence of the Newton method is obtained.
2. If $\delta_k \neq 0$ and $\|\delta_k\|$ is not convergent to zero, then there is no guarantee that the error for the Newton method will converge to zero.
3. If $\|\Delta_k\| \neq 0$, then the convergence of the Newton method is slowed down from quadratic to linear, or to superlinear if $\|\Delta_k\| \to 0$.

Therefore, we see that the inaccuracy evaluation of the Hessian of the minimizing function is not so important. It is the accuracy of the evaluation of the gradient which is more important. This is the motivation for the development of the quasi-Newton methods or, for example, the methods in which the Hessian is approximated as a diagonal matrix, (Nazareth, 1995; Dennis & Wolkowicz, 1993; Zhu, Nazareth, & Wolkowicz, 1999; Leong, Farid, & Hassan, 2010, 2012; Andrei, 2018e, 2019c, 2019d).

Some disadvantages of the Newton method are as follows:

1. *Lack of global convergence.* If the initial point is not sufficiently close to the solution, i.e., it is not within the *region of convergence*, then the Newton method may diverge. In other words, the Newton method does not have the *global convergence property*. This is because, far away from the solution, the search direction (1.41) may not be a valid descent direction even if $g_k^T d_k < 0$, a unit stepsize might not give a descent in minimizing the function values. The remedy is to use the globalization strategies. The first one is the line search which alters

the magnitude of the step. The second one is the trust-region which modifies both the stepsize and the direction.

2. *Singular Hessian.* The second difficulty is when the Hessian $\nabla^2 f(x_k)$ becomes singular during the progress of iterations, or becomes nonpositive definite. When the Hessian is singular at the solution point, then the Newton method loses its quadratic convergence property. In this case, the remedy is to select a positive definite matrix M_k in such a way that $\nabla^2 f(x_k) + M_k$ is sufficiently positive definite and solve the system $(\nabla^2 f(x_k) + M_k)d_k = -g_k$. The *regularization* term M_k is typically chosen by using the spectral decomposition of the Hessian, or as $M_k = \max\{0, -\lambda_{\min}(\nabla^2 f(x_k))\}I$, where $\lambda_{\min}(\nabla^2 f(x_k))$ is the smallest eigenvalue of the Hessian. Another method for modifying the Newton method is to use the modified Cholesky factorization see Gill and Murray (1974), Gill, Murray, and Wright (1981), Schnabel and Eskow (1999), Moré and Sorensen (1984).
3. *Computational efficiency.* At each iteration, the Newton method requires the computation of the Hessian matrix $\nabla^2 f(x_k)$, which may be a difficult task, especially for large-scale problems and for finding the solution of a linear system. One possibility is to replace the analytic Hessian by a finite difference approximation see Sun and Yuan (2006). However, this is costly because n additional evaluations of the minimizing function are required at each iteration. To reduce the computational effort, the quasi-Newton methods may be used. These methods generate approximations to the Hessian matrix using the information gathered from the previous iterations. To avoid solving a linear system for the search direction computation, variants of the quasi-Newton methods which generate approximations to the inverse Hessian may be used. Anyway, when run, the Newton method is the best.

1.4.3 Quasi-Newton Methods

These methods were introduced by Davidon (1959) and developed by Broyden (1970), Fletcher (1970), Goldfarb (1970), Shanno (1970), Powell (1970) and modified by many others. A deep analysis of these methods was presented by Dennis and Moré (1974, 1977).

The idea underlying the quasi-Newton methods is to use an approximation to the inverse Hessian instead of the true Hessian required in the Newton method (1.42). Many approximations to the inverse Hessian are known, from the simplest one where it remains fixed throughout the iterative process to more sophisticated ones that are built by using the information gathered during the iterations.

The search directions in quasi-Newton methods are computed as

$$d_k = -H_k g_k, \tag{1.45}$$

where $H_k \in \mathbb{R}^{n\times n}$ is an approximation to the inverse Hessian. At the iteration k, the approximation H_k to the inverse Hessian is updated to achieve H_{k+1} as a new approximation to the inverse Hessian in such a way that H_{k+1} satisfies a particular equation, namely the secant equation, which includes the second order information. The most used equation is the *standard secant equation*:

$$H_{k+1} y_k = s_k, \tag{1.46}$$

where $s_k = x_{k+1} - x_k$ and $y_k = g_{k+1} - g_k$.

Given the initial approximation H_0 to the inverse Hessian as an arbitrary symmetric and positive definite matrix, the most known quasi-Newton updating formulae are the BFGS (Broyden–Fletcher–Goldfarb–Shanno) and DFP (Davidon–Fletcher–Powell) updates:

$$H_{k+1}^{BFGS} = H_k - \frac{s_k y_k^T H_k + H_k y_k s_k^T}{y_k^T s_k} + \left(1 + \frac{y_k^T H_k y_k}{y_k^T s_k}\right)\frac{s_k s_k^T}{y_k^T s_k}, \tag{1.47}$$

$$H_{k+1}^{DFP} = H_k - \frac{H_k y_k y_k^T H_k}{y_k^T H_k y_k} + \frac{s_k s_k^T}{y_k^T s_k}. \tag{1.48}$$

The BFGS and DFP updates can be linearly combined, thus obtaining the *Broyden class* of quasi-Newton update formula

$$\begin{aligned} H_{k+1}^{\phi} &= \phi H_{k+1}^{BFGS} + (1-\phi) H_{k+1}^{DFP} \\ &= H_k - \frac{H_k y_k y_k^T H_k}{y_k^T H_k y_k} + \frac{s_k s_k^T}{y_k^T s_k} + \phi v_k v_k^T, \end{aligned} \tag{1.49}$$

where ϕ is a real parameter and

$$v_k = \sqrt{y_k^T H_k y_k}\left(\frac{s_k}{y_k^T s_k} - \frac{H_k y_k}{y_k^T H_k y_k}\right). \tag{1.50}$$

The main characteristics of the Broyden class of update are as follows (Sun & Yuan, 2006). If H_k is positive definite and the line search ensures that $y_k^T s_k > 0$, then H_{k+1}^{ϕ} with $\phi \geq 0$ is also a positive definite matrix and therefore, the search direction $d_{k+1} = -H_{k+1}^{\phi} g_{k+1}$ is a descent direction. For a strictly convex quadratic objective function, the search directions of the Broyden class of quasi-Newton method are conjugate directions. Therefore, the method possesses the quadratic termination property. If the minimizing function f is convex and $\phi \in [0, 1]$, then the Broyden class of the quasi-Newton methods is globally and locally superlinear convergent (Sun & Yuan, 2006). Intensive numerical experiments showed that among the quasi-Newton update formulae of the Broyden class, the BFGS is the top performer (Xu & Zhang, 2001).

It is worth mentioning that similar to the quasi-Newton approximations to the inverse Hessian $\{H_k\}$ satisfying the secant Equation (1.46), the quasi-Newton approximations to the (direct) Hessian $\{B_k\}$ can be defined, for which the following equivalent version of the standard secant Equation (1.46) is satisfied

$$B_{k+1}s_k = y_k. \tag{1.51}$$

In this case, the search direction can be obtained by solving the linear algebraic system (the *quasi-Newton system*)

$$B_k d_k = -g_k. \tag{1.52}$$

Now, to determine the BFGS and DFP updates of the (direct) Hessian, the following inverse must be computed: $(H_{k+1}^{BFGS})^{-1}$ and $(H_{k+1}^{DFP})^{-1}$ respectively. For this, the Sherman–Morrison formula is used (see Appendix A).

Therefore, using Sherman–Morrison formula from (1.47) to (1.48) the corresponding update of B_k is as follows:

$$B_{k+1}^{BFGS} = B_k - \frac{B_k s_k s_k^T B_k}{s_k^T B_k s_k} + \frac{y_k y_k^T}{y_k^T s_k}, \tag{1.53}$$

$$B_{k+1}^{DFP} = B_k + \frac{(y_k - B_k s_k)y_k^T + y_k(y_k - B_k s_k)^T}{y_k^T s_k} - \frac{(y_k - B_k s_k)^T s_k}{(y_k^T s_k)^2} y_k y_k^T. \tag{1.54}$$

The convergence of the quasi-Newton methods is proved under the following *classical assumptions*: the function f is twice continuously differentiable and bounded below; the level set $S = \{x \in \mathbb{R}^n : f(x) \le f(x_0)\}$ is bounded; the gradient $g(x)$ is Lipschitz continuous with constant $L > 0$, i.e., $\|g(x) - g(y)\| \le L\|x - y\|$, for any $x, y \in \mathbb{R}^n$.

In the convergence analysis, a key requirement for a line search algorithm like (1.4) is that the search direction d_k is a direction of *sufficient descent*, which is defined as

$$\frac{g_k^T d_k}{\|g_k\|\|d_k\|} \le -\varepsilon, \tag{1.55}$$

where $\varepsilon > 0$. This condition bounds the elements of the sequence $\{d_k\}$ of the search directions from being arbitrarily close to the orthogonality to the gradient. Often, the line search methods are so that d_k is defined in a way that satisfies the sufficient descent condition (1.55), even though an explicit value for $\varepsilon > 0$ is not known.

Theorem 1.13 *Suppose that $\{B_k\}$ is a sequence of bounded and positive definite symmetric matrices whose condition number is also bounded, i.e., the smallest eigenvalue is bounded away from zero. If d_k is defined to be the solution of the system* (1.52)*, then $\{d_k\}$ is a sequence of sufficient descent directions.*

Proof Let B_k be a symmetric positive definite matrix with eigenvalues $0 < \lambda_1^k \le \lambda_2^k \le \cdots \le \lambda_n^k$. Therefore, from (1.52) it follows that

$$\|g_k\| = \|B_k d_k\| \le \|B_k\| \|d_k\| = \lambda_n^k \|d_k\|. \tag{1.56}$$

From (1.52), using (1.56) we have

$$-\frac{g_k^T d_k}{\|g_k\| \|d_k\|} = \frac{d_k^T B_k d_k}{\|g_k\| \|d_k\|} \ge \lambda_1^k \frac{\|d_k\|^2}{\|g_k\| \|d_k\|} = \lambda_1^k \frac{\|d_k\|}{\|g_k\|} \ge \lambda_1^k \frac{\|d_k\|}{\lambda_n^k \|d_k\|} = \frac{\lambda_1^k}{\lambda_n^k} > 0.$$

The quality of the search direction d_k can be determined by studying the angle θ_k between the steepest descent direction $-g_k$ and the search direction d_k. Hence, applying this result to each matrix in the sequence $\{B_k\}$, we get

$$\cos \theta_k = -\frac{g_k^T d_k}{\|g_k\| \|d_k\|} \ge \frac{\lambda_1^k}{\lambda_n^k} \ge \frac{1}{M}, \tag{1.57}$$

where M is a positive constant. Observe that M is a positive constant and it is well defined since the smallest eigenvalue of matrices B_k in the sequence $\{B_k\}$ generated by the algorithm is bounded away from zero. Therefore, the search directions $\{d_k\}$ generated as solutions of (1.52) form a sequence of sufficient descent directions. ♦

The main consequence of this theorem on how to modify the quasi-Newton system defining the search direction d_k is to ensure that it is a solution of a system that has the same properties as B_k.

A global convergence result for the BFGS method was given by Powell (1976a). Using the trace and the determinant to measure the effect of the two rank-one corrections on B_k in (1.53), he proved that if f is convex, then for any starting point x_0 and any positive definite starting matrix B_0, the BFGS method gives $\liminf_{k \to \infty} \|g_k\| = 0$. In addition, if the sequence $\{x_k\}$ converges to a solution point at which the Hessian matrix is positive definite, then the rate of convergence is superlinear. The analysis of Powell was extended by Byrd, Nocedal, and Yuan (1987) to the Broyden class of quasi-Newton methods.

With Wolfe line search, BFGS approximation is always positive definite, so the line search works very well. It behaves "almost" like Newton in the limit (convergence is superlinear). DFP has the interesting property that, for a quadratic objective, it simultaneously generates the directions of the conjugate gradient method while constructing the inverse Hessian. However, DFP is highly sensitive to inaccuracies in line searches.

1.4.4 Modifications of the BFGS Method

In the following, some modifications of the BFGS updating method, both subject to its updating formula and subject to the line search conditions, are going to be presented. Intensive numerical experiments on minimizing functions with different dimensions and complexities showed that the BFGS method may require a large number of iterations or function and gradient evaluations on certain problems (Gill & Leonard, 2001). The sources of the inefficiency of the BFGS method may be caused by a poor initial approximation to the Hessian or, more importantly, by the ill-conditioning of the Hessian approximations along the iterations. To improve the efficiency and the robustness of the BFGS method and to overcome the difficulties, some modified versions of it were given. All these modified BFGS methods can be classified into three large classes: the scaling of the BFGS update matrix, the BFGS update with modified secant equation and the modified BFGS method using different line search conditions for stepsize computation. The scaling of the BFGS update has two developments: *sizing*, i.e., multiplying by an appropriate scalar the approximate Hessian matrix before it is updated in the BFGS method [Contreras and Tapia (1993), Oren and Luenberger (1974), Oren and Spedicato (1976), Shanno and Phua (1978), Yabe, Martínez, and Tapia (2004)], and the *proper scaling of the terms* on the right hand side of the BFGS updating formula with positive factors [Biggs (1971, 1973), Oren (1972), Liao (1997), Nocedal and Yuan (1993), Andrei (2018c, 2018d, 2018f)]. The purpose of the BFGS update with *modified secant equation* is to approximate the curvature of the objective function along the search direction more accurately than the standard secant equation does [Yuan (1991), Yuan and Byrd (1995), Al-Baali (1998), Zhang, Deng, and Chen (1999), Zhang and Xu (2001), Wei, Yu, Yuan, and Lian (2004), Zhu and Wen (2006), Yabe, Ogasawara, and Yoshino (2007), Al-Baali and Grandinetti (2009), Yuan and Wei (2010), Wu and Liang (2014), Arzam, Babaie-Kafaki, and Ghanbari (2017)]. The BFGS methods with *new line search conditions* for stepsize computation try to ensure the global convergence by modifying the Wolfe line search conditions [Wan, Huang, and Zheng (2012), Wan, Teo, Shen, and Hu (2014), Yuan, Wei, and Lu (2017), Yuan, Sheng, Wang, Hu, and Li (2018), Dehmiry, 2019)].

Scaling the Terms on the Right-Hand Side of the BFGS Update

From (1.53) we see that BFGS update involves two corrections matrices, each of rank-one. Therefore, by the interlocking eigenvalue theorem of Wilkinson (1965), the first rank-one correction matrix which is subtracted decreases the eigenvalues, i.e., it shifts the eigenvalues to the left. On the other hand, the second rank-one matrix which is added shifts the eigenvalues to the right. More exactly, two important tools in the analysis of the properties and of the convergence of the BFGS method are the trace and the determinant of the standard B_{k+1} given by (1.53). The trace of a matrix is exactly the sum of its eigenvalues. The determinant of a matrix is the product of its eigenvalues. By direct computation from (1.53), we get (see Appendix A)

$$tr(B_{k+1}) = tr(B_k) - \frac{\|B_k s_k\|^2}{s_k^T B_k s_k} + \frac{\|y_k\|^2}{y_k^T s_k}.$$

On the other hand,

$$\det(B_{k+1}) = \det(B_k) \frac{y_k^T s_k}{s_k^T B_k s_k}.$$

As it is known, the efficiency of the BFGS method is dependent on the structure of the eigenvalues of the approximation to the Hessian matrix (Nocedal, 1992). Powell (1987) and Byrd, Liu, and Nocedal (1992) emphasized that the BFGS method actually suffers more from the large eigenvalues than from the small ones. Observe that the second term on the right hand side of $tr(B_{k+1})$ is negative. Therefore, it produces a shift of the eigenvalues of B_{k+1} to the left. Thus, the BFGS method is able to correct large eigenvalues. On the other hand, the third term on the right hand side of $tr(B_{k+1})$ being positive produces a shift of the eigenvalues of B_{k+1} to the right. If this term is large, B_{k+1} may have large eigenvalues, too. Therefore, a correction of the eigenvalues of B_{k+1} can be achieved by scaling the corresponding terms in (1.53) and this is the main motivation for which the scaled BFGS methods is used. There must be a balance between these eigenvalue shifts, otherwise the Hessian approximation could either approach singularity or become arbitrarily large, thus ruining the convergence of the method. The scaling procedures of the BFGS update (1.53) with one or two parameters know the following developments.

1. *One parameter scaling the third term on the right hand side of the BFGS update*. In this case, the general scaling BFGS updating formula is:

$$B_{k+1} = B_k - \frac{B_k s_k s_k^T B_k}{s_k^T B_k s_k} + \gamma_k \frac{y_k y_k^T}{y_k^T s_k}, \tag{1.58}$$

where γ_k is a positive parameter. For the selection of the scaling factor γ_k in (1.58), the following procedures have been considered in literature.

1.1. *Scaling BFGS with Hermite interpolation conditions* (Biggs, 1971, 1973). If the objective function is cubic along the line segment connecting x_{k-1} and x_k and if the Hermite interpolation is used on the same line between x_{k-1} and x_k, Biggs (1971) proposed the following value for the scaling factor γ_k :

$$\gamma_k = \frac{6}{y_k^T s_k}(f(x_k) - f(x_{k+1}) + s_k^T g_{k+1}) - 2. \tag{1.59}$$

For one-dimensional problems, Wang and Yuan (1992) showed that the scaling BFGS (1.58) with γ_k given by (1.59) and without line search is r-linearly convergent.

1.2. *Scaling BFGS with a simple interpolation condition* (Yuan, 1991). By using a simple interpolation condition on the quadratic approximation of the minimizing function *f*, the value for the scaling parameter in (1.58) suggested by Yuan (1991) is

$$\gamma_k = \frac{2}{y_k^T s_k}(f(x_k) - f(x_{k+1}) + s_k^T g_{k+1}). \tag{1.60}$$

Powell (1986a) showed that the scaling BFGS update (1.58) with γ_k given by (1.60) is globally convergent for convex functions with inexact line search. However, for general nonlinear functions, the inexact line search does not involve the positivity of γ_k. In these cases, Yuan restricted γ_k in the interval $[0.01, 100]$ and proved the global convergence of this variant of the scaling BFGS method.

1.3. *Spectral scaling BFGS* (Cheng & Li, 2010). In this update, the scaling parameter γ_k in (1.58) is computed as:

$$\gamma_k = \frac{y_k^T s_k}{\|y_k\|^2}, \tag{1.61}$$

which is obtained as solution of the problem: $\min\|s_k - \gamma_k y_k\|^2$. Observe that γ_k given by (1.61) is exactly one of the spectral stepsizes introduced by Barzilai and Borwein (1988). Therefore, the scaling BFGS method (1.58) with γ_k given by (1.61) is viewed as the spectral scaling BFGS method. It is proved that this spectral scaling BFGS method with Wolfe line search is globally convergent and r-linearly convergent for convex optimization problems. Cheng and Li (2010) presented the computational evidence that their spectral scaling BFGS algorithm is top performer versus the standard BFGS and also versus the scaling BFGS algorithms by Al-Baali (1998), Yuan (1991), and Zhang and Xu (2001).

1.4. *Scaling BFGS with diagonal preconditioning and conjugacy condition* (Andrei, 2018a). Andrei (2018a) introduced another scaling BFGS update given by (1.58), in which the scaling parameter γ_k is computed in an adaptive manner as

$$\gamma_k = \min\left\{\frac{y_k^T s_k}{\|y_k\|^2 + \beta_k}, 1\right\}, \tag{1.62}$$

where $\beta_k > 0$ for all $k = 0, 1, \ldots$. Since under the Wolfe line search conditions (1.12) and (1.13) $y_k^T s_k > 0$ for all $k = 0, 1, \ldots$, it follows that γ_k given by (1.62) is bounded away from zero, i.e., $0 < \gamma_k \leq 1$. If γ_k is selected as in (1.62), where $\beta_k > 0$ for all $k = 0, 1, \ldots$, then the large eigenvalues of B_{k+1} given by (1.58) are shifted to the left (Andrei, 2018a). Intensive numerical experiments showed that this scaling BFGS algorithm with $\beta_k = |s_k^T g_{k+1}|$ is the best one, being more efficient and more robust versus the standard BFGS algorithm as well as versus some other scaling BFGS algorithms, including the versions of Biggs (1971), (1973), Yuan (1991), and Cheng and Li (2010).

Andrei (2018a) gives the following theoretical justification for selecting the parameter γ_k as in (1.62) with $\beta_k = |s_k^T g_{k+1}|$. To have a good algorithm, we need $\gamma_k I$ to be a diagonal preconditioner of $\nabla^2 f(x_{k+1})$ that reduces the condition number to the inverse of $\nabla^2 f(x_{k+1})$. Such matrix $\gamma_k I$ should be a rough approximation to the inverse of $\nabla^2 f(x_{k+1})$. Therefore, γ_k can be computed to minimize $\|s_k - \gamma_k y_k\|^2$. On the other hand, for nonlinear functions, as known, the classical conjugacy condition used by Hestenes and Stiefel (1952) for quadratic functions which incorporate the second-order information is $d_{k+1}^T y_k = -s_k^T g_{k+1}$. Therefore, in this algorithm, $\gamma_k I$ is selected to be a diagonal preconditioner of $\nabla^2 f(x_{k+1})$ and also to minimize the conjugacy condition, i.e., γ_k is selected to minimize a combination of these two conditions:

$$\min\{\|s_k - \gamma_k y_k\|^2 + \gamma_k^2 |s_k^T g_{k+1}|\}.$$

2. *One parameter scaling the first two terms of the BFGS update* [Oren and Luenberger (1974), Nocedal and Yuan (1993)]. This scaling BFGS update was introduced by Oren and Luenberger (1974) in their study on self-scaling variable metric algorithms for unconstrained optimization and was defined as

$$B_{k+1} = \delta_k \left[B_k - \frac{B_k s_k s_k^T B_k}{s_k^T B_k s_k} \right] + \frac{y_k y_k^T}{y_k^T s_k}, \tag{1.63}$$

where δ_k is a positive parameter. Oren and Luenberger (1974) suggested

$$\delta_k = \frac{y_k^T s_k}{s_k^T B_k s_k} \tag{1.64}$$

as being one of the best factors, since it simplifies the analysis of the eigenvalues structure of the inverse Hessian approximation. Furthermore, Nocedal and Yuan (1993) presented a deep analysis of this scaling quasi-Newton method and showed that even if the corresponding algorithm with inexact line search is superlinear convergent on general functions it is computationally expensive as regards the stepsize computation.

3. *Two parameters scaling the terms on the right-hand side of the BFGS update* [Liao (1997, Andrei (2018c), (2018d), (2018f)]. In these methods, the scaling parameters of the terms on the right hand side of the BFGS update are selected to modify the structure of the eigenvalues of the iteration matrix B_{k+1}, mainly to cluster them and to shift the large ones to the left. The following two approaches are known.

   3.1. *Scaling the first two terms on the right-hand side of the BFGS update with a positive parameter and the third one with another positive parameter* (Andrei, 2018c). Motivated by the idea of changing the structure of the eigenvalues of the BFGS approximation to the Hessian matrix, Andrei (2018c) proposed a *double parameter scaling BFGS method* in which the updating of the approximation Hessian matrix B_{k+1} is computed as

$$B_{k+1} = \delta_k \left[B_k - \frac{B_k s_k s_k^T B_k}{s_k^T B_k s_k} \right] + \gamma_k \frac{y_k y_k^T}{y_k^T s_k}, \tag{1.65}$$

where δ_k and γ_k are positive parameters. In this scaling BFGS method, the parameter δ_k is selected to cluster the eigenvalues of B_{k+1}. On the other hand, γ_k is determined to reduce the large eigenvalues of B_{k+1}, i.e., to shift them to the left, thus obtaining a better distribution of the eigenvalues:

$$\gamma_k = \min \left\{ \frac{y_k^T s_k}{\|y_k\|^2 + |s_k^T g_{k+1}|}, 1 \right\}, \tag{1.66}$$

and

$$\delta_k = \frac{n - \gamma_k \dfrac{\|y_k\|^2}{y_k^T s_k}}{n - \dfrac{\|B_k s_k\|^2}{s_k^T B_k s_k}}. \tag{1.67}$$

Theorem 1.14 *If the stepsize* α_k *is determined by the standard Wolfe line search* (1.12) *and* (1.13), B_k *is positive definite and* $\gamma_k > 0$, *then* B_{k+1} *given by* (1.65) *is also positive definite.* ♦

For general nonlinear functions, this scaling BFGS algorithm with inexact line search is globally convergent under the very reasonable condition that the scaling parameters are bounded. Intensive numerical experiments using over 80 unconstrained optimization test problems of different structures and complexities showed that this double parameter scaling BFGS update is more efficient than the standard BFGS algorithm and also than some other well-known scaling BFGS algorithms, including those by Biggs (1971), (1973), Cheng and Li (2010), Liao (1997), Nocedal and Yuan (1993), and Yuan (1991).

3.2. *Scaling the first two terms on the right-hand side of the BFGS update with a positive parameter and the third one with another positive parameter using the measure function of Byrd and Nocedal* (Andrei, 2018d). In this method, the BFGS update is scaled as in (1.65), where parameters δ_k and γ_k are computed to minimize the measure function $\varphi(.)$ of Byrd and Nocedal (1989). Minimizing the function $\varphi(B_{k+1}) = tr(B_{k+1}) - \ln(\det(B_{k+1}))$, subject to the parameters δ_k and γ_k, where B_{k+1} is given in (1.65), the following values are obtained:

$$\delta_k = \frac{n-1}{tr(B_k) - \frac{\|B_k s_k\|^2}{s_k^T B_k s_k}}. \tag{1.68}$$

$$\gamma_k = \frac{y_k^T s_k}{\|y_k\|^2}. \tag{1.69}$$

Theorem 1.15 *If the stepsize* α_k *is determined by the standard Wolfe line search* (1.12) *and* (1.13), *then the scaling parameters* δ_k *and* γ_k *given by* (1.68) *and* (1.69) *respectively are the unique global solution of the problem* $\min_{\delta_k > 0, \gamma_k > 0} \varphi(B_{k+1})$. ♦

Intensive numerical experiments in Andrei (2018d) proved that this scaling procedure of the BFGS with two parameters is more efficient and more robust than the other scaling procedures including those of Biggs (1971), (1973), Cheng and Li (2010), Yuan (1991), Nocedal and Yuan (1993), Liao (1997), and Andrei (2018c), (2018d).

3.3. *Scaling the last terms on the right-hand side of the BFGS update with two positive parameters* Liao (1997). Liao (1997) introduced the two parameter scaling BFGS method as

$$B_{k+1} = B_k - \delta_k \frac{B_k s_k s_k^T B_k}{s_k^T B_k s_k} + \gamma_k \frac{y_k y_k^T}{y_k^T s_k}, \tag{1.70}$$

and proved that this method corrects the large eigenvalues better than the standard BFGS method given by (1.53). In other words, it was proved that this scaling BFGS method has a strong self-correcting property with respect to the determinant (Liao, 1997). In Liao's method, the parameters scaling the terms in the BFGS update are computed in an adaptive way subject to the values of a positive parameter as

$$(\delta_k, \gamma_k) = \begin{cases} \left(\dfrac{s_k^T B_k s_k}{s_k^T B_k s_k + y_k^T s_k}, \dfrac{y_k^T s_k}{s_k^T B_k s_k + y_k^T s_k} \right), & \text{if } \dfrac{s_k^T B_k s_k}{s_k^T B_k s_k + y_k^T s_k} \geq \tau_k, \\ (\tau_k, 1), & \text{otherwise}, \end{cases} \tag{1.71}$$

where $0 < \tau_k < 1$. (Liao proposed $\tau_k = \exp(-1/k^2)$.) Liao proved that the scaling BFGS method given by (1.70) and (1.71) with the Wolfe line search generates iterations which converge superlinearly to the optimal solution. Limited numerical experiments with Liao's scaling BFGS method proved that this is competitive with the standard BFGS method and it corrects large eigenvalues better than the standard BFGS method. However, subject to other scaling BFGS updates, the scaling BFGS update by Liao is less efficient and less robust see Andrei (2018d).

BFGS with modified secant equation

The standard secant Equation (1.46) or its equivalent form (1.51) uses only the gradient information, without referring to the function values. Therefore, to obtain better approximations for the Hessian, the idea was to modify the secant equation in order to include more available information about the minimizing function f. Wei, Li, and Qi (2006a) and Wei, Yu, Yuan, and Lian (2004) proposed the following modified secant equation. Suppose that the objective function f is smooth enough. From the Taylor series, we obtain

$$f_k = f_{k+1} - s_k^T g_{k+1} + \frac{1}{2} s_k^T \nabla^2 f(x_{k+1}) s_k - \frac{1}{6} s_k^T (T_{k+1} s_k) s_k + O(\|s_k\|^4), \quad (1.72)$$

where

$$s_k^T (T_{k+1} s_k) s_k = \sum_{i,j,l}^{n} \frac{\partial^3 f(x_{k+1})}{\partial x^i \partial x^j \partial x^l} s_k^i s_k^j s_k^l. \quad (1.73)$$

After some simple algebraic manipulations, from (1.72) we get

$$\begin{aligned} s_k^T \nabla^2 f(x_{k+1}) s_k &= s_k^T y_k + 2(f_k - f_{k+1}) + s_k^T (g_k + g_{k+1}) \\ &\quad + \frac{1}{3} s_k^T (T_{k+1} s_k) s_k + O(\|s_k\|^4). \end{aligned}$$

Therefore, neglecting the terms which include the tensor, the following approximation is obtained:

$$s_k^T \nabla^2 f(x_{k+1}) s_k = s_k^T y_k + \vartheta_k,$$

where

$$\vartheta_k = 2(f_k - f_{k+1}) + s_k^T (g_k + g_{k+1}). \quad (1.74)$$

With these, the following *modified secant equation* is obtained

$$B_{k+1} s_k = z_k, \quad z_k = y_k + \frac{\vartheta_k}{s_k^T u_k} u_k, \quad (1.75)$$

where $u_k \in \mathbb{R}^n$ is a vector satisfying $s_k^T u_k \neq 0$, see: Yuan (1991), Yuan and Byrd (1995), Babaie-Kafaki (2011).

Another modified secant equation was suggested by Zhang, Deng, and Chen (1999). From the Taylor series, we get

$$s_k^T g_k = s_k^T g_{k+1} - s_k^T \nabla^2 f(x_{k+1}) s_k + \frac{1}{2} s_k^T (T_{k+1} s_k) s_k + O(\|s_k\|^4). \tag{1.76}$$

Now, combining (1.72) and (1.76) and cancelling the terms including the tensor, it follows that

$$s_k^T \nabla^2 f(x_{k+1}) s_k = s_k^T y_k + 3\vartheta_k + O(\|s_k\|^4),$$

where ϑ_k is defined in (1.74). Therefore, the following *modified secant equation* can be obtained

$$B_{k+1} s_k = w_k, \; w_k = y_k + \frac{3\vartheta_k}{s_k^T u_k} u_k, \tag{1.77}$$

where $u_k \in \mathbb{R}^n$ is a vector satisfying $s_k^T u_k \neq 0$.

The theoretical advantages of these modified secant Equations (1.75) and (1.77) can be seen from the following theorem, which shows the accuracy of these modified secant equations versus the standard secant Equation (1.51) see: Zhang, Deng and Chen (1999), Wei, Li and Qi (2006a).

Theorem 1.16 *If the function f is sufficiently smooth and* $\|s_k\|$ *is small enough, then the following estimating relations hold:*

$$s_k^T \nabla^2 f(x_{k+1}) s_k - s_k^T y_k = \tfrac{1}{2} s_k^T (T_{k+1} s_k) s_k + O(\|s_k\|^3),$$
$$s_k^T \nabla^2 f(x_{k+1}) s_k - s_k^T z_k = \tfrac{1}{3} s_k^T (T_{k+1} s_k) s_k + O(\|s_k\|^4),$$
$$s_k^T \nabla^2 f(x_{k+1}) s_k - s_k^T w_k = O(\|s_k\|^4),$$

where T_{k+1} *is the tensor of f at* x_{k+1} *defined in* (1.73). ♦

In this context, we can see that for quadratic objective functions, $\vartheta_k = 0$ and therefore the modified secant Equations (1.75) and (1.77) reduce to the standard secant equation. Concerning the vector u_k, it can usually be selected as $u_k = s_k$ or $u_k = y_k$ provided that the line search satisfies the Wolfe conditions (1.12) and (1.13). To get positive definite quasi-Newton approximations for the Hessian based on the modified secant Equations (1.75) or (1.77), we should have $s_k^T z_k > 0$ and $s_k^T w_k > 0$, respectively. To overcome this difficulty, a simple procedure is to replace ϑ_k in (1.75) and (1.77) by $\max\{0, \vartheta_k\}$.

For nonconvex objective functions, Li and Fukushima (2001a), (2001b) proposed a new modified BFGS (called *cautious* BFGS), for which the local and the global superlinear convergence were proved. The method is based on the following modified secant equation

$$B_{k+1}s_k = \bar{y}_k, \quad \bar{y}_k = y_k + h_k\|g_k\|^r s_k, \tag{1.78}$$

where r is a positive constant and h_k is defined as

$$h_k = C + \max\{0, -\frac{y_k^T s_k}{\|s_k\|^2}\}\|g_k\|^{-r},$$

for a positive constant C. In the cautious BFGS method, the update of B_k is defined as:

$$B_{k+1} = \begin{cases} B_k - \dfrac{B_k s_k s_k^T B_k}{s_k^T B_k s_k} + \dfrac{y_k y_k^T}{y_k^T s_k}, & \text{if } \dfrac{y_k^T s_k}{\|s_k\|^2} \geq \varepsilon\|g_k\|^\gamma, \\ B_k, & \text{otherwise,} \end{cases} \tag{1.79}$$

where ε and γ are positive constants. For the modified secant Equation (1.78), $\bar{y}_k^T s_k > 0$ independently by the line search conditions and by the objective function convexity. This ensures heredity of positive definiteness of the corresponding BFGS update (Guo, Liu, & Wang, 2008).

Recently, Babaie-Kafaki (2012), (2013), (2014) and Babaie-Kafaki and Ghanbari (2014a) proposed scaled memoryless BFGS methods with modified secant Equations (1.75), (1.77), or (1.78) which satisfy the sufficient descent property $d_k^T g_k \leq -c\|g_k\|^2$, where c is a positive constant. A new approach using the polynomial interpolation of the data from the most recent m steps in modified secant equations was developed by Ford and Moghrabi (1994), (1996a), (1996b) and by Ford, Narushima, and Yabe (2008).

BFGS with modified line search

Usually, the BFGS method is implemented by using the Wolfe line search (1.12) and (1.13). This is important since it ensures that $y_k^T s_k > 0$ for any $k = 1, \ldots,$ thus conserving the positive definiteness of the BFGS approximations along the iterations. Suppose that the gradient of the minimizing function is Lipschitz continuous with constant L. Let L_k be an approximation of L. Wan, Huang, and Zheng (2012) proposed the cautious BFGS method (1.79) with the following modification of the Armijo line search procedure. Set $\beta_k = -g_k^T d_k/(L_k\|d_k\|^2)$. "*Find the stepsize* α_k *as the largest component in the set* $\{\beta_k, \beta_k\rho, \beta_k\rho^2, \ldots\}$ *so that the inequality*

$$f(x_k + \alpha_k d_k) \leq f(x_k) + \sigma\alpha_k(g_k^T d_k - \frac{1}{2}\alpha_k\mu L_k\|d_k\|^2) \tag{1.80}$$

holds, where $\sigma \in (0,1)$, $\mu \in [0,\infty)$, *and* $\rho \in (0,1)$ *are given constants.*" Under classical assumptions, if $\|B_k s_k\| \leq a_1\|s_k\|$ and $a_2\|s_k\|^2 \leq s_k^T B_k s_k$, then $\liminf_{k\to\infty}\|g_k\| = 0$, where a_1 and a_2 are positive constants.

Another interesting modification of the strong Wolfe line search (1.14) was given by Wan, Teo, Shen, and Hu (2014). In the Wolfe line search, the choice of σ

can affect the cost of finding a stepsize satisfying (1.12) and (1.13) or (1.12) and (1.14). For a larger value of σ, the cost of searching a stepsize decreases. Therefore, it seems reasonable to select a sufficiently large value for σ. On the other hand, for a large value for σ, the obtained stepsize might be far away from the optimal one, i.e., the one obtained by the exact line search (1.9). With the value of $s_k^T y_k$ far away from $-s_k^T g_k$, it is clear that $g(x_k + \alpha_k d_k)^T d_k$ is far away from zero. In other words, a large value for σ may give rise to a stepsize, which is not a good approximation to the one obtained by the exact line search. To overcome this difficulty, Wan, Teo, Shen, and Hu (2014) replaced the strong Wolfe condition (1.14) by:

$$-(1+\sigma^U)s_k^T g_k \geq s_k^T y_k \geq -(1-\sigma^L)s_k^T g_k, \tag{1.81}$$

where σ^L and σ^U are two sufficiently small constants satisfying: $0 \leq \sigma^L \leq \sigma < 1$ and $0 \leq \sigma^U \leq \sigma$. See also Al-Baali and Grandinetti (2009). As above, it is proved that, if $\|B_k s_k\| \leq b_1 \|s_k\|$ and $b_2 \|s_k\|^2 \leq s_k^T B_k s_k$, where B_k is the BFGS update, then $\liminf_{k\to\infty} \|g_k\| = 0$, where b_1 and b_2 are positive constants. Numerical experiments reported by the authors show that this variant of BFGS with modified line search (1.81) is competitive versus standard BFGS or the cautious BFGS (1.79).

Dai (2003a) presented a numerical example showing that the standard BFGS method fails for nonconvex functions under the Wolfe line search. Therefore, the question is: "*are there any other inexact line searches that possess the global convergence of the BFGS method for general functions*?" A positive answer was given by Yuan, Wei, and Lu (2017) and Yuan, Sheng, Wang, Hu, and Li (2018). They presented the following modified Wolfe line search:

$$f(x_k + \alpha_k d_k) \leq f_k + \rho \alpha_k g_k^T d_k + \alpha_k \min\left\{-\rho_1 g_k^T d_k, \rho \frac{\alpha_k}{2} \|d_k\|^2\right\}, \tag{1.82}$$

$$g(x_k + \alpha_k d_k)^T d_k \geq \sigma g_k^T d_k + \min\left\{-\rho_1 g_k^T d_k, \rho \alpha_k \|d_k\|^2\right\}, \tag{1.83}$$

where $\rho \in (0, 1/2)$, $\rho_1 \in (\rho/2, \rho)$ and $\sigma \in (\rho, 1)$. Under classical assumptions if $\|B_k s_k\| \leq b_1 \|s_k\|$ and $b_2 \|s_k\|^2 \leq s_k^T B_k s_k$, where B_k is the BFGS update (1.53), then for the corresponding algorithm with modified line search given by (1.82) and (1.83), $\liminf_{k\to\infty} \|g_k\| = 0$, where b_1 and b_2 are positive constants.

An improvement of the global convergence of the BFGS method with Yuan–Wei–Lu line search (1.82) and (1.83) was presented by Dehmiry (2019):

$$f(x_k + \alpha_k d_k) \leq f_k + \rho \alpha_k g_k^T d_k - \rho \frac{\alpha_k^2}{2\beta_k} \|d_k\|^2, \tag{1.84}$$

$$g(x_k + \alpha_k d_k)^T d_k \geq \sigma g_k^T d_k - \frac{\rho \alpha_k}{\beta_k} \|d_k\|^2, \tag{1.85}$$

$$\alpha_k < -\frac{\beta_k(1-\sigma)}{\varepsilon_0\beta_k + \rho}\frac{g_k^T d_k}{\|d_k\|^2}, \tag{1.86}$$

where $\rho \in (0, 1/2)$,$\sigma \in (\rho, 1)$, ε_0 is a small parameter ($\varepsilon_0 = 10^{-6}$) and $\{\beta_k\}$ is an arbitrary increasing sequence of positive numbers so that $\lim_{k\to\infty} \beta_k/k = +\infty$. Like in Byrd and Nocedal (1989), Dhemiry (2019) proved that there exist constants $b_1 > b_2 > 0$ such that $\|B_k s_k\| \leq b_1\|s_k\|$ and $b_2\|s_k\|^2 \leq s_k^T B_k s_k$ for at least $t/2$ values of $k \in \{1,\ldots,t\}$ with any positive integer t. Therefore, the sequence $\{g_k\}$ generated by the corresponding algorithm with modified line search given by (1.84)–(1.86), satisfies $\liminf_{k\to\infty}\|g_k\| = 0$. Numerical experiments with BFGS algorithm where the stepsize is determined by the modified Yuan–Wei–Lu line search (1.84)–(1.86) show that it is more efficient and more robust versus the algorithm with Yuan–Wei–Lu line search (1.82) and (1.83) and versus the Li and Fukushima (2001b) algorithm.

1.4.5 Quasi-Newton Methods with Diagonal Updating of the Hessian

A relative recent idea to generate simple minimization algorithms for the unconstrained optimization in the frame of the quasi-Newton methods is to approximate the Hessian of the minimizing function by a diagonal matrix with positive diagonal elements. This approach was introduced by Gill and Murray (1979) and discussed by Gilbert and Lemaréchal (1989). The search direction is computed as

$$d_{k+1} = -B_{k+1}^{-1} g_{k+1}, \tag{1.87}$$

where $B_{k+1} = diag(b_{k+1}^1,\ldots,b_{k+1}^n)$ is a positive definite diagonal matrix retaining only the diagonal elements of the BFGS update matrix (1.53):

$$b_{k+1}^i = b_k^i - \frac{(b_k^i)^2(s_k^i)^2}{\sum_{i=1}^n b_k^i (s_k^i)^2} + \frac{(y_k^i)^2}{y_k^T s_k}, \quad i = 1,\ldots,n. \tag{1.88}$$

This diagonal-updating approach uses only $O(n)$ storage. If we assume that $y_k^T s_k > 0$, then B_{k+1} is positive definite.

Another quasi-Newton algorithm with diagonal approximation to the Hessian is using the weak quasi-Newton equation $s_k^T B_{k+1} s_k = s_k^T y_k$ and was introduced and studied by Dennis and Wolkowicz (1993). The update proposed by Dennis and Wolkowicz is as follows

$$B_{k+1} = B_k + \frac{s_k^T y_k - s_k^T B_k s_k}{(s_k^T B_k s_k)^2} B_k s_k s_k^T B_k, \tag{1.89}$$

where B_k is positive definite. The condition $s_k^T y_k > 0$ implies that B_{k+1} in (1.89) is also positive definite. If B_k is taken to be a positive definite diagonal matrix $B_k = diag(b_k^1, \ldots, b_k^n)$, then (1.89) can be restricted to update only the diagonal elements of B_{k+1} as

$$b_{k+1}^i = b_k^i + \frac{s_k^T y_k - \sum_{i=1}^n b_k^i (s_k^i)^2}{\left(\sum_{i=1}^n b_k^i (s_k^i)^2\right)^2} (b_k^i)^2 (s_k^i)^2, \; i = 1, \ldots, n, \tag{1.90}$$

yielding a positive definite diagonal matrix. The search direction in this algorithm is computed as in (1.87), where the diagonal elements of B_{k+1} are computed as in (1.90).

The diagonal quasi-Newton approximation algorithm presented by Zhu, Nazareth, and Wolkowicz (1999) is as follows. Suppose that B_k is a positive definite diagonal matrix and B_{k+1}, which is also diagonal, is the updated version of B_k. The algorithm requires that the updated B_{k+1} satisfies the weak quasi-Newton secant equation and the deviation between B_k and B_{k+1} is minimized under a variational principle. The search direction is computed as $d_{k+1} = -B_{k+1}^{-1} g_{k+1}$, where the diagonal elements of $B_{k+1} = diag(b_{k+1}^1, \ldots, b_{k+1}^n)$ are computed as

$$b_{k+1}^i = b_k^i + \frac{s_k^T y_k - \sum_{i=1}^n b_k^i (s_k^i)^2}{\sum_{i=1}^n (s_k^i)^4} (s_k^i)^2, \; i = 1, \ldots, n. \tag{1.91}$$

In the same way of developments, Andrei (2019c) presented a new diagonal quasi-Newton updating method, where the elements of the diagonal matrix approximating the Hessian are determined by minimizing both the size of the change from the previous estimate and the trace of the update, subject to the weak secant equation. The search direction in this algorithm is computed as

$$d_{k+1}^i = -\frac{g_{k+1}^i}{b_{k+1}^i}, \; i = 1, \ldots n, \tag{1.92}$$

and

$$b_{k+1}^i = b_k^i + \frac{s_k^T y_k + s_k^T s_k - \sum_{i=1}^n b_k^i (s_k^i)^2}{\sum_{i=1}^n (s_k^i)^4} (s_k^i)^2 - 1, \quad i = 1, \ldots n. \tag{1.93}$$

Another approach using the same paradigm of diagonal quasi-Newton updating method was given by Andrei (2018e), where the diagonal elements are determined by minimizing the measure function of Byrd and Nocedal. The components of the search direction in this approach are computed as:

$$d_{k+1}^i = -g_{k+1}^i(1+\lambda(s_k^i)^2),\ i = 1,\ldots,n, \tag{1.94}$$

where

$$\lambda = \begin{cases} r+\theta, & \text{if } \ \bar{\lambda}<r, \\ \bar{\lambda} & \text{if } \ \bar{\lambda}\geq r, \end{cases} \text{ and } \bar{\lambda} = \frac{t(s_k^T g_{k+1}) - y_k^T g_{k+1}}{\sum_{i=1}^n (y_k^i g_{k+1}^i (s_k^i)^2)},$$

and

$$r = -1/(s_k^j)^2 = \max_{i=1,\ldots,n;s_k^i \neq 0} \{-1/(s_k^i)^2\}.$$

t is a positive parameter and θ is a small perturbation (for example $\theta = 1$).

Andrei (2019d) has recently presented a new quasi-Newton method, in which the Hessian of the function f is approximated as a positive definite diagonal matrix. In this method, the approximation Hessian B_{k+1} is a diagonal matrix computed as:

$$B_{k+1} = Y_k S_k^{-1}, \tag{1.95}$$

where $Y_k = diag(y_k^1,\ldots,y_k^n)$ and $S_k = diag(s_k^1,\ldots,s_k^n)$, y_k^i, $i = 1,\ldots n$, being the components of the vector y_k and s_k^i, $i = 1,\ldots n$, being the components of the vector s_k. In other words,

$$b_{k+1}^i = \frac{y_k^i}{s_k^i} = \frac{g_{k+1}^i - g_k^i}{x_{k+1}^i - x_k^i} = \frac{g^i(x_k + \alpha_k d_k) - g^i(x_k)}{\alpha_k d_k^i},\ i = 1,\ldots n, \tag{1.96}$$

where g_k^i, is the i-th component of the gradient in x_k and d_k^i is the i-th component of the search direction d_k. Therefore, in this approach, the element b_{k+1}^i may be considered as an approximation of the second order derivative of function f, corresponding to the i-th diagonal element of the Hessian computed in x_{k+1} by a *scaled forward finite differences directional derivative scheme.*

Intensive numerical experiments with these diagonal quasi-Newton methods reported in Andrei (2018e), (2019d) proved that the algorithm in which the diagonal elements of the approximation to the Hessian are computed by a scaled forward finite differences directional derivative scheme is competitive among the algorithms in this class.

1.4.6 Limited-Memory Quasi-Newton Methods

Limited-memory quasi-Newton methods are dedicated to solving large-scale unconstrained optimization problems whose Hessian matrix cannot be computed and stored at a reasonable cost (Nocedal, 1980). Instead of storing an $n \times n$ approximation to the Hessian, these methods save only a few vectors which can be used to represent the approximation implicitly. Among different limited-memory methods, the best known and used one is L-BFGS, which is based on the BFGS update. The main idea behind this method is to use the curvature information from only the most recent iterations in order to construct the Hessian approximation.

As known, each step of the BFGS method is computed as

$$x_{k+1} = x_k - \alpha_k H_k g_k, \; k = 0, 1, \ldots, \tag{1.97}$$

where α_k is the stepsize, $g_k = \nabla f(x_k)$ and H_k is the approximation to the inverse Hessian updated by the formula

$$H_{k+1} = V_k^T H_k V_k + \rho_k s_k s_k^T, \tag{1.98}$$

where

$$\rho_k = \frac{1}{y_k^T s_k}, \; V_k = I - \rho_k y_k s_k^T. \tag{1.99}$$

Since the inverse Hessian H_k is generally a dense matrix, the limited-memory BFGS method will implicitly store a modified version of H_k by storing only a certain number (say m) of the vector pairs $\{s_i, y_i\}$ used in the updating formulae (1.98) and (1.99). After the new iterate has been computed, the oldest vector pair in the set of pairs $\{s_i, y_i\}$ is replaced by the new pair $\{s_k, y_k\}$ obtained from the current iteration (Nocedal & Wright, 2006). Therefore, the set of vector pairs includes curvature information from the m most recent iterations. It is worth mentioning that only a small number of vector pairs, between 3 and 11, needs to be used for solving large-scale optimization problems. The product $H_k g_k$ is computed during the updating process by performing a sequence of inner products and vector summations involving only g_k and the pairs $\{s_i, y_i\}$. Nocedal and Wright (2006, page 178) present the two-loop recursion L-BFGS algorithm to compute the product $H_k g_k$ in (1.97). At iteration k, the current iterate is x_k and the set of vector pairs is given by $\{s_i, y_i\}$ for $i = k - m, \ldots, k - 1$. Selecting some initial Hessian approximation H_k^0, the two-loop recursion algorithm for the search direction computation is as follows (Liu & Nocedal, 1989).

Algorithm 1.6 L-BFGS algorithm

Set $q = g_k$. The *first loop* is:
For $i = k-1, k-2, \ldots, k-m$ compute: $\alpha_i = \rho_i s_i^T q$, $q = q - \alpha_i y_i$ End for.
Set $r = H_k^0 q$.

The *second loop* is:
For $i = k-m, k-m+1, \ldots, k-1$ compute: $\beta = \rho_i y_i^T r$, $r = r + s_i(\alpha_i - \beta)$. End for. Set $H_k g_k = r$. ♦

Usually, H_k^0 is chosen as a diagonal matrix. Often, $H_k^0 = \gamma_k I$, where $\gamma_k = (s_{k-1}^T y_{k-1})/\|y_{k-1}\|^2$. The parameter γ_k is a scaling factor that attempts to estimate the size of the true Hessian along the most recent search direction. This selection of γ_k ensures that the search direction is well scaled and therefore, as a by product, the stepsize $\alpha_k = 1$ is accepted in most iterations (Gilbert & Lemaréchal, 1989). In the implementation of Liu and Nocedal (1989), L-BFGS is one of the best algorithms (and computing programs) for solving large-scale unconstrained optimization problems.

1.4.7 Truncated Newton Methods

The main idea behind the *truncated Newton methods* introduced by Dembo, Eisenstat, and Steihaug (1982) and analyzed by Dembo and Steihaug (1983) and Deuflhard (1990) is to approximate the solution of the Newton system $\nabla^2 f(x_k) d_k = -g_k$ for the search direction d_k. Since far away from a local minimum the objective function cannot be well approximated by a quadratic model, it follows that it is not necessary to spend too much time on computing an accurate Newton search vector. Therefore, for $k = 0, 1, \ldots$ the search direction d_k is determined so that

$$\left\|\nabla^2 f(x_k) d_k + g_k\right\| \leq \eta_k \|g_k\|, \tag{1.100}$$

after which a new estimation $x_{k+1} = x_k + \alpha_k d_k$ is computed. The problem with this approach is how to choose the sequence $\{\eta_k\}$ convergent to zero to emphasize the reduction of the residual

$$r_k = \left\|\nabla^2 f(x_k) d_k + g_k\right\|. \tag{1.101}$$

The local convergence of the truncated Newton method is obtained simply by ensuring that the sequence $\{\eta_k\}$ is bounded away from 1.

Theorem 1.17 *Suppose that $\nabla^2 f(x)$ exists and is continuous in a neighborhood of a minimizer x^*, with $\nabla^2 f(x^*)$ positive definite. Consider the iteration $x_{k+1} = x_k + d_k$, where d_k satisfies $\|r_k\| \leq \eta_k \|g_k\|$ and assume that $\eta_k \leq \eta$ for some* *constant* $\eta \in (0,1)$. *If the starting point* x_0 *is sufficiently close to* x^*, *then the sequence* $\{x_k\}$ *generated by the truncated Newton method converges to* x^* *and satisfies*

$$\left\|\nabla^2 f(x^*)(x_{k+1} - x^*)\right\| \leq \hat{\eta}\left\|\nabla^2 f(x^*)(x_k - x^*)\right\|,$$

for some constant $\hat{\eta}$ *with* $\eta < \hat{\eta} < 1$. ♦

The best selection of the sequence $\{\eta_k\}$ is unknown. The only requirement is $\eta_k \to 0$. If $\eta_k \to 0$, then the rate of convergence of the truncated Newton method is superlinear. If in addition, $\nabla^2 f(x)$ is Lipschitz continuous for x near x^*, and if $\eta_k = O(\|\nabla f(x_k)\|)$, then the convergence is quadratic.

In the practical implementation of the method, two selections are used: $\eta_k = 1/k$ or $\eta_k = \|g_k\|$. Dembo and Steihaug (1983) showed that if

$$\eta_k = \min\left\{\frac{1}{k}, \|g_k\|^t\right\}, \tag{1.102}$$

where $0 < t \leq 1$, then the rate of convergence of the truncated Newton method is of order $O(1+t)$. The superlinear convergence is obtained if, for example, $\eta_k = \min\{1/2, \sqrt{\|g_k\|}\}$. Moreover, the choice $\eta_k = \min\{1/2, \|g_k\|\}$ would yield quadratic convergence (Nocedal & Wright, 2006).

Observe that the truncated Newton method has two imbedded loops. The first *exterior* loop implements the Newton algorithm. The second *inner* loop implements the mechanism for an approximate solution to the Newton system. Two implementations of this method are known. In the first one, given by Schlick and Fogelson (1992a), (1992b), the Hessian matrix is computed by finite differences and an approximate solution to the preconditioned Newton system is computed. In the second one, given by Nash (1985), a BFGS approximation of the Hessian is computed and an approximate solution to the Newton system is determined.

The numerical study on the performances of the limited-memory BFGS method L-BFGS (Liu, & Nocedal, 1989) and on the truncated-Newton TN (Nash, 1985) was given by Nash and Nocedal (1991). Some of their conclusions are as follows. L-BFGS and TN use different principles to compute the search direction. L-BFGS uses a fixed, low-cost formula requiring no extra derivative information. TN uses an elaborate, variable-cost iteration with partial second-derivative information. Both use cubic interpolation to obtain the strong Wolfe conditions. The numerical experiments showed that neither algorithm is clearly superior to the other. In terms of CPU computing time, neither algorithm is a clear winner: the higher iteration cost of TN is compensated by a much lower iteration count, on average. The performance of these algorithms appears to be correlated with the degree of nonlinearity: for quadratic and approximately quadratic problems, TN outperforms L-BFGS. For highly nonlinear problems, L-BFGS is the best. In terms of function evaluation, L-BFGS is preferable to TN for more highly nonlinear problems. However, TN almost always requires many fewer iterations than L-BFGS, and therefore, if the number of gradient evaluations in the inner iteration could be significantly reduced, TN would be competitive or more efficient than L-BFGS. Subject to the distribution of the eigenvalues of the Hessian, there is no clear correlation between the success of the methods and the eigenvalue structure. For problems with ill-conditioned Hessians, TN seems to be better. Clustering of eigenvalues at the solution does not seem to benefit one method more than the other.

1.4.8 Conjugate Gradient Methods

Having a very simple iteration formula and very modest memory requirements, conjugate gradient methods are very well suited for solving large-scale unconstrained optimization problems. The method was introduced by Hestenes and Stifel (1952) for minimizing convex quadratic functions with symmetric and positive definite matrices. This is known as *linear conjugate gradient* method, which is presented and analyzed in Chapter 2 of this book. Later on, Fletcher and Reeves (1964) extended the method for solving nonlinear unconstrained optimization problems. The search direction in these methods is computed as

$$d_{k+1} = -g_{k+1} + \beta_k d_k, \ k = 0, 1, \ldots, \tag{1.103}$$

where $d_0 = -g_0$ and β_k is a scalar called the *conjugate gradient parameter*. There is a large variety of nonlinear conjugate gradient methods and in the following we shall give a general overview of them. Some of these methods will be presented and developed along the chapter of the book by showing their theoretical properties and their efficiency and robustness for solving large-scale nonlinear unconstrained optimization problems. Different conjugate gradient methods correspond to different choices for the conjugate gradient parameter β_k. Reviews of these methods can be found in: Nazareth (2001), Hager and Zhang (2006b), Narushima and Yabe (2014), Babaie-Kafaki (2016). Good survey and new developments highlighting the linear–nonlinear conjugate gradient connection with different optimization techniques and application domains are described in The Proceedings of the AMS-IMS-SIAM Joint Summer Research Conference on "Linear and Nonlinear Conjugate Gradient-Related Methods" (SIAM, Philadelphia, 1996) edited by Adams and Nazareth (1996).

When the minimizing function f is a strictly convex quadratic function and the line search is exact, then all the nonlinear conjugate gradient algorithms are equivalent. However, for general functions, their behavior may be quite different (Powell, 1984a), (Dai & Ni, 2003), (Andrei, 1995, 2007d, 2013d). It should be said that the search directions of the linear conjugate gradient methods are conjugate directions. However, for nonlinear conjugate gradient methods, the search directions may not be conjugate directions. Concerning the convergence, Crowder and Wolfe (1969) established that the nonlinear conjugate gradient method without restarts is only linearly convergent. If the method is restarted along the negative gradient after n steps, then McCormick and Ritter (1974) established the n-step quadratic convergence rate.

Many conjugate gradient methods were proposed in literature. Some of them are known as *standard conjugate gradient* methods: HS (Hestenes & Stiefel, 1952), FR (Fletcher & Reeves, 1964), PRP (Polak & Ribiére, 1969; Polyak, 1969), CD (Fletcher, 1987), LS (Liu & Storey, 1991) and DY (Dai & Yuan, 1999). These methods are classified by the formula for β_k computation in (1.103) and are discussed in Chapter 4 of this book.

Other methods proposed are known as the *hybrid conjugate gradient* methods and the *parameterized conjugate gradient* methods. The hybrid methods are based on the *projection* idea: (Touati-Ahmed & Storey, 1990; Hu & Storey, 1991; Gilbert & Nocedal, 1992; Dai & Yuan, 2001a; Yabe & Takano, 2004). Others are based on the idea of *convex combination* of the standard conjugate gradient methods: (Andrei, 2008b, 2008d, 2008e, 2009a, 2010a; Liu & Li, 2014). The hybrid and the parameterized conjugate gradient methods are to be presented in Chapter 6 of this book.

New conjugate gradient methods were suggested as being based on different ingredients in order to improve their convergence to the solution and to reduce jamming. Thus, the following methods can be mentioned: *modified of the classical schemes*: (Dai & Liao, 2001; Hager & Zhang; 2005; Andrei, 2009b; Dai & Kou, 2013); *memoryless conjugate gradient methods BFGS preconditioned* (Perry, 1976; Shanno, 1978a, 1978b; Andrei, 2007a, 2017b); *conjugate gradient method with guaranteed descent* (Hager & Zhang, 2005, 2006a); *spectral conjugate gradient methods* (Birgin & Martínez, 2001); *scaled conjugate gradient methods* (Andrei, 2007a, 2007b, 2007c, 2008a); *three-term conjugate gradient methods* (Beale, 1972; Nazareth, 1977; Andrei, 2007a, 2011a, 2013a, 2013b, 2015a), (Babaie-Kafaki & Ghanbari, 2014a); *conjugate gradient methods with sufficient descent* (Andrei, 2008c); *conjugate gradient methods with modified secant equation* (Andrei, 2008b); *conjugate gradient method with finite difference Hessian/vector product approximation* (Andrei, 2009d). Some of these conjugate gradient methods will be presented and analyzed in the chapters of this book.

The relationship between the conjugate gradient methods and the quasi-Newton methods was given by Perry (1976), who noticed that (1.103) with HS conjugate gradient parameter can be rewritten as $d_{k+1} = -Q_{k+1}g_{k+1}$, where Q_{k+1} is a nonsymmetric matrix. Later on, Shanno (1978a), using the idea of Perry, discovered that *the conjugate gradient methods are precisely the BFGS quasi-Newton method, where the approximation to the inverse Hessian is restarted as the identity matrix at every step*. Thus, no significant storage is used to develop a better approximation to the inverse Hessian. This was the starting point for the *memoryless conjugate gradient methods BFGS preconditioned* and for the *self-scaling conjugate gradient methods*. This idea was also discussed by Nazareth (1975, 1979) and Buckley (1978a). An excellent survey concerning the relationship between the conjugate gradient methods and the quasi-Newton methods was given by Nazareth (1986).

Different modifications of the HS conjugate gradient method were given by: Dai and Liao (2001), Hager and Zhang (2005, 2006a), and Dai and Kou (2013). Thus, the conjugate gradient method with guaranteed descent and the conjugate gradient method with new conjugacy condition were later developed (Andrei, 2012, 2013c). Babaie-Kafaki and Ghanbari (2014b) studied some suitable choices of parameters incorporated into the Dai–Liao method. However, selection of an optimal parameter in the Dai–Liao method is still an open problem (Andrei, 2011b). The combination of the search directions of the conjugate gradient methods with the two-point stepsize gradient algorithms of Barzilai and Borwein (1988) and the use of the secant equation generated the spectral conjugate gradient method, proposed by Birgin and Martínez (2001). This was later developed by Andrei (2007a) as a double quasi-Newton update scheme (for applications see Wong and Protas 2013). The concept of three-term conjugate gradient method was originally introduced by Beale (1972) and has been used ever since. As some conjugate gradient methods such as HS, FR and PRP cannot generally guarantee the descent condition $g_k^T d_k < 0$, where g_k is the gradient of the minimizing function *f*, Zhang, Zhou, and Li (2006a, 2006b, 2007) proposed the three-term versions of these conjugate gradient methods in order to overcome this drawback. Furthermore, Narushima, Yabe, and Ford (2011) proposed a three-term conjugate gradient method that involves the scaled conjugate gradient methods. Based on the memoryless BFGS quasi-Newton method, Andrei (2007a, 2011a, 2013a, 2014) proposed different three-term conjugate gradient methods that generate descent directions under the Wolfe line search. Nonlinear conjugate gradient methods are a class of important methods for solving unconstrained optimization problems. Surveys on these methods are presented by: Hager and Zhang (2006b), Dai (2010, 2011), Narushima and Yabe (2014) and Babaie-Kafaki (2016).

1.4.9 Trust-Region Methods

Trust-region methods do not use line search. They generate the steps with the help of a quadratic model of the objective function. These methods define a region around the current point within which they trust the model to be an adequate representation of the minimizing function and choose the step to be an approximate minimizer of the model in this region. Therefore, the trust-region methods choose the search direction and the stepsize simultaneously. If a step is not acceptable, then the size of the region is reduced and a new minimizer is computed. For solving the problem (1.1), at each iteration a trust-region method computes a trial step s_k as solution of the following *trust-region subproblem*:

$$\min_{d \in \mathbb{R}^n} g_k^T d + \frac{1}{2} d^T B_k d = m_k(d) \text{ subject to } \|d\| \le \Delta_k, \tag{1.104}$$

where B_k is an $n \times n$ symmetric matrix which approximates the Hessian of function f at x_k and $\Delta_k > 0$ is a *trust-region radius*. In (1.104), $m_k(.)$ is the quadratic *model* of f around x_k and $\|.\|$ is the Euclidian norm. If B_k is an approximation to the Hessian, then the difference between $m_k(d)$ and $f(x_k + d)$ is of order $O(\|d\|^2)$, which is small when d is small. If $B_k = \nabla^2 f(x_k)$, then the approximation error in the model m_k is of order $O(\|d\|^3)$, so the model is accurate when $\|d\|$ is small. The size of the trust-region is critical to the effectiveness of each step. Thus, the trust-region approach requires solving a sequence of subproblems (1.104) in which the objective and the constrained are both quadratic. A key ingredient in a trust-region method is the strategy for choosing the trust-region radius Δ_k at each iteration. A trust-region method uses the ratio

$$r_k = \frac{Ared_k}{Pred_k} = \frac{f(x_k) - f(x_k + s_k)}{m_k(0) - m_k(s_k)} \tag{1.105}$$

which decides whether the trial step s_k is acceptable and how the next trust-region radius is chosen. The numerator of r_k is called the *actual reduction* and the denominator of r_k is the *predicted reduction*. The idea of this method is as follows. Since the step s_k is obtained by minimizing the model m_k over a region that includes $s = 0$, it follows that the predicted reduction will always be nonnegative. Therefore, if r_k in (1.105) is negative, the new objective value $f(x_k + s_k)$ is greater than the current value $f(x_k)$, so the step must be rejected. On the other hand if r_k is close to 1, then there is a good agreement between the model m_k and the function f over this step, so it is safe to expand the trust-region region for the next iteration. If r_k is positive but significantly smaller than 1, the trust region is not altered, but if it is close to zero or negative, then the trust region is shrinked by reducing the trust-region radius Δ_k for the next iteration (Nocedal and Wright, 2006).

The algorithm solves the trust-region subproblem (1.104) to get the solution s_k. The solution and the trust-region parameter are updated as:

$$x_{k+1} = \begin{cases} x_k, & \text{if } r_k \le \tau_0, \\ x_k + s_k, & \text{otherwise} \end{cases} \qquad \Delta_{k+1} = \begin{cases} [\tau_3 \|s_k\|,\ \tau_4 \Delta_k], & \text{if } r_k < \tau_2, \\ [\Delta_k,\ \tau_1 \Delta_k], & \text{otherwise} \end{cases}$$

where τ_i, $i = 0, 1, \ldots, 4$ are so that $0 < \tau_3 < \tau_4 < 1 < \tau_1$, $0 \le \tau_0 \le \tau_2 < 1$, $\tau_2 > 0$. The approximation of the Hessian is updated to get B_{k+1}. Typical values of the constants τ_i, $i = 0, \ldots 4$, are $\tau_0 = 0$, $\tau_1 = 2$, $\tau_2 = \tau_3 = 0.25$ and $\tau_4 = 0.5$ (Moré, 1983; Powell, 1984b). If $\|g_k\| \le \varepsilon$, then a solution has been obtained. The values of the above constants make no difference in the convergence proofs of the trust-region methods. However, whether $\tau_0 > 0$ or $\tau_0 = 0$, it will lead to different convergence results and require different techniques in the proof. The proof when $\tau_0 = 0$ was given by Powell (1984b). The proof when $\tau_0 > 0$ and the sequence $\{\|B_k\|\}$ is bounded was given by Yuan (1998). Details on the trust-region method can be found in Nocedal and Wright (2006) and Sun and Yuan (2006) and in the landmark monograph by Conn, Gould, and Toint, (2000).

1.4.10 p-Regularized Methods

For solving (1.1) many known algorithms employ a quadratic model of the minimizing function *f*. The Newton and the quasi-Newton methods use a second-order Taylor series expansion of the function f with an explicit or an approximated Hessian matrix, respectively. The algorithms corresponding to these methods are efficient and robust for solving problems that satisfy certain assumptions and exhibit superlinear rates of local convergence on those problems. However, to ensure the global convergence for the Newton method and to see its computational complexity, it requires modifications in order to secure a sufficient descent in the values of function *f*. This can be achieved by trust-region methods. As was mentioned above, the main idea is to compute a trial step by minimizing the second-order Taylor's expansion of *f* over a trust-region ball centered at the current iterate x. Thus, the following trust-region subproblem is obtained:

$$\min_{x\in\mathbb{R}^n}\{c^T x + \frac{1}{2}x^T Bx\}, \text{ subject to} \tag{1.106}$$

$$\|x\|^2 \le \Delta,$$

where B is the Hessian of f at the current iterate. Since the trust region ball $\{x : \|x\|^2 \le \Delta\}$ is a compact set, it follows that the trust-region subproblem (1.106) always has a global minimizer. This problem can be solved via a semi-definite program followed by a rank-one decomposition procedure. Due to compactness it is not necessary to assume that B is positive definite. A merit function is then used to determine whether the global minimizer of (1.106) is to be accepted or rejected, followed by an update for the next subproblem. Detailed discussions on trust region methods are given by Conn, Gould, and Toint (2000) and by Yuan (2015).

A relatively new approach is the p-regularized methods. In these methods the trust region ball $\{x : \|x\|^2 \le \Delta\}$ is replaced with a higher-order regularization term weighted by a positive parameter σ. Specifically, the *p-regularized subproblem* is the following unconstrained minimization problem:

$$\min_{x\in\mathbb{R}^n}\{h(x) \equiv c^T x + \frac{1}{2}x^T Bx + \frac{\sigma}{p}\|x\|^p\}, \tag{1.107}$$

where $p > 2$ is an integer and $\sigma > 0$ is the *regularization parameter*. The regularization term $(\sigma/p)\|x\|^p$ in (1.107) determines that $h(x)$ is a coercive function (see Appendix A), that is $\lim_{\|x\|\to+\infty} h(x) = +\infty$, i.e., (1.107) can always attain its global minimum even for nonpositive definite matrix B. Therefore, a local approximation (1.107) of f is constructed and solved. If the global minimizer of (1.107) gives a reduction in the value of *f*, then it is accepted, otherwise σ is increased in order to force the regularization. Usually the p-regularized subproblem

(1.107) is solved by seeking the unique root of a *secular equation* as described by Gould, Robinson and Thorne (2010).

The most common choice to regularize the quadratic approximation of function f is the p-regularized subproblem (1.107) with $p = 3$, known as the *cubic regularization*. The cubic model is a global upper estimate for the minimizing function and therefore enforces global performance guarantees of the resulting method. The cubic regularization was introduced by Griewank (1981) in order to develop an algorithm based on the Newton's method that is affine-invariant and convergent to second-order critical points. (A second-order critical point of f, is a point $\bar{x} \in \mathbb{R}^n$ satisfying $\nabla f(\bar{x}) = 0$ and $\nabla^2 f(\bar{x})$ semipositive definite.) Nesterov and Polyak (2006) re-introduced cubic regularization of Newton's method for unconstrained optimization and established its global complexity for certain classes of problems. The cubic regularization was further developed and documented by Cartis, Gould and Toint (2011a, 2011b) who proposed the adaptive cubic regularization algorithm. This algorithm uses an approximate Hessian of the minimizing function and finds an approximate minimizer of the cubic model which reduces the computational effort while retaining the local and the global convergence properties established by Griewank (1981) and by Nesterov and Polyak (2006), including the worst-case complexity results. Recently, a comparison for the numerical effectiveness between the p-regularized subproblem with $p > 2$ and trust-region subproblem was given by Gould, Robinson and Thorne (2010). A theoretical presentation and application of the p-regularized subproblems for $p > 2$ was given by Hsia, Sheu and Yuan (2017). The main result on the p-regularized subproblem is given by the following theorem.

Theorem 1.18 *The point x^* is a global minimizer of the p-regularized subproblem* (1.107) *for $p > 2$ if and only if*

$$(B + \sigma \|x^*\|^{p-2} I)x^* = -c, \quad \text{and} \quad B + \sigma \|x^*\|^{p-2} I \geq 0. \tag{1.108}$$

Moreover, the l_2 norm of all the global minimizers are equal. ♦

A characterization of the global minimizers of the p-regularized subproblem (1.107) is given by the following results.

Proposition 1.4 *Assume that $\bar{x}$ is a local minimizer of the p-regularized subproblem* (1.107) *with $p > 2$. Then*

$$\nabla h(\bar{x}) = (B + \sigma \|\bar{x}\|^{p-2} I)\bar{x} + c = 0, \tag{1.109}$$

$$\nabla^2 h(\bar{x}) = (B + \sigma \|\bar{x}\|^{p-2} I) + \sigma(p-2)\|\bar{x}\|^{p-4} \bar{x}\bar{x}^T \geq 0, \tag{1.110}$$

where ∇h and $\nabla^2 h$ represent the gradient and the Hessian of h, respectively. ♦

The next theorem shows that a local minimizer $\bar{x}$ is global if and only if $B + \sigma \|\bar{x}\|^{p-2} I \geq 0$, see: Gould, Robinson, and Thorne (2010), Hsia, Sheu, and Yuan (2017).

Theorem 1.19 *The point x^* is a global minimizer of the p-regularized subproblem* (1.107) *for $p > 2$ if and only if it is a critical point satisfying $\nabla h(x^*) = 0$ and $B + \sigma \|x^*\|^{p-2} I \geq 0$. Moreover, the l_2 norms of all the global minimizers are equal.* ♦

In the following let us characterize the global minimizers of the p-regularized subproblem (1.107) assuming that B is a diagonal matrix, i.e., $B = diag(\mu_1, \ldots, \mu_n)$, where $\mu_1 = \cdots = \mu_k < \mu_{k+1} \leq \cdots \leq \mu_n$, are the eigenvalues of B and the smallest eigenvalue has multiplicity k. Observe that if B is not a diagonal matrix, then its spectral decomposition $B = Q\Lambda Q^T$ may be used, where $Q = [q_1, \ldots, q_n] \in \mathbb{R}^{n \times n}$ is an orthogonal matrix with columns q_i, $i = 1, \ldots, n$, as eigenvectors of B and $\Lambda = diag(\mu_1, \ldots, \mu_n)$, where μ_i are the eigenvalues of B (see Appendix A). Let $y = Q^T x$. Then $\|y\| = \|Q^T x\| = \|x\|$ and therefore a diagonal version of the p-regularized subproblem in terms of y is obtained.

Theorem 1.20 *The set of the global minimizers of the p-regularized subproblem* (1.107) *with B in the diagonal form is either a singleton or a k-dimensional sphere centered at*

$$\left(0, \ldots, 0, \frac{-c_{k+1}}{\mu_{k+1} - \mu_1}, \ldots \frac{-c_n}{\mu_n - \mu_1}\right) \tag{1.111}$$

with the radius

$$\sqrt{\left(\frac{\mu_1}{\sigma}\right)^{2/(p-2)} - \sum_{i=k+1}^{n} \left(\frac{c_i}{\mu_i - \mu_1}\right)^2}. \tag{1.112}$$

Proof Let x^* be the global minimizer of (1.107) and define $z^* = \|x^*\|^{p-2} \geq 0$. By Theorem 1.19 it follows that $\mu_i + \sigma z^* \geq 0$ for any $i = 1, \ldots, n$. Now, if $B + \sigma z^* I$ is nonsingular, then $z^* \in (\max\{-\mu_1/\sigma, 0\}, +\infty)$ and the global minimizer x^* is uniquely defined by

$$x_i^* = \frac{-c_i}{\mu_i + \sigma z^*}, \quad i = 1, \ldots, n. \tag{1.113}$$

Observe that in (1.113) z^* is unknown. However, by summing all $(x_i^*)^2$, it follows that z^* is necessarily a nonnegative root of the following secular function

$$s(z) \equiv \sum_{i=1}^{n} \frac{c_i^2}{(\mu_i - \sigma z)^2} - z^{2/(p-2)}, \tag{1.114}$$

on the interval $z \in I_z \equiv (\max\{-\mu_1/\sigma, 0\}, +\infty)$. Since $s(z)$ is strictly decreasing on I_z, the secular function $s(z)$ has a unique root on I_z which must be z^*.

On the other hand, if $B+\sigma z^* I$ is singular, then $z^* = -\mu_1/\sigma$. In this case $c_1^2 + \ldots + c_k^2 = 0$ and $\mu_i + \sigma z^* > 0$. for $i = k+1, \ldots, n$, such that

$$\hat{x}^* = \left(0, \ldots, 0, \frac{-c_{k+1}}{\mu_{k+1} - \mu_1}, \ldots, \frac{-c_n}{\mu_n - \mu_1}\right)^T \tag{1.115}$$

is one trivial solution of $(B - \mu_1 I)x^* = -c$. Similarly, by summing all $(\hat{x}_i^*)^2$ in (1.115) again a secular function is obtained as

$$\hat{s}(z) \equiv \sum_{i=k+1}^{n} \frac{c_i^2}{(\mu_i + \sigma z)^2} - z^{2/(p-2)}, \tag{1.116}$$

where $z \in I_{\hat{z}} \equiv [-\mu_1/\sigma, +\infty)$. Observe that $\hat{s}(z)$ is also strictly decreasing on $I_{\hat{z}}$. Therefore, if $\hat{s}(-\mu_1/\sigma) = 0$, then $z^* = -\mu_1/\sigma$ is the unique root of $\hat{s}(z)$ on $I_{\hat{z}}$. Thus $\hat{x}^*$ defined by (1.115) is the unique global minimizer of (1.107).

If $\hat{s}(-\mu_1/\sigma) < 0$, then (1.116) has no solution and the trivial solution $\hat{x}^*$ to $(B - \mu_1 I)x^* = -c$ does not satisfies $z^* = -\mu_1/\sigma = \|\hat{x}^*\|^{p-2}$. Therefore, any x^* satisfying

$$(x_1^*)^2 + \cdots + (x_k^*)^2 + \sum_{i=k+1}^{n} \frac{c_i^2}{(\mu_i - \mu_1)^2} = \left(\frac{-\mu_1}{\sigma}\right)^{2/(p-2)} \tag{1.117}$$

is a global minimizer of (1.107). Namely, the global minimum solution set forms a k-dimensional sphere defined as in (1.111) and (1.112).

If $\hat{s}(-\mu_1/\sigma) > 0$, then the secular function given by (1.116) has no solution. Therefore (1.117) cannot hold for any x^*, thus obtaining a contradiction that (1.107) has no global minimizer. ♦

The above developments present the necessary and sufficient optimality conditions for the global minimizers of the p-regularized subproblem (1.107). It emphasizes a closed-form expression for the global minimum set of (1.107). In Chapter 11 these theoretical developments are used to obtain a conjugate gradient algorithm with subspace minimization based on regularization model of the minimizing function.

1.5 Test Problems and Applications

In order to see the performances of the algorithms discussed in this book and to make comparisons among them, a number of 80 large-scale unconstrained optimization test problems, as well as five real large-scale applications are used. The unconstrained optimization test problems selected in this set, which we call the *UOP collection*, have different structures and complexities. The name of these problems and the initial points are given in Table 1.1. The algebraic expression of these problems is presented in Appendix B.

Table 1.1 The UOP collection of unconstrained optimization test problems

No.	Problem name	Initial point
1	Extended Freudenstein & Roth	[0.5, −2, 0.5, −2, …, 0.5, −2]
2	Extended White & Holst	[−1.2, 1, −1.2, 1, …, −1.2, 1]
3	Tridiagonal White & Holst	[−1.2, 1, −1.2, 1, …, −1.2, 1]
4	Extended Beale	[1, 0.8, 1, 0.8, …, 1, 0.8]
5	Extended Powell	[3, −1, 0, 1, …, 3, −1, 0, 1]
6	Extended Maratos	[0.1, 0.1, …, 0.1]
7	Extended Cliff	[0.001, 0.001, …, 0.001]
8	Extended Woods	[−3, −1, −3, −1, … −3, −1, −3, −1]
9	Extended Hiebert	[5.001, 5.001, … ,5.001]
10	Extended Rosenbrock	[−1.2, 1, −1.2, 1, …, −1.2, 1]
11	Generalized Rosenbrock	[−1.2, 1, −1.2, 1, …, −1.2, 1]
12	Extended Himmelblau—HIMMELBC	[1, 1, … 1]
13	HIMMELBG	[1.5, 1.5 ,…, 1.5]
14	HIMMELBH	[0.8, 0.8, …, 0.8]
15	Extended Trigonometric ET1	[0.2, 0.2, …, 0.2]
16	Extended Trigonometric ET2	[0.2, 0.2, …, 0.2]
17	Extended Block-Diagonal BD1	[1, 1, …, 1]
18	Extended Tridiagonal-1	[2, 2, …, 2]
19	Extended Three Exponential Terms	[0.1, 0.1, …, 0.1]
20	Generalized Tridiagonal-1	[2, 2, …, 2]
21	Generalized Tridiagonal-2	[−1, −1, …, −1]
22	Tridiagonal Double Borded	[−1, −1, …, −1]
23	Broyden Pentadiagonal	[−1, −1, …, −1]
24	Extendeed PSC1	[3.,0.1,3.,0.1,…,3.,0.1]
25	Perturbed Quadratic PQ1	[1.,1.,…,1.]
26	Perturbed Quadratic PQ2	[0.5, 0.5, …, 0.5]
27	Almost Perturbed Quadratic	[0.5, 0.5, …, 0.5]
28	Almost Perturbed Quartic	[0.5, 0.5, …, 0.5]
29	Extended Penalty Function U52	[1/100, 2/100, …, n/100]
30	TR-Sum of Quadratics	[1.,1.,…,1.]
31	Quadratic Diagonal Perturbed	[0.5, 0.5, …, 0.5]
32	Full Hessian FH1	[1/n, 2/n, …, n/n]
33	Full Hessian FH2	[1, 1, …, 1]
34	Full Hessian FH3	[1, 1, …, 1]
35	Diagonal Full Borded	[0.001, 0.001, …, 0.001]
36	Diagonal Double Border Arrow Up	[0.4, 1, 0.4, 1, …, 0.4, 1]
37	QP1 Extended Quadratic Penalty	[1, 1, …, 1]
38	QP2 Extended Quadratic Penalty	[2, 2, …, 2]
39	QP3 Extended Quadratic Penalty	[1, 1, …, 1]
40	Staircase S1	[1, 1, …, 1]

(continued)

Table 1.1 (continued)

No.	Problem name	Initial point
41	Staircase S2	[1, 1, …, 1]
42	Staircase S3	[2, 2, …, 2]
43	NONDQUAR	[1, −1, 1, −1, …, 1, −1]
44	TRIDIA	[1, 1, …, 1]
45	ARWHEAD	[1, 1, …, 1]
46	NONDIA	[−0.01, −0.01, …, −0.01]
47	BDQRTIC	[1, 1, …, 1]
48	DQDRTIC	[3, 3, …, 3]
49	EG2	[0.001, 0.001, …, 0.001]
50	EG3	[0.02, 0.02, …, 0.02]
51	EDENSCH	[0, 0, …, 0]
52	FLETCHCR	[0.5, 0.5, …, 0.5]
53	ENGVAL1	[2, 2, …, 2]
54	DENSCHNA	[1, 1, …, 1]
55	DENSCHNB	[10, 10, …, 10]
56	DENSCHNC	[1, 1, …, 1]
57	DENSCHNF	[100, −100, …, 100, −100]
58	SINQUAD	[0, 0, …, 0]
59	DIXON3DQ	[−0.1, −0.1, …, −0.1]
60	BIGGSB1	[0.1, 0.1, …, 0.1]
61	PRODsin	[0.000001, …, 0.000001]
62	PROD1	[1, 1, …, 1]
63	PRODcos	[1, 0, …, 0]
64	PROD2	[0.00001, …, 0.00001]
65	DIXMAANA	[2, 2, …, 2]
66	DIXMAANB	[2, 2, …, 2]
67	DIXMAANC	[2, 2, …, 2]
68	DIXMAAND	[2, 2, …, 2]
69	DIXMAANL	[1, 1, …, 1]
70	ARGLINB	[0.01, 0.001, …, 0.01, 0.001]
71	VARDIM	[1−1/n, 1−2/n, …, 1−n/n]
72	DIAG-AUP1	[4, 4, …, 4]
73	ENGVAL8	[2, 2, …, 2]
74	QUARTIC	[2, 2, …, 2]
75	LIARWHD	[4, 4, …, 4]
76	NONSCOMP	[3, 3, …, 3]
77	Linear Perturbed	[2, 2, …, 2]
78	CUBE	[−1.2, 1.1, −1.2, 1.1,…, −1.2, 1.1]
79	HARKERP	[1, 2, …, n]
80	QUARTICM	[2, 2, …, 2]

In this collection, some problems are quadratic and some of them are highly nonlinear. The problems are presented in extended (separable) or generalized (chained) form. The Hessian for problems in extended form has a block-diagonal structure. On the other hand, the Hessian for problems in generalized form has a banded structure with small bandwidth, often being tri- or penta-diagonal. For some other optimization problems from this set, the corresponding Hessian has a sparse structure or it is a dense (full) matrix. All the test problems used in our numerical experiments are described in (Andrei, 2018g). The vast majority of the optimization problems included in our collection is taken from the CUTE (Bongartz, Conn, Gould, & Toint, 1995) collection. See also Gould, Orban, and Toint (2003).

The applications used in our numerical experiments are taken from the MINPACK-2 test problem collection Averick, Carter, and Moré (1991), Averick, Carter, Moré, and Xue (1992), Andrei (2013e, 2017c). MINPACK-2 contains applications from different fields, such as: elasticity, fluid dynamics, combustion, lubrication, molecular conformation, nondestructive testing, chemical cinetics, etc. In our numerical experiments we have considered only five MINPACK-2 real applications: A1—elastic–plastic torsion ($c = 5$), A2—pressure distribution in a journal bearing ($b = 10$, $\varepsilon = 0.1$), A3—optimal design with composite materials ($\lambda = 0.008$), A4—steady-state combustion ($\lambda = 5$) and A5—minimal surface with Enneper conditions.

Application A1 (Elastic–Plastic Torsion)

The description and the physical interpretation of the torsion problem are discussed in Glowinski (1984, pp.41–55). We follow the presentation of it from Averick, Carter, Moré, and Xue (1992). The elastic–plastic torsion problem arises from the determination of the stress field on an infinitely long cylindrical bar. The infinite-dimensional version of this problem is of the following form

$$\min\{q(v) : v \in K\},$$

where $q : K \to \mathbb{R}$ is a quadratic function

$$q(v) = \frac{1}{2}\int_D \|\nabla v(x)\|^2 dx - c\int_D v(x)dx$$

for some constant c and D is a bounded domain with smooth boundary. The convex set K is defined as

$$K = \left\{v \in H_0^1(D) : \ |v(x)| \leq dist(x, \partial D), \ x \in D\right\},$$

where $dist(., \partial D)$ is the distance function to the boundary of D and $H_0^1(D)$ is the Hilbert space of all functions with compact support in D, so that v and $\|\nabla v\|^2$ belong to $L^2(D)$.

A finite element approximation to the torsion problem is obtained by triangulating D and then by replacing the minimization of q over $H_0^1(D)$ with the minimization of q over the set of piecewise linear functions that satisfy the constraints specified by K. The finite element approximation thus gives rise to a finite-dimensional minimization problem whose variables are the values of the piecewise linear function at the vertices of the triangulation.

In (Averick, Carter, Moré, & Xue, 1992) a finite element approximation to a minimization problem with a quadratic q of the general form

$$q(v) = \frac{1}{2}\int_D w_q(x)\|\nabla v(x)\|^2 dx - \int_D w_l(x)v(x)dx$$

where $w_q : D \to \mathbb{R}$ and $w_l : D \to \mathbb{R}$ are functions defined on the rectangle D is described. In the torsion problem, $w_q = 1$ and $w_l = c$.

Let $D = (\xi_{1,l}, \xi_{1,u}) \times (\xi_{2,l}, \xi_{2,u})$ be a rectangle in $\mathbb{R}^2$. Vertices $z_{i,j} \in \mathbb{R}^2$ for a triangulation of D are obtained by choosing the grid spacing h_x and h_y and by defining the grid points:

$$z_{i,j} = (\xi_{1,l} + ih_x, \xi_{2,l} + jh_y),\ 0 \le i \le n_x + 1,\ 0 \le j \le n_y + 1,$$

so that $z_{n_x+1,n_y+1} = (\xi_{1,u}, \xi_{2,u})$. The triangulation consists of the triangular elements T_L with vertices at $z_{i,j}$, $z_{i+1,j}$ and $z_{i,j+1}$ and the triangular elements T_U with vertices $z_{i,j}$,$z_{i-1,j}$ and $z_{i,j-1}$.

A finite element approximation to the torsion problem is obtained by minimizing q over the space of piecewise linear functions v with values $v_{i,j}$ at $z_{i,j}$. The approximation to the integral

$$\int_D w_q(x)\|\nabla v(x)\|^2 dx$$

over the element T_L is the quadratic $q_{i,j}^L(v)$, where

$$q_{i,j}^L(v) = \mu_{i,j}\left\{\left(\frac{v_{i+1,j} - v_{i,j}}{h_x}\right)^2 + \left(\frac{v_{i,j+1} - v_{i,j}}{h_y}\right)^2\right\},$$
$$\mu_{i,j} = \frac{h_x h_y}{6}\left\{w_q(z_{i,j}) + w_q(z_{i+1,j}) + w_q(z_{i,j+1})\right\}.$$

Similarly, the approximation over the element T_U is the quadratic $q_{i,j}^U(v)$, where

$$q_{i,j}^U(v) = \lambda_{i,j}\left\{\left(\frac{v_{i-1,j} - v_{i,j}}{h_x}\right)^2 + \left(\frac{v_{i,j-1} - v_{i,j}}{h_y}\right)^2\right\},$$
$$\lambda_{i,j} = \frac{h_x h_y}{6}\left\{w_q(z_{i,j}) + w_q(z_{i-1,j}) + w_q(z_{i,j-1})\right\}.$$

Therefore, the finite element approximation to the problem leads to a quadratic programming problem of the following form

$$\min\{q(v) : v \in \Omega\},$$

where q is the quadratic function

$$q(v) = \frac{1}{2}\sum\left(q_{i,j}^L(v) + q_{i,j}^U(v)\right) - h_x h_y \sum w_l(z_{i,j})v_{i,j}.$$

Observe that in this formulation, the quadratic $q_{i,j}^L$ is defined only when $0 \le i \le n_x$ and $0 \le j \le n_y$, while $q_{i,j}^U$ is defined when $1 \le i \le n_x + 1$ and $1 \le j \le n_y + 1$. Besides, note that for the torsion problem, $w_q = 1$ and $w_l = c$, and the feasible set Ω, is $\Omega = \{v \in R^{n_x n_y} : |v_{i,j}| \le d_{i,j}\}$, where $d_{i,j}$ is the value of $dist(., \partial D)$ at vertices $z_{i,j}$.

Considering $D = (0,1) \times (0,1)$, $c = 5$ and $nx = 200$, $ny = 200$, then a minimization problem with 40,000 variables is obtained. The solution of this application is illustrated in Figure 1.1.

Numerical results for the elastic–plastic torsion problem are presented, for example, by O'Leary and Yang (1978), Elliott and Ockendon (1982), Moré and Toraldo, 1991), Andrei (2007e, 2007f, 2008h, 2008i, 2009g, 2009h).

Application A2 (Pressure Distribution in a Journal Bearing)

This problem consists in determining the pressure distribution in a thin film of lubricant between two circular cylinders (Averick, Carter, & Moré, 1991). The infinite-dimensional version of this problem is of the following form:

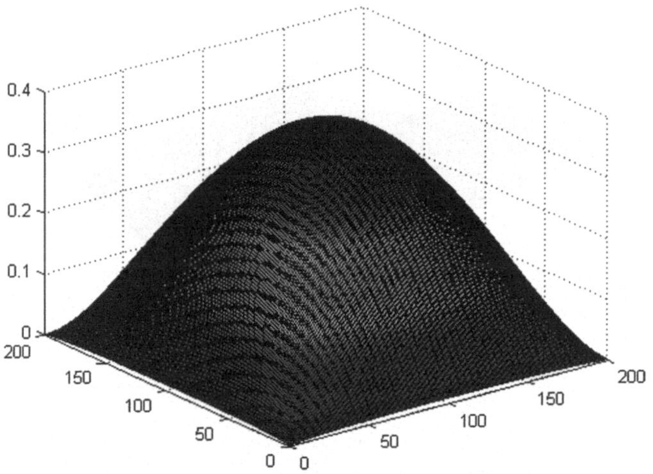

Figure 1.1 Solution of the application A1—Elastic–Plastic Torsion. $nx = 200$, $ny = 200$

$$\min\{q(v) : v \in K\},$$

$$q(v) = \frac{1}{2}\int_D w_q(x)\|\nabla v(x)\|^2 dx - \int_D w_l(x)v(x)dx$$

with

$$w_q(z_1, z_2) = (1 + \varepsilon \cos z_1)^3, \ w_l(z_1, z_2) = \varepsilon \sin z_1,$$

for some constant $\varepsilon \in (0, 1)$ and $D = (0, 2\pi) \times (0, 2b)$, where $b > 0$ is again an arbitrary constant. The convex set K is defined as

$$K = \{v \in H_0^1(D) : v \in D, \ v \geq 0\}.$$

A finite element approximation to the journal bearing problem is obtained as in the torsion problem. In the case of the journal bearing problem, $w_q(\xi_1, \xi_2) = (1 + \varepsilon \cos \xi_1)^3$ and $w_l(\xi_1, \xi_2) = \varepsilon \sin \xi_1$. The feasible set Ω is given by $\Omega = \{v \in \mathbb{R}^{n_x n_y} : v_{i,j} \geq 0\}$. Considering $b = 10$ and $\varepsilon = 0.1$, as well as a discretization $n_x \times n_y$ of the domain $D = (0, 2\pi) \times (0, 2b)$, where $nx = 200$ and $ny = 200$, then the solution of this application is represented in Figure 1.2.

Numerical results for this problem are given, for example, by Lin and Cryer (1985), Cimatti and Menchi (1978), Moré and Toraldo (1991), Andrei (2007e, 2007f, 2008h, 2008i, 2009g, 2009h).

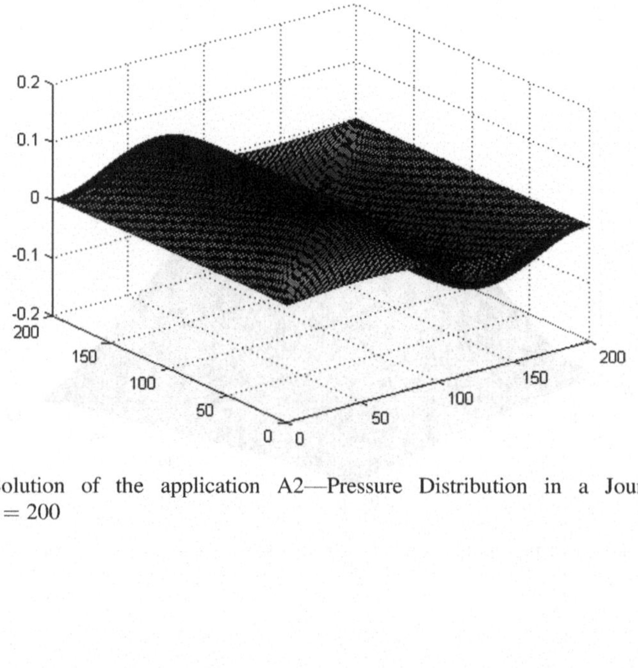

Figure 1.2 Solution of the application A2—Pressure Distribution in a Journal Bearing. $nx = 200, \ ny = 200$

Application A3 (Optimal Design with Composite Materials)
This application requires determining the placement of two elastic materials in the cross-section of a rod with maximal torsional rigidity. The formulation of this problem is given in (Averick, Carter, Moré, & Xue, 1992) and follows the presentation from (Goodman, Kohn, & Reyna, 1986).

Let $D \subset \mathbb{R}^2$ be a bounded domain and let $w < |D|$, where $|D|$ is the area of D. The solution of the optimal design problem is a subset Ω of D that solves the problem

$$min\{F(v,\Omega) : v \in H_0^1(D), |\Omega| = w\},$$

where

$$F(v,\Omega) = \int_D \left\{\frac{1}{2}\mu(x)\|\nabla v(x)\|^2 + v(x)\right\}dx,$$

and $\mu(x) = \mu_1$ for $x \in \Omega$, and $\mu(x) = \mu_2$ for $x \notin \Omega$. The reciprocals of the constants μ_1 and μ_2 are the shear moduli of the elastic materials in the rod. It is assumed that $\mu_1 < \mu_2$.

Goodman, Kohn, and Reyna (1986) formulate the optimal design problem in terms of a family of problems of the form

$$min\{f_\lambda(v) : v \in H_0^1(D)\},$$

where $f_\lambda : H_0^1(D) \to \mathbb{R}$ is the functional

$$f_\lambda(v) = \int_D \{\psi_\lambda(\|\nabla v(x)\|) + v(x)\}dx$$

and $\psi_\lambda : \mathbb{R} \to \mathbb{R}$ is a piecewise quadratic function. In this formulation, λ is a Lagrange multiplier associated with the optimal design problem. The piecewise quadratic $\psi_\lambda : \mathbb{R} \to \mathbb{R}$ is of the form

$$\psi_\lambda(t) = \begin{cases} \frac{1}{2}\mu_2 t^2, & 0 \le t \le t_1, \\ \mu_2 t_1(t - \frac{1}{2}t_1), & t_1 \le t \le t_2, \\ \frac{1}{2}\mu_1(t^2 - t_2^2) + \mu_2 t_1(t_2 - \frac{1}{2}t_1), & t_2 \le t, \end{cases}$$

with the breakpoints t_1 and t_2 defined by:

$$t_1 = \left(2\lambda\frac{\mu_1}{\mu_2}\right)^{1/2} \quad \text{and} \quad t_2 = \left(2\lambda\frac{\mu_2}{\mu_1}\right)^{1/2}.$$

The definition of these breakpoints implies that $\mu_1 t_2 = \mu_2 t_1$ and thus ψ_λ is continuously differentiable. The solution of the optimum design problem considered by Averick, Carter, Moré and Xue (1992) is the minimizing f_λ for a fixed value of λ, where $\mu_1 = 1$ and $\mu_2 = 2$, so that $t_1^2 = \lambda$ and $t_2^2 = 2\lambda$.

A finite element approximation to this problem is obtained by minimizing f_λ over the space of piecewise linear functions v with values v_{ij} at z_{ij}, where $z_{ij} \in \mathbb{R}^2$ are the vertices of a triangulation of D with grid spacings h_x and h_y. The values v_{ij} are obtained by solving the minimization problem

$$\min\left\{\sum \left(f_{i,j}^L(v) + f_{i,j}^U(v) + h_x h_y v_{i,j}\right) : v \in \mathbb{R}^n\right\},$$

where the functions $f_{i,j}^L$ and $f_{i,j}^U$ are defined by:

$$f_{i,j}^L(v) = \frac{h_x h_y}{2}\psi_\lambda\left(d_{i,j}^+(v)\right),\ f_{i,j}^U(v) = \frac{h_x h_y}{2}\psi_\lambda\left(d_{i,j}^-(v)\right),$$

with

$$d_{i,j}^{\pm}(v) = \left\{\left(\frac{v_{i\pm 1,j} - v_{i,j}}{h_x}\right)^2 + \left(\frac{v_{i,j\pm 1} - v_{i,j}}{h_y}\right)^2\right\}^{1/2}.$$

Observe that in this formulation, $f_{i,j}^L$ is defined only for $0 \le i \le n_x$ and $0 \le j \le n_y$, while $f_{i,j}^U$ is defined for $1 \le i \le n_x + 1$ and $1 \le j \le n_y + 1$.

Considering $\lambda = 0.008$, $\mu_1 = 1$ and $\mu_2 = 2$, then Figure 1.3 presents the solution of this application for $nx = 200$ and $ny = 200$.

Numerical results for this problem are given by Goodman, Kohn and Reyna (1986), Averick, Carter, Moré, and Xue (1992), Andrei (2007e, 2007f, 2008h, 2008i, 2009g, 2009h).

Figure 1.3 Solution of the application A3—Optimal Design with Composite Materials. $nx = 200$, $ny = 200$

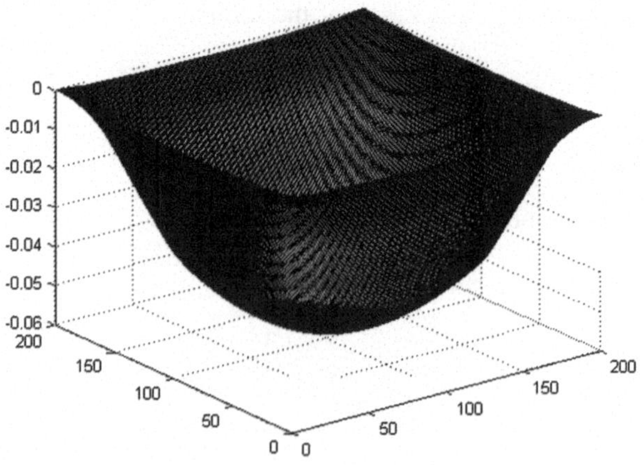

Application A4 (Steady-State Combustion)
This application is taken from (Averick, Carter, Moré, & Xue, 1992). The infinite-dimensional optimization problem is as follows

$$\min\{f_\lambda(v) : v \in H_0^1(D)\},$$

where $f_\lambda : H_0^1(D) \to \mathbb{R}$ is the functional

$$f_\lambda(v) = \int_D \left\{\frac{1}{2}\|\nabla v(x)\|^2 - \lambda \exp[v(x)]\right\} dx,$$

and $\lambda \geq 0$ is a known parameter. This problem is the variational formulation of the boundary value problem

$$\begin{aligned} -\Delta v(x) &= \lambda \exp[v(x)], & x &\in D, \\ v(x) &= 0, & x &\in \partial D, \end{aligned}$$

where Δ is the Laplacian operator. Aris (1975) and Bebernes and Eberly (1989) discuss this application in the context of combustion problems.

An interesting property of the variational Bratu problem is that f_λ is unbounded below for any $\lambda > 0$. This can be seen by noting that if v is any positive constant function, then $f_\lambda(\alpha v) \to -\infty$ as $\alpha \to \infty$. Another interesting property of the variational Bratu problem is that, if $\lambda_{FK} > 0$ is the Frank-Kamenetskii parameter, then f_λ has a unique minimizer for $\lambda \in [0, \lambda_{FK}]$, but no minimizers for $\lambda > \lambda_{FK}$. If D is the unit square, then $\lambda_{FK} \approx 6.81$, known as the Frank-Kamenetskii parameter.

A finite element approximation to this problem is obtained by minimizing f over the space of piecewise linear functions v with values v_{ij} at z_{ij}, where $z_{ij} \in \mathbb{R}^2$ are the vertices of a triangulation of D with grid spacings h_x and h_y. The values of v_{ij} are computed by solving the following minimization problem

$$\min\left\{\sum \left(f_{i,j}^L(v) + f_{i,j}^U(v)\right) : v \in \mathbb{R}^n\right\},$$

where

$$f_{i,j}^L(v) = \frac{h_x h_y}{4}\left\{\left(\frac{v_{i+1,j} - v_{i,j}}{h_x}\right)^2 + \left(\frac{v_{i,j+1} - v_{i,j}}{h_y}\right)^2 - \lambda \mu_{i,j}^L\right\},$$
$$\mu_{i,j}^L = \frac{2}{3}\left\{\exp(v_{i,j}) + \exp(v_{i+1,j}) + \exp(v_{i,j+1})\right\},$$

and

$$f_{i,j}^U(v) = \frac{h_x h_y}{4}\left\{\left(\frac{v_{i-1,j} - v_{i,j}}{h_x}\right)^2 + \left(\frac{v_{i,j-1} - v_{i,j}}{h_y}\right)^2 - \lambda \mu_{i,j}^U\right\},$$
$$\mu_{i,j}^U = \frac{2}{3}\left\{\exp(v_{i,j}) + \exp(v_{i-1,j}) + \exp(v_{i,j-1})\right\}.$$

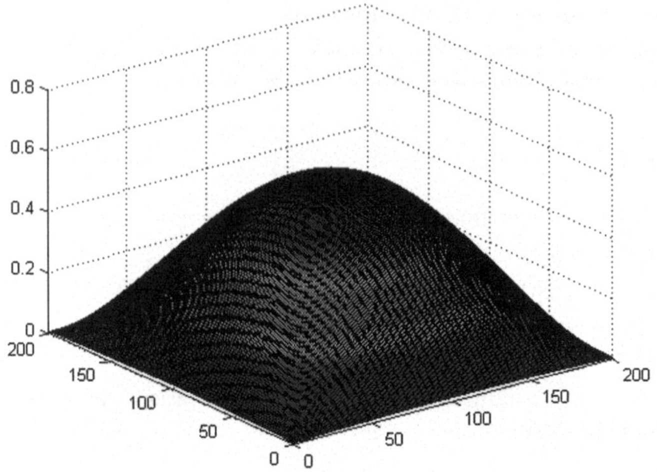

Figure 1.4 Solution of the application A4—Steady-State Combustion. $nx = 200$, $ny = 200$

In this formulation, $f_{i,j}^L$ is defined only for $0 \leq i \leq n_x$ and $0 \leq j \leq n_y$, while $f_{i,j}^U$ is defined when $1 \leq i \leq n_x + 1$ and $1 \leq j \leq n_y + 1$. Considering $\lambda = 5$, Figure 1.4 shows the solution of this application, where $nx = 200$ and $ny = 200$.

Numerical results for this problem are given by Averick, Carter, Moré, and Xue (1992) and Andrei (2007e, 2007f, 2008h, 2008i, 2009g, 2009h).

Application A5 (Minimal Surfaces with Enneper boundary conditions)
Determination of the surface with minimal area and given boundary values in a convex domain D is an infinite-dimensional optimization problem of the form

$$\min\{f(v) : v \in K\},$$

where $f : K \to \mathbb{R}$ is the functional

$$f(v) = \int_D (1 + \|\nabla v(x)\|^2)^{1/2} dx,$$

and the set K is defined by:

$$K = \{v \in H^1(D) : v(x) = v_D(x) \text{ for } x \in \partial D\}$$

for some boundary data function $v_D : \partial D \to \mathbb{R}$. The boundary function v_D uniquely defines the solution to the minimal surface problem. An interesting minimal surface, given by Enneper, is obtained by defining v_D on $D = (-1/2, 1/2) \times (-1/2, 1/2)$ by

$$v_D(\xi_1, \xi_2) = u^2 - v^2,$$

where u and v are the unique solutions to the nonlinear equations

$$\xi_1 = u + uv^2 - \frac{1}{3}u^3, \ \xi_2 = -v - u^2 v + \frac{1}{3}v^3.$$

A finite element approximation to the minimal surface problem is obtained by minimizing f over the space of piecewise linear functions v with values $v_{i,j}$ at $z_{i,j}$, where $z_{i,j} \in \mathbb{R}^2$ are the vertices of a triangulation of D with grid spacings h_x and h_y. The values $v_{i,j}$ are obtained by solving the minimization problem

$$\min\left\{\sum \left(f_{i,j}^L(v) + f_{i,j}^U(v)\right) : v \in \mathbb{R}^n\right\},$$

where the functions $f_{i,j}^L$ and $f_{i,j}^U$ are defined by

$$f_{i,j}^L(v) = \frac{h_x h_y}{2}\left\{1 + \left(\frac{v_{i+1,j} - v_{i,j}}{h_x}\right)^2 + \left(\frac{v_{i,j+1} - v_{i,j}}{h_y}\right)^2\right\}^{1/2},$$

$$f_{i,j}^U(v) = \frac{h_x h_y}{2}\left\{1 + \left(\frac{v_{i-1,j} - v_{i,j}}{h_x}\right)^2 + \left(\frac{v_{i,j-1} - v_{i,j}}{h_y}\right)^2\right\}^{1/2}.$$

Note that in this formulation, $f_{i,j}^L$ is defined only when $0 \le i \le n_x$ and $0 \le j \le n_y$, while $f_{i,j}^U$ is defined when $1 \le i \le n_x + 1$ and $1 \le j \le n_y + 1$. Figure 1.5 shows the solution of this application, where $nx = 200$ and $ny = 200$.

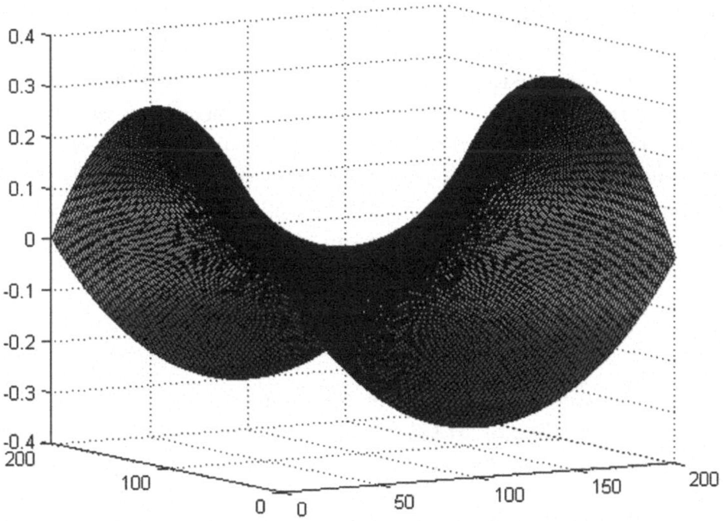

Figure 1.5 Solution of the application A5—minimal surfaces with Enneper boundary conditions. $nx = 200$, $ny = 200$

Numerical results for this problem are given by Nitsche (1989), Averick, Carter, Moré, and Xue (1992), Andrei (2007e, 2007f, 2008h, 2008i, 2009g, 2009h).

1.6 Numerical Experiments

Plenty of algorithms for unconstrained optimization are shown in this book. Some of them accumulate information gathered at previous iterations, while others use only local information available at the current iteration. Anyway, a good algorithm must have the following properties: *efficiency*—it should not requires excessive storage and computer time, *robustness*—it should solve a large variety of problems in its class, for any values of the initial point, *accuracy*—it should get a solution with precision without being affected by errors in the data or by arithmetic rounding errors that arise when the algorithm is implemented as a computer program.

To be reliable and to have a guarantee that they find a solution, some properties and their convergence results have to be discussed. However, this is not enough. The irony is that in optimization, algorithms for which there are very well established theoretical properties (descent, convergence and complexity) prove to have modest numerical performances. Having in view that the final test of a theory is its capacity to solve the problems and the applications which originated it, our efforts have been directed to see the numerical performances of the algorithms for solving 80 problems assembled in the UOP collection (Andrei, 2018g) (see Appendix B) and five applications from the MINPACK-2 collection (Averick, Carter, Moré, & Xue, 1992). The algorithms have been tested in the following environment.

Remark 1.1 All algorithms have been coded in double precision Fortran and compiled with f77 (default compiler settings) and run on an Intel Pentium 4, 1.8 GHz workstation. For each problem from the UOP collection, 10 numerical experiments with an increasing number of variables as $n = 1000, 2000, \ldots, 10000$ have been performed. Hence, 800 problems have been solved in this set of numerical experiments. The algorithms compared in these numerical experiments find local solutions. Therefore, the comparisons of the algorithms are given in the following context. Let f_i^{ALG1} and f_i^{ALG2} be the optimal value found by ALG1 and ALG2 for problem $i = 1, \ldots, 800$, respectively. We say that, in the particular problem i, the performance of ALG1 was better than the performance of ALG2 if

$$|f_i^{ALG1} - f_i^{ALG2}| < 10^{-3} \tag{1.118}$$

and if the number of iterations (#iter), or the number of function-gradient evaluations (#fg), or the CPU time of ALG1 was less than the number of iterations, or the number of function-gradient evaluations, or the CPU time corresponding to ALG2, respectively.

The iterations are stopped if the inequality $\|g_k\|_\infty \leq 10^{-6}$ is satisfied, where $\|.\|_\infty$ is the maximum absolute component of a vector. All algorithms implement the standard Wolfe line search (1.12) and (1.13), where $\rho = 0.0001$ and $\sigma = 0.8$. The maximum number of iterations was limited to 2000.

For the applications from the MINPACK-2 collection, $nx = 200$ and $ny = 200$. Therefore, these problems have 40,000 variables.

To compare the performances of algorithms, the Dolan and Moré (2002) performance profiles are used. For n_p problems and n_s solvers, the performance profile $P : \mathbb{R} \to [0, 1]$ is defined as follows. Let P and S be the set of problems and the set of solvers, respectively. For each problem $p \in P$ and for each solver $s \in S$, define $t_{p,s}$ = computing time (similarly for the number of iterations or the number of function and its gradient evaluations) required to solve problem p by solver s. The idea is to compare the performance of solver s on problem p with the best performance by any solver on this problem. Therefore, the *performance ratio* is defined by:

$$r_{p,s} = \frac{t_{p,s}}{\min_{s \in S}\{t_{p,s}\}}. \tag{1.119}$$

With this, the *performance profile* may be defined by:

$$P_s(\tau) = \frac{1}{n_p} size\{p \in P : r_{p,s} \leq \tau\}, \tag{1.120}$$

for $\tau > 0$, where for any set A, $size(A)$ stands for the number of elements in A. The performance profile $P_s : \mathbb{R} \to [0, 1]$ for a solver s is a nondecreasing, piecewise constant function, continuous from the right at each breakpoint. $P_s(\tau)$ is the probability for solver $s \in S$ so that the performance ratio $r_{p,s}$ is within a factor $\tau > 0$ of the best possible ratio. The function P_s is the cumulative distribution function for the performance ratio. Observe that $1 - P_s(\tau)$ is the fraction of problems that the solver cannot solve within a factor τ of the best solver. In our numerical experiments, in each figure, we used $n_p = 800$ and $\tau = 16$.

The performance profile plot of solvers may be computed as follows. Suppose that two solvers ($n_s = 2$) have to be compared subject to a given metric (which can be the number of iterations, the number of function and its gradient evaluations or the CPU computing time) for solving n_p problems. Consider an integer value for parameter τ. Firstly, out of n_p problems only the problems for which the criterion (1.118) is satisfied are retained. Let $\bar{n}_p$ be the number of problems satisfying the criterion (1.118). For each solver and for each problem, compute the performance ratio $r_{p,s}$, for $p = 1, \ldots, \bar{n}_p$ and $s = 1, 2$, corresponding to the metric selected. For each solver $s = 1, 2$ and for $i = 1, \ldots, \tau$, compute the performance profile $P_s(\tau)$.

The percentage of problems for which an algorithm is the best is given on the left side of the plot. On the other hand, the right side of the plot gives the percentage of the problems that are successfully solved. In other words, for a given algorithm, the plot for $\tau = 1$, represents the fraction of problems for which the algorithm was the most efficient over all algorithms. The plot for $\tau = \infty$ represents the fraction of problems solved by the algorithm irrespective of the required effort. Therefore, the plot for $\tau = 1$ is associated to the *efficiency* of the algorithm, while the plot for $\tau = \infty$ is associated to the *robustness* of the algorithm. ♦

In the previous sections we have seen that L-BFGS (Limited-Memory BFGS) and TN (Truncated Newton) are designed to solve large-scale unconstrained optimization problems. In the following, we shall present the performances of the L-BFGS in the implementation of Liu and Nocedal (1989) and of TN in the implementation of Nash (1985) for solving 80 unconstrained optimization test problems from our UOP collection with $n = 1000, \ldots, 10000$. Figure 1.6 presents the Dolan and Moré (2002) performance profiles of L-BFGS ($m = 5$) versus TN for solving this set of 800 unconstrained optimization test problems subject to iterations calls, function calls and CPU time metric, respectively.

On comparing L-BFGS versus TN (see Figure 1.6) subject to the number of iterations, we see that L-BFGS was better in 310 problems (i.e., it achieved the minimum number of iterations in 310 problems), while TN was better in 387 problems, etc. Out of 800 problems considered in this set of numerical experiments, only for 728 problems did the criterion (1.118) hold.

It is worth seeing that subject to the number of iterations and to the number of function calls, TN is top performer. However, subject to the CPU time metric, L-BFGS is more efficient and more robust. Both these algorithms are reliable for solving a large variety of large-scale unconstrained optimization problems. In our numerical experiments we noticed that the performances of these methods do not depend on the structure of the Hessian matrix of the problems.

Let us now present the performances of these algorithms for solving the applications from MINPACK-2 collection, described above. Tables 1.2 and 1.3 present the performances of L-BFGS ($m = 5$) and TN for solving these applications, where n is the number of variables ($n = nx \times ny$, $nx = 200$, $ny = 200$), *#iter* is the number of iterations to get the solution, *#fg* is the number of function and its gradient calls and *cpu* is the CPU time in seconds for solving the application.

Observe that subject to the number of iterations and to the function calls, TN is better than L-BFGS, but subject to the CPU time metric, both L-BFGS and TN have similar performances, L-BFGS being slightly faster.

Notes and References

Unconstrained optimization is very well consolidated with a lot of methods for which both their theoretical properties and their computational efficiency and robustness are known. In this presentation only the gradient methods have been discussed. Derivative-free methods are not considered. The vast majority of unconstrained optimization methods are line search methods. An exception is the trust-region method. Both of them generate steps using a quadratic model of the minimizing function. A line search method generates a descent search direction and then a step along this direction is taken, hoping that the value of the minimizing function is improved. Trust-region methods define a region around the current iteration within

Figure 1.6 Performance profiles of L-BFGS ($m = 5$) versus TN (Truncated Newton) based on: iterations calls, function calls, and CPU time, respectively

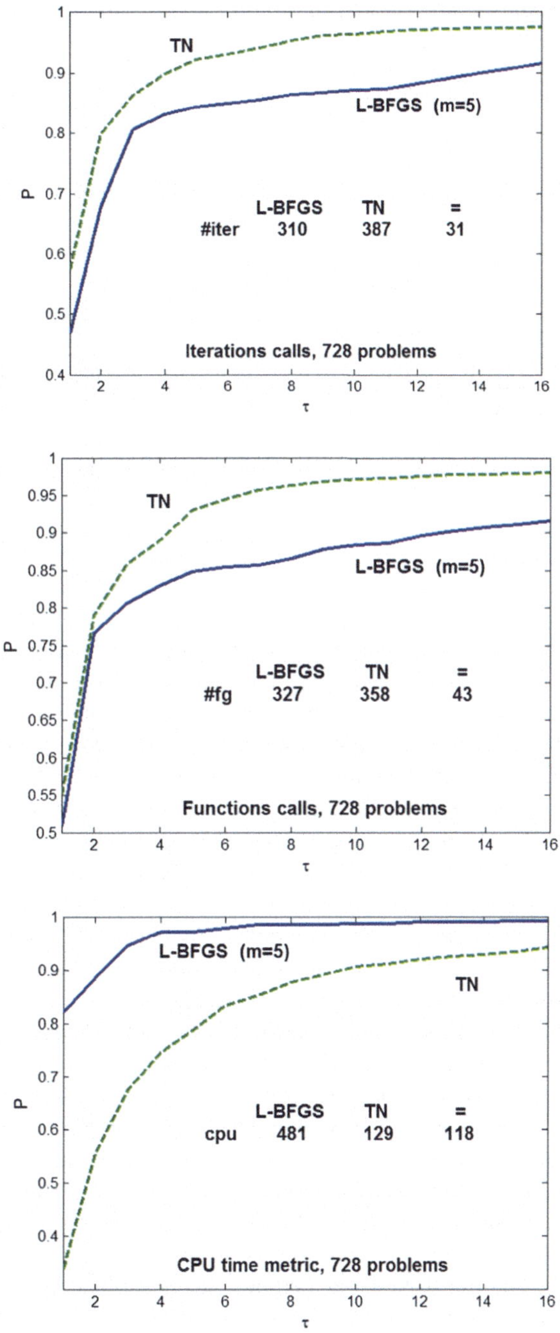

Table 1.2 Performances of L-BFGS ($m = 5$) for solving five applications from the MINPACK-2 collection

	n	*#iter*	*#fg*	*cpu*
A1	40,000	548	560	7.94
A2	40,000	1210	1258	19.01
A3	40,000	1558	1583	33.72
A4	40,000	847	883	30.02
A5	40,000	679	703	12.23
Total	–	4842	4987	102.92

Table 1.3 Performances of TN for solving five applications from the MINPACK-2 collection

	n	*#iter*	*#fg*	*cpu*
A1	40,000	14	320	10.35
A2	40,000	40	790	13.92
A3	40,000	53	1791	53.37
A4	40,000	26	501	21.87
A5	40,000	20	312	5.06
Total	–	153	3714	104.57

which they trust a model (quadratic) of the minimizing function and choose a step to be the approximate minimizer of the model in this region. For the stepsize computation, known as line search, we have presented the most important monotone and non-monotone methods. The most important are the Wolfe line search conditions. Modifications of these conditions are the *approximate Wolfe line search* (Hager & Zhang, 2005) and *improved Wolfe line search* (Dai & Kou, 2013). Both these modified Wolfe line search are of a paramount importance in conjugate gradient methods.

Fundamental is the steepest descent method. Its convergence is only linear, with a convergence ratio given by (1.38). The convergence of the Newton method is found in Dennis and Schnabel (1983). The most important result on the behavior of the Newton method is given by Theorem 1.12. A proof is found in (Kelley, 1995, 1999). Although the Newton method is the best in terms of its convergence near solution, it requires some modifications. The most important are the quasi-Newton, limited-memory quasi-Newton, truncated Newton methods and the p-regularized methods. These modified Newton methods are operational for solving large-scale problems, but they are only linear or superlinear convergent. Plenty of scaling BFGS methods are known. They try to improve the eigenvalue distribution of the approximated scaled Hessian. An important class of the modified Newton methods is that of the quasi-Newton methods with diagonal updating of the Hessian (Dennis & Wolkowicz, 1993; Zhu, Nazareth, & Wolkowicz, 1999; Andrei, 2018e, 2019c, 2019d). Their convergence is only linear.

The quasi-Newton methods are described in a lot of papers, but the synthetic study given by Dennis and Moré (1974, 1977) is a landmark. A new derivation of the BFGS quasi-Newton method using symmetric positive definite secant updates was developed by Dennis and Schnabel (1981). The convergence of the BFGS method can be found in Ge and Powell (1983) and Boggs and Tolle (1994). The global convergence of the BFGS method was established by Powell (1976a). Two important properties of BFGS method are the self-correcting and bounded deterioration properties. The *self-correcting property* with respect to the determinant and with respect to the trace of the BFGS quasi-Newton method was established by Nocedal (1992). The analysis of the quasi-Newton methods was based on the *bounded deterioration property* (Dennis & Moré, 1977). Let x^* be a stationary point. An updating formula for the Hessian approximation B_k satisfies the bounded deterioration property if there exists a constant $c > 0$ so that for all x_k and B_k the new x_{k+1} and B_{k+1} satisfy

$$\left\|B_{k+1} - \nabla^2 f(x^*)\right\| \leq \left\|B_k - \nabla^2 f(x^*)\right\|(1+\sigma) + c\sigma,$$

where $\sigma = \max\{\|x_k - x^*\|, \|x_{k+1} - x^*\|\}$. If a quasi-Newton method satisfies the bounded deterioration property and there exist the positive constants ε and δ so that $\|x_0 - x^*\| \leq \varepsilon$ and $\left\|B_0 - \nabla^2 f(x^*)\right\| \leq \delta$, then the sequence $\{x_k\}$ generated by the algorithm is well defined and converges q-linearly to x^*. In order to have q-superlinearly convergence of the algorithm, it is necessary to have the *consistency* of it. A quasi-Newton method is consistent if $\{x_k\}$ converges to x^*, then $\{B_k\}$ converges to $\nabla^2 f(x^*)$. The consistency condition is sufficient but not necessary.

In both BFGS and DFP updating formulae, the updated matrix B_{k+1} (and its inverse H_{k+1}) differs from previous B_k (or H_k) by a rank-two matrix. There is another rank-one update, known as the *symmetric-rank-one*, or SR1, update. The SR1 method does not guarantee that the updated matrix maintains positive definiteness. In fact, even for convex quadratic minimizing function, there may be steps on which there is no symmetric-rank-one update that satisfies the secant equation. Computational experiments by Conn, Gould, and Toint (1988) using both line search and trust-region approaches, showed that the SR1 method appears to be competitive with the BFGS method (Nocedal & Wright, 2006).

For large-scale problems, the limited-memory quasi-Newton and the truncated Newton methods are recommended. A compact representation of the limited-memory BFGS updating is described by Nocedal and Wright (2006). The best (free) software implementing these methods are L-BFGS (Liu & Nocedal, 1989), M1QN3 (Gilbert & Lemaréchal, 1989), TN/TNBC (Nash, 1985), TNPACK (Schlick & Fogelson 1992a, 1992b).

The conjugate gradient method was designed by Hestenes and Stiefel (1952) as an alternative to factorization methods for solving symmetric positive definite linear algebraic systems. Extension of this method for the nonlinear optimization was given by Fletcher and Reeves (1964). These methods are "matrix fee," involving only a small number of dot products. Papers on this subject are too many to be enumerated. The global convergence of nonlinear conjugate gradient methods received much attention and a lot of papers detail this aspect: Al-Baali (1985), Gilbert and Nocedal (1992), Hager and Zhang (2005). In a number of papers, with major intellectual effort, Dai (1997, 2001, 2002a, 2003b, 2010, 2011) and Dai and

Yuan (1996a, 1996b, 1996c, 1999, 2000) established fundamental results concerning the convergence of these methods. At present, the best software implementing conjugate gradient methods for solving large-scale unconstrained optimization problems are: CONMIN (Shanno & Phua, 1976), CG-DESCENT (Hager & Zhang, 2005), L-CG-DESCENT (Hager & Zhang, 2013) (for ill-conditioned problems), DESCON (Andrei, 2013c), CGOPT (Dai & Kou, 2013), which will be presented in the chapters of this book. The modern conjugate gradient algorithms CG-DESCENT, DESCON and CGOPT include advanced line searches based on modifications or improvements of the Wolfe line search conditions. For solving large-scale unconstrained optimization problems the recommended methods are: the limited-memory quasi-Newton, the truncated Newton and the conjugate gradient methods. As it is to be seen, out of these methods, the conjugate gradient methods prove to be more efficient and more robust. The conjugate gradient algorithms for the nonconvex optimization and for problems with box constraints are discussed by Pytlak (2009).

One of the first presentations of the trust-region method was given by Winfield (1969) in his Ph.D. Thesis. The convergence of the trust-region method was given by Powell (1975). The general theory of trust-region methods is in (Byrd, Schnabel, & Schultz, 1985, 1988). A survey on trust-region methods is found in Dennis and Schnabel (1983, 1989) and Nocedal and Wright (2006). The monograph of Conn, Gould, and Toint (2000) is an exhaustive presentation of the state of the art in the trust-region method for both unconstrained and constrained nonlinear optimization.

The p-regularized methods have been introduced by Griewank (1981) and have been developed by many authors: Nesterov and Polyak (2006), Dollar, Gould, and Robinson (2009), Gould, Robinson, and Thorne (2010), Cartis, Gould, and Toint (2011a, 2011b), Gould, Porcelli, and Toint (2012), Bianconcini, Liuzzi, Morini, and Sciandrone (2013), Benson, and Shanno (2014, 2018), Bellavia, and Morini (2015), Bianconcini, and Sciandrone (2016). The theory of p-regularized subproblems including the necessary and sufficient optimality condition and a closed-form expression for the global minimum set of these subproblems was presented by Hsia, Sheu, and Yuan (2017). It is proved that the p-regularized subproblems have all properties that the trust-region subproblems have.

Chapter 2
Linear Conjugate Gradient Algorithm

The linear conjugate gradient algorithm is dedicated to minimizing convex quadratic functions (or solving linear algebraic systems of equations with positive definite matrices). This algorithm was introduced by Hestenes and Stiefel (1952). Let us consider the quadratic function

$$f(x) = \frac{1}{2}x^T Ax - b^T x, \tag{2.1}$$

where $A \in \mathbb{R}^{n \times n}$ is a symmetric and positive definite matrix, and $b \in \mathbb{R}^n$ is a known vector. From (2.1), we get

$$\nabla f(x) = Ax - b, \quad \nabla^2 f(x) = A. \tag{2.2}$$

Observe that the Hessian of function (2.1) is independent of x. Since the Hessian A is symmetric and positive definite, from the optimality conditions for a minimum of a differentiable function, it follows that there is a unique minimizer x^* of (2.1). From (2.2), observe that x^* is the solution of the linear system

$$Ax = b.$$

Having in view that the function f is quadratic, from the Taylor theorem, for all $t \in \mathbb{R}$ and all $y, z \in \mathbb{R}^n$, the following identity is obtained:

$$f(y + tz) = f(y) + t\nabla f(y)^T z + \frac{t^2}{2} z^T Az. \tag{2.3}$$

N. Andrei, *Nonlinear Conjugate Gradient Methods for Unconstrained Optimization*,
Springer Optimization and Its Applications 158,
https://doi.org/10.1007/978-3-030-42950-8_2

2.1 Line Search

The linear conjugate gradient algorithm is exactly a line search with a special choice of directions. Given the current approximation x_j to the minimum x^* as well as a direction vector d_j, a *line search algorithm* computes the next approximation x_{j+1} by using the following two steps:

1. Find the stepsize $\alpha_j = \arg\min f(x_j + \alpha d_j)$
2. Set $x_{j+1} = x_j + \alpha_j d_j$ ♦

Assuming that an initial point x_0 is given, then, applying k steps of the above line search method k iterates are obtained: $\{x_0, x_1, \ldots, x_{k-1}\}$. From (2.3), the stepsize α_j is computed as follows:

$$\alpha_j = \frac{-d_j^T r_j}{d_j^T A d_j}, \tag{2.4}$$

where $r_j = Ax_j - b$ is the *residual* at x_j.

Definition 2.1 The set of directions $\{d_0, \ldots, d_{k-1}\}$ is a set of conjugate directions if and only if $d_j^T A d_i = 0$ for all $i = 0, \ldots, k-1$, $j = 0, \ldots, k-1$ and $i \neq j$. ♦

Now, for all $k = 1, \ldots,$ the following vector space and affine space are introduced:

$$W_k = span\{d_0, \ldots, d_{k-1}\}, \tag{2.5}$$

$$U_k = x_0 + W_k = \{z \in \mathbb{R}^n : z = x_0 + w_k,\ w_k \in W_k\}. \tag{2.6}$$

Denote $W_0 = \{0\}$ and $U_0 = \{x_0\}$.

Proposition 2.1 *Assume that $d_i^T A d_j = 0$ for all $0 \leq j < i$, where i is a fixed integer, and that $\{x_0, \ldots, x_i\}$ are computed by the line search algorithm. Then*

$$d_i^T r_i = d_i^T \nabla f(y), \tag{2.7}$$

for all $y \in U_i$.

Proof Firstly, observe that, since $\{x_0, \ldots, x_i\}$ are computed by the line search algorithm, it follows that $x_i \in U_i$. If $y \in U_i$, then, from the definition of U_i, it follows that $x_i - y \in W_i$ and hence $d_i^T A(x_i - y) = 0$. Therefore,

$$d_i^T (r_i - \nabla f(y)) = d_i^T (Ax_i - b - Ay + b) = d_i^T A(x_i - y) = 0,$$

which proves (2.7). ♦

2.2 Fundamental Property of the Line Search Method with Conjugate Directions

Obviously, at every step, the line search algorithm minimizes $f(x)$ only in a fixed direction. However, if the directions are conjugate according to Definition 2.1, then a stronger result can be proved, as in Theorem 2.1 below: A choice of conjugate directions in the line search method leads to obtaining a minimizer x_k for the whole space U_k.

Theorem 2.1 *If the directions in the line search algorithm are conjugate and* $\{x_0, \ldots, x_k\}$ *are the iterates generated after k steps of the line search algorithm, then*

$$x_j = \arg\min_{x \in U_j} f(x),$$

for all $1 \le j \le k$.

Proof The theorem is proved by induction. For $k = 1$, the result is obtained from the definition of x_1 as a minimizer on U_1. Assume that for $k = i$,

$$x_j = \arg\min_{y \in U_j} f(y),$$

for all $1 \le j \le i$. We need to show that if $x_{i+1} = x_i + \alpha_i d_i$, then

$$x_{i+1} = \arg\min_{x \in U_{i+1}} f(x).$$

By definition of U_{i+1}, any $x \in U_{i+1}$ can be written as $x = y + \alpha d_i$, where $\alpha \in \mathbb{R}$ and $y \in U_i$. Now, using (2.3) and Proposition 2.1, it results

$$\begin{aligned} f(x) &= f(y + \alpha d_i) = f(y) + \alpha d_i^T \nabla f(y) + \frac{\alpha^2}{2} d_i^T A d_i \\ &= f(y) + \left[\alpha d_i^T \nabla f(x_i) + \frac{\alpha^2}{2} d_i^T A d_i \right]. \end{aligned} \tag{2.8}$$

Observe that (2.8) is a decoupled function. The first term of the right-hand side of (2.8) does not depend on α, and the second term does not depend on y. Therefore,

$$\min_{x \in U_{i+1}} f(x) = \min_{y \in U_i} f(y) + \min_{\alpha \in \mathbb{R}} \left[\alpha d_i^T r_i + \frac{\alpha^2}{2} d_i^T A d_i \right]. \tag{2.9}$$

But, the right-hand side of (2.9) is minimized when $y = x_i$ and

$$\alpha = \alpha_i = \frac{-d_i^T r_i}{d_i^T A d_i},$$

that is, the left-hand side of (2.9) is minimized exactly for $x_{i+1} = x_i + \alpha_i d_i$. In other words, x_{i+1} is the minimizer of f over the set $\{x : x = x_0 + span\{d_0, \ldots .d_i\}\}$. ♦

To show the significance of making the search directions mutually conjugate with respect to A, let us first state and prove a technical result involving only the first two iterations of the conjugate gradient algorithm.

Proposition 2.2 *After two iterations of the conjugate gradient method, the gradient* $g_2 = Ax_2 - b$ *satisfies*

$$d_1^T g_2 = d_0^T g_2 = 0.$$

Proof After the first iteration, the new point is x_1. Therefore, $g_1 = Ax_1 - b$. Since the line search is exact, we also have $d_0^T g_1 = 0$.

Now, consider iteration two. At this iteration, the algorithm will generate a point $x_2 = x_1 + \alpha d_1$ where $g_2 = Ax_2 - b$ and $d_1^T g_2 = 0$. But we have

$$d_0^T g_2 = d_0^T (Ax_1 + \alpha A d_1 - b) = d_0^T g_1 + \alpha d_0^T A d_1.$$

The first term, $d_0^T g_1$, in the right-hand side of the above equality is zero because of the line search on iteration one. The second term, $\alpha d_0^T A d_1$, is zero because d_0 and d_1 are conjugate with respect to A. ♦

This result shows that after two iterations, the gradient is orthogonal to both search directions d_0 and d_1. Similarly, the above result can be generalized to prove the following proposition.

Proposition 2.3 *After k iterations of the conjugate gradient method, the gradient* $g_k = Ax_k - b$ *satisfies*

$$d_j^T g_k = 0 \quad \text{for } j = 0, 1, \ldots, k-1.$$

♦

This proposition implies that after k iterations, the gradient g_k is restricted to the $(n-k)$-dimensional subspace orthogonal to the vectors $d_0, \ldots, d_{k-1}$. From this, the important *finite termination property* of the conjugate gradient method can be obtained.

Proposition 2.4 *The conjugate gradient method solves an* $n \times n$ *linear algebraic system* $Ax = b$ *in* n *iterations at the most.*

Proof Proposition 2.3 implies that, after n iterations, g_n is orthogonal to the n vectors $d_0, \ldots, d_{n-1}$. But this means that g_n must lie in a subspace of dimension zero and so $g_n = 0$, which proves that $Ax_n = b$. ♦

The finite termination property is only guaranteed when the calculations are exact. In practice, the conclusion of Proposition 2.3 may not be exactly satisfied when the iterations are performed in real arithmetic, which is subject to rounding errors. Hence, for solving some $n \times n$ linear algebraic systems, the conjugate gradient method needs a few more than n iterations.

2.3 The Linear Conjugate Gradient Algorithm

The result of Theorem 2.1, for generating conjugate directions, may now be used to present the linear conjugate gradient algorithm. To start with, a general recurrence relation that generates a set of conjugate directions is proved. Next, it is shown that this recurrence relation can be reduced to a simple expression. Finally, the linear conjugate gradient algorithm is described.

Proposition 2.5 *Let* $d_0 = -r_0$ *and for* $k = 1, 2, \ldots$ *set*

$$d_k = -r_k + \sum_{j=0}^{k-1} \frac{d_j^T A r_k}{d_j^T A d_j} d_j. \tag{2.10}$$

Then $d_j^T A d_m = 0$ *for all* $0 \leq m < j \leq k$.

Proof By induction, it is shown that (2.10) generates conjugate directions. For $k = 1$, $d_1^T A d_0 = 0$. Assume that for $k = i$, the vectors $\{d_0, \ldots, d_i\}$ are pairwise conjugate. We must show that $d_{i+1}^T A d_m = 0$ for all $m \leq i$. Consider $m \leq i$. Then

$$\begin{aligned} d_{i+1}^T A d_m &= -r_{i+1}^T A d_m + \sum_{j=0}^{i} \frac{d_j^T A r_{i+1}}{d_j^T A d_j} d_j^T A d_m \\ &= -r_{i+1}^T A d_m + \frac{d_m^T A r_{i+1}}{d_m^T A d_m} d_m^T A d_m = 0, \end{aligned}$$

which proves the proposition. ♦

Proposition 2.6 *Let* $\{d_0, \ldots, d_k\}$ *be the directions generated by* (2.10). *Then*

(i) $W_k = span\{r_0, \ldots, r_{k-1}\}$,
(ii) $r_m^T r_j = 0$, *for all* $0 \leq j < m \leq k$,
(iii) $d_k^T r_j = -r_k^T r_k$, *for all* $0 \leq j \leq k$,
(iv) *The direction* d_k *satisfies*

$$d_k = -r_k + \beta_{k-1} d_{k-1}, \tag{2.11}$$

where

$$\beta_{k-1} = \frac{r_k^T r_k}{r_{k-1}^T r_{k-1}}. \tag{2.12}$$

Proof Since $d_0 = -r_0$, (i) follows directly from (2.10).

To prove (ii), observe that for $0 \le j < m \le k$ and any $t \in \mathbb{R}$, $r_j \in W_{j+1} \subset W_m$ and hence $x_m + tr_j \in U_m$. Now, from Theorem 2.1, since x_m is the unique minimizer of function f over U_m, it follows that $t = 0$ is the unique minimizer of $f(x_m + tr_j)$. Therefore,

$$0 = \left.\frac{df(x_m + tr_j)}{dt}\right|_{t=0} = \nabla f(x_m)^T r_j = r_m^T r_j,$$

for $0 \le j < m \le k$, which proves (ii).

To prove (iii), the identity in (iii) is first proved for $j = k$. Now, from (i), it follows that r_k is orthogonal to each d_l for $l < k$. Therefore, if we take the inner product with r_k, the second term in the right side of (2.10) will vanish. But, this is exactly the identity in (iii). for $j = k$. If $j < k$, then $(x_k - x_j) \in W_k$, and hence $d_k^T A(x_k - x_j) = 0$. Therefore,

$$d_k^T (r_k - r_j) = d_k^T A(x_k - x_j) = 0.$$

To prove (iv), observe first that $\{r_0, \ldots, r_k\}$ form an orthogonal basis of W_{k+1}. Hence, we can write $d_k \in W_{k+1}$ as a linear combination of $\{r_0, \ldots, r_k\}$ and then apply (iii). Indeed

$$\begin{aligned} d_k &= \sum_{j=0}^{k} \frac{d_k^T r_j}{r_j^T r_j} r_j = -\sum_{j=0}^{k} \frac{r_k^T r_k}{r_j^T r_j} r_j = -r_k - \frac{r_k^T r_k}{r_{k-1}^T r_{k-1}} \sum_{j=0}^{k-1} \frac{r_{k-1}^T r_{k-1}}{r_j^T r_j} r_j \\ &= -r_k + \beta_{k-1} \sum_{j=0}^{k-1} \frac{d_{k-1}^T r_j}{r_j^T r_j} r_j = -r_k + \beta_{k-1} d_{k-1}. \end{aligned}$$

♦

Remark 2.1 From (2.4) and from Proposition 2.6 (iii), it follows that

$$\alpha_k = \frac{-d_k^T r_k}{d_k^T A d_k} = \frac{r_k^T r_k}{d_k^T A d_k}. \tag{2.13}$$

On the other hand, the residual vector r_{k+1} can be written as

$$r_{k+1} = Ax_{k+1} - b = Ax_k - b + \alpha_k Ad_k = r_k + \alpha_k Ad_k. \tag{2.14}$$

With these, using Proposition 2.3 and Remark 2.1, the following linear conjugate gradient algorithm can be presented.

Algorithm 2.1 Linear conjugate gradient

1.	Select an initial point x_0 and $\varepsilon > 0$ sufficiently small
2.	Set $r_0 = Ax_0 - b$, $d_0 = -r_0$ and $k = 0$
3.	If $\|r_k\| \le \varepsilon$, then stop, otherwise, continue with step 4
4.	Compute $\alpha_k = \frac{r_k^T r_k}{d_k^T A d_k}$, $x_{k+1} = x_k + \alpha_k d_k$, $r_{k+1} = r_k + \alpha_k A d_k$, $\beta_k = \frac{r_{k+1}^T r_{k+1}}{r_k^T r_k}$, $d_{k+1} = -r_{k+1} + \beta_k d_k$
5.	Set $k = k + 1$ and continue with step 3 ♦

The algorithm is simple and very easy to implement. The initial residual is the same as the first gradient search direction. If the initial solution x_0 is zero, then $r_0 = -b$ and $d_0 = b$. Observe that if A is not symmetric and positive definite, then the denominator in α_k may vanish, resulting in the breakdown of the iterations.

2.4 Convergence Rate of the Linear Conjugate Gradient Algorithm

In the following, an estimate for the convergence rate of the linear conjugate gradient algorithm is presented. For this, the error reduction in the linear conjugate gradient algorithm is discussed. After that, based on the Chebyshev polynomials, a convergence rate estimate is given.

Proposition 2.7 *The following relation holds:*

$$W_k = span\{r_0, \ldots, A^{k-1} r_0\}. \tag{2.15}$$

Proof For $k = 1$, (2.15) is true. Assume that (2.15) holds for $k = i$ and let us show that it holds for $k = i + 1$. From Proposition 2.6 (i), this would be equivalent to show that $r_i \in span\{r_0, \ldots, A^i r_0\}$. Observe that $r_{i-1} \in W_i$ and $d_{i-1} \in W_i$. By inductive assumption, we can write $r_{i-1} = R_{i-1}(A) r_0$ and $d_{i-1} = P_{i-1}(A) r_0$, where $R_{i-1}(.)$ and $P_{i-1}(.)$ are polynomials depending on the matrix A, of degree less than or equal to $i - 1$. Hence,

$$\begin{aligned} r_i &= r_{i-1} + \alpha_{i-1}Ad_{i-1} \\ &= R_{i-1}(A)r_0 + \alpha_{i-1}AP_{i-1}(A)r_0 \in span\{r_0, \ldots, A^i r_0\}, \end{aligned}$$

thus proving the proposition. ♦

In the following, let us present a general error estimate which connects $\|x^* - x_k\|_A$ and $\|x^* - x_0\|_A$, where for any $y \in \mathbb{R}^n$, $\|y\|_A^2 = y^T Ay$. For this, denote $\overline{P}_k$ as the set of polynomials of degree less than or equal to k.

Proposition 2.8 *The following estimate holds:*

$$\|x^* - x_k\|_A = \inf_{P \in \bar{P}_k, P(0)=1} \|P(A)(x^* - x_0)\|_A. \tag{2.16}$$

Proof Since r_k is orthogonal to W_k, it follows that for all $y \in W_k$,

$$(x^* - x_k)^T Ay = r_k^T y = 0. \tag{2.17}$$

Denoting $w_k = x_k - x_0 \in W_k$ and $e_0 = x^* - x_0$, from (2.17) we get

$$0 = (x^* - x_k)^T Ay = (e_0 - w_k)^T Ay,$$

for all $y \in W_k$. Therefore, $w_k = x_k - x_0$ is an A-orthogonal projection of e_0 on W_k. Thus,

$$\|e_0 - w_k\|_A = \min_{w \in W_k} \|e_0 - w\|_A.$$

But, from Proposition 2.7, it is known that $w = Q_{k-1}(A)r_0$, for a polynomial $Q_{k-1} \in \overline{P}_{k-1}$, where $\overline{P}_{k-1}$ is the set of polynomials of degree less than or equal to $k-1$. Also, $Ae_0 = -r_0$ and $e_0 - w = (I + Q_{k-1}(A)A)e_0$, and hence,

$$\|x^* - x_k\|_A = \|e_0 - w_k\|_A = \min_{P_k \in \overline{P}_k, P_k(0)=1} \|P_k(A)e_0\|_A, \tag{2.18}$$

which completes the proof. ♦

This convergence rate is rather general and does not take into account knowledge of the distribution of the eigenvalues of A. In order to refine the above results and to obtain a qualitative estimate on the right-hand side of (2.18), observe that for A symmetric and positive definite the following spectral decomposition can be written as follows (see Appendix A):

$$A = U\Lambda U^T,$$

where U is an orthogonal matrix whose columns are the eigenvectors of A, and Λ is a diagonal matrix with the positive eigenvalues of A, $\lambda_1 \leq \lambda_2 \leq \ldots \leq \lambda_n$ on the diagonal. Since $UU^T = U^T U = I$ by orthogonality of U, it follows that for any j,

$$A^j = U\Lambda^j U^T.$$

Therefore,

$$P_k(A) = UP_k(\Lambda)U^T.$$

Define $A^{1/2} = U\Lambda^{1/2}U^T$. Observe that $\|x\|_A^2 = x^T Ax = \left\|A^{1/2}x\right\|_2^2$. Hence, for any $x \in \mathbb{R}^n$,

$$\|P_k(A)x\|_A = \left\|A^{1/2}P_k(A)x\right\|_2 \leq \|P_k(A)\|_2 \left\|A^{1/2}x\right\|_2 \leq \|P_k(A)\|_2 \|x\|_A.$$

This together with (2.18) implies that for any polynomial $P_k(\lambda)$,

$$\|x^* - x_k\|_A = \min_{P_k \in \overline{P}_k, P_k(0)=1} \|P_k(A)e_0\|_A \leq \min_{P_k \in \overline{P}_k, P_k(0)=1} \rho(P_k(A))\|e_0\|_A,$$

where $\rho(P_k(A))$ is the spectral radius of $P_k(A)$. Since both A and $P_k(A)$ have the same eigenvectors, it follows that

$$\|x^* - x_k\|_A \leq \min_{P_k \in \overline{P}_k, P_k(0)=1} \max_{1 \leq j \leq n} \left|P_k(\lambda_j)\right| \|e_0\|_A, \tag{2.19}$$

where $\lambda_1 \leq \lambda_2 \leq \cdots \leq \lambda_n$ are the eigenvalues of A. The above inequality shows that minimizing the error in the linear conjugate gradient corresponds to minimizing the polynomial $P_k(\lambda)$ over the entire range of eigenvalues $[\lambda_1, \lambda_n]$. This can be accomplished via the Chebyshev polynomials.

The Chebyshev polynomials of the first kind on $[-1, 1]$ are defined as

$$T_k(\xi) = \cos(k \arccos(\xi)), \quad k = 0, 1, \ldots.$$

It is easy to see that $T_k(\xi)$ is a polynomial if the following trigonometric identities are used:

$$\cos(\alpha + \beta) = \cos\alpha \cos\beta - \sin\alpha \sin\beta,$$
$$\cos(\alpha + \beta) + \cos(\alpha - \beta) = 2\cos\alpha \cos\beta.$$

Let us denote $\theta = \arccos(\xi)$, then

$$\begin{aligned}
T_0(\xi) &= \cos(0\theta) = 1, \\
T_1(\xi) &= \cos(1\theta) = \xi, \\
T_2(\xi) &= \cos(2\theta) = \cos^2\theta - \sin^2\theta = 2\cos^2\theta - 1 = 2\xi^2 - 1, \\
T_{k+1}(\xi) + T_{k-1}(\xi) &= \cos((k+1)\theta) + \cos((k-1)\theta) \\
&= 2\cos(k\theta)\cos(\theta) = 2\xi T_k(\xi).
\end{aligned}$$

Therefore,

$$T_0(\xi) = 1, \; T_1(\xi) = \xi, \tag{2.20}$$

$$T_{k+1}(\xi) = 2\xi T_k(\xi) - T_{k-1}(\xi), \tag{2.21}$$

for any $\xi \in \mathbb{R}$.

From (2.21) for fixed ξ, it follows that

$$T_k(\xi) = c_1(\eta_1(\xi))^k + c_2(\eta_2(\xi))^k, \; k = 0, 1, \ldots,$$

where $\eta_1(\xi)$ and $\eta_2(\xi)$ are the roots of the characteristic equation

$$\eta^2 - 2\xi\eta + 1 = 0.$$

The constants c_1 and c_2 are determined from the initial conditions (2.20). Therefore,

$$T_k(\xi) = \frac{1}{2}\left[\left(\xi + \sqrt{\xi^2 - 1}\right)^k + \left(\xi - \sqrt{\xi^2 - 1}\right)^k\right]. \tag{2.22}$$

Figure 2.1 presents some Chebyshev polynomials. Observe that $|T_k(\xi)| \leq 1$ for all $\xi \in [-1, 1]$. The polynomial that minimizes (2.19) over the interval $[\lambda_1, \lambda_n]$ is

$$S_k(\lambda) = \left[T_k\left(\frac{\lambda_n + \lambda_1}{\lambda_n - \lambda_1}\right)\right]^{-1}\left[T_k\left(\frac{\lambda_n + \lambda_1 - 2\lambda}{\lambda_n - \lambda_1}\right)\right]. \tag{2.23}$$

To prove this, assume that there exists another polynomial of degree k, Q_k, that is better at minimizing (2.19) on the appropriate interval $[\lambda_1, \lambda_n]$, so that $Q_k(0) = 1$,

$$Q_k(\lambda) < \left[T_k\left(\frac{\lambda_n + \lambda_1}{\lambda_n - \lambda_1}\right)\right]^{-1}.$$

The polynomial $P_k - Q_k$ must have a zero at $\lambda = 0$ and at k zeros of the polynomials, meaning that this polynomial must have $k+1$ zeros, which is a contradiction. Therefore, S_k from (2.23) must be the minimizing polynomial on the interval $[\lambda_1, \lambda_n]$. Therefore, from (2.19), it results that

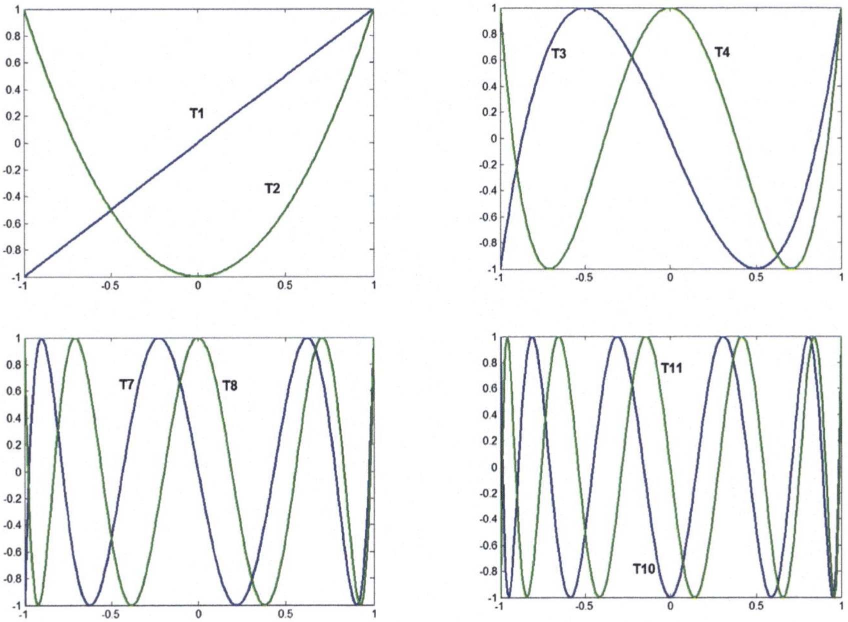

Figure 2.1 Some Chebyshev polynomials

$$\|x^* - x_k\|_A \leq \left[T_k\left(\frac{\lambda_n + \lambda_1}{\lambda_n - \lambda_1}\right)\right]^{-1} \|x^* - x_0\|_A. \tag{2.24}$$

Theorem 2.2 *The error after k iterations of the linear conjugate gradient algorithm can be bounded as follows:*

$$\|x^* - x_k\|_A \leq \frac{2}{\left(\frac{\sqrt{\kappa}+1}{\sqrt{\kappa}-1}\right)^k + \left(\frac{\sqrt{\kappa}-1}{\sqrt{\kappa}+1}\right)^k} \|x^* - x_0\|_A \leq 2\left(\frac{\sqrt{\kappa}-1}{\sqrt{\kappa}+1}\right)^k \|x^* - x_0\|_A, \tag{2.25}$$

where $\kappa = \kappa(A) = \lambda_n/\lambda_1$ *is the condition number of A.*

Proof The purpose is to calculate

$$\left[T_k\left(\frac{\lambda_n + \lambda_1}{\lambda_n - \lambda_1}\right)\right]^{-1}.$$

From (2.22), for $\xi = (\lambda_n + \lambda_1)/(\lambda_n - \lambda_1) = (\kappa + 1)/(\kappa - 1)$, we obtain

$$\xi \pm \sqrt{\xi^2 - 1} = \frac{\kappa + 1}{\kappa - 1} \pm \frac{2\sqrt{\kappa}}{\kappa - 1} = \frac{\kappa + 1 \pm 2\sqrt{\kappa}}{\kappa - 1} = \frac{(\sqrt{\kappa} \pm 1)^2}{(\sqrt{\kappa} - 1)(\sqrt{\kappa} + 1)} = \frac{\sqrt{\kappa} \pm 1}{\sqrt{\kappa} \mp 1}.$$

Therefore,

$$T_k\left(\frac{\lambda_n + \lambda_1}{\lambda_n - \lambda_1}\right) = \frac{1}{2}\left[\left(\frac{\sqrt{\kappa} + 1}{\sqrt{\kappa} - 1}\right)^k + \left(\frac{\sqrt{\kappa} - 1}{\sqrt{\kappa} + 1}\right)^k\right].$$

Hence,

$$\left[T_k\left(\frac{\lambda_n + \lambda_1}{\lambda_n - \lambda_1}\right)\right]^{-1} = \frac{2}{\left(\frac{\sqrt{\kappa}+1}{\sqrt{\kappa}-1}\right)^k + \left(\frac{\sqrt{\kappa}-1}{\sqrt{\kappa}+1}\right)^k} \leq 2\left(\frac{\sqrt{\kappa} - 1}{\sqrt{\kappa} + 1}\right)^k. \tag{2.26}$$

The proof is completed by substituting (2.26) in (2.24). ♦

Knowing only the largest and the smallest eigenvalues of A, the bound (2.25) is the best possible. Theorem 2.2 shows that the error $\|x^* - x_k\|_A$ is upper bounded by a sequence which is convergent to zero. Besides, the convergence is monotone, and this explains why the linear conjugate gradient algorithm is viewed as an iterative method. As an iterative method, the performance of the linear conjugate gradient algorithm depends both on b and on the spectrum of A [see Kelley (1995), Greenbaum (1997)]. The linear conjugate gradient will perform well if κ is near 1, and it may perform very poorly if κ is large. Geometrically, κ is large if the ellipsoidal level surfaces of the quadratic function f are very far from spherical.

If additional information about the eigenvalues of A in the interval $[\lambda_1, \lambda_n]$ is available, then the estimate (2.25) can be improved. Suppose, for example, that A has one eigenvalue much larger than the others, i.e., $\lambda_1 \leq \cdots \leq \lambda_{n-1} \ll \lambda_n$, that is $\lambda_n/\lambda_{n-1} \gg 1$. Consider a polynomial P_k that is the product of a linear factor which is zero at λ_n and the $(k-1)$ degree scaled and shifted Chebyshev polynomial on the interval $[\lambda_1, \lambda_{n-1}]$:

$$P_k(\lambda) = \frac{T_{k-1}\left(\frac{\lambda_{n-1} + \lambda_1 - 2\lambda}{\lambda_{n-1} - \lambda_1}\right)}{T_{k-1}\left(\frac{\lambda_{n-1} + \lambda_1}{\lambda_{n-1} - \lambda_1}\right)} \frac{\lambda_n - \lambda}{\lambda_n}. \tag{2.27}$$

Since the second factor in (2.27) is zero at λ_n and less than one in absolute value at each of the other eigenvalues, then the maximum absolute value of this polynomial in the entire spectrum $\{\lambda_1, \ldots, \lambda_n\}$ of A is less than the maximum absolute value of the first factor on $\{\lambda_1, \ldots, \lambda_{n-1}\}$. Therefore, using similar arguments as those in Theorem 2.2, it follows that

$$\|x^* - x_k\|_A \le 2\left(\frac{\sqrt{\kappa_{n-1}} - 1}{\sqrt{\kappa_{n-1}} + 1}\right)^{k-1} \|x^* - x_0\|_A, \tag{2.28}$$

where $\kappa_{n-1} = \frac{\lambda_{n-1}}{\lambda_1}$. A detailed study of the case of isolated eigenvalues is given by Axelsson and Lindskog (1986).

Similarly, if the matrix A has just a few large eigenvalues, say, $\lambda_1 \le \cdots \lambda_{n-m} \ll \lambda_{n-m+1} \le \cdots \le \lambda_n$, i.e., $\lambda_{n-m+1}/\lambda_{n-m} \gg 1$, then one can consider a polynomial P_k which is the product of an mth degree factor that is zero at each large eigenvalues and a scaled and shifted Chebyshev polynomial of degree $k - m$ on the interval $[\lambda_1, \lambda_{n-m}]$. Bounding the size of this polynomial, it results that

$$\|x^* - x_k\|_A \le 2\left(\frac{\sqrt{\kappa_{n-m}} - 1}{\sqrt{\kappa_{n-m}} + 1}\right)^{k-m} \|x^* - x_0\|_A, \tag{2.29}$$

where this time, $\kappa_{n-m} = \frac{\lambda_{n-m}}{\lambda_1}$ [See Greenbaum (1997), Van der Vorst (1993)].

It is generally true that if the eigenvalues of the matrix A occur in m distinct clusters, then the iterates of the linear conjugate gradient algorithm will approximately solve the problem in about m steps. This is illustrated in the following example.

Example 2.1 In order to see the influence of the distribution of the eigenvalues on the convergence of the linear conjugate gradient algorithm, let us consider a linear system $Ax = b$, where A is a diagonal matrix and b is selected in such a way that the solution of the system is always $[1, 1, \ldots, 1]$ (Andrei, 2009f). In Algorithm 2.1, let us consider $n = 1000$ and $\varepsilon = 10^{-6}$. In the first set of numerical experiments, the matrix A has four distributions of the eigenvalues: (a) $A = diag(1, 2, \ldots, 1000)$ for which $\kappa(A) = 1000$, (b) the diagonal elements are uniformly distributed in [0,1) with $\kappa(A) = 997.4945$, (c) the eigenvalues of A are distributed in ten intervals with $\kappa(A) = 19.0198$, and (d) the eigenvalues of A are distributed in five intervals with $\kappa(A) = 9.0099$. Figure 2.2 presents the norm of residuals corresponding to these four distributions of the eigenvalues.

In the second set of numerical experiments, the matrix A has 999 eigenvalues uniformly distributed in $[0, 1)$ and one large eigenvalue equal to 100. In this case, the condition number is 4448807.0435, and the linear conjugate gradient gives a solution in 185 iterations. Figure 2.3 presents the evolution of the norm of the residuals for this distribution of the eigenvalues.

Suppose that the eigenvalues of A consist of m large values and $n - m$ small eigenvalues clustered around 1. Defining $\tau = \lambda_{n-m} - \lambda_1$, then (2.29) tells us that after $m + 1$ steps of the linear conjugate gradient algorithm,

$$\|x^* - x_{m+1}\|_A \approx \tau \|x^* - x_0\|_A.$$

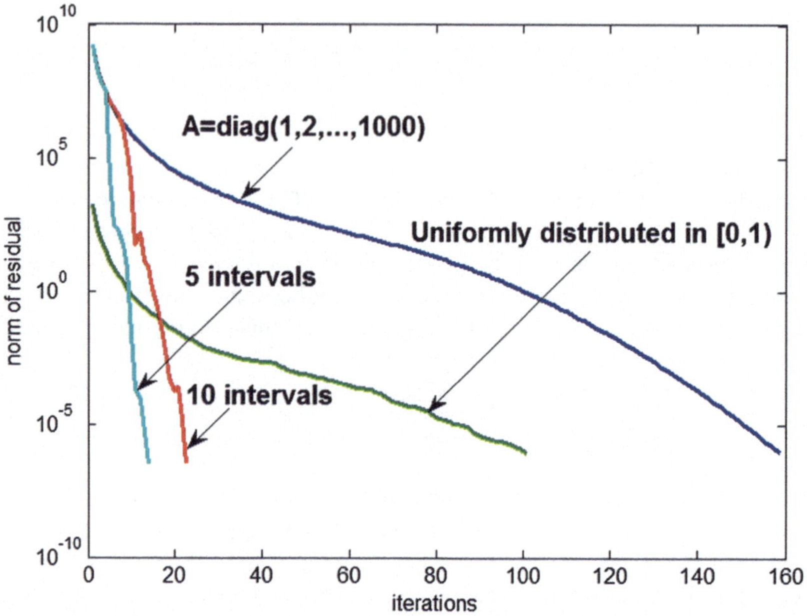

Figure 2.2 Performance of the linear conjugate gradient algorithm for solving the linear system $Ax = b$, where: a) $A = diag(1, 2, \ldots, 1000)$, b) the diagonal elements of A are uniformly distributed in [0,1), c) the eigenvalues of A are distributed in 10 intervals, and d) the eigenvalues of A are distributed in 5 intervals

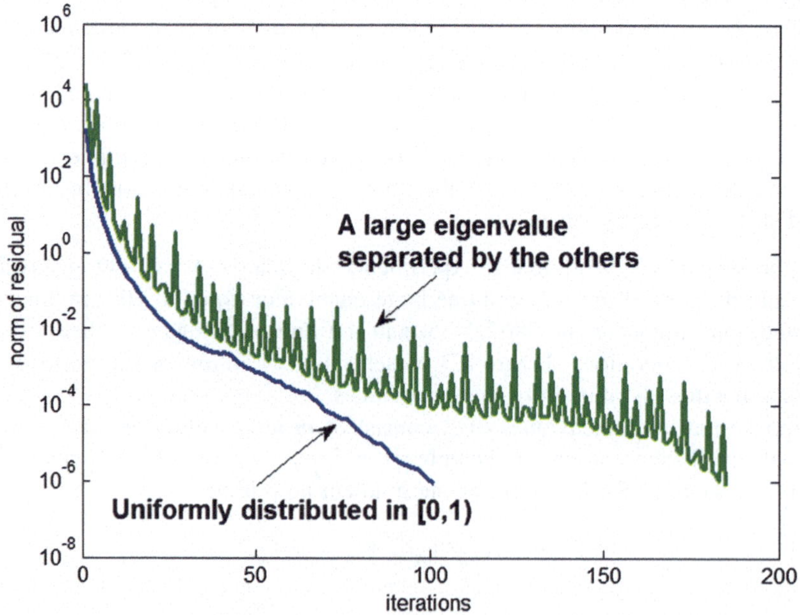

Figure 2.3 Performance of the linear conjugate gradient algorithm for solving the linear system $Ax = b$, where the matrix A has a large eigenvalue separated from others, which are uniformly distributed in [0,1)

Therefore, for a small value of τ, i.e., for small well-clustered eigenvalues, it follows that the iterates of the algorithm provide a good estimate of the solution after only $m+1$ steps.

Example 2.2 Let us consider the linear algebraic system $Ax = b$, where

$$A = \begin{bmatrix} 2 & -1 & & & \\ -1 & 2 & \ddots & & \\ & \ddots & \ddots & \ddots & \\ & & \ddots & 2 & -1 \\ & & & -1 & 2 \end{bmatrix}, \quad \text{and} \quad b = \begin{bmatrix} 1 \\ 0 \\ \vdots \\ 0 \\ 1 \end{bmatrix},$$

obtained from the finite difference numerical method to discretize the one-dimensional Poisson equation (Andrei, 2000, 2009f).

For $n = 1000$, the linear conjugate gradient algorithm gives a solution in 500 iterations. Figure 2.4 shows the evolution of the error $\|b - Ax_k\|$ along the iterations for obtaining a solution with accuracy less than or equal to 10^{-8}.

The eigenvalues of A are

$$\lambda_i = 2\left(1 - \cos\frac{\pi i}{n+1}\right).$$

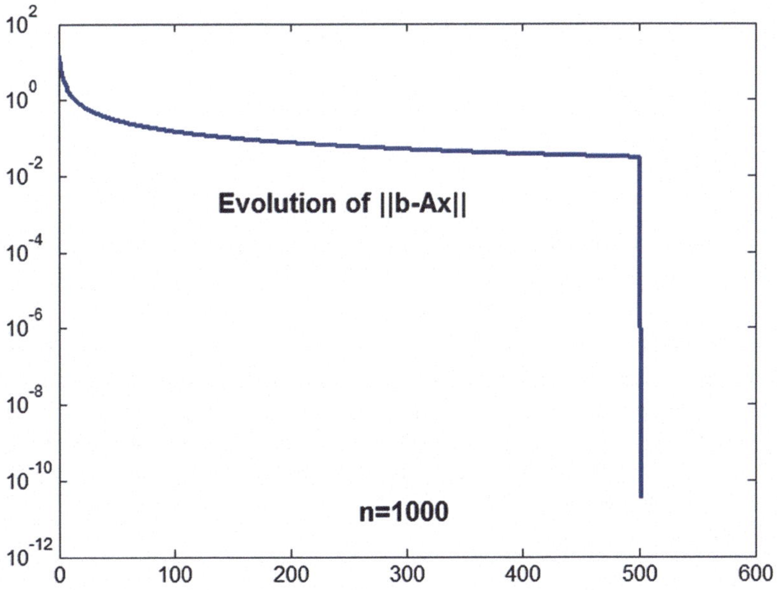

Figure 2.4 Evolution of the error $\|b - Ax_k\|$

Observe that the maximum eigenvalue is $\lambda_n = 2\left(1 - \cos\frac{\pi n}{n+1}\right) \approx 4$. On the other hand, the minimum eigenvalue is $\lambda_1 = 2\left(1 - \cos\frac{\pi}{n+1}\right)$. For i sufficiently small,

$$\lambda_i = 2\left(1 - \cos\frac{\pi i}{n+1}\right) \approx 2\left(1 - \left(1 - \frac{\pi^2 i^2}{2(n+1)^2}\right)\right) = \left(\frac{\pi i}{n+1}\right)^2.$$

Therefore, A is positive definite, and for n large, the conditioning number of A is

$$\kappa(A) = \frac{\lambda_n}{\lambda_1} \approx \frac{4(n+1)^2}{\pi^2}.$$

The behavior of the linear conjugate gradient algorithm illustrated in Figure 2.4 has a *plateau,* where for a large number of iterations, the error $\|b - Ax_k\|$ decreases very slowly. This behavior is typical of the linear conjugate gradient algorithm. Greenbaum and Strakoš (1992) demonstrated that finite precision conjugate gradient computations for solving a symmetric positive definite linear system $Ax = b$ behave very similarly to the exact algorithms applied to any of a certain class of larger matrices. This class consists of matrices $\widetilde{A}$ that have lots of eigenvalues spread throughout tiny intervals around the eigenvalues of A. The width of these intervals is a modest multiple of the machine precision times the norm of A. This analogy appears to hold unless the algorithms are run for huge numbers of steps. See also Naiman, Babuska and Elman (1997).

Example 2.3 In the following, let us consider the linear algebraic system $Ax = b$, where

$$A = \begin{bmatrix} B & -I & & & \\ -I & B & -I & & \\ & \ddots & \ddots & \ddots & \\ & & -I & B & -I \\ & & & -I & B \end{bmatrix}, \quad \text{and} \quad B = \begin{bmatrix} 4 & -1 & & & \\ -1 & 4 & \ddots & & \\ & \ddots & \ddots & \ddots & \\ & & \ddots & 4 & -1 \\ & & & -1 & 4 \end{bmatrix},$$

obtained from the finite difference numerical method to discretize the two-dimensional Poisson equation (Andrei, 2000, 2009f). Matrix A has n_2 blocks on the main diagonal, where each block $B \in \mathbb{R}^{n_1 \times n_1}$. Hence, $A \in \mathbb{R}^{n \times n}$, where $n = n_1 n_2$. The right-hand side b is chosen so that the solution of the system $Ax = b$ is $x^* = [1, 1, \ldots, 1]$. Considering n = 10,000, the evolution of error $\|b - Ax_k\|$ computed by the linear conjugate gradient algorithm for five different values of n_1 and n_2 is presented in Figure 2.5.

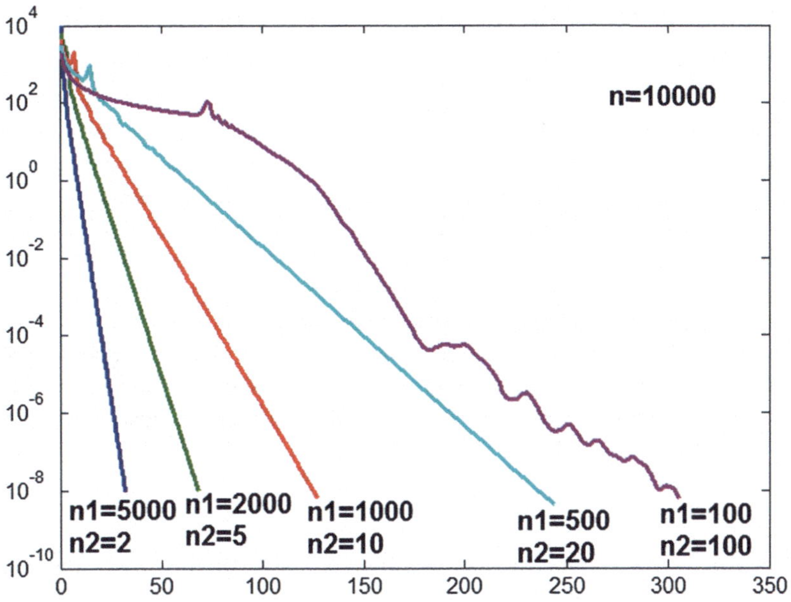

Figure 2.5 Evolution of the error $\|b - Ax_k\|$ of the linear conjugate gradient algorithm for different numbers (n_2) of blocks on the main diagonal of matrix A

The eigenvalues of A are as follows:

$$\lambda_{ij} = 4\sin^2\left(\frac{i\pi}{2(n_1+1)}\right) + 4\sin^2\left(\frac{j\pi}{2(n_2+1)}\right), \quad i = 1, \ldots, n_1, \; j = 1, \ldots, n_2.$$

The maximum eigenvalue is $\lambda_{\max} = 8$, while the minimum eigenvalue is $\lambda_{\min} = 8\sin^2(\pi/2)$. Therefore, A is positive definite and its condition number is

$$\frac{\lambda_{\max}}{\lambda_{\min}} \approx \frac{4}{\pi^2}.$$

From Figure 2.5, for $n_1 = 5000$ and $n_2 = 2$, that is when there are only two blocks on the main diagonal of A, the linear conjugate gradient algorithm needs only 31 iterations. Therefore, the convergence is faster. On the other hand, when $n_2 = 100$, i.e., there are 100 blocks on the main diagonal of matrix A, then the algorithm needs 304 iterations. In other words, the smaller the number of blocks on the main diagonal of matrix A, the faster the convergence.

2.5 Comparison of the Convergence Rate of the Linear Conjugate Gradient and of the Steepest Descent

As we have already seen in Proposition 2.4, the linear conjugate gradient algorithm has a quadratic (finite) termination property, i.e., for convex quadratic functions, the linear conjugate gradient algorithm with exact line search terminates after n iterations. In (2.24), (2.25), (2.28) and (2.29), some formulae for the convergence rates of the linear conjugate gradient algorithm have been presented, showing that the convergence rate of linear conjugate gradient algorithms is not worse than one of the steepest descent algorithm, that is, it is not worse than linear.

In the following, let us have a comparison between the linear conjugate gradient algorithm and the steepest descent algorithm subject to the reduction of the function values along the iterations (Sun & Yuan, 2006). Consider the quadratic function

$$f(x) = \frac{1}{2}x^T Ax, \tag{2.30}$$

where $A \in \mathbb{R}^{n\times n}$ is symmetric and positive definite. In this case, the explicit expression for the stepsize is

$$\alpha_k = -\frac{d_k^T Ax_k}{d_k^T Ad_k} = -\frac{d_k^T g_k}{d_k^T Ad_k}. \tag{2.31}$$

Therefore,

$$\begin{aligned} f(x_{k+1}) &= \frac{1}{2}x_{k+1}^T Ax_{k+1} \\ &= \frac{1}{2}(x_k + \alpha_k d_k)^T A(x_k + \alpha_k d_k) \\ &= \frac{1}{2}x_k^T Ax_k - \frac{1}{2}\frac{(g_k^T d_k)^2}{d_k^T Ad_k}. \end{aligned} \tag{2.32}$$

Now, for the steepest descent algorithm $d_k = -g_k$ and from (2.32),

$$f(x_{k+1}^{SD}) = \frac{1}{2}x_k^T Ax_k - \frac{1}{2}\frac{\|g_k\|^4}{g_k^T Ag_k}. \tag{2.33}$$

On the other hand, for the linear conjugate gradient algorithm $d_k = -g_k + \beta_{k-1}d_{k-1}$ and from (2.32),

$$f(x_{k+1}^{CG}) = \frac{1}{2}x_k^T Ax_k - \frac{1}{2}\frac{\|g_k\|^4}{d_k^T Ad_k}. \tag{2.34}$$

Since

$$\begin{aligned} d_k^T A d_k &= (-g_k + \beta_{k-1} d_{k-1})^T A (-g_k + \beta_{k-1} d_{k-1}) \\ &= g_k^T A g_k + \beta_{k-1}^2 d_{k-1}^T A d_{k-1} \\ &\leq g_k^T A g_k, \end{aligned}$$

it follows that

$$f(x_{k+1}^{CG}) \leq f(x_{k+1}^{SD}).$$

Therefore, the linear conjugate gradient algorithm reduces the value of the minimizing function f at least as much as the steepest descent algorithm. Since the steepest descent algorithm has a linear rate of convergence, it follows that the linear conjugate gradient algorithm has a convergence rate that is not worse than the linear rate. From (2.34), it results that for the linear conjugate gradient algorithm, the objective function is strictly decreased along the iterations.

2.6 Preconditioning of the Linear Conjugate Gradient Algorithms

Preconditioning is a technique for the acceleration of the conjugate gradient method. The idea is to change the variables from x to $\hat{x}$ via a nonsingular matrix $C \in \mathbb{R}^{n \times n}$, that is

$$\hat{x} = Cx. \tag{2.35}$$

The quadratic function f given by (2.1) is transformed as

$$\hat{f}(\hat{x}) = \frac{1}{2}\hat{x}^T (C^{-T} A C^{-1}) \hat{x} - (C^{-T} b)^T \hat{x}. \tag{2.36}$$

To minimize function $\hat{f}$, Algorithm 2.1 can be used, or equivalently, the following linear system

$$(C^{-T} A C^{-1}) \hat{x} = C^{-T} b \tag{2.37}$$

can be solved. The convergence rate of Algorithm 2.1 will depend on the structure of the eigenvalues of the matrix $C^{-T}AC^{-1}$ rather than on those of A. Therefore, the purpose of the preconditioning is to choose C in such a way so that the structure of the eigenvalues of the matrix $C^{-T}AC^{-1}$ is better for the convergence theory presented above. There are several possibilities to choose the preconditioning matrix C. For example, C can be selected so that the condition number of $C^{-T}AC^{-1}$ should be much smaller than the condition number of A. In this case, the constant in (2.25) is much smaller. Another possibility is to choose the matrix C in such a way so that the eigenvalues of $C^{-T}AC^{-1}$ should be clustered. In this case, according to the above discussion, the number of iterations to find a good approximate solution is not much larger than the number of clusters.

Practical implementation of the preconditioned linear conjugate gradient does not require the transformation (2.35) explicitly, but rather the matrix $M = C^TC$, which is a symmetric and positive definite matrix, as it is presented in Algorithm 2.2.

Algorithm 2.2 Preconditioned linear conjugate gradient

1.	Select an initial point x_0, the preconditioner M and $\varepsilon > 0$ sufficiently small
2.	Set $r_0 = Ax_0 - b$. Solve the system $My_0 = r_0$. Set $d_0 = -y_0$ and $k = 0$
3.	If $\|r_k\| \leq \varepsilon$, then stop, otherwise continue with step 4
4.	Compute $\alpha_k = \dfrac{r_k^T y_k}{d_k^T A d_k}$; $x_{k+1} = x_k + \alpha_k d_k$; $r_{k+1} = r_k + \alpha_k A d_k$; Solve $My_{k+1} = r_{k+1}$; $\beta_k = \dfrac{r_{k+1}^T y_{k+1}}{r_k^T y_k}$; $d_{k+1} = -y_{k+1} + \beta_k d_k$
5.	Set $k = k + 1$ and continue with step 3 ♦

The main difference between the preconditioned and the un-preconditioned linear conjugate gradient methods is the need to solve the linear system of the form $My = r$ in step 4 of the Algorithm 2.2.

Selection of the preconditioner M is not easy. Effective preconditioners are based on a deep insight into the structure of the problem. Selection of M is based on different objectives, like the effectiveness of M to get a good structure of the eigenvalues of the preconditioned problem, inexpensive storage and computation of M, inexpensive solution of $My = r$, etc. Often, the preconditioner is defined in such a way so that the system $My = r$ should be a simplified version of the original system $Ax = b$. One simple preconditioner is the Jacoby preconditioning, where M is the inverse of the diagonal elements of A. Other preconditioners are based on the stationary iterative methods, such as the symmetric Gauss–Seidel preconditioner. Another approach is to apply a sparse Cholesky factorization to matrix A and discard small elements of the factors and/or allow only a fixed amount of factors storage. Such preconditioners are called *incomplete Cholesky factorization*. The idea is as follows: instead of computing the exact Cholesky factor L that satisfies $A = LL^T$, an approximation factor $\overline{L}$ is computed so that $A = \overline{L}\overline{L}^T + E$, where E is small. Therefore, with this incomplete factorization, $A \approx \overline{L}\overline{L}^T$. Now, by choosing $C = \overline{L}^T$, it results that $M = \overline{L}\overline{L}^T$ and

$$C^{-T}AC^{-1} = \overline{L}^{-1}A\overline{L}^{-T} \approx I,$$

showing that the eigenvalue distribution of $C^{-T}AC^{-1}$ is favorable. In this case, solving the system $My = r$ is reduced to solving two triangular systems [see: Axelsson (1994), Golub and Van Loan (1996), Kelley (1995), Nocedal and Wright (2006)].

For the preconditioned linear conjugate gradient algorithm, from (2.32) observe that

$$f(x_{k+1}^{CG}) = \frac{1}{2}x_k^T A x_k - \frac{1}{2}\frac{(g_k^T v_k)^2}{d_k^T A d_k},$$

where $v = M^{-1}g_k$. Therefore, the convergence rate for the preconditioned linear conjugate gradient algorithm is also linear.

Notes and References

Linear conjugate gradient methods are characterized by low memory requirements and strong local and global convergence properties. These methods were introduced by Hestenes and Stiefel (1952) for efficiently solving linear algebraic systems with positive definite matrices. The survey by Golub and O'Leary (1989) gives a history of the conjugate gradient and Lanczos algorithms from 1948–1976. A thorough discussion on the linear conjugate gradient method is given by Golub and Van Loan (1996) or by Luenberger (1984). There is a vast literature on these methods: Crowder and Wolfe (1969), Reid (1971), Beale (1972), Cohen (1972), McCormick and Ritter (1974), Concus and Golub (1976), Hestenes (1980), Al-Baali (1985), Axelsson and Lindskog (1986), Strakoš (1991), Nocedal (1992), Axelsson (1980, 1994), Axelsson and Barker (2001), Arnold (2001), Saad (2003), Wen Shen (2008), Andrei (1995, 2000, 2009f). Our presentation is based on Luenberger (1984), Greenbaum (1997), Arnold (2001), Wen Shen (2008) and Luenberger and Ye (2016).

Chapter 3
General Convergence Results for Nonlinear Conjugate Gradient Methods

For solving the nonlinear unconstrained optimization problem

$$\min f(x), \tag{3.1}$$

where $f : \mathbb{R}^n \to \mathbb{R}$ is a continuously differentiable function, any nonlinear conjugate gradient algorithm generates the sequence $\{x_k\}$ of the form

$$x_{k+1} = x_k + \alpha_k d_k, \tag{3.2}$$

where α_k is the stepsize obtained by line search and d_k is the search direction computed by

$$d_{k+1} = -g_{k+1} + \beta_k d_k, \tag{3.3}$$

for $k \geq 0$, where β_k is the conjugate gradient parameter and $g_k = \nabla f(x_k)$. In conjugate gradient methods, $d_0 = -g_0$.

A popular strategy for stepsize determination, which plays a key role in the efficiency of the unconstrained optimization algorithms, consists in accepting a positive stepsize α_k satisfying the *standard Wolfe* line search conditions

$$f(x_k + \alpha_k d_k) \leq f(x_k) + \rho \alpha_k d_k^T g_k, \tag{3.4}$$

$$\nabla f(x_k + \alpha_k d_k)^T d_k \geq \sigma d_k^T g_k, \tag{3.5}$$

where $0 < \rho < \sigma < 1$. Often, the *strong Wolfe* line search is used in the implementation of conjugate gradient methods. These are given by (3.4) and

$$\left| \nabla f(x_k + \alpha_k d_k)^T d_k \right| \leq -\sigma d_k^T g_k, \tag{3.6}$$

N. Andrei, *Nonlinear Conjugate Gradient Methods for Unconstrained Optimization*, Springer Optimization and Its Applications 158,
https://doi.org/10.1007/978-3-030-42950-8_3

where again $0 < \rho < \sigma < 1$. Observe that if $\sigma = 0$, then the strong Wolfe line search reduces to the exact line search. Dai and Yuan (1999, 2001a) proved that the standard Wolfe line search (3.4) and (3.5) ensures the convergence and therefore can be successfully used in the current implementations in computing programs of the conjugate gradient methods.

Nonlinear conjugate gradient methods have a very nice theory, with a lot of important results on their convergence. This is the main argument for which these methods are intensely used in solving practical unconstrained optimization applications. This chapter is dedicated to presenting the main convergence theorems of the nonlinear conjugate gradient methods, assuming that their search directions are descent. A short presentation of the types of convergence of the sequences generated by the optimization algorithms is firstly discussed. After that the concept of nonlinear conjugate gradient is detailed, continuing with the convergence of the conjugate gradient method under the strong Wolfe line search and then under the standard Wolfe line search.

3.1 Types of Convergence

Starting from an initial point x_0, every unconstrained optimization method generates a sequence $\{x_k\}$ of points which hopefully converges to a solution of the problem. The objective of the convergence analysis of unconstrained optimization algorithms is to study the properties of the sequence $\{x_k\}$ subject to its convergence to a solution of the problem or to a stationary point, to see the convergence rate of the sequence and to compare the convergence performances of different algorithms. The convergence rate is a local characterization of an algorithm subject to its effectiveness to solve a problem. By local convergence methods, we mean that the initial point x_0 is *close* to a local minimizer x^* of the problem at which the sufficient optimality conditions hold.

The sequence $\{x_k\}$ converges to a point x^* if

$$\lim_{k \to \infty} \|x_k - x^*\| = 0. \tag{3.7}$$

However, in practical situations, the solution x^* is not known, and therefore, it is not possible to use (3.7) as a test of convergence. A possibility to see the convergence of $\{x_k\}$ is to compute the limit

$$\lim_{k \to \infty} \|x_k - x_{k-1}\| = 0. \tag{3.8}$$

Unfortunately, criterion (3.8) cannot guarantee the convergence of $\{x_k\}$. Therefore, the global convergence study of the unconstrained optimization algorithms tries to prove the following limit

$$\lim_{k \to \infty} \|g_k\| = 0, \tag{3.9}$$

which ensures that x_k is close to the set of the stationary points where $g_k = \nabla f(x_k) = 0$, *or* the limit

$$\liminf_{k \to \infty} \|g_k\| = 0, \tag{3.10}$$

which ensures that at least a subsequence of $\{x_k\}$ is close to the set of stationary points. In other words, if the iterates $\{x_k\}$ stay in a boundary region, then (3.9) says that every cluster point of $\{x_k\}$ will be a stationary point of f, while (3.10) means that there exists at least one cluster point which is a stationary point of f (see Appendix A). Denote $x_k \to x^*$ which means that the sequence $\{x_k\}$ converges to x^*, i.e., (3.7) holds. In the following, the *q-convergence*, which means *quotient-convergence* and the *r-convergence*, which means *root-convergence*, are introduced. More details can be found, for example, in Ortega and Rheinboldt (1970), Potra (1989), Sun and Yuan (2006), Cătinaş (2019).

Definition 3.1 Let $\{x_k\}$ be a sequence from $\mathbb{R}^n$ and $x^* \in \mathbb{R}^n$. Then, we say that:

1. $x_k \to x^*$ q-quadratically if $x_k \to x^*$ and there exists $K > 0$ which is independent of the iterative number k, so that

$$\|x_{k+1} - x^*\| \le K\|x_k - x^*\|^2. \tag{3.11}$$

2. $x_k \to x^*$ q-superlinearly with q-order $p > 1$ if $x_k \to x^*$ and there exists $K > 0$ which is independent of the iterative number k, so that

$$\|x_{k+1} - x^*\| \le K\|x_k - x^*\|^p. \tag{3.12}$$

3. $x_k \to x^*$ q-superlinearly if

$$\lim_{k \to \infty} \frac{\|x_{k+1} - x^*\|}{\|x_k - x^*\|} = 0. \tag{3.13}$$

4. $x_k \to x^*$ q-linearly with q-factor $\sigma \in (0, 1)$ if

$$\|x_{k+1} - x^*\| \le \sigma\|x_k - x^*\| \tag{3.14}$$

for k sufficiently large. ♦

Often, p is called *q-order* and K is *q-factor*. Notice that the q-linear convergence $(p = 1)$ means that asymptotically the point x_{k+1} approximates x^* with $\log_{10} K$ more correct digits than x_k. Therefore, in this case, the number of correct digits grows linearly subject to the number of iterations. On the other hand, for q-superlinearly convergence of q-order p, the number of additional correct digits asymptotically grows by a factor of p, i.e., the number of correct digits is exponential subject to the number of iterations, and therefore the convergence is fast.

A q-superlinearly convergent sequence is also q-linearly convergent with q-factor σ for any $\sigma > 0$. A q-quadratically convergent sequence is q-superlinearly convergent with q-order of 2.

The motivation for introducing the q-convergence rate was to compare the speed of convergence of the algorithms. We see that the q-convergence rate depends more on p and less on K. Suppose that there are two sequences $\{x_k\}$ and $\{\hat{x}_k\}$, where their q-order and q-factor are $\{p, K\}$ and $\{\hat{p}, \hat{K}\}$, respectively. If $p > \hat{p}$, then the sequence with q-order p converges faster than the sequence with q-order $\hat{p}$. In other words, q-quadratically convergent sequences eventually converge faster than q-superlinearly and q-linearly convergent sequences. When $p = \hat{p}$, i.e., the sequences have the same q-order of convergence, if $K < \hat{K}$, then the sequence $\{x_k\}$ is faster than $\{\hat{x}_k\}$. Usually, we are interested in q-superlinearly and q-quadratically convergent sequences. In this case, we say that the sequence has a rapid convergence rate.

Definition 3.2 An iterative method for computing x^* is said to be locally (q-quadratically, q-superlinearly, q-linearly, etc.) convergent if the iterates x_k generated by the method converge to x^* (q-quadratically, q-superlinearly, q-linearly, etc.), provided that the initial point of the iterative method is sufficiently close to x^*. ♦

Another measure of the convergence rate of a sequence $\{x_k\}$, which is weaker than the q-convergence rate, is the r-convergence rate. The motivation of introducing the r-convergence rate is as follows. There are situations when the accuracy of the iteration can be improved by means of some procedures that are external to the algorithm, for example, by evaluating the objective function and its gradient with an increasing accuracy. In such cases, there is no guarantee that the accuracy of the iteration is monotonically increasing, but only that the accuracy of the results is improving at a rate determined by the improving accuracy in the function-gradient evaluations. Another situation is when there are some sequences which still converge reasonably fast, but whose speed of convergence is variable. The concept of r-convergence rate captures these situations (Kelley, 1999).

Definition 3.3 Let $\{x_k\}$ be a sequence from $\mathbb{R}^n$ and $x^* \in \mathbb{R}^n$. Let

$$r_\delta = \begin{cases} \limsup_{k\to\infty} \|x_k - x^*\|^{1/k}, & \text{if } \delta = 1, \\ \limsup_{k\to\infty} \|x_k - x^*\|^{1/\delta^k}, & \text{if } \delta > 1. \end{cases} \tag{3.15}$$

If $r_1 = 0$, then $\{x_k\}$ is said to be r-superlinearly convergent to x^*.

If $0 < r_1 < 1$, then $\{x_k\}$ is said to be r-linearly convergent to x^*.

If $r_1 = 1$, then $\{x_k\}$ is said to be r-sublinearly convergent to x^*.

Similarly, if $r_2 = 0$, $0 < r_2 < 1$ and $r_2 \geq 1$, respectively, then $\{x_k\}$ is said to be r-superquadratically, r-quadratically, and r-subquadratically convergent to x^*, respectively. ♦

Observe that when $\{x_k\}$ is convergent to x^*, then there is always an index $k_0 \geq 0$ so that $0 \leq \|x_k - x^*\| < 1$, for all $k \geq k_0$. Hence, for all $\delta \geq 1$, we have $0 \leq r_\delta \leq 1$. In this case, δ is called r-order and r_δ is r-factor. The higher the r-order, the faster the sequence $\{x_k\}$ converges. When two sequences have the same r-order, the smaller the r-factor, faster the corresponding sequence converges (Sun & Yuan, 2006).

Another definition to the r-convergence rate, which presents the connection with the q-convergence rate, is as follows.

Definition 3.4 Let $\{x_k\}$ be a sequence from $\mathbb{R}^n$ and $x^* \in \mathbb{R}^n$. The sequence $\{x_k\}$ converges to x^* r-(quadratically, superlinearly, linearly) if there is a sequence $\{\gamma_k\}$ from $\mathbb{R}$ converging q-(quadratically, superlinearly, linearly) to zero so that

$$\|x_k - x^*\| \leq \gamma_k.$$

The sequence $\{x_k\}$ converges r-superlinearly with r-order $p > 1$ if the sequence $\{\gamma_k\}$ converges to zero q-superlinearly with q-order p. ♦

In general, in analyzing the convergence of algorithms, the q-convergence rate is used. An algorithm with q-superlinear or q-quadratic rate of convergence is considered a good one. Therefore, we are interested in designing superlinear or quadratically convergent algorithms.

3.2 The Concept of Nonlinear Conjugate Gradient

The idea behind the conjugate gradient approach for minimizing quadratic functions is to use search directions which do not interfere with one another. As in Definition 2.1, given a symmetric and positive definite matrix A, the set of vectors $\{d_0, \ldots, d_{k-1}\}$ is a set of conjugate directions (or "A conjugate", or even "A orthogonal") if and only if $d_i^T A d_j = 0$ for all $i \neq j$, where $i = 0, \ldots, k-1$ and $j = 0, \ldots, k-1$. Observe that $d_i^T A d_i > 0$ for all i, since A is positive definite.

In the conjugate gradient algorithm, the search direction d_{k+1} is computed as a linear combination

$$d_{k+1} = -g_{k+1} + \beta_k d_k \tag{3.16}$$

of the current negative gradient and the last search direction in such a way that d_{k+1} is A conjugate to all the previous search directions.

For quadratic functions, the conjugate gradient parameter β_k is chosen so that $d_{k+1}^T A d_k = 0$, which determines

$$\beta_k = \frac{g_{k+1}^T A d_k}{d_k^T A d_k}. \tag{3.17}$$

The next approximation to the minimum is x_{k+1}. This is the unique minimum of f along the line $l(\alpha) = x_k + \alpha d_k$, which is given by

$$x_{k+1} = x_k + \alpha_k d_k, \tag{3.18}$$

where

$$\alpha_k = -\frac{d_k^T g_k}{d_k^T A d_k}. \tag{3.19}$$

This is the *traditional conjugate gradient algorithm* which works for quadratic functions, since it explicitly uses the matrix A, both in (3.17) and (3.19). In Chapter 2, we have presented an equivalent expression for the linear conjugate gradient, (see Algorithm 2.1) which numerically is more stable and where the stepsize and the conjugate gradient parameter indirectly depend on the matrix A.

Now, the problem is *how the algorithm could be modified so that the quadratic nature of the minimizing function f should not appear explicitly into the algorithm, i.e., in* (3.17) *and* (3.19), *but remain unchanged if f is a quadratic function*. As in Algorithm 2.1, the initial search direction is $d_0 = -g_0$. Observe that the initial search direction does not involve the matrix A. Therefore, for nonlinear functions, the initial search direction could be very well considered as the gradient of the minimizing function at the initial point. In step 4 of Algorithm 2.1, two crucial elements of the algorithm are computed: the stepsize α_k and the parameter β_k. To determine α_k, a line search from x_k in direction d_k is executed. For the convex quadratic functions, this line search determines an explicit stepsize (3.19), which is the unique minimum of the function f in direction d_k. This is the advantage of quadratic functions, there is an explicit formula for stepsize computation. For nonlinear functions, this formula for stepsize computation can be replaced by a general line search procedure (Armijo, Goldstein, Wolfe, etc.). This will change nothing in the quadratic case, but generalizes the algorithm to nonlinear functions. For the computation of the conjugate gradient parameter β_k, from (3.17), we need to compute Ad_k. There are several possibilities to modify the algorithm in order to eliminate the explicit mention of matrix A in (3.17).

One way is to see from (3.18) that $x_{k+1} - x_k = \alpha_k d_k$. Therefore, for quadratic functions,

$$g_{k+1} - g_k = (Ax_{k+1} - b) - (Ax_k - b) = \alpha_k A d_k.$$

With this, since $Ad_k = (g_{k+1} - g_k)/\alpha_k$, from (3.17) it follows that

$$\beta_k = \frac{g_{k+1}^T y_k}{d_k^T y_k}, \tag{3.20}$$

where $y_k = g_{k+1} - g_k$.

If f is a quadratic function, then the definitions (3.17) and (3.20) are equivalent, but the new definition (3.20) can be used for *any* differential function. Observe that (3.20) is *exactly the Hestenes and Stiefel formula* for the conjugate gradient parameter computation.

Another way to eliminate A from (3.17) is as follows. Suppose that f is quadratic. From (3.20), it follows that $d_k^T y_k = d_k^T g_{k+1} - d_k^T g_k$. But, from Proposition 2.3, $d_k^T g_{k+1} = 0$. Therefore, $d_k^T y_k = -d_k^T g_k$. Now, observe that $d_k = -g_k + \beta_{k-1} d_{k-1}$, so that since $d_{k-1}^T g_k = 0$, it results that

$$d_k^T y_k = -(-g_k^T + \beta_{k-1} d_{k-1}^T) g_k = g_k^T g_k.$$

Therefore, from (3.20), another formula for β_k computation is

$$\beta_k = \frac{g_{k+1}^T y_k}{g_k^T g_k}. \tag{3.21}$$

Observe that if f is quadratic, (3.21) is equivalent with (3.17). Hence, (3.21) generalizes the computation of the conjugate gradient parameter to the nonquadratic case. Formula (3.21) is *exactly the Polak–Ribière–Polyak formula* for the conjugate gradient parameter computation.

Furthermore, it is very easy to see that if f is a quadratic function, then $g_{k+1}^T g_k = 0$, and in this case, from (3.21),

$$\beta_k = \frac{g_{k+1}^T g_{k+1}}{g_k^T g_k}, \tag{3.22}$$

which generalizes the computation of the conjugate gradient parameter to the nonquadratic case. Observe that (3.22) is *exactly the Fletcher-Reeves formula* for the conjugate gradient parameter computation.

Formulae (3.20), (3.21), and (3.22) generalize the conjugate gradient algorithm to *any* nonlinear differential function. They require only the gradient evaluations. For quadratic functions they are equivalent, but they are not equivalent for nonlinear functions.

For general nonlinear functions, the matrix A should be replaced by the Hessian. Since $y_k = g_{k+1} - g_k$, from the mean value theorem it follows that

$$d_{k+1}^T y_k = \alpha_k d_{k+1}^T \nabla^2 f(x_k + \tau \alpha_k d_k) d_k, \tag{3.23}$$

where $\tau \in (0, 1)$. Therefore, from (3.23), for nonlinear functions, it is reasonable to consider the conjugacy condition as

$$d_{k+1}^T y_k = 0, \tag{3.24}$$

or as $d_{k+1}^T y_k = t(g_{k+1}^T s_k)$, where $t > 0$ is a parameter and $s_k = x_{k+1} - x_k$, given by Dai and Liao (2001). For example, the Hestenes–Stiefel method has the property that the conjugacy condition (3.24) always holds, independent of the line search.

In general, not all the nonlinear conjugate gradient algorithms satisfy the conjugacy condition (3.24). However, *characteristic for nonlinear conjugate gradient algorithms is that the search direction is computed as a linear combination between the negative gradient and the last search direction, where the conjugate gradient parameter* β_k *is determined in such a way that the descent condition or the sufficient descent condition hold.* Of course, as it is to be seen, to improve the convergence of the algorithms, this linear combination between the negative gradient and the last search direction can be modified in such a way as to take into account, for example, the second-order information of the minimizing function, or some other ingredients which accelerate the convergence. In conjugate gradient methods, the descent or the sufficient descent conditions are crucial. The descent condition $d_k^T g_k < 0$, (if $g_k \neq 0$) is more general and leads to general convergence results for line search algorithms under the exact or inexact line search (and not only for conjugate gradient algorithms). The sufficient descent condition $d_k^T g_k \leq -c\|g_k\|^2$, for some positive constant c, is fundamental in conjugate gradient methods.

3.3 General Convergence Results for Nonlinear Conjugate Gradient Methods

As it has already been seen in (3.2), $d_0 = -g_0$. Selection $d_0 = -g_0$ is critical in conjugate gradient algorithms. A very important property of the linear conjugate gradient method is that it terminates after at most n iterations if $f(x)$ is a convex quadratic function and if the first search direction is $d_0 = -g_0$. However, for a general nonlinear function, it can be closely approximated by a quadratic function only after a certain number of iterations. Therefore, the local analysis cannot apply to show the quadratic termination because in this case $d_k \neq -g_k$ for $k > 1$, due to previous iterations. Crowder and Wolfe (1969) gave a three-dimensional example, showing that even for a strongly convex quadratic, the convergence rate is linear if the initial search direction is not the steepest descent direction. Powell (1976b) obtained a stronger result, showing that if the objective function is a convex quadratic one and if the initial search direction is an arbitrary descent direction, then either optimality is achieved in at most $n + 1$ iterations or the rate of convergence is only linear. Yuan (1993) gave a theoretical study, showing that the conjugate gradient method applied to convex quadratic functions always converges only linearly if the finite termination does not happen. Some details for this selection of d_0 are also given by Andrei (2011b).

A conjugate gradient algorithm given by (3.2) and (3.3) generates a sequence $\{x_k\}$. The interest is to see the conditions under which this sequence converges to the solution x^* of problem (3.1). *Since the algorithm given by* (3.2) *and* (3.3) *depends only on the parameter β_k, it follows that the interest is to see the values of this parameter for which the algorithm is convergent.*

An important requirement for the optimization methods based on line search is that the search direction must be a descent one. The search direction d_k satisfies the descent property, i.e., it is a *descent direction*, if for all $k = 1, 2, \ldots$

$$g_k^T d_k < 0, \tag{3.25}$$

(see Definition 1.1) For conjugate gradient methods, from (3.3), it follows that:

$$g_{k+1}^T d_{k+1} = -\|g_{k+1}\|^2 + \beta_k g_{k+1}^T d_k. \tag{3.26}$$

Now, if the line search is exact, i.e., if $g_{k+1}^T d_k = 0$, then $g_{k+1}^T d_{k+1} = -\|g_{k+1}\|^2$. Therefore, d_{k+1} is a descent direction if $g_{k+1} \neq 0$. However, for the inexact line search this may not be true. Using the restart with $d_{k+1} = -g_{k+1}$, this situation can be corrected. The search direction d_k satisfies the sufficient descent property, i.e., it is a *sufficient descent direction* if

$$g_k^T d_k \leq -c\|g_k\|^2, \tag{3.27}$$

for all $k = 1, 2, \ldots$, where $c > 0$ is a constant.

The convergence properties of a line search method, like the nonlinear conjugate gradient, can be studied by measuring the effectiveness of the search direction and of the length of the step. The quality of a search direction d_k can be determined by studying the angle between the steepest descent direction $-g_k$ and the search direction d_k defined by

$$\cos\theta_k = \frac{-g_k^T d_k}{\|g_k\|\|d_k\|}. \tag{3.28}$$

To establish the general convergence results of any method of form (3.2) and (3.3), the following *basic assumptions* on the objective function are introduced.

Assumption CG

1. The level set $S = \{x \in \mathbb{R}^n : f(x) \leq f(x_0)\}$ is bounded, i.e., there exists *a* constant $B > 0$ so that $\|x\| \leq B$ for all x in the level set.
2. In some neighborhood N of the level set, f is continuously differentiable and its gradient is Lipschitz continuous, i.e., there exists a constant $L > 0$ so that

$$\|g(x) - g(y)\| \leq L\|x - y\|, \textit{for all } x, y \in N. \tag{3.29}$$

Note that these assumptions imply that there is a constant Γ so that $\|g(x)\| \leq \Gamma$, for all x from the level set S. The boundedness assumption of the level set is not necessary in all situations. Only the assumption that f is bounded below on the level set can be used for global convergence analysis. ♦

Under the Assumption CG, the following theorem, due to Zoutendijk (1970) and Wolfe (1969, 1971), is essential in proving the global convergence results of the unconstrained optimization algorithms, including conjugate gradient or Newton one.

Theorem 3.1 *Suppose that f is bounded below in $\mathbb{R}^n$ and that f is continuously differentiable in a neighborhood N of the level set $S = \{x \in \mathbb{R}^n : f(x) \le f(x_0)\}$. Assume also that the gradient is Lipschitz continuous, i.e., there exists a constant $L > 0$ so that* (3.29) *is satisfied for all $x, y \in N$. Consider any iteration of the form* (3.2), *where d_k is a descent direction and α_k satisfies the Wolfe line search conditions* (3.4) *and* (3.5). *Then,*

$$\sum_{k=1}^{\infty} \cos^2 \theta_k \|g_k\|^2 < \infty. \tag{3.30}$$

Proof From (3.5) it follows that

$$(g_{k+1} - g_k)^T d_k \ge (\sigma - 1) g_k^T d_k.$$

On the other hand, the Lipschitz continuity (3.29) results in

$$(g_{k+1} - g_k)^T d_k \le \alpha_k L \|d_k\|^2.$$

Therefore, the combination of these two relations gives

$$\alpha_k \ge \frac{(\sigma - 1)}{L} \frac{g_k^T d_k}{\|d_k\|^2}. \tag{3.31}$$

Now, using the first Wolfe condition (3.4) and (3.31) it results that

$$f_{k+1} \le f_k + \rho \frac{(\sigma - 1)}{L} \frac{(g_k^T d_k)^2}{\|d_k\|^2}. \tag{3.32}$$

From the definition (3.28) of $\cos \theta_k$, it follows that (3.32) can be written as

$$f_{k+1} \le f_k + c \cos^2 \theta_k \|g_k\|^2, \tag{3.33}$$

where $c = \rho(1 - \sigma)/L$. Summing (3.33) for $k \ge 1$ and having in view that f is bounded below, (3.30) is obtained. ♦

The relation (3.30) is called the *Zoutendijk condition,* and from (3.28), it can be rewritten as

$$\sum_{k=1}^{\infty} \frac{(g_k^T d_k)^2}{\|d_k\|^2} < \infty. \tag{3.34}$$

It is worth seeing how the Zoutendijk condition is used to obtain global convergence results (Nocedal, 1992). Suppose that the iteration (3.2) is so that

$$\cos\theta_k \geq \delta > 0, \tag{3.35}$$

for all k. Then, from (3.30), it follows that

$$\lim_{k\to\infty} \|g_k\| = 0. \tag{3.36}$$

In other words, *if the search direction generated by any unconstrained optimization method does not tend to be orthogonal to the gradient, then the corresponding sequence of gradients converges to zero.* For line search methods (3.2), the limit (3.36) is the best type of the global convergence result that can be obtained. We cannot guarantee that the method converges to the minimizers of function f, but only that it converges to stationary points. The implications of the Zoutendijk condition are as follows.

1. For steepest descent with Wolfe line search, $\cos\theta_k = 1$ for all k. Thus, the steepest descent method is globally convergent only if the stepsizes are adequately computed.
2. Consider the Newton-type methods, where the search direction is computed as $d_k = -B_k^{-1} g_k$, where B_k is a nonsingular symmetric matrix, ($B_k = I$, $B_k = \nabla^2 f(x_k)$ or B_k is a symmetric and positive definite approximation of the Hessian $\nabla^2 f(x_k)$). Assuming that the condition number of the matrices B_k is uniformly bounded, i.e., for any k, $\|B_k\|\|B_k^{-1}\| \leq \Delta$, where $\Delta > 0$ is a constant, then from (3.28) it follows that

$$\begin{aligned}
\cos\theta_k &= -\frac{g_k^T d_k}{\|g_k\|\|d_k\|} = \frac{g_k^T B_k^{-1} g_k}{\|g_k\|\|B_k^{-1} g_k\|} \\
&\geq \frac{1}{\|g_k\|} \frac{\|g_k\|^2}{\|B_k\|} \frac{1}{\|B_k^{-1}\|\|g_k\|} \\
&= \frac{1}{\|B_k\|\|B_k^{-1}\|} \geq \frac{1}{\Delta}.
\end{aligned}$$

   Hence, $\cos\theta_k \geq 1/\Delta$, i.e., it is bounded away from 0. Therefore, the Newton method or the quasi-Newton methods are globally convergent if the matrices B_k are positive definite (descent condition), if their condition number is bounded and if the line search satisfies the Wolfe conditions. Observe that condition (3.35) is crucial for obtaining these results.

3. For conjugate gradient methods, it is not possible to show the limit (3.36), but only a weaker result, that is

$$\liminf_{k\to\infty} \|g_k\| = 0. \tag{3.37}$$

This type of result is also obtained from the Zoutendijk condition. Indeed, suppose that (3.37) does not hold, i.e., the gradients remain bounded away from zero. In other words, suppose that there exists a constant $\gamma > 0$ so that for any k

$$\|g_k\| \geq \gamma. \tag{3.38}$$

In this case, again from the Zoutendijk condition (3.30), it follows that

$$\cos\theta_k \to 0. \tag{3.39}$$

Therefore, the algorithm can only fail in the sense of (3.38) if the sequence $\{\cos\theta_k\}$ converges to zero. Therefore, to establish (3.37), it suffices to show that a subsequence $\{\cos\theta_{k_j}\}$ of the sequence $\{\cos\theta_k\}$ is bounded away from zero.

Let us now present some conditions on β_k, which determine the convergence of the conjugate gradient algorithms. Suppose that $\beta_k \geq 0$ and the search direction d_k is a descent direction, i.e., $g_k^T d_k < 0$. At this moment, we are interested in finding a β_k which produces a descent direction d_{k+1}, i.e., a direction which satisfies

$$g_{k+1}^T d_{k+1} = -\|g_{k+1}\|^2 + \beta_k g_{k+1}^T d_k < 0. \tag{3.40}$$

Proposition 3.1 *Suppose that* $\beta_k \geq 0$. *If*

$$\beta_k \leq \frac{\|g_{k+1}\|^2}{d_k^T y_k}, \tag{3.41}$$

then d_{k+1} *is a descent direction for function f.*

Proof Since $g_k^T d_k < 0$ and $\beta_k \geq 0$ for all $k \geq 1$, we can write:

$$\begin{aligned} -\|g_{k+1}\|^2 + \beta_k g_{k+1}^T d_k &= -\|g_{k+1}\|^2 + \beta_k g_{k+1}^T d_k - \beta_k g_k^T d_k + \beta_k g_k^T d_k \\ &= -\|g_{k+1}\|^2 + \beta_k d_k^T y_k + \beta_k g_k^T d_k \\ &\leq -\|g_{k+1}\|^2 + \beta_k d_k^T y_k, \end{aligned} \tag{3.42}$$

where $y_k = g_k - g_{k-1}$. Therefore, the nonpositivity of (3.42) is sufficient to show that condition (3.40) holds. Hence,

$$\|g_{k+1}\|^2 \geq \beta_k d_k^T y_k, \tag{3.43}$$

proving the proposition. ♦

The following theorem, proved by Dai and Yuan (2001a), shows how β_k selected to satisfy (3.41) determines the convergence of the conjugate gradient algorithm.

Theorem 3.2 *Suppose that the Assumption CG holds. Let $\{x_k\}$ be the sequence generated by the algorithm (3.2) and (3.3), where β_k satisfies (3.41). Then, the algorithm either determines a stationary point or converges in the sense that*

$$\liminf_{k\to\infty} \|g_k\| = 0.$$

Proof From (3.3), we get $d_{k+1} + g_{k+1} = \beta_k d_k$. Squaring both sides of this relation, it results that

$$\|d_{k+1}\|^2 = \beta_k^2 \|d_k\|^2 - 2g_{k+1}^T d_{k+1} - \|g_{k+1}\|^2. \tag{3.44}$$

From (3.44), dividing both sides by $(g_{k+1}^T d_{k+1})^2$, it follows that

$$\begin{aligned}
\frac{\|d_{k+1}\|^2}{(g_{k+1}^T d_{k+1})^2} &= \frac{(\beta_k)^2 \|d_k\|^2}{(g_{k+1}^T d_{k+1})^2} - 2\frac{g_{k+1}^T d_{k+1}}{(g_{k+1}^T d_{k+1})^2} - \frac{\|g_{k+1}\|^2}{(g_{k+1}^T d_{k+1})^2} \\
&= \frac{(\beta_k)^2 \|d_k\|^2}{(g_{k+1}^T d_{k+1})^2} - \frac{2}{g_{k+1}^T d_{k+1}} - \frac{\|g_{k+1}\|^2}{(g_{k+1}^T d_{k+1})^2} \\
&= \frac{(\beta_k)^2 \|d_k\|^2}{(g_{k+1}^T d_{k+1})^2} - \left(\frac{1}{\|g_{k+1}\|} + \frac{\|g_{k+1}\|}{g_{k+1}^T d_{k+1}}\right)^2 + \frac{1}{\|g_{k+1}\|^2} \\
&\leq \frac{(\beta_k)^2 \|d_k\|^2}{(g_{k+1}^T d_{k+1})^2} + \frac{1}{\|g_{k+1}\|^2} = \left(\frac{(\beta_k)^2 (g_k^T d_k)^2}{(g_{k+1}^T d_{k+1})^2}\right) \frac{\|d_k\|^2}{(g_k^T d_k)^2} + \frac{1}{\|g_{k+1}\|^2}.
\end{aligned}$$

Now, using (3.3) again,

$$\frac{\|d_{k+1}\|^2}{(g_{k+1}^T d_{k+1})^2} \leq \left(\frac{(\beta_k)^2 (g_k^T d_k)^2}{\left(-\|g_{k+1}\|^2 + \beta_k g_{k+1}^T d_k\right)^2}\right) \frac{\|d_k\|^2}{(g_k^T d_k)^2} + \frac{1}{\|g_{k+1}\|^2}. \tag{3.45}$$

Therefore, from the above relations and from (3.41),

$$\left(\underbrace{-\|g_{k+1}\|^2+\beta_k g_{k+1}^T d_k}_{<0}+\underbrace{\beta_k g_k^T d_k}_{<0}\right)\left(\underbrace{-\|g_{k+1}\|^2+\beta_k g_{k+1}^T d_k-\beta_k g_k^T d_k}_{\le 0}\right)\ge 0.$$

Hence,

$$\frac{(\beta_k)^2(g_k^T d_k)^2}{\left(-\|g_{k+1}\|^2+\beta_k g_{k+1}^T d_k\right)^2}\le 1. \tag{3.46}$$

From (3.45) and (3.46), the following inequality is obtained

$$\frac{\|d_{k+1}\|^2}{(g_{k+1}^T d_{k+1})^2}\le\frac{\|d_k\|^2}{(g_k^T d_k)^2}+\frac{1}{\|g_{k+1}\|^2}. \tag{3.47}$$

Observe that

$$\frac{\|d_0\|^2}{(g_0^T d_0)^2}=\frac{1}{\|g_0\|^2}.$$

Therefore, from (3.47), it follows that

$$\frac{\|d_k\|^2}{(g_k^T d_k)^2}\le\sum_{i=0}^{k}\frac{1}{\|g_i\|^2} \tag{3.48}$$

for all k. Now, if the theorem is not true, then there exists a constant $\gamma>0$ so that $\|g_k\|\ge\gamma$ for all k, i.e., from (3.48)

$$\frac{(g_k^T d_k)^2}{\|d_k\|^2}\ge\frac{\gamma^2}{k+1},$$

which implies

$$\sum_{k=0}^{\infty}\frac{(g_k^T d_k)^2}{\|d_k\|^2}=\infty,$$

thus contradicting the Zoutendijk condition. ♦

Gilbert and Nocedal (1992) proved the convergence of the conjugate gradient method under the sufficient descent condition (3.27). In fact, this condition (3.27) is often implied or required in many convergence analyses of the conjugate gradient methods, for example, see: Hestenes and Stiefel (1952), Al-Baali (1985), Touati-Ahmed and Storey (1990), Hu and Storey (1991), Grippo and Lucidi (1997).

3.3.1 Convergence Under the Strong Wolfe Line Search

The following theorem shows that if β_k is chosen to satisfy the condition (3.41) for all k, then under the strong Wolfe line search conditions (3.4) and (3.6) the direction (3.3) satisfies the sufficient descent condition (3.27).

Theorem 3.3 *Suppose that x_0 is an initial point and the Assumption CG holds. Let $\{x_k\}$ be the sequence generated by the conjugate gradient algorithm* (3.2) *and* (3.3). *If β_k is so that*

$$\|g_{k+1}\|^2 \geq \beta_k d_k^T y_k,$$

$$\beta_k \geq 0$$

and the stepsize α_k satisfies the strong Wolfe conditions (3.4) *and* (3.6), *then the conjugate gradient algorithm satisfies the sufficient descent condition* (3.27) *with* $c = 1/(1+\sigma)$.

Proof The proof is given by induction as follows. Observe that $\sigma > 0$ implies that $-1 < -1/(1+\sigma)$. Therefore, for $k = 1$, the conclusion holds, since

$$g_0^T d_0 = -\|g_0\|^2 \leq -\frac{1}{1+\sigma}\|g_0\|^2 = -c\|g_0\|^2,$$

under $c = 1/(1+\sigma)$. Suppose that (3.27) holds for some $k \geq 0$. Then, from the second strong Wolfe condition (3.6), it follows that

$$l_k \equiv \frac{g_{k+1}^T d_k}{g_k^T d_k} \in [-\sigma, \sigma] \text{ and } d_k^T y_k > 0. \tag{3.49}$$

Hence,

$$l_k - 1 = \frac{g_{k+1}^T d_k}{g_k^T d_k} - 1 = \frac{g_{k+1}^T d_k - g_k^T d_k}{g_k^T d_k} = \frac{d_k^T y_k}{g_k^T d_k} \neq 0. \tag{3.50}$$

By (3.3), it results that

$$g_{k+1}^T d_{k+1} = g_{k+1}^T(-g_{k+1} + \beta_k d_k) = -\|g_{k+1}\|^2 + \beta_k g_{k+1}^T d_k.$$

Following the sign of $g_{k+1}^T d_k$, the following two cases are in order.

1. The case $g_{k+1}^T d_k \leq 0$. Hence,

$$g_{k+1}^T d_{k+1} = -\|g_{k+1}\|^2 + \beta_k g_{k+1}^T d_k \leq -\|g_{k+1}\|^2 \leq -\frac{1}{1+\sigma}\|g_{k+1}\|^2. \tag{3.51}$$

2. The case $g_{k+1}^T d_k > 0$. Using the conditions on β_k and (3.50), it follows that

$$\begin{aligned} g_{k+1}^T d_{k+1} &= -\|g_{k+1}\|^2 + \beta_k g_{k+1}^T d_k \\ &\leq -\|g_{k+1}\|^2 + \frac{\|g_{k+1}\|^2}{d_k^T y_k} g_{k+1}^T d_k = \left(-1 + \frac{g_{k+1}^T d_k}{d_k^T y_k}\right)\|g_{k+1}\|^2 \\ &= \left(\frac{g_k^T d_k}{d_k^T y_k}\right)\|g_{k+1}\|^2 = \frac{1}{l_k - 1}\|g_{k+1}\|^2. \end{aligned}$$

But

$$\frac{1}{1+\sigma} \leq \frac{1}{1-l_k} \leq \frac{1}{1-\sigma}. \tag{3.52}$$

From (3.52),

$$g_{k+1}^T d_{k+1} \leq \frac{1}{l_k - 1}\|g_{k+1}\|^2 \leq -\frac{1}{1+\sigma}\|g_{k+1}\|^2. \tag{3.53}$$

From (3.51) and (3.53), it follows that at the iteration $k+1$ the sufficient descent condition is satisfied with $c = 1/(1+\sigma)$. ♦

Another powerful result for general conjugate gradient methods with strong Wolfe line search, related to the Zoutendijk condition, is the following theorem (Dai, Han, Liu, Sun, Yin & Yuan, 1999).

Theorem 3.4 *Suppose that the Assumption CG holds. Consider any general conjugate gradient method given by* (3.2) *and* (3.3), *where the stepsize* α_k *is determined by the strong Wolfe line search* (3.4) *and* (3.6). *Then, either*

$$\liminf_{k\to\infty}\|g_k\| = 0, \tag{3.54}$$

or

$$\sum_{k=1}^{\infty}\frac{\|g_k\|^4}{\|d_k\|^2} < \infty. \tag{3.55}$$

Proof From (3.3), for all $k \geq 1$,

$$d_{k+1} + g_{k+1} = \beta_k d_k. \tag{3.56}$$

Squaring both sides of (3.56) results in

$$\|d_{k+1}\|^2 = -\|g_{k+1}\|^2 - 2g_{k+1}^T d_{k+1} + \beta_k^2 \|d_k\|^2. \tag{3.57}$$

Since d_k satisfies the descent condition $g_k^T d_k < 0$, from (3.57) it follows that

$$\|d_{k+1}\|^2 \geq -\|g_{k+1}\|^2 + \beta_k^2 \|d_k\|^2. \tag{3.58}$$

From (3.3),

$$g_{k+1}^T d_{k+1} - \beta_k g_{k+1}^T d_k = -\|g_{k+1}\|^2,$$

which, with the line search condition (3.6), shows that

$$\left|g_{k+1}^T d_{k+1}\right| + \sigma|\beta_k|\left|g_k^T d_k\right| \geq \|g_{k+1}\|^2. \tag{3.59}$$

Applying the inequality $(a + \sigma b)^2 \leq (1 + \sigma^2)(a^2 + b^2)$ for all $a, b, \sigma \geq 0$, with $a = \left|g_{k+1}^T d_{k+1}\right|$ and $b = |\beta_k|\left|g_k^T d_{k,}\right|$, from (3.59), we can write

$$(g_{k+1}^T d_{k+1})^2 + \beta_k^2 (g_k^T d_k)^2 \geq c\|g_{k+1}\|^4, \tag{3.60}$$

where $c = (1 + \sigma^2)^{-1}$.

Therefore, from (3.58) and (3.60), it follows that

$$\begin{aligned}
\frac{(g_{k+1}^T d_{k+1})^2}{\|d_{k+1}\|^2} + \frac{(g_k^T d_k)^2}{\|d_k\|^2} &= \frac{1}{\|d_{k+1}\|^2}\left[(g_{k+1}^T d_{k+1})^2 + \frac{\|d_{k+1}\|^2}{\|d_k\|^2}(g_k^T d_k)^2\right] \\
&\geq \frac{1}{\|d_{k+1}\|^2}\left[(g_{k+1}^T d_{k+1})^2 + \beta_k^2 (g_k^T d_k)^2 - \frac{(g_k^T d_k)^2}{\|d_k\|^2}\|g_{k+1}\|^2\right] \\
&\geq \frac{1}{\|d_{k+1}\|^2}\left[c\|g_{k+1}\|^4 - \frac{(g_k^T d_k)^2}{\|d_k\|^2}\|g_{k+1}\|^2\right].
\end{aligned} \tag{3.61}$$

Now, if (3.54) is not true, then (3.61) and the Zoutendijk condition (3.34) imply that the following inequality

$$\frac{(g_{k+1}^T d_{k+1})^2}{\|d_{k+1}\|^2} + \frac{(g_k^T d_k)^2}{\|d_k\|^2} \geq \frac{c}{2}\frac{\|g_{k+1}\|^4}{\|d_{k+1}\|^2} \tag{3.62}$$

holds for k sufficiently large.

With these, the inequality (3.55) follows from (3.62) and (3.61). ♦

The following corollary immediately follows from the above theorem.

Corollary 3.1 *Suppose that the Assumption CG holds and consider any general conjugate gradient method given by* (3.2) *and* (3.3), *where the stepsize* α_k *is determined by the strong Wolfe line search* (3.4) *and* (3.6). *If*

$$\sum_{k=1}^{\infty} \frac{\|g_k\|^t}{\|d_k\|^2} = \infty, \tag{3.63}$$

for any $t \in [0, 4]$, *then the method converges in the sense that* (3.54) *is true.*

Proof If (3.54) is not true, then from Theorem 3.4 it follows that

$$\sum_{k=1}^{\infty} \frac{\|g_k\|^4}{\|d_k\|^2} < \infty. \tag{3.64}$$

Because $\|g_k\|$ is bounded away from zero and $t \in [0, 4]$, it is easy to see that (3.64) contradicts (3.63). Therefore, the corollary is true. ♦

The above corollary shows that if a conjugate gradient method fails to converge, then the length of the search direction will converge to infinity. Observe that in the above developments, the sufficient descent condition is not assumed. This corollary is very useful for proving the global convergence of some conjugate gradient methods without assuming the sufficient descent condition, as can be seen in the next chapter, where the standard conjugate gradient methods are to be discussed. The general outline of the proof is as follows: assume that the convergence relation (3.54) does not hold, then it follows that $\sum_{k=1}^{\infty} \|g_k\|^2 / \|d_k\|^2 = \infty$ or $\sum_{k=1}^{\infty} 1/\|d_k\|^2 = \infty$, which by Corollary 3.1 implies that (3.54) holds, thus giving a contradiction.

The following theorem, which introduces the *Nocedal condition*, presents a general convergence result for any conjugate gradient method (3.2) and (3.3) under the strong Wolfe line search (3.4) and (3.6). Mainly, the theorem says that if $\|d_k\|^2$ is at most linearly increasing, i.e., if $\|d_k\|^2 \le c_1 k + c_2$ for all k, where c_1 and c_2 are some constants, and then a conjugate gradient method with strong Wolfe line search is globally convergent. The theorem is proved by Dai (2011). See also (Nocedal, 1996).

Theorem 3.5 *Suppose that the Assumption CG holds. Consider any conjugate gradient method* (3.2) *and* (3.3) *with* d_k *satisfying* $g_k^T d_k < 0$ *and with strong Wolfe line search* (3.4) *and* (3.6). *Then, the method is globally convergent if*

$$\sum_{k=1}^{\infty} \frac{1}{\|d_k\|^2} = \infty. \tag{3.65}$$

Proof From (3.26) and (3.6), it follows that

$$\left|g_{k+1}^T d_{k+1}\right| + \sigma|\beta_k|\left|g_k^T d_k\right| \geq \|g_{k+1}\|^2. \tag{3.66}$$

Applying the following inequality $(a+\sigma b)^2 \leq (1+\sigma^2)(a^2+b^2)$ for all $a, b, \sigma \geq 0$, with $a = \left|g_{k+1}^T d_{k+1}\right|$ and $b = |\beta_k|\left|g_k^T d_k\right|$, (3.66) yields to

$$(g_{k+1}^T d_{k+1})^2 + \beta_k^2 (g_k^T d_k)^2 \geq c\|g_{k+1}\|^4, \tag{3.67}$$

where $c = (1+\sigma^2)^{-1}$ is a positive constant.

Now, from (3.3), it follows that $d_{k+1} + g_{k+1} = \beta_k d_k$, for all $k \geq 1$. Squaring both sides of this equality, the following is obtained

$$\|d_{k+1}\|^2 = -\|g_{k+1}\|^2 - 2g_{k+1}^T d_{k+1} + \beta_k^2 \|d_k\|^2.$$

Since $g_{k+1}^T d_{k+1} < 0$, it follows that

$$\|d_{k+1}\|^2 \geq -\|g_{k+1}\|^2 + \beta_k^2 \|d_k\|^2. \tag{3.68}$$

Hence, from (3.67) and (3.68),

$$\begin{aligned}
\frac{(g_{k+1}^T d_{k+1})^2}{\|d_{k+1}\|^2} + \frac{(g_k^T d_k)^2}{\|d_k\|^2} &= \frac{1}{\|d_{k+1}\|^2}\left[(g_{k+1}^T d_{k+1})^2 + \frac{\|d_{k+1}\|^2}{\|d_k\|^2}(g_k^T d_k)^2\right] \\
&\geq \frac{1}{\|d_{k+1}\|^2}\left[(g_{k+1}^T d_{k+1})^2 + \beta_k^2 (g_k^T d_k)^2 - \frac{(g_k^T d_k)^2}{\|d_k\|^2}\|g_{k+1}\|^2\right] \\
&\geq \frac{1}{\|d_{k+1}\|^2}\left[c\|g_{k+1}\|^4 - \frac{(g_k^T d_k)^2}{\|d_k\|^2}\|g_{k+1}\|^2\right].
\end{aligned} \tag{3.69}$$

Assume that (3.37) is not true and there exists a constant $\gamma > 0$ so that for any k

$$\|g_k\| \geq \gamma. \tag{3.70}$$

Observe that the Zoutendijk condition (3.34) implies that $g_k^T d_k / \|d_k\|$ tends to zero. Therefore, by (3.69) and (3.70), for sufficiently large k, it results that

$$\frac{(g_{k+1}^T d_{k+1})^2}{\|d_{k+1}\|^2} + \frac{(g_k^T d_k)^2}{\|d_k\|^2} \geq c \frac{\|g_{k+1}\|^4}{\|d_{k+1}\|^2}. \tag{3.71}$$

Thus, by the Zoutendijk condition and (3.70), we must have

$$\sum_{k=1}^{\infty} \frac{1}{\|d_k\|^2} \leq \frac{1}{\gamma^2} \sum_{k=1}^{\infty} \frac{\|g_k\|^2}{\|d_k\|^2} < \infty,$$

which is a contradiction to the assumption (3.65). Therefore, the convergence relation (3.37) holds. ♦

The theorem says that the iterations of the conjugate gradient method can fail in the sense of (3.70) only if $\|d_k\| \to \infty$ rapidly enough. More exactly, the sequence of the gradient norms $\|g_k\|$ can be bounded away from zero only if

$$\sum_{k=1}^{\infty} \frac{1}{\|d_k\|^2} < \infty. \tag{3.72}$$

In the following, we shall provide a condition on β_k which is sufficient for the global convergence of the general conjugate gradient method with strong Wolfe line search (Dai, 2010). By Theorem 3.5, we know that if (3.65) holds, and then the conjugate gradient method is convergent in the sense of (3.37). Otherwise, we have (3.72), which gives

$$\lim_{k \to \infty} \|d_k\| = \infty. \tag{3.73}$$

Therefore, from the Assumption CG, it follows that

$$\|g_k\| \leq \Gamma, \tag{3.74}$$

for some $\Gamma > 0$ and for all $k \geq 0$. Now, from (3.3) and the above relations (3.73) and (3.74), it follows that

$$\|d_{k+1}\| \approx |\beta_k| \|d_k\|. \tag{3.75}$$

Therefore, if the scalars β_k are so that

$$\sum_{k=1}^{\infty} \prod_{j=0}^{k} \beta_j^{-2} = \infty, \tag{3.76}$$

it is possible to establish the Nocedal condition (3.65), and then by Theorem 3.5, a contradiction to (3.72) is obtained. The following theorem details these discussions.

Theorem 3.6 *Suppose that the Assumption CG holds. Consider a conjugate gradient method given by* (3.2) *and* (3.3), *where the search direction* d_k *is descent, i.e.,* $d_k^T g_k < 0$. *Consider that the stepsize is determined by the strong Wolfe line search conditions* (3.4) *and* (3.6). *If* β_k *satisfies* (3.76), *then* $\liminf_{k\to\infty} \|g_k\| = 0$.

Proof Observe that (3.3) can be written as

$$d_{k+1} + g_{k+1} = \beta_k d_k. \tag{3.77}$$

Squaring both sides of (3.77) leads to

$$\|d_{k+1}\|^2 = -2g_{k+1}^T d_{k+1} - \|g_{k+1}\|^2 + \beta_k^2 \|d_k\|^2. \tag{3.78}$$

But,

$$-2g_{k+1}^T d_{k+1} - \|g_{k+1}\|^2 \leq \frac{(g_{k+1}^T d_{k+1})^2}{\|g_{k+1}\|^2}. \tag{3.79}$$

Therefore, from (3.78) and (3.79), it results that

$$\|d_{k+1}\|^2 \leq \frac{(g_{k+1}^T d_{k+1})^2}{\|g_{k+1}\|^2} + \beta_k^2 \|d_k\|^2. \tag{3.80}$$

Having in view the definition of $\cos\theta_k$ in (3.28), from (3.80) it follows that

$$\begin{aligned} \|d_{k+1}\|^2 &\leq (1 - \cos^2\theta_{k+1})^{-1} \beta_k^2 \|d_k\|^2 \leq \cdots \\ &\leq \left(\prod_{j=1}^{k+1} (1 - \cos^2\theta_j)^{-1} \right) \left(\prod_{j=0}^{k} \beta_j^2 \right) \|d_0\|^2. \end{aligned} \tag{3.81}$$

Suppose that $\liminf_{k\to\infty} \|g_k\| \neq 0$. Therefore, there exists a constant $\gamma > 0$ so that

$$\|g_k\| \geq \gamma, \text{ for all } k \geq 0. \tag{3.82}$$

Hence, from the Zoutendijk condition (3.34), from the definition of $\cos\theta_k$ in (3.28) and (3.82), it follows that

$$\sum_{k=1}^{\infty} \cos^2\theta_k < \infty. \tag{3.83}$$

With this, (3.83) implies that

$$\prod_{j=1}^{k+1} (1 - \cos^2 \theta_j) \geq c, \tag{3.84}$$

where $c > 0$ is a constant. Hence,

$$\frac{1}{\|d_{k+1}\|^2} \geq c \left(\prod_{j=0}^{k} \beta_j^{-2} \right) \|d_0\|^{-2}. \tag{3.85}$$

From (3.85) and (3.76), it follows that (3.65) holds. Thus, by Theorem 3.5, $\liminf_{k \to \infty} \|g_k\| = 0$. But this, together with (3.82) give a contradiction, thus proving the theorem. ♦

Theorem 3.6 shows that the global convergence of any conjugate gradient algorithm is obtained if the conjugate gradient parameters β_k satisfy the condition (3.76) and the stepsize is determined by the strong Wolfe line search. Mainly, it is based on the Zoutendijk condition. It is worth mentioning that in Theorem 3.6, it is the descent condition (3.25) which is used and not the sufficient descent condition (3.27).

3.3.2 Convergence Under the Standard Wolfe Line Search

Dai (2010) proved that the conclusion of Theorem 3.6 for the global convergence of any conjugate gradient method also holds under the standard Wolfe line search. This result is based on the following proposition proved by Dai and Yuan (2003).

Proposition 3.2 *Consider any conjugate gradient method* (3.2) *and* (3.3). *Define* ϕ_k *and* t_k *as follows*

$$\phi_k^2 = \begin{cases} \|g_k\|^2, & for\, k = 0, \\ \beta_0^2 \beta_1^2 \cdots \beta_{k-1}^2, & for\, k \geq 1 \end{cases} \tag{3.86}$$

and

$$t_k = \frac{\|d_k\|^2}{\phi_k^2}. \tag{3.87}$$

Then, for all $k \geq 0$,

$$t_k = -2 \sum_{i=0}^{k} \frac{g_i^T d_i}{\phi_i^2} - \sum_{i=0}^{k} \frac{\|g_i\|^2}{\phi_i^2}. \tag{3.88}$$

Proof Since $d_0 = -g_0$ and $\phi_0^2 = \|g_0\|^2$, (3.88) holds for $k = 0$. For $k \geq 1$, dividing (3.78) by ϕ_{k+1}^2 and using the definitions of ϕ_k and t_k, it follows that

$$t_{k+1} = t_k - 2\frac{g_{k+1}^T d_{k+1}}{\phi_{k+1}^2} - \frac{\|g_{k+1}\|^2}{\phi_{k+1}^2}. \tag{3.89}$$

Summing (3.89) results in

$$t_{k+1} = t_0 - 2\sum_{i=1}^{k+1}\frac{g_i^T d_i}{\phi_i^2} - \sum_{i=1}^{k+1}\frac{\|g_i\|^2}{\phi_i^2}. \tag{3.90}$$

Since $t_0 = \|g_0\|^2/\phi_0^2$, it follows that (3.90) is equivalent to (3.88). Therefore, (3.88) holds for any $k \geq 0$. ♦

To show the sufficiency of the condition (3.76) on β_k, the following proposition is needed.

Proposition 3.3 *Suppose that* $\{a_i\}$ *and* $\{b_i\}$ *are two real positive number sequences, satisfying*:

$$b_k \leq c_1 + c_2\sum_{i=1}^{k} a_i, \textit{ for all } k,$$

where c_1 *and* c_2 *are positive constants. If the sum* $\sum_{k\geq 1} a_k$ *is divergent, then* $\sum_{k\geq 1} a_k/b_k$ *is also divergent.*

Proof Let $S_k = \sum_{i=1}^{k} a_i$, for any $k \geq 1$. Now, $\sum_{i\geq 1} a_i = \infty$ implies that $\lim_{k\to\infty} S_k = \infty$. Observe that the sequence $\{S_k\}$ is increasing. Let $c = c_1/c_2$. There exists $k_0 \geq 1$ so that $S_k \geq c$ for $k \geq k_0$. Note that

$$\sum_{k=k_0}^{\infty}\frac{a_k}{b_k} \geq \sum_{k=k_0}^{\infty}\frac{a_k}{c_1 + c_2 S_k} = \frac{1}{c_2}\sum_{k=k_0}^{\infty}\frac{a_k}{c + S_k} \geq \frac{1}{c_2}\sum_{k=k_0}^{\infty}\frac{a_k}{S_k + S_k} = \frac{1}{2c_2}\sum_{k=k_0}^{\infty}\frac{a_k}{S_k}.$$

Let

$$A(k, r) = \sum_{i=1}^{r}\frac{a_{k+i}}{S_{k+i}}, \text{ for } r \geq 1.$$

It is easy to see that for every $k \geq 1$ there exists $r \geq 1$ so that $A(k, r) \geq \frac{1}{2}$. Indeed,

$$A(k,r) \geq \sum_{i=1}^{r} \frac{a_{k+i}}{S_{k+r}} = \frac{S_{k+r} - S_k}{S_{k+r}} = 1 - \frac{S_k}{S_{k+r}}.$$

But,

$$\lim_{k\to\infty} A(k,r) \geq \lim_{k\to\infty} \left(1 - \frac{S_k}{S_{k+r}}\right) = 1.$$

Hence, there exists $r \geq 1$ so that $A(k,r) \geq 1/2$.

Therefore, there is a sequence $\{r_k\}$ with $r_k \geq 1$ so that $k_1 = k_0 + r_1$, $k_2 = k_1 + r_2, \ldots$ and $A(k_0, r_1) \geq \frac{1}{2}$, $A(k_1, r_2) \geq \frac{1}{2}, \ldots$ Hence,

$$\sum_{k=k_0}^{\infty} \frac{a_k}{S_k} = \sum_{i=0}^{\infty} A(k_i, r_{i+1}) \geq \sum_{i=0}^{\infty} \frac{1}{2} = \infty.$$ ♦

The following theorem, proved by Dai (2010), shows that the condition (3.76) on β_k is *sufficient* for the global convergence of any conjugate gradient method (3.2) and (3.3).

Theorem 3.7 *Suppose that the Assumption CG holds. Consider a conjugate gradient method given by* (3.2) *and* (3.3), *where the search direction* d_k *is descent, i.e.,* $d_k^T g_k < 0$ *and the stepsize is determined by the standard Wolfe line search conditions* (3.4) *and* (3.5). *If* β_k *satisfies* (3.76), *then* $\liminf_{k\to\infty} \|g_k\| = 0$.

Proof Define ϕ_k^2 as in (3.86). Then, from (3.76) it follows that

$$\sum_{k\geq 1} \frac{1}{\phi_k^2} = \infty. \tag{3.91}$$

Now, using (3.79) in (3.88), we can write

$$t_k \leq \sum_{i=0}^{k} \frac{(g_i^T d_i)^2}{\|g_i\|^2 \phi_i^2}. \tag{3.92}$$

Since $t_{k+1} \geq 0$, from (3.88), it follows that

$$-2\sum_{i=0}^{k} \frac{g_i^T d_i}{\phi_i^2} \geq \sum_{i=0}^{k} \frac{\|g_i\|^2}{\phi_i^2}. \tag{3.93}$$

But, $\left(\|g_{k+1}\|^2 + 2(g_{k+1}^T d_{k+1})\right)^2 \geq 0$. Hence, for any k,

$$-4g_{k+1}^T d_{k+1} - \|g_{k+1}\|^2 \le 4\frac{(g_{k+1}^T d_{k+1})^2}{\|g_{k+1}\|^2}. \tag{3.94}$$

Hence, from (3.93) and (3.94),

$$4\sum_{i=0}^{k}\frac{(g_i^T d_i)^2}{\|g_i\|^2\phi_i^2} \ge -4\sum_{i=0}^{k}\frac{g_i^T d_i}{\phi_i^2} - \sum_{i=0}^{k}\frac{\|g_i\|^2}{\phi_i^2} \ge \sum_{i=0}^{k}\frac{\|g_i\|^2}{\phi_i^2}. \tag{3.95}$$

Now, let us proceed by contradiction and assume that (3.82) holds. Then, by (3.95) and (3.91) it follows that

$$\sum_{k\ge 1}\frac{(g_k^T d_k)^2}{\|g_k\|^2\phi_k^2} = \infty. \tag{3.96}$$

From (3.96), (3.87) and Proposition 3.3,

$$\sum_{k=1}^{\infty}\frac{(g_k^T d_k)^2}{\|g_k\|^2\phi_k^2}\frac{1}{t_k} = \sum_{k=1}^{\infty}\frac{(g_k^T d_k)^2}{\|g_k\|^2\|d_k\|^2} = \sum_{k=1}^{\infty}\cos^2\theta_k = \infty, \tag{3.97}$$

which contradicts (3.83). This contradiction shows that (3.76) is true. ♦

Observe that the proof of this theorem is based on the Zoutendijk condition. Theorem 3.7 provides the condition (3.76) on β_k which is sufficient for the global convergence of the general conjugate gradient method with standard Wolfe line search. Besides, notice that only the descent condition $g_k^T d_k < 0$, for any k, is used in Theorem 3.7. This is an important contribution to the general theory of conjugate gradient methods, since very often, in the implementation of the conjugate gradient algorithms, only the standard Wolfe line search conditions are used.

In the following, let us show that the condition (3.76) is also *necessary* for the global convergence of the general conjugate gradient methods (3.2) and (3.3). This was proved by Dai (2010). It represents a remarkable effort for establishing the necessary and sufficient conditions on the conjugate gradient parameters β_k, that ensure the convergence of the general conjugate gradient algorithms. We follow his developments. Let us assume that the line search is exact, i.e.,

$$g_{k+1}^T d_k = 0, \quad \text{for all } k. \tag{3.98}$$

Besides, assume that the iterations of the algorithm fall into a region where the function f is quadratic with the unit Hessian

$$f(x) = \frac{1}{2}x^T x.$$

From (3.3) with (3.98), it follows that

$$g_{k+1}^T d_{k+1} = -\|g_{k+1}\|^2, \tag{3.99}$$

$$\|d_{k+1}\|^2 = \|g_{k+1}\|^2 + \beta_k^2 \|d_k\|^2, \tag{3.100}$$

for any k.

Since f is quadratic with the unit Hessian, from (3.2) and (3.3), it follows that

$$g_{k+1} = g_k + \alpha_k d_k. \tag{3.101}$$

Therefore, under the exact line search (3.98), from (3.101),

$$\alpha_k = -\frac{g_k^T d_k}{\|d_k\|^2} = \frac{\|g_k\|^2}{\|d_k\|^2}. \tag{3.102}$$

Squaring both sides of (3.101) and using the above relations (3.99), (3.100), and (3.102) lead to

$$\begin{aligned} \|g_{k+1}\|^2 &= \|g_k\|^2 + 2\alpha_k g_k^T d_k + \alpha_k^2 \|d_k\|^2 \\ &= \|g_k\|^2 \left(1 - \frac{\|g_k\|^2}{\|d_k\|^2}\right) = \beta_{k-1}^2 \frac{\|d_{k-1}\|^2}{\|d_k\|^2} \|g_k\|^2. \end{aligned} \tag{3.103}$$

Therefore, by recursion of (3.103),

$$\|g_{k+1}\|^2 = \left(\prod_{j=0}^{k-1} \beta_j^2\right) \frac{\|d_0\|^2}{\|d_k\|^2} \|g_1\|^2. \tag{3.104}$$

With ϕ_k and t_k defined in Proposition 3.2, from (3.104), it follows that

$$\|g_{k+1}\|^2 = \frac{1}{t_k} \|d_0\|^2 \|g_1\|^2. \tag{3.105}$$

On the other hand, from (3.88), since $g_i^T d_i = -\|g_i\|^2$ for any i, it follows that

$$t_k = \sum_{i=0}^{k} \frac{\|g_i\|^2}{\phi_i^2}. \tag{3.106}$$

But, from (3.103), the sequence $\{\|g_k\|^2\}$ is monotonically decreasing. Thus,

$$\|g_k\| \leq \|g_1\|, \text{ for all } k \geq 1. \tag{3.107}$$

Therefore, if (3.76) is false, i.e., if

$$\sum_{k=1}^{\infty} \prod_{j=0}^{k} \beta_j^{-2} < \infty,$$

then from the definition of ϕ_k^2, (3.106) and (3.107), it follows that there is a positive constant M, so that

$$t_k \le M. \tag{3.108}$$

Hence, from (3.108) and (3.105), it follows that

$$\|g_{k+1}\|^2 \ge \frac{\|d_0\|^2 \|g_1\|^2}{M}, \text{ for all } k,$$

which means that the iterations cannot approach the unique minimizer $x^* = 0$ of the quadratic function with the unit Hessian considered above. Therefore, by contrast, if (3.76) is true, from the definition of ϕ_k and (3.106), we have that $t_k \to \infty$, that is, from (3.105), $\lim_{k\to\infty} \|g_k\| = 0$. In other words, using this simple example, it follows that any conjugate gradient method given by (3.2) and (3.3) converges globally if and only if the condition (3.76) on β_k holds. ♦

The above theorems present the necessary and sufficient condition on the conjugate gradient parameter β_k, namely (3.76), for the global convergence of any general conjugate gradient method under the standard Wolfe line search.

In the following, two properties which establish some conditions on the parameter β_k to ensure the convergence of the corresponding conjugate gradient method are discussed. The first one is due to Gilbert and Nocedal (1992). The second was developed by Dai (2010).

Property (*) Gilbert and Nocedal (1992)
To prove the convergence of the conjugate gradient method, Gilbert and Nocedal (1992) introduced the so-called Property (*). The idea is that in addition to $\beta_k \ge 0$, it requires that β_k be small when the step $s_k = x_k - x_{k-1}$ is small. Formally, this property is as follows:

Property (*) *Consider any conjugate gradient method* (3.2) *and* (3.3). *Suppose that for all* $k \ge 0$, $0 < \gamma \le \|g_k\| \le \Gamma$. *Under this assumption, we say that the method has Property* (*) *if there exist constants* $b > 1$ *and* $\lambda > 0$ *so that for all* k

$$|\beta_k| \le b$$

and

$$\|s_k\| \le \lambda \Rightarrow |\beta_k| \le \frac{1}{2b}.$$ ♦

Gilbert and Nocedal (1992) proved that if the gradients are bounded away from zero and if the method has Property (*), then a fraction of the steps cannot be too small. Therefore, Property (*) determines the convergence of conjugate gradient methods.

Theorem 3.8 *Suppose that the Assumption CG holds and consider any conjugate gradient method* (3.2) *and* (3.3) *with the following properties:* $\beta_k \geq 0$ *for all k, the Zoutendijk, the sufficient descent and the Property* (*) *conditions hold. Then,* $\liminf_{k\to\infty} \|g_k\| = 0$.

The proof is given by Gilbert and Nocedal (1992) and is based on the fact that if the gradients are bounded away from zero and if the method has Property (*), then a fraction of the steps cannot be too small. It is also based on the bounding of $\sum_{k\geq 1} \|u_k - u_{k-1}\|^2$, where $u_k = d_k/\|d_k\|$. ♦

Under the Assumption CG, the Polak–Ribiére–Polyak (PRP) and the Hestenes–Stiefel (HS) have Property (*). If β_k has Property (*), so have $|\beta_k|$ and $\beta_k^+ = \max\{0, \beta_k\}$. Therefore, many other choices of $\beta_k > 0$ lead to algorithms with Property (*).

Property (#) Dai (2010)
This property was introduced by Dai (2010) as a generalization of Property (*). The idea was to relax the boundedness of β_k subject to a positive and uniformly bounded sequence.

Property (#) *Consider a conjugate gradient method given by* (3.2) *and* (3.3) *and suppose that for all* $k \geq 0$, $0 < \gamma \leq \|g_k\| \leq \Gamma$. *Under this assumption, we say that the method has Property (#) if there exist a positive and uniformly bounded sequence* $\{\varphi_k\}$ *and the constants* $b \geq 1$ *and* $\lambda > 0$, *so that for all k,*

$$|\beta_k| \leq b\frac{\varphi_k}{\varphi_{k-1}}$$

and

$$\|s_k\| \leq \lambda \Rightarrow |\beta_k| \leq \frac{1}{b}\frac{\varphi_k}{\varphi_{k-1}}.$$ ♦

Property (#) has Property (*) as its special case, i.e., if Property (*) holds, then Property (#) must be true with $\varphi_k = 1$. The following theorem shows how Theorem 3.7 and Property (#) may be used to analyze the global convergence of the conjugate gradient methods.

Theorem 3.9 *Suppose that the Assumption CG holds and consider any conjugate gradient method* (3.2) *and* (3.3), *where* β_k *has Property (#) with* $b = 1$. *Suppose that the search direction* d_k *is a descent direction, i.e.,* $g_k^T d_k < 0$ *for all k. If the* *stepsize* α_k *satisfies the standard Wolfe conditions* (3.4) *and* (3.5), *then* $\liminf_{k\to\infty}\|g_k\| = 0$.

Proof Suppose that there exists some constant $\gamma > 0$ so that $\|g_k\| \geq \gamma$, for all $k \geq 0$. By Property (#) with $b = 1$, it follows that

$$\prod_{j=2}^{k}\beta_j^2 = \frac{\varphi_k^2}{\varphi_1^2},$$

which with the boundedness of $\{\varphi_k\}$ implies that

$$\prod_{j=2}^{k}\beta_j^2 \leq c_1 + c_2 k,$$

where c_1 and c_2 are some nonnegative constants. Therefore, by Theorem 3.7, it follows that $\liminf_{k\to\infty}\|g_k\| = 0$, thus contradicting that $\|g_k\| \geq \gamma$. This contradiction proves the theorem. ♦

Observe the difference between the convergence of the conjugate gradient methods with Property (*) and with Property (#). In Theorem 3.8, the convergence is proved if Property (*) and the sufficient descent condition are satisfied. On the other hand, in Theorem 3.9, the convergence is proved if Property (#) and only the descent conditions are satisfied.

Using Property (*), Gilbert and Nocedal (1992) proved the convergence of Hestenes–Stiefel (HS) and Polak–Ribiére–Polyak (PRP) conjugate gradient algorithms. On the other hand, using Property (#), Dai (2010) proved the convergence of Fletcher-Reeves (FR), Polak–Ribiére–Polyak (PRP), Dai–Yuan (DY), as well as of the hybrid methods FR-PRP and DY-HS conjugate gradient methods introduced in Chapter 6.

3.4 Criticism of the Convergence Results

The most important ingredient in proving the global convergence of conjugate gradient methods is the *Zoutendijk condition*, originally given by Zoutendijk (1970) and Wolfe (1969, 1971).

Normally, the direction d_k is chosen so that it is a descent direction, namely $d_k^T g_k < 0$, if $g_k \neq 0$. Using the first standard Wolfe line search condition (3.4), it follows that

$$\alpha_k \geq c\frac{-d_k^T g_k}{\|d_k\|^2}, \tag{3.109}$$

where c is some positive constant (see Proposition 1.2). Since the search direction is descent, it follows that $\alpha_k > 0$. Besides, using (3.109), from (3.4), it results that

$$f(x_k) - f(x_{k+1}) \geq \rho c \frac{(-d_k^T g_k)^2}{\|d_k\|^2}. \tag{3.110}$$

Therefore, if $\{f(x_k)\}$ is bounded below, then from (3.110), it follows that

$$\sum_{k=1}^{\infty} \frac{(-d_k^T g_k)^2}{\|d_k\|^2} < \infty. \tag{3.111}$$

But,

$$\cos^2 \theta_k = \frac{(-d_k^T g_k)^2}{\|d_k\|^2 \|g_k\|^2},$$

where θ_k is the angle between the steepest descent direction and the search direction d_k. Therefore, from (3.111),

$$\sum_{k=1}^{\infty} \|g_k\|^2 \cos^2 \theta_k < \infty, \tag{3.112}$$

if $f(x)$ is bounded below. But (3.112) is exactly the Zoutendijk condition (3.34), which implies the convergence of the method.

Therefore, from the above developments, the following *general convergence results* may be established for the *line search algorithms* [see Wolfe (1969, 1971), Zoutendijk (1970)].

Theorem 3.10 *Let $\{x_k\}$ be the sequence generated by a line search algorithm under the exact line search or under any inexact line search which satisfies* (3.110). *If*

$$\sum_{k=1}^{\infty} \cos^2 \theta_k = \infty,$$

then the sequence $\{x_k\}$ is convergent in the sense that $\liminf_{k\to\infty} \|g_k\| = 0$.

Furthermore, if there exists a positive constant η so that $\cos^2 \theta_k \geq \eta$ for all k, then the sequence is convergent in the sense that $\lim_{k\to\infty} \|g_k\| = 0$. ♦

In conjugate gradient algorithms, it is assumed that the parameter β_k is chosen so that the sufficient descent condition $d_k^T g_k \leq -c\|g_k\|^2$ is satisfied for some positive constant c. For example, Theorem 3.8, proved by Gilbert and Nocedal (1992), shows that if the sufficient descent condition and Property (*) are satisfied, then under the Assumption CG, the general conjugate gradient algorithm is convergent, that is $\liminf\|g_k\| = 0$. On the other hand, Theorem 3.9, proved by Dai (2010), shows that if the descent condition and Property (#) are satisfied, then under the Assumption CG, the general conjugate gradient algorithm is convergent in the sense that $\liminf\|g_k\| = 0$.

The proofs on the convergence of the conjugate gradient methods are mainly based on the estimation of

$$\sum_{k=1}^{\infty}\cos^2\theta_k = \sum_{k=1}^{\infty}\frac{1}{\|d_k\|^2}\left(\frac{(d_k^T g_k)^2}{\|g_k\|^2}\right). \tag{3.113}$$

If $(d_k^T g_k)^2/\|g_k\|^2$ is bounded away from zero, then from Theorem 3.10, it follows that the condition

$$\sum_{k=1}^{\infty}\frac{1}{\|d_k\|^2} = \infty, \tag{3.114}$$

known as the Nocedal condition, implies the convergence of the conjugate gradient method. Therefore, in the convergence analysis of a conjugate gradient method, a widely used technique is to derive a contradiction by establishing the above relation if there is a positive constant γ so that $\|g_k\| \geq \gamma$, for any k. Observe that under the boundedness of $\|g_k\|$, the boundedness of $(d_k^T g_k)^2/\|g_k\|^2$ is equivalent to the sufficient descent condition (3.27).

Convergence results on conjugate gradient methods obtained by Dai and Yuan (1996c) show that the sufficient descent condition $d_k^T g_k \leq -c\|g_k\|^2$ is *not always necessary* [see also (Yuan, 1998)]. Instead, this condition has to be satisfied in the mean sense, i.e., the mean value of $-d_k^T g_k$, over every two consecutive iterations, has to be bounded away from zero. In other words, the sufficient condition $d_k^T g_k \leq -c\|g_k\|^2$ can be replaced by

$$\frac{(d_k^T g_k)^2}{\|g_k\|^4} + \frac{(d_{k+1}^T g_{k+1})^2}{\|g_{k+1}\|^4} \geq c, \text{ for any } k \geq 1. \tag{3.115}$$

Theorem 3.11 *If* $\{f(x_k)\}$ *is bounded below,* $\{\beta_k\}$ *is bounded and* (3.114) *holds, then the sequence* $\{x_k\}$ *generated by the conjugate gradient* (3.2) *and* (3.3) *under the strong Wolfe line search* (3.4) *and* (3.6) *converges in the sense that* $\liminf_{k\to\infty}\|g_k\| = 0$.

Proof From (3.3),

$$\frac{g_{k+1}^T d_{k+1}}{\|g_{k+1}\|^2} = -1 + \beta_k\frac{g_{k+1}^T d_k}{\|g_{k+1}\|^2}.$$

Using the strong Wolfe line search condition (3.6), it follows that

$$1 = -\frac{g_{k+1}^T d_{k+1}}{\|g_{k+1}\|^2} + \beta_k \frac{g_{k+1}^T d_k}{\|g_{k+1}\|^2} \leq -\frac{g_{k+1}^T d_{k+1}}{\|g_{k+1}\|^2} + \sigma|\beta_k| \frac{\|g_k\|^2}{\|g_{k+1}\|^2} \frac{|d_k^T g_k|}{\|g_k\|^2}$$
$$\leq \sqrt{1 + \frac{\sigma^2 \beta_k^2 \|g_k\|^2}{\|g_{k+1}\|^2}} \times \sqrt{\frac{(d_k^T g_k)^2}{\|g_k\|^4} + \frac{(d_{k+1}^T g_{k+1})^2}{\|g_{k+1}\|^4}}.$$

The above inequality and the assumptions imply that there exists a positive constant c so that (3.115) holds. Therefore, from Theorem 3.10 and (3.115), it follows that

$$\sum_{k=1}^{\infty} \min\left\{\frac{1}{\|d_{2k-1}\|^2}, \frac{1}{\|d_{2k}\|^2}\right\} < \infty, \tag{3.116}$$

which shows that

$$\max\{\|d_{2k-1}\|, \|d_{2k}\|\} \to \infty.$$

Using (3.3) and the boundedness of $\|g_k\|$ ($\|g_k\| \leq \gamma$, $\gamma > 0$ for any k), it results that

$$\|d_{2k}\| \leq \gamma + |\beta_{2k-1}| \|d_{2k-1}\|.$$

From (3.116) and the boundedness of β_k, it follows that

$$\sum_{k=1}^{\infty} \frac{1}{\|d_{2k}\|^2} < \infty.$$

Now, repeating the above analysis but this time with the indices $2k-1$ and $2k$ replaced by $2k$ and $2k+1$, respectively, it can be proved that

$$\sum_{k=1}^{\infty} \frac{1}{\|d_{2k+1}\|^2} < \infty.$$

Therefore, it follows that

$$\sum_{k=1}^{\infty} \frac{1}{\|d_k\|^2} < \infty,$$

which contradicts the assumption in the theorem. ♦

The above theorem, proved by Yuan (1998), shows that an essential technique for proving the convergence of conjugate gradient methods is to try to obtain some bounds on the increasing rate of $\|d_k\|$ so that (3.114) holds. To estimate the bounds on $\|d_k\|$, a direct way is to use (3.3) recursively. Therefore, this implies some inequality on β_k. In other words, the convergence results of the conjugate gradient methods are established under certain inequalities on β_k, which is quite normal because the search direction in conjugate gradient methods depends on the parameter β_k. In this presentation, such conditions on β_k are given by (3.41) or (3.76).

Numerical study. In the convergence analysis on the conjugate gradient methods, both the standard Wolfe line search and the strong Wolfe line search conditions are used. In order to see the effectiveness of the Wolfe line search conditions and to make a comparison between the standard Wolfe line search and the strong Wolfe line search, let us consider the following numerical experiment using the set of 80 unconstrained optimization test problems from our UOP collection (Andrei, 2018 g). We report the numerical results obtained with a Fortran implementation of the Hestenes and Stiefel (HS) conjugate gradient method with both the standard Wolfe line search and with the strong Wolfe line search. The standard Wolfe line search uses the implementations of Shanno (1983) with some additional minor modifications by Andrei (1995) (see Figure 5.1). The strong Wolfe line search uses the implementations of Moré and Thuente (1994).

The search direction in the HS conjugate gradient method is determined as in (3.3), where the parameter β_k is computed as $\beta_k = g_{k+1}^T y_k / d_k^T y_k$. For each test problem, ten numerical experiments have been considered with the number of variables $n = 1000, 2000, \ldots, 10000$. Therefore, a number of 800 unconstrained optimization test problems have been solved. The comparisons of the algorithms are given in the context of Remark 1.1.

Figure 3.1 shows the Dolan and Moré (2002) performance profile of the HS conjugate gradient algorithm with standard Wolfe line search versus HS with strong Wolfe line search. Out of 800 problems considered in this numerical experiment, only for 760 problems does the criterion (1.118) hold. The left side of Figure 3.1 (small values of τ) gives the percentage of the test problems, out of 760, for which an algorithm is more successful (faster); the right side (large values of τ) gives the percentage of the test problems that were successfully solved by each of the algorithms. Observe that the HS with standard Wolfe line search outperforms the HS with strong Wolfe line search in the vast majority of problems and the differences are substantial.

When comparing HS with standard Wolfe line search versus HS with strong Wolfe line search subject to the number of iterations (see Table 3.1), we notice that HS with standard Wolfe line search was better in 351 problems (i.e., it achieved the minimum number of iterations in 351 problems). HS with strong Wolfe line search was better in 220 problems and they achieved the same number of iterations in 99 problems. As for the CPU time, we see that HS with standard Wolfe line search was better in 352 problems and HS with strong Wolfe line search was better in 114 problems, etc.

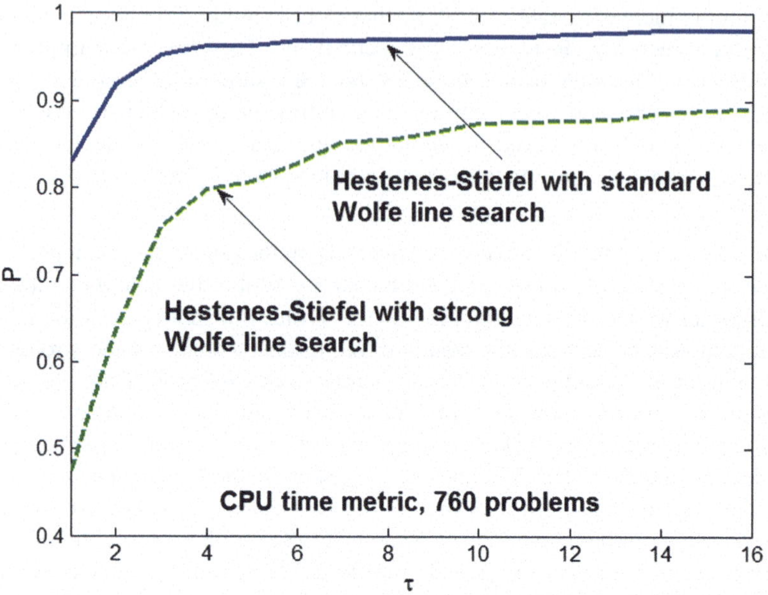

Figure 3.1 Performance profiles of Hestenes–Stiefel conjugate gradient with standard Wolfe line search versus Hestenes–Stiefel conjugate gradient with strong Wolfe line search, based on CPU time

Table 3.1 Performances of Hestenes–Stiefel conjugate gradient with standard Wolfe line search versus Hestenes–Stiefel conjugate gradient with strong Wolfe line search

	Standard Wolfe	Strong Wolfe	=
#iter	351	220	99
#fg	467	201	2
cpu	352	114	204

It is obvious that the conjugate gradient with standard Wolfe line search is top performer versus the conjugate gradient with strong Wolfe line search subject to all three metrics: the number of iterations, the number of function, and its gradient evaluations and the CPU time. The theory behind the conjugate gradient method makes it far more strongly dependent on the use of the line search than on any other minimization techniques (Newton or quasi-Newton).

Notes and References

The convergence of the conjugate gradient methods is a result of major intellectual efforts given by a number of scientists including: Hestenes and Steifel (1952), Zoutendijk (1970), Wolfe (1969, 1971), Powell (1976b, 1984a), Al-Baali (1985), Nocedal (1992), Gilbert and Nocedal (1992), Dai (2010, 2011), Dai and Yuan

(1996c, 1999, 2001a, 2003), etc. The convergence analysis of conjugate gradient algorithms is based on the Zoutendijk condition. The central result on the convergence of conjugate gradient algorithms is based on the growth in the length of the search direction. If the level set is bounded and the minimizing function is twice continuously differentiable on the level set, then the convergence can fail in the sense that $\|g_k\| \geq \gamma > 0$ for all k, where γ is a constant only if $\|d_k\| \to \infty$ rapidly enough. More exactly, the sequence of gradient norms $\|g_k\|$ can be bounded away from zero only if $\sum_{k \geq 0} 1/\|d_k\|^2 < \infty$. This result, known as the Nocedal condition, is crucial and is used as a basis for global convergence analysis on nonlinear conjugate gradient methods (see Theorem 3.5). Analysis of the convergence of conjugate gradient algorithms with inexact line search shows that unless the conjugate gradient parameter β_k is carefully chosen, the length of the search direction d_k can grow without bound, thus determining the failure of the algorithm. Therefore, some conditions must be imposed on β_k. The purpose of this chapter was to present the most important conditions on β_k. Such conditions on β_k include: $\beta_k \geq 0$, Property (*) of Gilbert and Nocedal or Property (#) of Dai, or the condition given by (3.76).

Concerning the rate of convergence of conjugate gradient algorithms, Crowder and Wolfe (1969) showed that under the exact line search, the rate of convergence is linear and show by constructing an example where q-superlinear convergence is not achievable. Cohen (1972) and Burmeister (1973) established that for general nonlinear functions, the rate of convergence is *n-step quadratic*, i.e., $\|x_{k+n} - x^*\| = O\left(\|x_k - x^*\|^2\right)$. Ritter (1980) shows that the rate of convergence is actually *superquadratic*, i.e., $\|x_{k+n} - x^*\| = o\left(\|x_k - x^*\|^2\right)$.

Using small problems, Powell (1983) gives a slightly better result and reports some numerical evidence on the convergence rate observed in practice. He also summarizes the rate of convergence results for asymptotically exact line searches, such as those obtained by Baptist and Stoer (1977) and Stoer (1977). Under the assumption that the search directions are uniformly linearly independent, Schuller (1974) reports faster rates of convergence.

Chapter 4
Standard Conjugate Gradient Methods

The purpose of this chapter is to present the standard conjugate gradient algorithms as well as their convergence for solving unconstrained optimization problems. The standard conjugate gradient methods are listed in Table 4.1. These methods have simple algebraic expressions and the convergence results are particularizations or specializations of the general results on the convergence of conjugate gradient algorithms presented in Chap. 3 of this book. In general, the convergence results of these algorithms are based on the Assumption CG, as well as on some other natural hypotheses on the line search.

A rapid inspection of Table 4.1 shows that except for the method by Daniel (1967), which requires the evaluation of the Hessian at each iteration, the numerator of the update parameter β_k is either $\|g_{k+1}\|^2$ or $g_{k+1}^T y_k$ and the denominator is either $\|g_k\|^2$ or $d_k^T y_k$ or $-d_k^T g_k$. Here, $y_k = g_{k+1} - g_k$. Mainly, these two choices for the numerator and the three choices for the denumerator lead to six different choices for β_k.

If function f is strongly convex quadratic and the line search is exact, then in theory, all the above choices for the update parameter β_k presented in Table 4.1 are equivalent. For nonquadratic objective functions, each choice for β_k leads to algorithms with different numerical performances (number of iterations, number of function, and its gradient evaluations or CPU time). Therefore, in the following, the global convergence properties of the standard conjugate gradient methods with the numerator $\|g_{k+1}\|^2$ for the update parameter β_k (FR, CD, and DY) and with $g_{k+1}^T y_k$ in the numerator of β_k (HS, PRP, and LS) will be separately presented. As a general remark, the convergence theory for the methods with numerator $\|g_{k+1}\|^2$ is better developed than the theory for the methods with numerator $g_{k+1}^T y_k$ of β_k. However, the methods with $g_{k+1}^T y_k$ in numerator of β_k perform better in practice than the methods with $\|g_{k+1}\|^2$ in numerator of β_k. The general algorithm for the standard conjugate gradient methods is as follows.

N. Andrei, *Nonlinear Conjugate Gradient Methods for Unconstrained Optimization*,
Springer Optimization and Its Applications 158,
https://doi.org/10.1007/978-3-030-42950-8_4

Table 4.1 Choices of β_k in standard conjugate gradient methods

$\beta_k^{HS} = \dfrac{g_{k+1}^T y_k}{d_k^T y_k}$	Original *linear* conjugate gradient method of Hestenes and Stiefel (1952)
$\beta_k^{FR} = \dfrac{g_{k+1}^T g_{k+1}}{g_k^T g_k}$	First *nonlinear* conjugate gradient method by Fletcher and Reeves (1964)
$\beta_k^{D} = \dfrac{g_{k+1}^T \nabla^2 f(x_k) d_k}{d_k^T \nabla^2 f(x_k) d_k}$	Proposed by Daniel (1967). This conjugate gradient method requires evaluation of the Hessian
$\beta_k^{PRP} = \dfrac{g_{k+1}^T y_k}{g_k^T g_k}$	Proposed by Polak and Ribière (1969) and by Polyak (1969)
$\beta_k^{PRP+} = \max\left\{0, \dfrac{g_{k+1}^T y_k}{g_k^T g_k}\right\}$	Proposed by Powell (1984a) and analyzed by Gilbert and Nocedal (1992)
$\beta_k^{CD} = \dfrac{g_{k+1}^T g_{k+1}}{-d_k^T g_k}$	Proposed by Fletcher (1987) known as CD (conjugate descent) method
$\beta_k^{LS} = \dfrac{g_{k+1}^T y_k}{-d_k^T g_k}$	Proposed by Liu and Storey (1991)
$\beta_k^{DY} = \dfrac{g_{k+1}^T g_{k+1}}{d_k^T y_k}$	Proposed by Dai and Yuan (1999)

Algorithm 4.1 General nonlinear conjugate gradient

1.	Choose an initial point x_0 and $\varepsilon > 0$ sufficiently small. Set $d_0 = -g_0$ and $k = 0$
2.	Test a criterion for stopping the iterations. If this test is satisfied, then stop; otherwise continue with step 3
3.	Using the Wolfe line search conditions, determine the stepsize α_k
	Compute $x_{k+1} = x_k + \alpha_k d_k$, f_{k+1}, g_{k+1} and $y_k = g_{k+1} - g_k$
4.	Compute the conjugate gradient parameter β_k
5.	Compute the search direction $d_{k+1} = -g_{k+1} + \beta_k d_k$
6.	Restart criterion. If $\left|g_{k+1}^T g_k\right| > 0.2\|g_{k+1}\|^2$ then set $d_{k+1} = -g_{k+1}$
7.	Set $k = k + 1$ and continue with step 2 ♦

Observe that in Algorithm 4.1, in step 6, a restart criterion is used. This is known as the Powell restart criterion, introduced by Powell (1977). Crowder and Wolfe (1969) proved that the standard conjugate gradient method without restart is at most linearly convergent. Yuan (1993) showed that the convergence rate of the conjugate gradient method without restart is exactly linear for uniformly convex quadratics. Cohen (1972) and McCormick and Ritter (1974) proved that the convergence rate of the conjugate gradient method may be improved from linear to n-step quadratic if the method is restarted with the negative gradient. Powell (1977) reported some numerical results with conjugate gradient methods, showing that the immediate reduction in the objective function with restart is usually smaller than the reduction without restart. Therefore, the current implementations of the conjugate gradient methods use this restart criterion. Some restart procedures for conjugate gradient methods can be found in Dai, Liao, and Li (2004).

4.1 Conjugate Gradient Methods with $\|g_{k+1}\|^2$ in the Numerator of β_k

In this section, let us discuss the methods: Fletcher-Reeves (FR), conjugate descent by Fletcher (CD), and Dai–Yuan (DY). As a general characterization of these methods versus some other choices for the update parameter β_k, we mention that the global convergence theorems only require the Lipschitz assumption, not the boundedness assumption.

Fletcher-Reeves method

This conjugate gradient method is defined by

$$x_{k+1} = x_k + \alpha_k d_k, \tag{4.1}$$

$$d_{k+1} = -g_{k+1} + \beta_k d_k, \tag{4.2}$$

where the conjugate gradient parameter is computed as

$$\beta_k^{FR} = \frac{\|g_{k+1}\|^2}{\|g_k\|^2}. \tag{4.3}$$

The first global convergence result for the FR method under the exact line search was given by Zoutendijk (1970). In other words, when the stepsize α_k is the exact solution of the problem $\min_{\alpha \geq 0} f(x_k + \alpha d_k)$, then the FR method is global convergent. As usual, let $S = \{x \in \mathbb{R}^n : f(x) \leq f(x_0)\}$ be the level set.

Theorem 4.1 *Suppose that $f : \mathbb{R}^n \to \mathbb{R}$ is continuously differentiable on the level set S and consider that the FR method is implemented with the exact line search. Then, the sequence $\{x_k\}$ generated by the algorithm has at least an accumulation point which is a stationary point of function f, that is:*

(i) If $\{x_k\}$ is a finite sequence, then x^* is a stationary point of f.
(ii) If $\{x_k\}$ is an infinite sequence, then this sequence has a limit point and any limit point is a stationary point.

Proof *(i)* If $\{x_k\}$ is a finite sequence, then from the condition of termination of iterations it follows that the final point x^* satisfies $\nabla f(x^*) = 0$, i.e., x^* is a stationary point of f.

(ii) If $\{x_k\}$ is an infinite sequence, then for any k, $\nabla f(x_k) \neq 0$. Since $d_{k+1} = -g_{k+1} + \beta_k d_k$ and the line search is exact, that is, $g_{k+1}^T d_k = 0$, it follows that

$$g_{k+1}^T d_{k+1} = -\|g_{k+1}\|^2 + \beta_k g_{k+1}^T d_k = -\|g_{k+1}\|^2 < 0, \tag{4.4}$$

i.e., d_k is a descent direction, $\{f(x_k)\}$ is a monotone decreasing sequence and consequently, $\{x_k\} \subset S$. Hence, $\{x_k\}$ is bounded and has a limit. Let x^* be the limit of $\{x_k\}$. Then, there exists a subsequence $\{x_k\}_{K_1}$ convergent to x^* , where K_1 is the set of indices corresponding to the subsequence from $\{x_k\}$. Since $\{x_k\}_{K_1} \subset \{x_k\}$, it follows that $\{f(x_k)\}_{K_1} \subset \{f(x_k)\}$. From the continuity of f, we know that for $k \in K_1$, (see Appendix A)

$$f(x^*) = f(\lim_{k\to\infty} x_k) = \lim_{k\to\infty} f(x_k) = f^*. \tag{4.5}$$

Similarly, $\{x_{k+1}\}$ is a bounded sequence. Therefore, there exists a subsequence $\{x_{k+1}\}_{K_2}$ convergent to $\bar{x}^*$, where K_2 is the set of indices corresponding to the subsequence from $\{x_{k+1}\}$. In this case,

$$f(\bar{x}^*) = f(\lim_{k\to\infty} x_{k+1}) = \lim_{k\to\infty} f(x_{k+1}) = f^*. \tag{4.6}$$

Hence,

$$f(\bar{x}^*) = f(x^*) = f^*. \tag{4.7}$$

Now, let us prove by contradiction that $\nabla f(x^*) = 0$. Suppose that $\nabla f(x^*) \neq 0$. Then, for any α small enough, it follows that

$$f(x^* + \alpha d^*) < f(x^*). \tag{4.8}$$

Since for any $\alpha > 0$,

$$f(x_{k+1}) = f(x_k + \alpha_k d_k) \leq f(x_k + \alpha d_k),$$

then, for $k \in K_2$, at limit for $k \to \infty$ from (4.8), it results that

$$f(\bar{x}^*) \leq f(x^* + \alpha d^*) < f(x^*), \tag{4.9}$$

which contradicts (4.7). Therefore, this shows that $\nabla f(x^*) = 0$, i.e., x^* is a stationary point of f. ♦

Powell (1977) showed that the FR method with exact line search is susceptible of jamming, i.e., along the iterations the algorithm could take many short steps without any significant progress to the minimum. The modest performances of the FR method can be explained by this jamming phenomenon, as it is detailed by Nocedal and Wright (2006).

The first global convergence result of the FR method under the inexact line search was given by Al-Baali (1985). Using the strong Wolfe line search

$$f(x_k + \alpha_k d_k) \leq f(x_k) + \rho \alpha_k d_k^T g_k, \tag{4.10}$$

$$\left|\nabla f(x_k + \alpha_k d_k)^T d_k\right| \leq -\sigma d_k^T g_k, \tag{4.11}$$

with $\sigma < 1/2$, Al-Baali proved that FR method generates sufficient descent conditions.

Theorem 4.2 *Suppose that the conjugate gradient FR is implemented with the strong Wolfe line search* (4.10) *and* (4.11), *where* $0 < \sigma < 1/2$. *Then, the FR method generates descent directions* d_k *satisfying the following inequalities:*

$$-\frac{1}{1-\sigma} \leq \frac{g_k^T d_k}{\|g_k\|^2} \leq \frac{2\sigma - 1}{1-\sigma}, \tag{4.12}$$

for all $k = 0, 1, \ldots$

Proof At first, notice that the function $u(t) = (2t-1)/(1-t)$ on the interval [0,1/2] is monotonically increasing and $u(0) = -1$ and $u(1/2) = 0$. Therefore, since $\sigma \in (0, 1/2)$, it follows that

$$-1 < \frac{2\sigma - 1}{1-\sigma} < 0. \tag{4.13}$$

The proof is given by induction as follows. For $k = 0$, the middle term in (4.12) is -1, so by using (4.13), both inequalities in (4.12) are satisfied. Now, assume that (4.12) holds for some $k \geq 1$. From the definition of β_k^{FR} given in (4.3) and (4.2), it follows that

$$\frac{g_{k+1}^T d_{k+1}}{\|g_{k+1}\|^2} = -1 + \frac{g_{k+1}^T d_k}{\|g_k\|^2}. \tag{4.14}$$

From the second Wolfe condition (4.11) and (4.14),

$$-1 + \sigma \frac{g_k^T d_k}{\|g_k\|^2} \leq \frac{g_{k+1}^T d_{k+1}}{\|g_{k+1}\|^2} \leq -1 - \sigma \frac{g_k^T d_k}{\|g_k\|^2}.$$

Substituting for the term $g_k^T d_k / \|g_k\|^2$ from the left-hand side of the induction hypothesis (4.12), it follows that

$$-1 - \frac{\sigma}{1-\sigma} \leq \frac{g_{k+1}^T d_{k+1}}{\|g_{k+1}\|^2} \leq -1 + \frac{\sigma}{1-\sigma},$$

which shows that (4.12) holds for $k+1$ as well. ♦

The theorem shows that for the FR conjugate gradient method, the sufficient descent condition $g_k^T d_k \leq -c\|g_k\|^2$ is satisfied. Therefore, from the Zoutendijk condition, it follows the global convergence of the FR method. Besides, the bounds on $g_k^T d_k$ in (4.12) impose a limit on how fast $\|d_k\|$ can grow along the iterations when the gradients are not small. For $\sigma = 1/2$, d_k is a descent direction, however, the analysis did not establish the sufficient descent.

The most important conclusion of Theorem 4.2 is that from (4.12) it follows that there are two positive constants c_1 and c_2 so that

$$c_1 \frac{\|g_k\|}{\|d_k\|} \leq \cos\theta_k \leq c_2 \frac{\|g_k\|}{\|d_k\|}. \tag{4.15}$$

Therefore, for the FR method or any method with $|\beta_k| \leq \beta_k^{FR}$, it follows that $\cos\theta_k$ is proportional to $\|g_k\|/\|d\|_k$, (Gilbert & Nocedal, 1992). This result may be used to prove the convergence of any conjugate gradient method with $|\beta_k| \leq \beta_k^{FR}$.

Theorem 4.3 *Suppose that the Assumption CG holds. Consider any conjugate gradient method* (4.1) *and* (4.2) *where* $|\beta_k| \leq \beta_k^{FR}$ *and where the stepsize is determined by the strong Wolfe line search* (4.10) *and* (4.11) *with* $0 < \rho < \sigma < 1/2$. *Then,* $\liminf_{k\to\infty} \|g_k\| = 0$.

Proof From the strong Wolfe line search (4.11) and Theorem 4.2, it results that

$$\left|g_{k+1}^T d_k\right| \leq -\sigma g_k^T d_k \leq \frac{\sigma}{1-\sigma}\|g_k\|^2. \tag{4.16}$$

Now, from (4.2), since $|\beta_k| \leq \beta_k^{FR}$ and having in view (4.16), it results that

$$\begin{aligned}
\|d_{k+1}\|^2 &\leq \|g_{k+1}\|^2 + 2|\beta_k|\left|g_{k+1}^T d_k\right| + \beta_k^2\|d_k\|^2 \\
&\leq \|g_{k+1}\|^2 + 2|\beta_k|\frac{\sigma}{1-\sigma}\|g_k\|^2 + \beta_k^2\|d_k\|^2 \\
&\leq \left(\frac{1+\sigma}{1-\sigma}\right)\|g_{k+1}\|^2 + \beta_k^2\|d_k\|^2.
\end{aligned} \tag{4.17}$$

Defining $\bar{\sigma} = (1+\sigma)/(1-\sigma)$ and applying (4.17) repeatedly and using the condition $|\beta_k| \leq \beta_k^{FR}$, it follows that

$$\begin{aligned}
&\|d_{k+1}\|^2 \leq \bar{\sigma}\|g_{k+1}\|^2 + \beta_k^2\|d_k\|^2 \\
&\quad \leq \bar{\sigma}(\|g_{k+1}\|^2 + \beta_k^2\|g_k\|^2 + \beta_k^2\beta_{k-1}^2\|g_{k-1}\|^2 + \cdots + \beta_k^2\beta_{k-1}^2\cdots\beta_1^2\|g_1\|^2) \\
&\quad + \beta_k^2\beta_{k-1}^2\cdots\beta_0^2\|d_0\|^2 \\
&\quad \leq \bar{\sigma}\|g_{k+1}\|^4\sum_{i=1}^{k+1}\frac{1}{\|g_i\|^2} + \frac{\|g_{k+1}\|^4}{\|g_0\|^2} \leq \bar{\sigma}\|g_{k+1}\|^4\sum_{i=0}^{k+1}\frac{1}{\|g_i\|^2}.
\end{aligned} \tag{4.18}$$

As it is known, from the Assumption CG, there is a constant Γ so that $\|g(x)\| \leq \Gamma$ for all $x \in S$. Now, let us proceed in a canonical manner to assume that $\|g_k\| \geq \gamma > 0$ for all k. From (4.18), this implies that

$$\|d_{k+1}\|^2 \leq \frac{\bar{\sigma}\Gamma^4}{\gamma^2}(k+2). \tag{4.19}$$

From the left inequality in (4.15) and the Zoutendijk condition, it follows that

$$\sum_{k \geq 0} \frac{\|g_k\|^4}{\|d_k\|^2} < \infty. \tag{4.20}$$

If the gradients are bounded away from zero, then (4.20) implies that

$$\sum_{k \geq 0} \frac{1}{\|d_k\|^2} < \infty.$$

But this is in contradiction with (4.19), thus proving the theorem. ♦

Remark 4.1 Intensive numerical experiments with the FR method with inexact line search showed that sometimes it slows down away from the solution, i.e., the steps become very small and this continues for a very large number of iterations until the method is restarted. The argument for this behavior of the FR method is as follows. Suppose that at iteration k a poor search direction is generated in the sense that it makes an angle of nearly 90° with $-g_k$, that is, $\cos\theta_k \approx 0$. By multiplying both sides of (4.12) by $\|g_k\|/\|d_k\|$ and using the definition (3.28) of $\cos\theta_k$, it follows that

$$\frac{1-2\sigma}{1-\sigma}\frac{\|g_k\|}{\|d_k\|} \leq \cos\theta_k \leq \frac{1}{1-\sigma}\frac{\|g_k\|}{\|d_k\|},$$

for all $k = 0, 1, \ldots$. From these inequalities, we see that $\cos\theta_k \approx 0$ if and only if $\|g_k\| \ll \|d_k\|$. Since d_k is almost orthogonal to the gradient, it follows that the step from x_k to x_{k+1} is small, that is, $x_{k+1} \approx x_k$. Therefore, $g_{k+1} \approx g_k$. Hence, $\beta_k^{FR} \approx 1$. By (4.15), it follows that $\|g_{k+1}\| \approx \|g_k\| \ll \|d_k\|$. Since $\beta_k^{FR} \approx 1$ from (4.2), we see that $\|d_{k+1}\| \approx \|d_k\| \gg \|g_{k+1}\|$, which by (4.15) implies that $\cos\theta_{k+1} \approx 0$. This situation can continue all over again and this explains the inefficiency of the FR method. ♦

The main difficulty in proving the global convergence of a conjugate gradient algorithm is to show that the search direction d_k is descent under the mild assumption on the line search conditions. If the line search is exact, i.e., $g_{k+1}^T d_k = 0$, then, d_{k+1} is exactly the steepest descent since $g_{k+1}^T d_{k+1} = -\|g_{k+1}\|^2$. Using this property and supposing that the

Assumption CG holds, Powell showed that the FR algorithm is globally convergent, i.e., $\liminf_{k\to\infty}\|g_k\| = 0$. Al-Baali extended this result, showing that the descent property holds for all k if α_k is determined by the strong Wolfe line search.

The following theorem shows that under the inexact line search, the search directions d_k generated by the FR method satisfy the descent condition $g_k^T d_k < 0$.

Theorem 4.4 *If for all k the stepsize α_k is determined by the strong Wolfe line search* (4.10) *and* (4.11), *where $\sigma \in (0, 1/2)$, then for the FR method the following inequalities hold*

$$-\sum_{j=0}^{k} \sigma^j \le \frac{g_k^T d_k}{\|g_k\|^2} \le -2 + \sum_{j=0}^{k} \sigma^j \tag{4.21}$$

for any k. As soon as $g_k \neq 0$ for all k, the descent property of d_k is satisfied, i.e.,

$$g_k^T d_k < 0. \tag{4.22}$$

Proof The theorem is proved by induction. For $k = 0$, $d_0 = -g_0$, $\sigma^0 = 1$, and therefore (4.21) and (4.22) are true. Now suppose that (4.21) and (4.22) are true for any $k \ge 0$. From $d_{k+1} = -g_{k+1} + \beta_k d_k$ and $\beta_k = g_{k+1}^T g_{k+1}/g_k^T g_k$ (Fletcher-Reeves updating formula), it follows that

$$\frac{g_{k+1}^T d_{k+1}}{\|g_{k+1}\|^2} = -1 + \frac{g_{k+1}^T d_k}{\|g_k\|^2}. \tag{4.23}$$

From the second strong Wolfe condition $|g_{k+1}^T d_k| \le -\sigma g_k^T d_k$ and (4.22),

$$-1 + \sigma\frac{g_k^T d_k}{\|g_k\|^2} \le \frac{g_{k+1}^T d_{k+1}}{\|g_{k+1}\|^2} \le -1 - \sigma\frac{g_k^T d_k}{\|g_k\|^2}. \tag{4.24}$$

From (4.21), it follows that

$$-\sum_{j=0}^{k+1} \sigma^j = -1 - \sigma\sum_{j=0}^{k} \sigma^j \le \frac{g_{k+1}^T d_{k+1}}{\|g_{k+1}\|^2} \le -2 + \sum_{j=0}^{k+1} \sigma^j,$$

showing that (4.21) holds for $k+1$. On the other hand, since

$$\frac{g_{k+1}^T d_{k+1}}{\|g_{k+1}\|^2} \le -2 + \sum_{j=0}^{k+1} \sigma^j \tag{4.25}$$

and

$$\sum_{j=0}^{k+1} \sigma^j < \sum_{j=0}^{\infty} \sigma^j = \frac{1}{1-\sigma}, \tag{4.26}$$

where $\sigma \in [0, 1/2)$, form $1 - \sigma > 1/2$, it results that

$$-2 + \sum_{j=0}^{k+1} \sigma^j < 0.$$

Therefore, from (4.22), $g_{k+1}^T d_{k+1} < 0$ is obtained, proving the theorem. ♦

With this result, the global convergence of the FR method with inexact line search can be proved as in the following theorem (Al-Baali, 1985).

Theorem 4.5 *Assume that the function $f : \mathbb{R}^n \to \mathbb{R}$ is twice continuously differentiable and the level set S is bounded. Suppose that α_k is determined by the strong Wolfe conditions (4.10) and (4.11), where $\rho < \sigma < 1/2$. Then, the sequence $\{x_k\}$ generated by the Fletcher-Reeves method is global convergent, i.e.,*

$$\liminf_{k\to\infty} \|g_k\| = 0. \tag{4.27}$$

Proof From the second strong Wolfe condition (4.11) and from the relations (4.21) and (4.26), it follows that

$$\left|g_{k+1}^T d_k\right| \leq -\sigma g_k^T d_k \leq \sigma \|g_k\|^2 \sum_{j=0}^{k} \sigma^j \leq \frac{\sigma}{1-\sigma} \|g_k\|^2.$$

From the relations $d_{k+1} = -g_{k+1} + \beta_k d_k$ and $\beta_k = g_{k+1}^T g_{k+1} / g_k^T g_k$ (Fletcher-Reeves formula), it results that

$$\begin{aligned}
\|d_{k+1}\|^2 &= \|g_{k+1}\|^2 - 2\beta_k^{FR} g_{k+1}^T d_k + (\beta_k^{FR})^2 \|d_k\|^2 \\
&\leq \|g_{k+1}\|^2 + \frac{2\sigma}{1-\sigma} \|g_{k+1}\|^2 + (\beta_k^{FR})^2 \|d_k\|^2 \\
&= \left(\frac{1+\sigma}{1-\sigma}\right) \|g_{k+1}\|^2 + (\beta_k^{FR})^2 \|d_k\|^2.
\end{aligned}$$

But, from the Fletcher-Reeves updating Formula (4.3), it is clear that

$$(\beta_k^{FR})^2 (\beta_{k-1}^{FR})^2 \ldots (\beta_{k-i}^{FR})^2 = \frac{\|g_{k+1}\|^4}{\|g_{k-i}\|^4}.$$

Therefore, by recurrence, the following is obtained:

$$\|d_{k+1}\|^2 \le \left(\frac{1+\sigma}{1-\sigma}\right)\|g_{k+1}\|^4\left(\sum_{j=0}^{k+1}\|g_j\|^{-2}\right). \tag{4.28}$$

Now, let us prove (4.27) by contradiction. Suppose that (4.27) is not true. Then, there exists a positive constant $\varepsilon > 0$ so that for all k sufficiently large,

$$\|g_k\| \ge \varepsilon > 0. \tag{4.29}$$

Since on the level set S, the gradient g_k is upper bounded, from (4.28), it follows that

$$\|d_k\|^2 \le c_1 k, \tag{4.30}$$

where c_1 is a positive constant. Hence, from (4.21) and (4.26),

$$\begin{aligned}\cos\theta_k &= -\frac{g_k^T d_k}{\|g_k\|\|d_k\|} \ge \left(2-\sum_{j=0}^{k}\sigma^j\right)\frac{\|g_k\|}{\|d_k\|} \\ &\ge \left(\frac{1-2\sigma}{1-\sigma}\right)\frac{\|g_k\|}{\|d_k\|}.\end{aligned} \tag{4.31}$$

Since $\sigma < 1/2$, introducing (4.29) and (4.30) in (4.31) it results that

$$\sum_k \cos^2\theta_k \ge \left(\frac{1-2\sigma}{1-\sigma}\right)^2\sum_k \frac{\|g_k\|^2}{\|d_k\|^2} \ge c_2\sum_k\frac{1}{k}, \tag{4.32}$$

where c_2 is a positive constant. Therefore, the series $\sum_k \cos^2\theta_k$ is divergent. Let M be an upper bound of $\|\nabla^2 f(x)\|$ on the level set S. Then,

$$g_{k+1}^T d_k = (g_k + \alpha_k\nabla^2 f(x_k)d_k)^T d_k \le g_k^T d_k + \alpha_k M\|d_k\|^2.$$

From the second strong Wolfe condition, (4.11) written as

$$\sigma g_k^T d_k \le g_{k+1}^T d_k \le -\sigma g_k^T d_k,$$

it follows that

$$\alpha_k \ge -\frac{1-\sigma}{M\|d_k\|^2} g_k^T d_k. \tag{4.33}$$

Introducing this value of α_k from (4.33) in the first Wolfe condition (4.10), it results that

$$f(x_{k+1}) \le f(x_k) - \frac{(1-\sigma)\rho}{M}\left(\frac{g_k^T d_k}{\|d_k\|}\right)^2 = f(x_k) - c_3\|g_k\|^2\cos^2\theta_k,$$

where $c_3 = \frac{(1-\sigma)\rho}{M} > 0$. Since f is lower bounded, it follows that the series $\sum_k \|g_k\|^2\cos^2\theta_2$ is convergent. Hence, from (4.29), the series $\sum_k \cos^2\theta_2$ is convergent. But this contradicts (4.32), proving the theorem. ♦

The global convergence proof of Al-Baali was extended to the case $\sigma = 1/2$ by Liu, Han, and Yin (1995). This result was further extended by Dai and Yuan (1996a), showing that in consecutive FR iterations, at least one iteration satisfies the sufficient descent condition. In other words,

$$\max\left\{\frac{-g_k^T d_k}{\|g_k\|^2}, \frac{-g_{k-1}^T d_{k-1}}{\|g_{k-1}\|^2}\right\} \ge \frac{1}{2}.$$

In the same avenue of research, Dai and Yuan (2000) showed that the FR method with strong Wolfe line search may not yield a direction of descent when $\sigma > 1/2$, even for the functions $f(x) = \lambda\|x\|^2$, where $\lambda > 0$ is a constant. Therefore, the constraints $\sigma \le 1/2$ must be imposed in order to ensure descent. In the current implementation of the Wolfe line search, good performances are obtained when σ is chosen to be close to 1. Hence, the constraint $\sigma \le 1/2$ imposed to ensure the descent character of the search direction represents a significant restriction in the choice of the line search parameters.

When $\sigma > 1/2$ and $g_k^T d_k > 0$, Dai and Yuan (1996b) showed that $-d_k$ can be used as a search direction; if $g_k^T d_k = 0$, then the line search can be skipped by setting $x_{k+1} = x_k$. If the gradient is bounded, i.e., if there exists a constant $\Gamma > 0$ so that $\|g_k\| \le \Gamma$, then under the Lipschitz assumption, the FR method with standard Wolfe line search and with the above special adjustments when $g_k^T d_k \ge 0$ is globally convergent.

In Dai and Yuan (1996a), the FR method with strong Wolfe line search is relaxed to the generalized Wolfe line search (curvature condition) (1.20). The main result is that the global convergence is obtained when $\sigma_1 + \sigma_2 \le 1$. For a strong Wolfe line search, $\sigma_1 = \sigma_2 = \sigma$, case in which the constraint $\sigma_1 + \sigma_2 \le 1$ implies $\sigma \le 1/2$. Therefore, the condition $\sigma_1 + \sigma_2 \le 1$ is weaker than the strong Wolfe constraint $\sigma \le 1/2$. A possibility is to take σ_1 close to 1 by taking σ_2 close to 0. Some details are given by Hager and Zhang (2006b).

CD Method

The conjugate descent (CD) method elaborated by Fletcher (1987) is very close to the FR method. The conjugate gradient parameter β_k is computed as

$$\beta_k^{CD} = \frac{\|g_{k+1}\|^2}{-d_k^T g_k}. \tag{4.34}$$

Under the exact line search, $\beta_k^{CD} = \beta_k^{FR}$. The difference between CD and FR is that in CD with strong Wolfe line search the sufficient descent condition $g_k^T d_k \leq -c\|g_k\|^2$ holds. In this case, the constraint $\sigma \leq 1/2$ that arose in the FR method is not necessary for the CD method. For a line search that satisfies the generalized Wolfe conditions $\sigma_1 d_k^T g_k \leq d_k^T g_{k+1} \leq -\sigma_2 d_k^T g_k$, with $\sigma_1 < 1$ and $\sigma_2 = 0$, it can be shown that $0 \leq \beta_k^{CD} \leq \beta_k^{FR}$. Therefore, from the analysis given by Al-Baali (1985) or by Theorem 3.4, the global convergence is achieved. On the other hand, if $\sigma_1 \geq 1$, or $\sigma_2 > 0$, then Dai and Yuan (1996c) constructed numerical examples where $\|d_k\|^2$ increases exponentially, and therefore the CD method converges to a point where the gradient does not vanish. In particular, the CD method may not converge to a stationary point under the strong Wolfe line search. Details are given by Hager and Zhang (2006b).

Dai–Yuan Method

In the DY method, elaborated by Dai and Yuan (1999), the conjugate gradient parameter β_k is computed as

$$\beta_k^{DY} = \frac{\|g_{k+1}\|^2}{d_k^T y_k}. \tag{4.35}$$

Under the standard Wolfe line search, the DY method always generates descent directions. Besides, when the Lipschitz assumption holds, then the DY is globally convergent.

To get (4.35), let us suppose that the current search direction d_k is descent, i.e., $d_k^T g_k < 0$ and we are interested in finding a β_k for which the new search direction d_{k+1} is also descent. This requires that

$$-\|g_{k+1}\|^2 + \beta_k g_{k+1}^T d_k < 0.$$

Assume that $\beta_k > 0$ and denote $\tau_k = \|g_{k+1}\|^2 / \beta_k$. Then, the above inequality is equivalent to $\tau_k > g_{k+1}^T d_k$. Therefore, we can consider $\tau_k = d_k^T y_k$, which gives the DY updating Formula (4.35).

A new representation for β_k^{DY} can immediately be obtained. From (4.2) and (4.35), the following can be written

$$g_{k+1}^T d_{k+1} = \frac{\|g_{k+1}\|^2}{d_k^T y_k} g_k^T d_k = \beta_k^{DY} g_k^T d_k. \tag{4.36}$$

Therefore, β_k^{DY},

$$\beta_k^{DY} = \frac{g_{k+1}^T d_{k+1}}{g_k^T d_k}. \tag{4.37}$$

Observe that (4.35) is well defined because the Wolfe line search implies that $d_k^T y_k > 0$. If the line search in this method is exact, then the DY method is the same as the FR method. The convergence of this method is given by the following theorem.

Theorem 4.6 *Suppose that the initial point x_0 satisfies the Assumption CG and let $\{x_k\}$ be the sequence generated by the general Algorithm (4.1) and (4.2), where β_k is computed as in (4.37). Then, the algorithm either terminates at a stationary point of f, or converges in the sense*

$$\liminf_{k \to \infty} \|g_k\| = 0. \tag{4.38}$$

Proof If the algorithm does not terminate after a finite number of iterations, then $\|g_k\| > 0$ for all k. Firstly, let us show that the search directions are descent, i.e.,

$$g_k^T d_k < 0 \tag{4.39}$$

for all k. For $k = 1$, the above inequality (4.39) is satisfied. Now let us prove it for all $k > 1$ by induction. Assume that (4.39) holds for k. From the second Wolfe line search condition,

$$\nabla f(x_k + \alpha_k d_k)^T d_k \geq \sigma d_k^T g_k, \tag{4.40}$$

it follows that

$$d_k^T y_k \geq (\sigma - 1) d_k^T g_k > 0. \tag{4.41}$$

From (4.36), it follows that (4.39) holds for $k + 1$. Hence, the search direction given by β_k^{DY} is descent for all values of k.

Now, (4.2) can be written as

$$d_{k+1} + g_{k+1} = \beta_k d_k. \tag{4.42}$$

Squaring both sides of (4.42), it follows that

$$\|d_{k+1}\|^2 = \beta_k^2 \|d_k\|^2 - 2g_{k+1}^T d_{k+1} - \|g_{k+1}\|^2. \tag{4.43}$$

Dividing both sides of (4.43) by $(g_{k+1}^T d_{k+1})^2$ and applying (4.37),

$$\begin{aligned}
\frac{\|d_{k+1}\|^2}{(g_{k+1}^T d_{k+1})^2} &= \frac{\|d_k\|^2}{(g_k^T d_k)^2} - \frac{2}{g_{k+1}^T d_{k+1}} - \frac{\|g_{k+1}\|^2}{(g_{k+1}^T d_{k+1})^2} \\
&= \frac{\|d_k\|^2}{(g_k^T d_k)^2} - \left(\frac{1}{\|g_{k+1}\|} + \frac{\|g_{k+1}\|}{g_{k+1}^T d_{k+1}}\right)^2 + \frac{1}{\|g_{k+1}\|^2} \\
&\le \frac{\|d_k\|^2}{(g_k^T d_k)^2} + \frac{1}{\|g_{k+1}\|^2}.
\end{aligned} \tag{4.44}$$

But, $\|d_0\|^2/(g_0^T d_0) = 1/\|g_0\|^2$, then (4.44) shows that

$$\frac{\|d_k\|^2}{(g_k^T d_k)^2} \le \sum_{i=0}^{k} \frac{1}{\|g_i\|^2}, \tag{4.45}$$

for all k. If the theorem is not true, then there exists a positive constant $c > 0$ so that

$$\|g_k\| \ge c, \tag{4.46}$$

for all k. Therefore, from (4.45) and (4.46), it follows that

$$\frac{\|d_k\|^2}{(g_k^T d_k)^2} \le \frac{k}{c^2},$$

which implies that

$$\sum_{k \ge 0} \frac{(g_k^T d_k)^2}{\|d_k\|^2} = \infty. \tag{4.47}$$

Observe that the relation (4.47) contradicts the Zoutendijk condition (3.34). This contradiction proves the theorem. ♦

A remarkable property of the DY method, called the *self-adjusting property*, which is independent of the line search and the convexity of the objective function, was established by Dai (2001). The self-adjusting property says that: *if the search direction* d_k *is descent, i.e.,* $g_k^T d_k < 0$ *for all* k, *but* $\lim_{k\to\infty}\|g_k\| \neq 0$, *then the sufficient descent condition* $g_k^T d_k \leq -c\|g_k\|^2$ *must hold for most of the iterations.*

Assume that $g_k \neq 0$, for all $k \geq 0$, as otherwise a stationary point has been found. Let us define the following quantities:

$$q_k = \frac{\|d_k\|^2}{(g_k^T d_k)^2} \text{ and } r_k = -\frac{g_k^T d_k}{\|g_k\|^2}. \tag{4.48}$$

Observe that the quantity q_k represents the size of d_k, while r_k shows the descent degree of d_k. If $r_k > 0$, then d_k is a descent direction. Moreover, if $r_k \geq c$, where $c > 0$ is a constant, then a sufficient descent condition $g_k^T d_k \leq -c\|g_k\|^2$ holds. As it has already been seen, the DY method (4.35) is equivalent to (4.37). From (4.36) observe that since $g_{k+1}^T d_{k+1} < 0$, it follows that $g_k^T d_k < 0$ if $d_k^T y_k > 0$, and since $d_0^T g_0 = -\|g_0\|^2 < 0$, a direct consequence of (4.37) is that the DY method produces descent directions at every iteration.

From the proof of Theorem 4.6, it follows that

$$\frac{\|d_{k+1}\|^2}{(g_{k+1}^T d_{k+1})^2} = \frac{\|d_k\|^2}{(g_k^T d_k)^2} - \frac{2}{g_{k+1}^T d_{k+1}} - \frac{\|g_{k+1}\|^2}{(g_{k+1}^T d_{k+1})^2}. \tag{4.49}$$

Having in view the definitions of q_k and r_k introduced in (4.48), the following relation is obtained

$$q_{k+1} = q_k + \frac{2}{r_{k+1}}\frac{1}{\|g_{k+1}\|^2} - \frac{1}{r_{k+1}^2}\frac{1}{\|g_{k+1}\|^2}. \tag{4.50}$$

From (4.50) observe that if d_{k+1} is a descent direction, then the second term on the right side of (4.50) increases the value of q_k, whereas the third term decreases the value of q_k. Considering these two terms together, it follows that q_k increases if and only if $r_{k+1} \geq 1/2$. If r_{k+1} is close to zero, then q_k will be significantly reduced, since the order of $1/r_{k+1}$ in the second term is only one but its order in the third term is two. This remark, together with the fact that $q_{k+1} \geq 0$ for all k, implies that in case q_k is very small, then r_{k+1} must be relatively large. This small observation offers the possibility to get a fine estimation of the lower bound of r_k. For this, the following assumption must be introduced: there exist the positive constants γ and Γ, so that

$$0 < \gamma \leq \|g_k\| \leq \Gamma, \tag{4.51}$$

for all $k \geq 0$. The following theorem can be proved (see Dai (2001)).

Theorem 4.7 *Consider the method (4.1) and (4.2) where* $\beta_k = \beta_k^{DY}$ *and* d_k *is a descent direction. If (4.51) holds, then there exist the positive constants* δ_1, δ_2 *and* δ_3, *so that*

$$-g_k^T d_k \geq \frac{\delta_1}{\sqrt{k}}, \tag{4.52}$$

$$\|d_k\|^2 \geq \frac{\delta_2}{k}, \tag{4.53}$$

$$r_k \geq \frac{\delta_3}{\sqrt{k}} \tag{4.54}$$

hold for any $k \geq 0$.

Proof Summing (4.50) over the iterates and having in view that $d_0 = -g_0$, it results that

$$q_k = \sum_{i=0}^{k} \frac{1}{\|g_i\|^2}\left(\frac{2}{r_i} - \frac{1}{r_i^2}\right). \tag{4.55}$$

But, since $q_k \geq 0$, from (4.55) it follows that

$$\frac{1}{\|g_k\|^2}\left(-\frac{2}{r_k} + \frac{1}{r_k^2}\right) \leq \sum_{i=1}^{k-1} \frac{1}{\|g_i\|^2}\left(\frac{2}{r_i} - \frac{1}{r_i^2}\right). \tag{4.56}$$

This, together with (4.51) and the fact that

$$\frac{2}{r_i} - \frac{1}{r_i^2} \leq 1 \tag{4.57}$$

imply that

$$\frac{1}{r_k^2} - \frac{2}{r_k} - \frac{\Gamma^2}{\gamma^2}(k-1) \leq 0. \tag{4.58}$$

Since $r_k > 0$ and $\gamma/\Gamma < 1$, (4.58) shows that

$$\frac{1}{r_k} \leq 1 + \sqrt{1 + \frac{\Gamma^2}{\gamma^2}(k-1)} = 1 + \frac{\Gamma}{\gamma}\sqrt{\frac{\gamma^2}{\Gamma^2} + (k-1)} \leq 1 + \frac{\Gamma}{\gamma}\sqrt{k} \leq \frac{2\Gamma}{\gamma}\sqrt{k}. \tag{4.59}$$

Therefore, (4.54) holds with $\delta_3 = \gamma/2\Gamma$. But, $-g_k^T d_k = \|g_k\|^2 r_k$ and $\|d_k\| \geq \|g_k\| r_k$. Hence, from (4.54) and (4.51) it follows that (4.52) and (4.53) hold with $\delta_1 = \delta_3\gamma^2$ and $\delta_2 = \delta_3^2\gamma^2$, respectively, which completes the proof of the theorem. ♦

Observe that (4.54) does not imply that the sufficient descent condition holds. However, under the same assumptions, Dai (2001) proved that the sufficient descent condition must hold for most of the iterations.

Theorem 4.8 *Consider the method (4.1) and (4.2) where $\beta_k = \beta_k^{DY}$. If the DY method is implemented with a line search for which the corresponding search directions are descent and if there are the positive constants γ and Γ so that $\gamma \le \|g_k\| \le \Gamma$ for all $k \ge 0$, then for any $p \in (0,1)$ there exist the constants $\delta_4, \delta_5, \delta_6 > 0$, so that, for any k, the relations*

$$-g_i^T d_i \ge \delta_4, \tag{4.60}$$

$$\|d_i\|^2 \ge \delta_5, \tag{4.61}$$

$$r_i \ge \delta_6 \tag{4.62}$$

hold for at least $\lfloor pk \rfloor$ indices $i \in [0,k]$, where $\lfloor r \rfloor$ represents the largest integer smaller or equal to r.

Proof For any $p \in (0,1)$, $\delta_6 > 0$ is chosen so small, so that

$$\bar{\delta} \equiv \frac{1}{\delta_6^2} - \frac{2}{\gamma\delta_6} \ge \frac{p\Gamma^2}{(1-p)\gamma^2}. \tag{4.63}$$

For this δ_6 and for any k, let us define the set of indices

$$I_k = \{i \in [1,k] \ : \ r_i \ge \delta_6\} \tag{4.64}$$

and denote $|I_k|$ to be the number of elements in this set. From (4.50), (4.51) and since $q_k \ge 0$, it follows that

$$\sum_{i \in [1,k]-I_k} \left(-\frac{2}{r_i} + \frac{1}{r_i^2}\right) \le \frac{\Gamma^2}{\gamma^2} \sum_{i \in I_k} \left(\frac{2}{r_i} - \frac{1}{r_i^2}\right). \tag{4.65}$$

Therefore, from (4.57) and the definition of I_k, it results that

$$\bar{\delta}(k - |I_k|) \le \frac{\Gamma^2}{\gamma^2} |I_k|, \tag{4.66}$$

where $\bar{\delta}$ is defined in (4.63). Now, (4.66) and (4.63) imply that

$$|I_k| \ge \frac{\bar{\delta}\gamma^2}{\bar{\delta}\gamma^2 + \Gamma^2} k \ge pk \ge \lfloor pk \rfloor. \tag{4.67}$$

Therefore, for any $p \in (0, 1)$, if $\delta_6 > 0$ is chosen to satisfy (4.63), $\delta_4 = \delta_6 \gamma^2$ and $\delta_5 = \delta_6^2 \gamma^2$, then, from (4.67), (4.51) and $-g_k^T d_k = \|g_k\|^2 r_k$ and $\|d_k\| \geq \|g_k\| r_k$, the conclusion of the theorem is obtained. ♦

The theorem is important. It shows that independent of the line search and the convexity of the objective function, the Dai–Yuan conjugate gradient method has the self-adjusting property, i.e., it generates search directions which satisfy the sufficient descent condition *for most iterations.*

Remark 4.2 It is worth emphasizing that in the context of the quasi-Newton BFGS method, supposing that f is uniformly convex, then Byrd and Nocedal (1989) proved that for any $p \in (0, 1)$ there exists some positive constant c, so that for any $k \geq 0$, the relation:

$$\cos \theta_i = \frac{-g_i^T d_i}{\|g_i\| \, \|d_i\|} \geq c$$

holds for at least $\lfloor pk \rfloor$ values of $i \in [1, k]$. The main difference between these two above results is that Theorem 4.8 does not need to assume the uniform convexity of the objective function f. Besides, Theorem 4.8 refers to the sufficient descent condition, not to the angle between $-g_k$ and d_k. ♦

Dai and Yuan established a convergence result applicable to any method for which β_k can be expressed as a ratio

$$\beta_k = \frac{\Phi_{k+1}}{\Phi_k}. \tag{4.68}$$

The FR method corresponds to the choice $\Phi_k = \|g_k\|^2$. Similarly, from (4.37), the DY method has the form (4.68), where $\Phi_k = g_k^T d_k$. This structure of β_k^{FR} and β_k^{DY} gives the possibility to prove some important properties of these methods. The idea is that these conjugate gradient methods, despite their diversity, can be analyzed in a unified way. Dai and Yuan (2001b, 2003) established the following result.

Theorem 4.9 *Consider any iterative method* (4.1) *and* (4.2)*, where* β_k *is selected as in* (4.68)*,* d_k *satisfies the descent condition* $g_k^T d_k < 0$*, and the Lipschitz continuity holds. If the Zoutendijk condition holds and if*

$$\sum_{k=0}^{\infty} \frac{(g_k^T d_k)^2}{\Phi_k^2} = \infty \quad or \quad \sum_{k=0}^{\infty} \frac{\|g_k\|^2}{\Phi_k^2} = \infty \quad or \quad \sum_{k=1}^{\infty} \prod_{i=1}^{k} \beta_i^{-2} = \infty,$$

then the iterates are global convergent. ♦

A corollary of this theorem is that the DY method is globally convergent when it is implemented with the standard Wolfe line search, since

$$\sum_{k=0}^{N} \frac{(g_k^T d_k)^2}{\Phi_k^2} = N \text{ when } \Phi_k = g_k^T d_k.$$

Similarly, the FR method is globally convergent when it is implemented with the strong Wolfe line search with $\sigma \leq 1/2$, since, in this case

$$\sum_{k=0}^{N} \frac{\|g_k\|^2}{\Phi_k^2} = N \text{ when } \Phi_k = \|g_k\|^2.$$

Dai and Yuan (2003) proved that under the Assumptions CG, if for all k the search direction d_k is descent and α_k satisfies the Wolfe line search conditions and if

$$\sum_{k \geq 0} \frac{\|g_k\|^2}{\Phi_k^2} = \infty,$$

then $\liminf_{k\to\infty} \|g_k\| = 0$.

4.2 Conjugate Gradient Methods with $g_{k+1}^T y_k$ in the Numerator of β_k

In this section, let us consider the conjugate gradient methods: Hestenes–Stiefel (HS), Polak–Ribière–Polyak (PRP), and Liu–Storey (LS). Even if for the methods with $\|g_{k+1}\|^2$ in the numerator of β_k a strong convergence theory has been developed, these methods have modest performances, mainly because of the jamming phenomenon. Namely, these methods begin to take small steps without making significant progress to the minimum. On the other hand, HS, PRP, and LS methods, having $g_{k+1}^T y_k$ in the numerator of β_k, possess a built-in restart feature that addresses the jamming. When the step $s_k = x_{k+1} - x_k$ is small (i.e., close to the minimum), then the factor $y_k = g_{k+1} - g_k$ in the numerator of β_k tends to zero. Hence, β_k becomes small and the new search direction d_{k+1} is essentially the steepest descent direction $-g_{k+1}$. This property of the PRP method is important for its analysis. Such a method that has this property is said to have Property (*). These methods automatically adjust β_k in order to avoid jamming. Therefore, the numerical performance of these methods is better than the numerical performance of the methods with $\|g_{k+1}\|^2$ in the numerator of β_k .

Polak–Ribière–Polyak method

The Polak–Ribière–Polyak conjugate gradient method is defined by (4.1) and (4.2), where the conjugate gradient parameter is computed as

$$\beta_k^{PRP} = \frac{g_{k+1}^T y_k}{g_k^T g_k}. \tag{4.69}$$

If f is strongly convex and the line search is exact, then Polak and Ribière (1969) and Polyak (1969) established the global convergence of the PRP method. Powell (1977) proved that for a general nonlinear function f, if (i) the stepsize $s_k = x_{k+1} - x_k$ tends to zero, (ii) the line search is exact, and (iii) Lipschitz continuity holds, then the PRP method is globally convergent. On the other hand, Powell (1984a), in a laborious paper, constructed a counter-example with three variables and showed that the PRP method may cycle infinitely without approaching any solution. Therefore, the assumption that the stepsize tends to zero is needed for convergence.

Later on, under the assumption that the search direction is a descent direction, Yuan (1993) established the global convergence of the PRP method for strongly convex objective functions and the Wolfe line search. However, Dai (1997), in his Ph.D. Thesis, presented an example which showed that even when the objective function is strongly convex and $\sigma \in (0,1)$ is sufficiently small, then the PRP method may still fail by generating an ascent search direction. Dai, Han, Liu, Sun, Yin, and Yuan (1999) constructed an example showing that the boundedness of the level set is necessary for the convergence of the PRP method even if the line search is exact. Therefore, the convergence of the PRP method is not certain. However, this method proved to be one of the most efficient for solving large-scale unconstrained optimization problems.

In order to prove the convergence of the PRP method, we need a technical result presented in the following proposition.

Proposition 4.1 *Let $\nabla f(x)$ be uniformly continuous on the level set S. Consider the angle θ_k between d_k and $-\nabla f(x_k)$, where d_k is a descent direction. If*

$$\theta_k \le \frac{\pi}{2} - \mu, \text{for such } \mu > 0, \tag{4.70}$$

then $\nabla f(x_k) = 0$ for certain k; or $f(x_k) \to -\infty$; or $\nabla f(x_k) \to 0$.

Proof Suppose that for all k, $\nabla f(x_k) \neq 0$ and that $f(x_k)$ is lower bounded. Since $\{f(x_k)\}$ is monotone decreasing, this sequence has a limit. Hence,

$$f(x_k) - f(x_{k+1}) \to 0. \tag{4.71}$$

Suppose, by contradiction that $\nabla f(x_k) \to 0$ is not true. Then, there exists an $\varepsilon > 0$ so that $\|\nabla f(x_k)\| \ge \varepsilon$. Therefore,

$$-\frac{\nabla f(x_k)^T d_k}{\|d_k\|} = \|\nabla f(x_k)\| \cos\theta_k \geq \varepsilon \sin\mu \equiv \varepsilon_1 \tag{4.72}$$

Observe that

$$\begin{aligned} f(x_k + \alpha d_k) &= f(x_k) + \alpha \nabla f(\xi_k)^T d_k \\ &= f(x_k) + \alpha \nabla f(x_k)^T d_k + \alpha[\nabla f(\xi_k) - \nabla f(x_k)]^T d_k \\ &\leq f(x_k) + \alpha\|d_k\|\left(\frac{\nabla f(x_k)^T d_k}{\|d_k\|} + \|\nabla f(\xi_k) - \nabla f(x_k)\|\right), \end{aligned} \tag{4.73}$$

where ξ_k is on the line segment connecting x_k and $x_k + \alpha d_k$. Since $\nabla f(x)$ is uniformly continuous on S, there exists $\bar{\alpha}$ so that when $0 \leq \alpha\|d_k\| \leq \bar{\alpha}$, then

$$\|\nabla f(\xi_k) - \nabla f(x_k)\| \leq \frac{1}{2}\varepsilon_1. \tag{4.74}$$

From (4.71)–(4.74) it follows that

$$f\left(x_k + \bar{\alpha}\frac{d_k}{\|d_k\|}\right) \leq f(x_k) + \bar{\alpha}\left(\frac{\nabla f(x_k)^T d_k}{\|d_k\|} + \frac{1}{2}\varepsilon_1\right) \leq f(x_k) - \frac{1}{2}\bar{\alpha}\varepsilon_1.$$

Hence,

$$f(x_{k+1}) \leq f\left(x_k + \bar{\alpha}\frac{d_k}{\|d_k\|}\right) \leq f(x_k) - \frac{1}{2}\bar{\alpha}\varepsilon_1,$$

which contradicts (4.71). Therefore, $\nabla f(x_k) \to 0$, thus completing the proof. ♦

With these, for strongly convex functions and under the exact line search, the following result on the convergence of the PRP method can be proved.

Theorem 4.10 *Consider $f : \mathbb{R}^n \to \mathbb{R}$ twice continuously differentiable and suppose that the level set S is bounded. Assume that there exists a positive constant $m > 0$ so that for $x \in S$ and any $y \in \mathbb{R}^n$*

$$m\|y\|^2 \leq y^T \nabla^2 f(x) y. \tag{4.75}$$

Then, the sequence $\{x_k\}$ generated by the PRP method with exact line search converges to the unique minimum x^* of function f.

Proof From Proposition 4.1, it follows that it is enough to prove (4.70), that is, there exists a positive constant $\omega > 0$, so that

$$-g_{k+1}^T d_{k+1} \geq \omega \|g_{k+1}\| \|d_{k+1}\|, \tag{4.76}$$

i.e., $\cos\theta_k \geq \omega > 0$. Then, from Proposition 4.1 observe that $g_k \to 0$ and $g(x^*) = 0$. From (4.75), it follows that $\{x_k\} \to x^*$, which is the unique minimum of function f. Since the line search is exact, from $d_{k+1} = -g_{k+1} + \beta_k d_k$ and the fact that $g_{k+1}^T d_k = 0$, it follows that că $g_{k+1}^T d_{k+1} = -\|g_{k+1}\|^2$. Then, (4.76) is equivalent to

$$\frac{\|g_{k+1}\|}{\|d_{k+1}\|} \geq \omega. \tag{4.77}$$

Now, having in view that $d_{k+1} = -g_{k+1} + \beta_k d_k$ and the line search is exact, from (2.4), the following value for α_k is obtained

$$\alpha_k = -\frac{g_k^T d_k}{d_k^T A_k d_k} = \frac{\|g_k\|^2}{d_k^T A_k d_k}, \tag{4.78}$$

where

$$A_k = \int_0^1 \nabla^2 f(x_k + t\alpha_k d_k) \mathrm{d}t. \tag{4.79}$$

Using the mean value theorem (see Appendix A), from (4.79),

$$g_{k+1} - g_k = \nabla f(x_k + \alpha_k d_k) - \nabla f(x_k) = \alpha_k A_k d_k. \tag{4.80}$$

Therefore, β_k^{PRP} can be expressed as

$$\begin{aligned} \beta_k^{PRP} &= \frac{g_{k+1}^T (g_{k+1} - g_k)}{g_k^T g_k} = \alpha_k \frac{g_{k+1}^T A_k d_k}{\|g_k\|^2} \\ &= \frac{g_{k+1}^T A_k d_k}{d_k^T A_k d_k}. \end{aligned} \tag{4.81}$$

Since the level set S is bounded, there exists a positive constant $M > 0$, so that for $x \in S$ and any $y \in \mathbb{R}^n$,

$$y^T A(x) y \leq M \|y\|^2. \tag{4.82}$$

Using the above relations, we get the following bound for β_k^{PRP}:

$$|\beta_k^{PRP}| \leq \frac{\|g_{k+1}\| \|A_k d_k\|}{m \|d_k\|^2} \leq \frac{M}{m} \frac{\|g_{k+1}\|}{\|d_k\|}. \tag{4.83}$$

Hence,

$$\begin{aligned}\|d_{k+1}\| &\leq \|g_{k+1}\| + \left|\beta_k^{PRP}\right| \|d_k\| \\ &\leq \|g_{k+1}\| + \frac{M}{m}\|g_{k+1}\| = \left(1 + \frac{M}{m}\right)\|g_{k+1}\|,\end{aligned}$$

that is,

$$\frac{\|g_{k+1}\|}{\|d_{k+1}\|} \geq \left(1 + \frac{M}{m}\right)^{-1}, \tag{4.84}$$

showing that (4.77) holds, where $\omega = m/(m+M)$. ♦

Powell (1984a, 1986b) introduced the PRP+ method as:

$$\beta_k^{PRP+} = \max\{0, \beta_k^{PRP}\}. \tag{4.85}$$

Later on, Gilbert and Nocedal (1992) proved the global convergence of the conjugate gradient methods with nonnegative β_k under inexact line search. The analysis of Gilbert and Nocedal applies to a class of conjugate gradient algorithms which have the Property (*).

Proposition 4.2 *Suppose that the Assumption CG holds. Consider the method (4.1) and (4.2) with $\beta_k \geq 0$ and with any line search satisfying both the sufficient descent condition $g_k^T d_k \leq -c\|g_k\|^2$ and the Zoutendijk condition (3.34). If $\|g_k\| \geq \gamma$ for all k and some positive constant $\gamma > 0$, then $d_k \neq 0$ and*

$$\sum_{k \geq 1} \|u_k - u_{k-1}\|^2 < \infty, \tag{4.86}$$

where $u_k = d_k/\|d_k\|$.

Proof Observe that $d_k \neq 0$, otherwise, the sufficient descent condition $g_k^T d_k \leq -c\|g_k\|^2$ implies $g_k = 0$. Therefore, u_k is well defined. Now, let us define

$$r_k \equiv \frac{-g_k}{\|d_k\|} \text{ and } \delta_k \equiv \frac{\beta_k \|d_k\|}{\|d_{k+1}\|}. \tag{4.87}$$

Observe that from (4.2), for $k \geq 1$,

$$u_{k+1} = r_{k+1} + \delta_k u_k. \tag{4.88}$$

Since $\|u_{k+1}\| = \|u_k\|$, from (4.88), it follows that

$$\|r_{k+1}\| = \|u_{k+1} - \delta_k u_k\| = \|\delta_k u_{k+1} - u_k\|. \tag{4.89}$$

Using the triangle inequality and (4.89), it results that

$$\begin{aligned}\|u_{k+1} - u_k\| &\leq \|(1+\delta_k)u_{k+1} - (1+\delta_k)u_k\| \\ &\leq \|u_{k+1} - \delta_k u_k\| + \|\delta_k u_{k+1} - u_k\| \\ &= 2\|r_{k+1}\|.\end{aligned} \tag{4.90}$$

From (3.28), using the sufficient descent condition $g_k^T d_k \leq -c\|g_k\|^2$, the following estimation is obtained

$$\cos\theta_k \geq c\frac{\|g_k\|}{\|d_k\|}. \tag{4.91}$$

From (4.91), using the Zoutendijk condition and (4.87), it follows that

$$\sum_{k\geq 1}\frac{\|g_k\|^4}{\|d_k\|^2} = \sum_{k\geq 1}\|r_k\|^2\|g_k\|^2 < \infty. \tag{4.92}$$

Since $\|g_k\| \geq \gamma$ it results that

$$\sum_{k\geq 1}\|r_k\|^2 < \infty,$$

which together with (4.90) complete the proof of the proposition. ♦

This proposition shows that the condition (4.86) does not imply the convergence of the sequence $\{u_k\}$, but it shows that the search directions u_k changes slowly, asymptotically (Gilbert & Nocedal, 1992).

Proposition 4.2 applies to any choice of $\beta_k \geq 0$. Therefore, to prove the convergence, we additionally need to require that β_k be small when the step $s_k = x_k - x_{k-1}$ is small. Formally, this property was called Property (*). It can be presented as:

Property (*) *Consider a method given by (4.1) and (4.2) and suppose that* $0 < \gamma \leq \|g_k\| \leq \Gamma$ *for all* $k \geq 0$. *The method has Property (*) if there exist the constants* $b > 1$ *and* $\lambda > 0$ *so that for all* k

$$|\beta_k| \leq b \tag{4.93}$$

and

$$\|s_k\| \leq \lambda \textit{ implies } |\beta_k| \leq \frac{1}{2b}. \tag{4.94}$$

It is easy to show that under the Assumption CG, the PRP method has Property (*). Indeed, using the constants γ and Γ, then $b = 2\Gamma^2/\gamma^2$ and $\lambda = \gamma^2/(2L\Gamma b)$ can be chosen to have Property (*). From (4.69),

$$|\beta_k^{PRP}| \le \frac{\|g_{k+1}\|(\|g_{k+1}\| + \|g_k\|)}{\|g_k\|^2} \le \frac{2\Gamma^2}{\gamma^2} \equiv b,$$

and when $\|s_k\| \le \lambda$, from the Lipschitz condition, it results that

$$|\beta_k^{PRP}| \le \frac{\|g_{k+1}\|\|y_k\|}{\|g_k\|^2} \le \frac{\Gamma L\lambda}{\gamma^2} = \frac{1}{2b}.$$

The following result shows that if the gradients are bounded away from zero and if the method has Property (*), then a fraction of the steps cannot be too small. Let N be the set of positive integers. Then, for $\lambda > 0$ define

$$K^\lambda = \{i \in N : i \ge 1, \|s_i\| > \lambda\},$$

i.e., the set of integers corresponding to the steps that are larger than λ. Besides, let us introduce groups of Δ consecutive iterations and define

$$K_{k,\Delta}^\lambda = \{i \in N : k \le i \le k + \Delta - 1, \ \|s_i\| > \lambda\}.$$

Let $\left|K_{k,\Delta}^\lambda\right|$ be the number of the elements of $K_{k,\Delta}^\lambda$. The following propositions are extracted from Gilbert and Nocedal (1992) and Dai, Han, Liu, Sun, Yin, and Yuan (1999). (Proposition 4.3 follows from Proposition 4.2.)

Proposition 4.3 *Suppose that the Assumption CG holds and consider any method of the form* (4.1) *and* (4.2) *for which the search direction* d_k *is descent. If* $\beta_k \ge 0$, *then* $d_k \ne 0$ *and*

$$\|u_{k+1} - u_k\| \le 2\frac{\|g_{k+1}\|}{\|d_{k+1}\|}, \tag{4.95}$$

where $u_k = d_k/\|d_k\|$. ♦

Proposition 4.4 *Suppose that the Assumption CG holds and consider any method of the form* (4.1) *and* (4.2) *for which the search direction* d_k *is decent. Assume that the method has Property* (*) *and that*

$$\sum_{k=1}^{\infty} \frac{1}{\|d_k\|^2} < +\infty. \tag{4.96}$$

If $\|g_k\| \geq \gamma$ *for any* $k \geq 1$, *then there exists* $\lambda > 0$ *so that for any* $\Delta \in N$ *and any index* k_0, *there is a greater index* $k > k_0$, *so that*

$$\left|K_{k,\Delta}^{\lambda}\right| > \frac{\Delta}{2}.$$

♦

The following theorem is a general global convergence result of the conjugate gradient methods with Property (*). In the following, we shall apply this result for the PRP+ method given by (4.85). The proof of the theorem was given by Gilbert and Nocedal (1992). A variant was given by Dai, Han, Liu, Sun, Yin, and Yuan, (1999).

Theorem 4.11 *Suppose that the Assumption CG holds. Consider the method* (4.1) *and* (4.2) *with the following properties: (i)* $\beta_k \geq 0$; *(ii) the strong Wolfe line search conditions* (4.10) *and* (4.11) *hold for all k; (iii) the descent condition* $g_k^T d_k < 0$ *holds for all k; (iv) Property (*) holds. Then, the method converges in the sense that*

$$\liminf_{k \to \infty} \|g_k\| = 0.$$

Proof Suppose that the theorem is not true, that is, there exists a positive constant γ so that for all k,

$$\|g_k\| \geq \gamma. \tag{4.97}$$

Since $\beta_k \geq 0$ and d_k is a descent direction, from Proposition 4.3, it follows that

$$\|u_{k+1} - u_k\| \leq 2\frac{\|g_{k+1}\|}{\|d_{k+1}\|}$$

for all $k \geq 1$. The above inequality (4.97) and Theorem 3.4 imply that

$$\sum_{k=0}^{\infty} \|u_{k+1} - u_k\|^2 \leq \frac{4}{\gamma^2}\sum_{k=0}^{\infty} \frac{\|g_k\|^4}{\|d_k\|^2} < +\infty. \tag{4.98}$$

Denote $s_k = x_{k+1} - x_k$. For any two indices l and k with $l \geq k$,

$$\begin{aligned} x_l - x_{k-1} &= \sum_{i=k}^{l} \|s_{i-1}\| u_{i-1} \\ &= \sum_{i=k}^{l} \|s_{i-1}\| u_{k-1} + \sum_{i=k}^{l} \|s_{i-1}\| (u_{i-1} - u_{k-1}). \end{aligned}$$

Since $\|u_k\| = 1$, it follows that

$$\sum_{i=k}^{l} \|s_{i-1}\| \leq \|x_l - x_{k-1}\| + \sum_{i=k}^{l} \|s_{i-1}\| \, \|u_{i-1} - u_{k-1}\|. \tag{4.99}$$

From the Assumption CG there exists a positive constant B, so that for any $k \geq 1$, $\|x_k\| \leq B$. Hence,

$$\sum_{i=k}^{l} \|s_{i-1}\| \leq 2B + \sum_{i=k}^{l} \|s_{i-1}\| \, \|u_{i-1} - u_{k-1}\|. \tag{4.100}$$

By Corollary 3.1, it can be assumed that (4.96) holds. Thus, the conditions of Proposition 4.4 are satisfied.

Let $\lambda > 0$ be given by Proposition 4.4 and define $\Delta \equiv \lceil 8B/\lambda \rceil$. Then, by (4.98), there is an index $k_0 \geq 1$ so that

$$\sum_{i \geq k_0} \|u_{i+1} - u_i\|^2 \leq \frac{1}{4\Delta}. \tag{4.101}$$

With this Δ and k_0, Proposition 4.4 gives an index $k \geq k_0$, so that

$$\left| K_{k,\Delta}^{\lambda} \right| > \frac{\Delta}{2}. \tag{4.102}$$

Next, for any index $i \in [k, k + \Delta - 1]$, using the Cauchy–Schwarz inequality and (4.101), it follows that

$$\begin{aligned} \|u_i - u_{k-1}\| &\leq \sum_{j=k}^{i} \|u_j - u_{j-1}\| \\ &\leq (i - k + 1)^{1/2} \left(\sum_{j=k}^{i} \|u_j - u_{j-1}\|^2 \right)^{1/2} \\ &\leq \Delta^{1/2} \left(\frac{1}{4\Delta} \right)^{1/2} = \frac{1}{2}. \end{aligned} \tag{4.103}$$

Using this relation and (4.102) in (4.100) with $l = k + \Delta - 1$, it results that

$$2B \geq \frac{1}{2} \sum_{i=k}^{k+\Delta-1} \|s_{i-1}\| > \frac{\lambda}{2} \left| K_{k,\Delta}^{\lambda} \right| > \frac{\lambda \Delta}{4}.$$

Hence, $\Delta < 8B/\lambda$, which contradicts the definition of Δ . ♦

To prove the global convergence of a conjugate gradient method, condition *(i)* is essential. Gilbert and Nocedal (1992) pointed out that the PRP method can produce negative values for β_k^{PRP} even for strongly convex objective functions. But, β_k^{PRP+} satisfies this condition, therefore the PRP+ method is globally convergent.

It is worth saying that for the global convergence of the PRP method, the boundedness of the level set is essential, even if the line search is exact. Dai, Han, Liu, Sun, Yin, and Yuan (1999) presented an example which shows that under the Assumption CG and $\beta_k^{PR} \geq 0$ for all $k, \|g_k\| = 1$ for any k.

The PRP+ method was introduced to rectify the convergence failure of the PRP method when implemented with Wolfe line search. Another approach for rectifying the convergence failure is to retain Formula (4.69), but modify the line search. This was done by Grippo and Lucidi (1997), who proposed a new line search based on the Armijo condition of the following form

$$\alpha_k = \max\left\{\lambda^j \frac{\tau|g_k^T d_k|}{\|d_k\|^2}\right\}, \tag{4.104}$$

where $j \geq 0$ is the smallest integer with the properties

$$f(x_{k+1}) \leq f(x_k) - \rho\alpha_k^2\|d_k\|^2$$

and

$$-c_1\|g_{k+1}\|^2 \leq g_{k+1}^T d_{k+1} \leq c_2\|g_{k+1}\|^2,$$

where $0 < c_2 < 1 < c_1$, $0 < \lambda < 1$ and $\tau > 0$ are constants. Using this new line search, Grippo and Lucidi proved the global convergence of the PRP method.

Another interesting result was given by Dai and Yuan (2000). They proved that the PRP method is globally convergent, although the line search uses a constant stepsize $\alpha_k = \eta < 1/4L$, where L is the Lipschitz constant associated with ∇f. In the same context, Sun and Zhang (2001) gave a global convergence result of the PRP method, in which the stepsize is computed as

$$\alpha_k = -\delta\frac{g_k^T d_k}{d_k^T Q_k d_k}, \tag{4.105}$$

where Q_k is a positive definite matrix with smallest eigenvalue $\nu_{\min} > 0$, $\delta \in (0, \nu_{\min}/L)$ and L is the Lipschitz constant for ∇f. For these stepsize choices (4.104) and (4.105), the search directions are no longer conjugate when f is quadratic. Therefore, in their essence, these methods should be viewed as steepest descent methods rather than conjugate gradient methods (Hager & Zhang, 2006b).

The results of Al-Baali inspired Gilbert and Nocedal (1992) to consider the following globally convergent modification of the PRP method. They defined β_k in the PRP method by

$$\beta_k = \begin{cases} -\beta_k^{FR} & \text{if} \quad \beta_k^{PRP} < -\beta_k^{FR}, \\ \beta_k^{PRP} & \text{if} \quad \left|\beta_k^{PRP}\right| \leq \beta_k^{FR}, \\ \beta_k^{FR} & \text{if} \quad \beta_k^{PRP} > \beta_k^{FR}, \end{cases} \tag{4.106}$$

showing that this variant of the conjugate gradient algorithm does not have the drawback of the FR algorithm, i.e., if $x_{k+1} \approx x_k$, then $\beta_k \approx 0$ instead of being close to one. The global convergence of this variant of the conjugate gradient algorithm is based on Theorem 4.2.

Remark 4.3 Considering the situation presented in Remark 4.1, notice that the PRP method behaves quite differently from the FR method. If $g_{k+1} \approx g_k$, then $\beta_k^{PRP} \approx 0$. Therefore, by (4.2) and (4.15), it follows that $\cos\theta_{k+1} \gg \cos\theta_k$. Thus, the PRP method would recover from that situation and this is the explanation for a good behavior of the PRP method.

It is interesting to see the behavior of the method given by (4.106) in the situation presented in Remark 4.1. Here, $\beta_k^{FR} \approx 1$ and $\beta_k^{PRP} \approx 0$. Therefore, the method (4.106) will set $\beta_{k+1} = \beta_{k+1}^{PRP}$, thus avoiding the inefficiency of the FR method. ♦

Hestenes–Stiefel Method

The HS conjugate gradient method is defined by (4.1) and (4.2), where the conjugate gradient parameter is computed as

$$\beta_k^{HS} = \frac{g_{k+1}^T y_k}{d_k^T y_k}. \tag{4.107}$$

The HS method has the property that the conjugacy condition

$$d_{k+1}^T y_k = 0 \tag{4.108}$$

is always satisfied, independent of the line search used. For the exact line search, $\beta_k^{HS} = \beta_k^{PRP}$. Therefore, the global convergence properties of the HS method are similar to the global convergence properties of the PRP method. In particular, by Powell's example (1984a), the HS method with exact line search may not converge for general nonlinear functions. If the search directions satisfy the sufficient descent condition, and if the standard Wolfe line search is used, and then the HS method satisfies Property (*). Indeed, assume that the descent condition $g_k^T d_k \leq -c\|g_k\|^2$ and the second Wolfe condition (4.40) are satisfied. Suppose that $0 < \gamma \leq \|g_k\| \leq \Gamma$, for all $k \geq 1$. Then,

$$\begin{aligned} d_k^T y_k &= d_k^T g_{k+1} - d_k^T g_k \\ &\geq -(1-\sigma) d_k^T g_k \\ &\geq (1-\sigma) c \|g_k\|^2 \\ &\geq (1-\sigma) c\gamma^2. \end{aligned}$$

Using this in (4.107), it follows that

$$|\beta_k^{HS}| \leq \frac{2\Gamma^2}{(1-\sigma)c\gamma^2} \equiv b.$$

Now, define $\lambda = (1-\sigma)c\gamma^2/(2L\Gamma b)$. Using the Lipschitz condition, if $\|s_k\| \leq \lambda$, then

$$|\beta_k^{HS}| \leq \frac{\Gamma L \lambda}{(1-\sigma)c\gamma^2} = \frac{1}{2b}.$$

Similar to the PRP+ method, if β_k^{HS+} is defined as $\beta_k^{HS+} = \max\{0, \beta_k^{HS}\}$, then by Theorem 4.10 it follows that the HS+ method is globally convergent.

Liu–Storey Method

This conjugate gradient method is defined by (4.1) and (4.2), where the conjugate gradient parameter is computed as

$$\beta_k^{LS} = \frac{g_{k+1}^T y_k}{-d_k^T g_k}. \tag{4.109}$$

For the exact line search, the LS method is identical to the PRP method. Liu and Storey (1991) studied this method, proving its global convergence. The techniques developed for the analysis the PRP method may be applied to the LS method.

4.3 Numerical Study

In this section, some numerical results with standard conjugate gradient methods are presented. For this, let us consider 80 unconstrained optimization problems from the UOP collection (Andrei, 2018g), each with the number of variables $n = 1000, 2000, \ldots, 10000$. The comparisons of algorithms are given in the context of Remark 1.1. The maximum number of iterations was limited to 2000.

Figures 4.1 and 4.2 present the Dolan and Moré's performance profiles of the standard conjugate gradient methods. The tables inside the plots show the performances of the algorithms subject to the number of iterations (*#iter*), the number of function and its gradient evaluations (*#fg*) and subject to the CPU time metric (*cpu*)

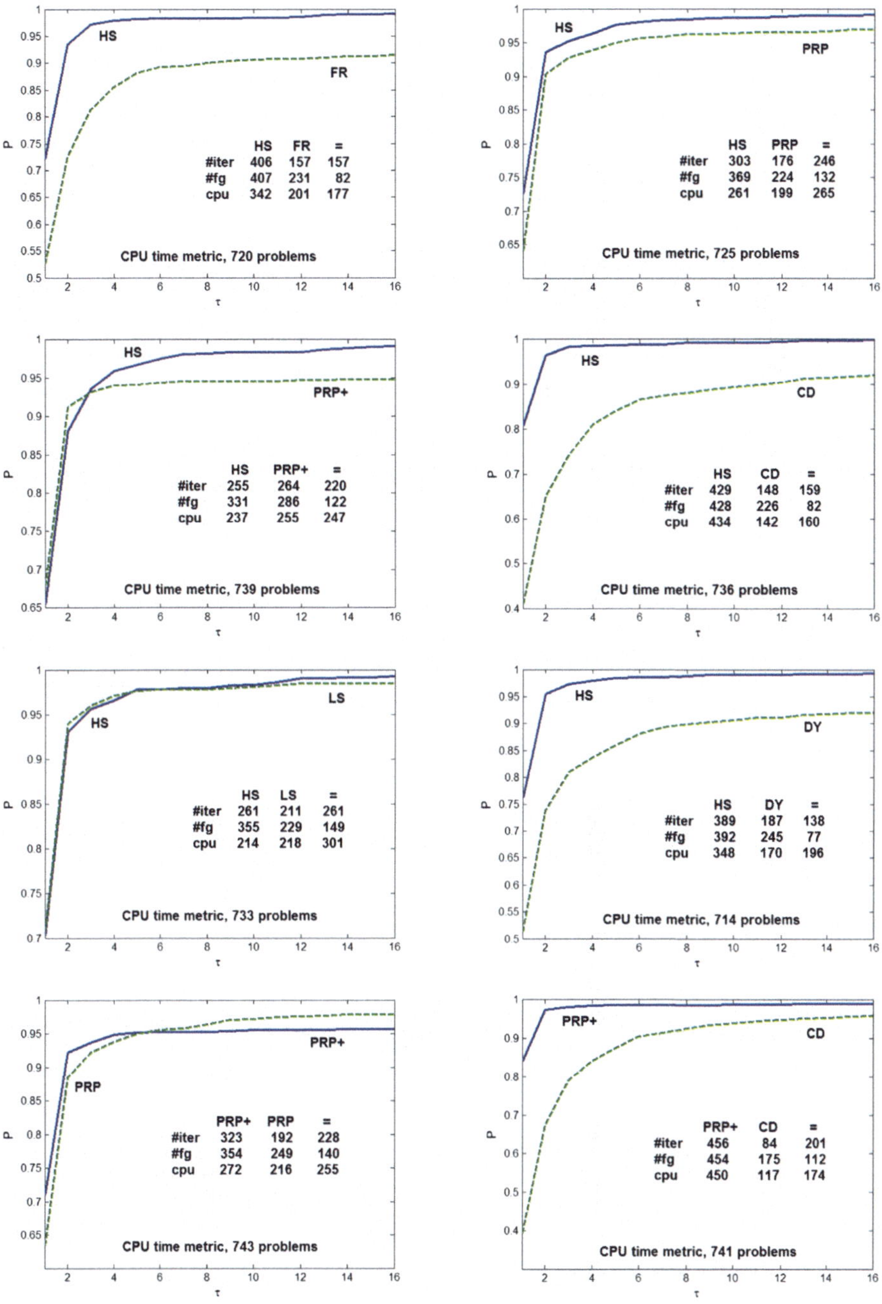

Figure 4.1 Performance profiles of the standard conjugate gradient methods

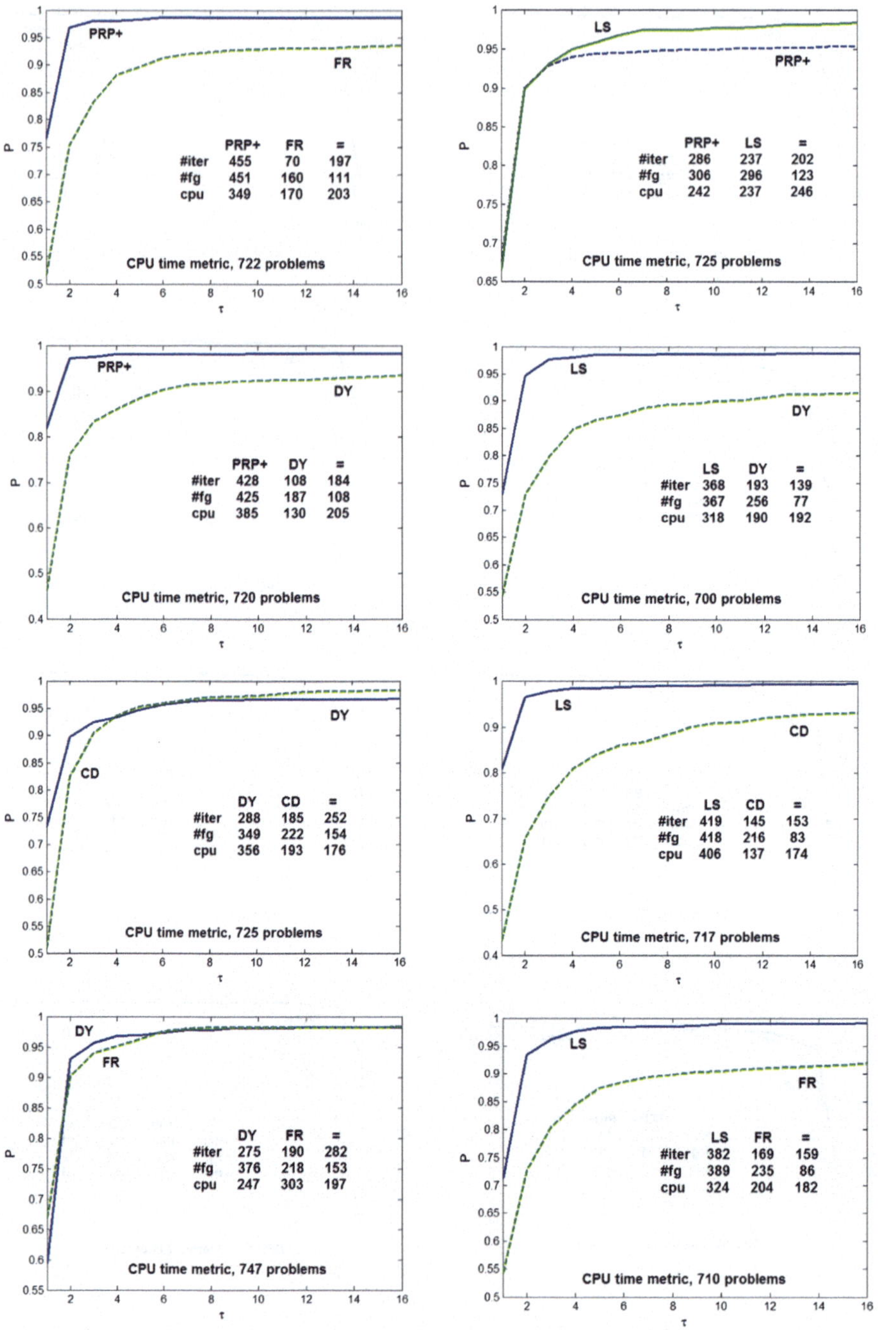

Figure 4.2 Performance profiles of the standard conjugate gradient methods

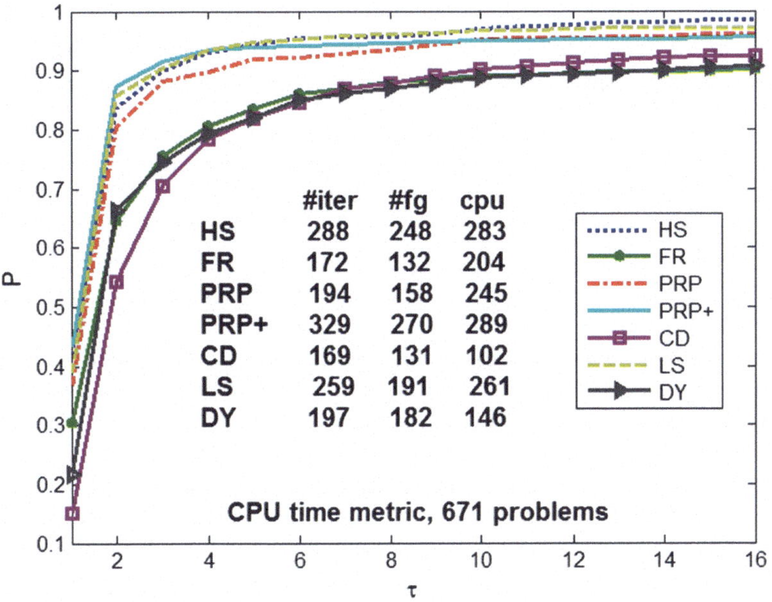

Figure 4.3 Performance profiles of seven standard conjugate gradient methods

in seconds. When comparing HS versus FR (see Figure 4.1) subject to the number of iterations, we see that HS was better in 406 problems (i.e., it achieved the minimum number of iterations in 406 problems). FR was better in 157 problems and they achieved the same number of iterations in 157 problems, etc. Out of 800 problems considered in this numerical experiment, only for 720 problems does the criterion (1.118) hold. From Figure 4.1, we see that subject to the CPU time metric, HS is the fastest. Close to HS are PRP+ and LS. Observe that HS is much better than FR, CD, and DY. One explanation is that the HS method satisfies the conjugacy condition (4.108), independent of the line search. From Figures 4.1 and 4.2, we notice that PRP+ is more efficient than FR, PRP, CD, and DY. Close to PRP+ is LS, but LS is slightly more robust. From Figure 4.2, we see that LS is better than FR, CD, and DY. Close to CD is DY. At least for this set of 800 unconstrained optimization problems, HS and PRP+ are the best methods. HS, PRP+, and LS methods automatically adjust β_k to avoid jamming. Figure 4.3 presents a global comparison among the standard conjugate gradient methods. *Here, there is an experimental confirmation of the classification of conjugate gradient methods in two classes according to the formula for* β_k *computation: with* $\|g_{k+1}\|^2$ *or with* $g_{k+1}^T y_k$ *at the numerator of* β_k.

We see that HS, PRP, PRP+ , and LS (all with $g_{k+1}^T y_k$ at the numerator of β_k) are more efficient and more robust subject to the CPU time metric than FR, CD, and DY (all with $\|g_{k+1}\|^2$ at the numerator of β_k).

Despite the strong convergence theory that has been developed for methods with $\|g_{k+1}\|^2$ in the numerator of β_k, these methods have modest numerical performances in comparison with the methods with $g_{k+1}^T y_k$ at the numerator of β_k.

In the second set of numerical experiments, let us present the performances of standard conjugate gradient methods for solving five applications from the MINPACK-2 collection (Averick, Carter, Moré, & Xue, 1992), described in Chap. 1 of this book, where $nx = 200$ and $ny = 200$. The conjugate gradient algorithms implement the same stopping criterion $\|g_k\|_\infty \leq \varepsilon_g$, where $\varepsilon_g = 10^{-6}$ and the stepsize are computed by the Wolfe line search (4.10) and (4.40) with $\rho = 0.0001$ and $\sigma = 0.9$.

Tables 4.2, 4.3, and 4.4 present the performances of the standard conjugate gradient methods for solving these applications. In these tables, n is the number of variables of the application, *#iter* is the number of iterations, *#fg* the number of function, and its gradient evaluations and *cpu* are the CPU time (in seconds) for solving the application. The last line in these tables contains the total number of iterations, and the total number of function and its gradient evaluations and the total CPU time for solving these five applications on each conjugate gradient method. Observe that for solving all five large-scale applications, CD, HS, and PRP required the minimum time. For example, CD needs 137.09 s, HS 139.65, and PRP 139.42 s. PRP+ and LS have comparable performances. However, FR and DY need the longest time for solving all these five applications. Details on the performances of conjugate gradient algorithms for solving the applications from the MINPACK-2 collection are found in (Andrei, 2006b).

Comparisons of the performances of the standard conjugate gradient methods presented in Tables 4.2, 4.3, and 4.4 versus the performances of the L-BFGS ($m = 5$) and TN presented in Tables 1.2 and 1.3, respectively, show that both L-BFGS ($m = 5$) and TN are faster than CD, HS, and PRP. This is because L-BFGS and TN use the second-order information given by an approximation of the Hessian of the applications.

Table 4.2 Performances of HS, FR, and PRP for solving five applications from the MINPACK-2 collection

		HS			FR			PRP		
	n	*#iter*	*#fg*	*cpu*	*#iter*	*#fg*	*cpu*	*#iter*	*#fg*	*cpu*
A1	40,000	359	457	8.89	1082	1164	24.05	601	857	5.73
A2	40,000	1286	1650	34.65	2698	2885	35.40	1036	1477	12.84
A3	40,000	1510	1730	45.55	20001	25625	411.04	3001	4035	65.31
A4	40,000	841	1069	41.32	2070	2167	83.06	809	1148	43.91
A5	40,000	583	714	9.24	1830	1931	24.65	668	927	11.63
Total	–	4579	5620	139.65	27681	33772	578.20	6115	8444	139.42

Table 4.3 Performances of PRP+ and CD for solving five applications from the MINPACK-2 collection

		PRP+			CD		
	n	*#iter*	*#fg*	*cpu*	*#iter*	*#fg*	*cpu*
A1	40,000	604	862	6.67	944	1097	12.80
A2	40,000	2335	1901	21.05	1887	2138	32.17
A3	40,000	3219	4306	69.73	2148	2287	37.52
A4	40,000	915	1314	50.17	1032	1148	43.99
A5	40,000	810	1125	14.09	759	836	10.61
Total	–	7883	9508	161.71	6770	7506	137.09

Table 4.4 Performances of LS and DY for solving five applications from the MINPACK-2 collection

		LS			DY		
	n	*#iter*	*#fg*	*cpu*	*#iter*	*#fg*	*cpu*
A1	40,000	642	886	10.51	464	488	6.08
A2	40,000	1085	1526	19.01	1031	1062	14.72
A3	40,000	2398	3120	50.76	8393	8423	141.57
A4	40,000	1455	2033	77.50	886	909	35.22
A5	40,000	591	821	10.25	2382	2410	31.61
Total	–	6171	8386	168.03	13156	13292	229.20

Notes and References

Table 4.1 presents the standard conjugate gradient methods. When the minimizing function f is quadratic, then all these methods give the same conjugate gradient parameter β_k, i.e., the same search direction. However, when f is not quadratic, the parameters β_k are different, thus yielding different search directions. Obviously, when f is not quadratic, the search directions d_k and d_{k-1} are not truly conjugate because there is not a constant Hessian $\nabla^2 f$ for them to be conjugate.

The first nonlinear conjugate gradient algorithm was proposed by Fletcher and Reeves (1964). The global convergence of the FR algorithm for general nonlinear problems under the inexact line search was established by Al-Baali (1985). His arguments and techniques for proving the global convergence became a standard, influencing research on nonlinear conjugate gradients algorithms, suggesting that the global convergence of these algorithms require some assumptions on parameters ρ and σ in the stepsize selection procedure based on the Wolfe conditions. Later on, Dai and Yuan (1996a, 2000) weakened the condition on σ and proposed the curvature condition (1.20), known as the generalized Wolfe line search, under which the FR method is global convergent. The proof of Dai and Yuan is similar to that given by Al-Baali. The main points of the proof are the ingredients which show that $\|d_k\|^2$ has a linear growth rate and that it leads to a contradiction to the Zoutendijk condition.

The FR algorithm has one of the strongest convergence properties among all standard conjugate gradient algorithms. However, it requires the strong Wolfe conditions for the line search. Dai and Yuan (1999) relaxed this assumption and proposed a new conjugate gradient algorithm, known as the DY method, which is globally convergent under the Wolfe line search without assuming that $\sigma < 1/2$.

It is interesting to notice that, although the conjugate gradient algorithms with $g_{k+1}^T y_k$ at the numerator of β_k, like HS, PRP, and LS, are more efficient and more robust, much effort has been made in the analysis of the FR algorithm. This is motivated by the interesting results obtained by Al-Baali, which gave the possibility to analyze the conjugate gradient algorithms with $\|g_{k+1}\|^2$ at the numerator of β_k.

In practice, the standard conjugate gradient algorithms usually need more iterations than a quasi-Newton method. The rate of convergence of conjugate gradient methods is n-step quadratic, which means that

$$\|x_{k+n} - x^*\| = O(\|x_k - x^*\|^2)$$

for k sufficiently large (Shanno, 1980). This implies that the convergence will be slower than for the Newton and the quasi-Newton methods. However, in spite of having slower convergence, the standard conjugate gradient methods have some advantages over the Newton and the quasi-Newton methods. Since the standard conjugate gradient methods do not use any matrices, for large-scale problems they require less computer memory. Moreover, the number of multiplications per iteration is of order $O(n)$ compared to $O(n^3)$ for the Newton method and $O(n^2)$ for the quasi-Newton method. Thus, although they may do more iterations than the quasi-Newton or the Newton methods, their overhead cost per iteration may be significantly less.

Chapter 5
Acceleration of Conjugate Gradient Algorithms

It is common knowledge that in conjugate gradient algorithms, the search directions tend to be poorly scaled and consequently the line search must perform more function evaluations in order to obtain a suitable stepsize α_k. In conjugate gradient methods, the stepsizes differ from 1 in a very unpredictable way. They can be larger or smaller than 1, depending on how the problem is scaled. This is in very sharp contrast to the Newton and the quasi-Newton methods, including the limited-memory quasi-Newton methods, which accept the unit stepsize most of the time along the iterations and therefore they usually require only few function evaluations per search direction.

The purpose of this chapter is to present an acceleration of conjugate gradient algorithms. The idea is to modify the stepsize α_k (computed by means of the Wolfe line search) through a positive parameter η_k in a multiplicative manner, in such a way as to improve the behavior of these algorithms (Andrei, 2009c). We show that the resulting algorithm is linear convergent, but the reduction in the function value is significantly improved. An acceleration of the gradient descent algorithm with backtracking for unconstrained optimization is given in Andrei (2006a).

For solving the unconstrained optimization problems

$$\min_{x \in \mathbb{R}^n} f(x), \tag{5.1}$$

where $f : \mathbb{R}^n \to \mathbb{R}$ is a continuously differentiable function bounded from below, starting from an initial guess $x_0 \in \mathbb{R}^n$, a nonlinear conjugate gradient method generates a sequence $\{x_k\}$ as

$$x_{k+1} = x_k + \alpha_k d_k, k = 0, 1, \ldots, \tag{5.2}$$

N. Andrei, *Nonlinear Conjugate Gradient Methods for Unconstrained Optimization*, Springer Optimization and Its Applications 158,
https://doi.org/10.1007/978-3-030-42950-8_5

where $\alpha_k > 0$ is obtained by line search and the directions d_k are generated as

$$d_{k+1} = -g_{k+1} + \beta_k d_k, k = 0, 1, \ldots, \tag{5.3}$$

where $d_0 = -g_0$. In (5.3), β_k is known as the conjugate gradient parameter and $g_k = \nabla f(x_k)$. The search direction d_k, assumed to be a descent one, plays the main role in these methods. On the other hand, the stepsize α_k guarantees the global convergence in some cases and is crucial in efficiency. Different conjugate gradient algorithms correspond to different choices for the scalar parameter β_k. For example, Table 4.1 presents a number of standard conjugate gradient algorithms which differ only in their formula for the β_k computation. In this chapter, we develop a procedure for modifying the stepsize determined by the Wolfe line search conditions in such a way as to improve the reduction of the minimizing function values along the iterations.

5.1 Standard Wolfe Line Search with Cubic Interpolation

For implementing the algorithm (5.2), one of the crucial elements is the stepsize computation. Many procedures have been suggested. In the *exact line search*, the step α_k is selected as

$$\alpha_k = \arg\min_{\alpha > 0} f(x_k + \alpha d_k), \tag{5.4}$$

where d_k is a descent direction. In some very special cases (quadratic problems, for example), it is possible to compute the step α_k analytically, but for the vast majority of cases it is computed to approximately minimize f along the ray

$$\{x_k + \alpha d_k \ : \ \alpha \geq 0\},$$

or at least to sufficiently reduce f. In practice, as we mentioned, the most used are the *inexact procedures*. Plenty of inexact line search procedures were proposed: Goldstein (1965), Armijo (1966), Wolfe (1969, 1971), Powell (1976a), Dennis and Schnabel (1983), Fletcher (1987), Potra and Shi (1995), Lemaréchal (1981), Shanno (1983), Moré and Thuente (1990), Hager and Zhang (2005), and many others. The line search in conjugate gradient algorithms is often based either on the Goldstein's conditions

$$\delta_1 \alpha_k g_k^T d_k \leq f(x_k + \alpha_k d_k) - f(x_k) \leq \delta_2 \alpha_k g_k^T d_k, \tag{5.5}$$

where $0<\delta_2<\frac{1}{2}<\delta_1<1$ and $\alpha_k>0$, or on the standard Wolfe conditions

$$f(x_k+\alpha_k d_k)-f(x_k)\leq \rho\alpha_k g_k^T d_k, \tag{5.6}$$

$$g_{k+1}^T d_k \geq \sigma g_k^T d_k, \tag{5.7}$$

where $0<\rho<1/2\leq\sigma<1$. The first Wolfe condition (5.6) is called the *sufficient reduction* of the minimizing function values, which ensures a sufficient reduction of function values. The second Wolfe condition (5.7) is the *curvature condition,* which ensures unacceptable short stepsizes. In current implementations of the conjugate gradient algorithms, the standard Wolfe line search is used. Of course, some other advanced line search procedures, like the approximate Wolfe line search of Hager and Zhang (2005), or the improved Wolfe line search of Dai and Kou (2013) may be used.

Any line search procedure requires an initial estimate α_0 of the stepsize and generates a sequence $\{\alpha_k\}$ that converges to a stepsize satisfying the conditions imposed by the user (e.g., Armijo, Goldstein, Wolfe, etc.), or it informs that such a stepsize does not exist. Mainly, line search procedures have two phases: a *bracketing phase,* that finds an interval $[a,b]$ containing an acceptable value for the stepsize, and a *selection phase,* that locates the final stepsize. The selection phase reduces the bracketing interval during the search and usually interpolates some values of the function and its gradient obtained from the earlier steps of searching to guess the location of the minimum.

In the following, let us describe *a variant* of a line search procedure which is simple enough to generate safeguarded stepsizes satisfying the standard Wolfe conditions (5.6) and (5.7) (see Shanno, 1983). Suppose that we are at the iteration k. To have a simple interpretation of the procedure and a clear description, a Fortran version of it is presented in Figure 5.1. The inputs of this procedure are: n the number of variables, $x=x_k$ a vector with the current values of variables, $f=f(x_k)$ the value of the minimizing function in x, $d=d_k$ the current search direction, $gtd=\nabla f(x_k)^T d_k$ the scalar product of the current gradient and the search direction, and $dnorm=\|d_k\|$ the l_2-norm of the search direction. The outputs of the procedure are: $alpha=\alpha_k$ the stepsize satisfying the standard Wolfe line search conditions, $xnew=x_{k+1}=x_k+\alpha_k d_k$ the new point, $fnew=f(x_{k+1})$ the function value in new point, $gnew=\nabla f(x_{k+1})$ the gradient of the minimizing function in the new point, *fgcnt* the number of function and its gradient calls, *lscnt* indicates that the line search procedure performed a number of iterations, and *lsflag* indicates that the number of iterations in the line search procedure is greater than a prespecified threshold.

In Figure 5.1, **`max$ls`** is the maximum number of iterations in the line search procedure, **`epsm`** is the epsilon machine and **`evalfg(n, xnew, fnew, gnew)`** is the subroutine implementing the algebraic expressions of the minimizing function and its gradient. In input, this subroutine has: **`n`** as the number of variables and

```
      subroutine LineSearch (n,x,f,d,gtd,dnorm,alpha,xnew,fnew,gnew,
     +                       fgcnt,lscnt,lsflag)

C     SCALAR ARGUMENTS
      integer n,fgcnt,lscnt,lsflag
      double precision f,gtd,dnorm,alpha,fnew
C     ARRAY ARGUMENTS
      double precision x(n),d(n),xnew(n),gnew(n)
C     LOCAL SCALARS
      integer i,lsiter, max$ls
      double precision alphap,alphatemp,fp,dp,gtdnew,a,b
      common/acca/epsm

      lsflag = 0
* Maximum number of iterations in LineSearch is max$ls (now is 20)
      max$ls=20

      alphap = 0.0d0
      fp     = f
      dp     = gtd
      do i = 1,n
        xnew(i) = x(i) + alpha * d(i)
      end do
c1
      call evalfg(n,xnew,fnew,gnew)
      fgcnt = fgcnt + 1

      gtdnew = 0.0d0
      do i = 1,n
        gtdnew = gtdnew + gnew(i) * d(i)
      end do

      lsiter = 0

 10   if ( alpha * dnorm .gt. 1.0d-30 .and. lsiter .lt. max$ls .and.
     +     .not. ( gtdnew .eq. 0.0d0 .and. fnew .lt. f ) .and.
     +     ( ( fnew .gt. f + 1.0d-04 * alpha * gtd .or.
     +     dabs( gtdnew / gtd ) .gt. 0.9d0 ) .or. ( lsiter .eq. 0 .and.
     +     dabs( gtdnew / gtd ) .gt. 0.5d0 ) ) ) then

 20       if ( alpha * dnorm .gt. 1.0d-30 .and. fnew .gt. f .and.
     +          gtdnew .lt. 0.0d0 ) then

              alpha = alpha / 3.0d0
              do i = 1,n
                xnew(i) = x(i) + alpha * d(i)
              end do
c2
              call evalfg(n,xnew,fnew,gnew)
              fgcnt = fgcnt + 1
              gtdnew = 0.0d0
              do i = 1,n
                gtdnew = gtdnew + gnew(i) * d(i)
              end do
              alphap = 0.0d0
              fp     = f
              dp     = gtd
              goto 20
          end if
```

Figure 5.1 Subroutine LineSearch which generates safeguarded stepsizes satisfying the standard Wolfe line search with cubic interpolation

```
          a = dp + gtdnew - 3.0d0 * ( fp - fnew ) / ( alphap - alpha )
          b = a ** 2 - dp * gtdnew
          if ( b .gt. epsm ) then
              b = sqrt( b )
          else
              b = 0.0d0
          end if

          alphatemp = alpha - ( alpha - alphap ) * ( gtdnew + b - a ) /
     +                ( gtdnew - dp + 2.0d0 * b )

          if ( gtdnew / dp .le. 0.0d0 ) then
              if ( 0.99d0 * dmax1( alpha, alphap ) .lt. alphatemp .or.
     +            alphatemp .lt. 1.01d0 * dmin1( alpha, alphap ) ) then
                  alphatemp = ( alpha + alphap ) / 2.0d0
              end if
          else
              if ( gtdnew .lt. 0.0d0 .and.
     +            alphatemp .lt. 1.01d0 * dmax1( alpha, alphap ) ) then
                  alphatemp = 2.0d0 * dmax1( alpha, alphap )
              end if
              if ( ( gtdnew .gt. 0.0d0 .and.
     +            alphatemp .gt. 0.99d0 * dmin1( alpha, alphap ) ) .or.
     +            alphatemp .lt. 0.0d0 ) then
                  alphatemp = dmin1( alpha, alphap ) / 2.0d0
              end if
          end if

          alphap = alpha
          fp     = fnew
          dp     = gtdnew
          alpha = alphatemp
          do i = 1,n
            xnew(i) = x(i) + alpha * d(i)
          end do
c3
          call evalfg(n,xnew,fnew,gnew)
          fgcnt = fgcnt + 1
          gtdnew = 0.0d0
          do i = 1,n
            gtdnew = gtdnew + gnew(i) * d(i)
          end do

          lsiter = lsiter + 1

      goto 10
      end if

      if ( lsiter .ge. max$ls ) then
          lsflag = 1
      end if
      if ( lsiter .ne. 0 ) then
          lscnt = lscnt + 1
      end if

      return
      end
```

Figure 5.1 (continued)

xnew as the new point. In output, it computes: **fnew** as the value of function f in the new point and **gnew** as the gradient of f in the new point.

We see that a line search procedure is complicated and to be reliable it must incorporate a lot of features. Firstly, observe that the standard Wolfe conditions are implemented in a complicated form, which takes into consideration both the ratio between the rate of decrease of f in the direction d_k at the new point and the rate of decrease in the direction d_k at the current point x_k, and also some precautions to avoid too small or too large values of the stepsize. Observe that in the selection phase of the procedure, the cubic interpolation is used. Cubic interpolation provides a good model for the minimizing function in the searching interval. Suppose we have an interval $[\bar{a}, \bar{b}]$ containing the desirable stepsize and two previous stepsizes estimates α_{i-1} and α_i in this interval. We use a cubic function to interpolate the values, $\varphi_k(\alpha_{i-1})$, $\varphi_k(\alpha_i)$, $\varphi'_k(\alpha_{i-1})$, and $\varphi'_k(\alpha_i)$, where $\varphi_k(\alpha) = f(x_k + \alpha d_k)$. (Bulirsch and Stoer (1980) proved that such a cubic function always exists and is unique.) The minimizer of this cubic function in $[\bar{a}, \bar{b}]$, that is, a new estimation of the stepsize, is either at one of the endpoints or in the interior, case in which it is given by

$$\alpha_{i+1} = \alpha_i - (\alpha_i - \alpha_{i-1}) \left[\frac{\varphi'_k(\alpha_i) + b - a}{\varphi'_k(\alpha_i) - \varphi'_k(\alpha_{i-1}) + 2b} \right], \tag{5.8}$$

where

$$a = \varphi'_k(\alpha_{i-1}) + \varphi'_k(\alpha_i) - 3\frac{\varphi_k(\alpha_{i-1}) - \varphi_k(\alpha_i)}{\alpha_{i-1} - \alpha_i}, \tag{5.9}$$

$$b = \left(a^2 - \varphi'_k(\alpha_{i-1})\varphi'_k(\alpha_i)\right)^{1/2}. \tag{5.10}$$

In Figure 5.1, the new estimate α_{i+1} is computed as **alphatemp**. The interpolation process can be repeated by discarding the data at one of the stepsizes α_{i-1} or α_i and replacing it by $\varphi_k(\alpha_{i+1})$ and $\varphi'_k(\alpha_{i+1})$. Observe that the interpolation step that determines a new estimation to the stepsize is safeguarded in order to ensure that the new stepsize is not too close to the endpoints of the interval. Some more details may be found, for example, in (Dennis & Schnabel, 1983), (Shanno, 1983), or (Birgin & Martínez, 2001).

5.2 Acceleration of Nonlinear Conjugate Gradient Algorithms

In this section, let us present the accelerated conjugate gradient algorithms for solving the unconstrained optimization problem (5.1) (Andrei, 2009c). Suppose that function f is twice continuously differentiable. At the current iteration k, x_k, f_k, g_k, and $d_k = -g_k + \beta_{k-1} d_{k-1}$ are known, where β_{k-1} is computed according to the conjugate gradient algorithm we want to accelerate. Suppose that d_k is a descent direction. Now, by the Wolfe line search (5.6) and (5.7), the stepsize α_k can be computed, with which the following point $z = x_k + \alpha_k d_k$ is determined. The first Wolfe condition (5.6) shows that the stepsize $\alpha_k > 0$ satisfies

$$f(z) = f(x_k + \alpha_k d_k) \le f(x_k) + \rho \alpha_k g_k^T d_k.$$

With these, let us introduce the accelerated conjugate gradient algorithm by means of the following iterative scheme

$$x_{k+1} = x_k + \eta_k \alpha_k d_k, \tag{5.11}$$

where $\eta_k > 0$ is a parameter which follows to be determined in such a way so as to improve the algorithm. Now, we have

$$f(x_k + \alpha_k d_k) = f(x_k) + \alpha_k g_k^T d_k + \frac{1}{2} \alpha_k^2 d_k^T \nabla^2 f(x_k) d_k + o\left(\|\alpha_k d_k\|^2 \right).$$

On the other hand, for $\eta > 0$, it follows that

$$f(x_k + \eta \alpha_k d_k) = f(x_k) + \eta \alpha_k g_k^T d_k + \frac{1}{2} \eta^2 \alpha_k^2 d_k^T \nabla^2 f(x_k) d_k + o\left(\|\eta \alpha_k d_k\|^2 \right).$$

Therefore,

$$f(x_k + \eta \alpha_k d_k) = f(x_k + \alpha_k d_k) + \Psi_k(\eta), \tag{5.12}$$

where

$$\begin{aligned} \Psi_k(\eta) = {} & \frac{1}{2} (\eta^2 - 1) \alpha_k^2 d_k^T \nabla^2 f(x_k) d_k + (\eta - 1) \alpha_k g_k^T d_k \\ & + \eta^2 \alpha_k o\left(\alpha_k \|d_k\|^2 \right) - \alpha_k o\left(\alpha_k \|d_k\|^2 \right). \end{aligned} \tag{5.13}$$

Let us denote

$$\begin{aligned} a_k &\equiv \alpha_k g_k^T d_k \le 0, \\ b_k &\equiv \alpha_k^2 d_k^T \nabla^2 f(x_k) d_k, \\ \varepsilon_k &\equiv o\left(\alpha_k \|d_k\|^2 \right). \end{aligned}$$

Observe that $a_k \le 0$ since d_k is a descent direction and for convex functions $b_k \ge 0$. Therefore,

$$\Psi_k(\eta) = \frac{1}{2}(\eta^2 - 1)b_k + (\eta - 1)a_k + \eta^2 \alpha_k \varepsilon_k - \alpha_k \varepsilon_k. \tag{5.14}$$

But, $\Psi_k'(\eta) = (b_k + 2\alpha_k\varepsilon_k)\eta + a_k$ and $\Psi_k'(\eta_m) = 0$ where

$$\eta_m = -\frac{a_k}{b_k + 2\alpha_k\varepsilon_k}. \tag{5.15}$$

Observe that $\Psi_k'(0) = a_k \le 0$. Therefore, assuming that $b_k + 2\alpha_k\varepsilon_k > 0$, then $\Psi_k(\eta)$ is a convex quadratic function with minimum value in point η_m and

$$\Psi_k(\eta_m) = -\frac{(a_k + (b_k + 2\alpha_k\varepsilon_k))^2}{2(b_k + 2\alpha_k\varepsilon_k)} \le 0.$$

Considering $\eta = \eta_m$ in (5.12) and since $b_k \ge 0$, it follows that for every k,

$$f(x_k + \eta_m \alpha_k d_k) = f(x_k + \alpha_k d_k) - \frac{(a_k + (b_k + 2\alpha_k\varepsilon_k))^2}{2(b_k + 2\alpha_k\varepsilon_k)} \le f(x_k + \alpha_k d_k),$$

which is a possible improvement of the values of function f (when $a_k + (b_k + 2\alpha_k\varepsilon_k) \ne 0$).

Therefore, using this simple multiplicative modification of the stepsize α_k as $\eta_k\alpha_k$, where $\eta_k = \eta_m = -a_k/(b_k + 2\alpha_k\varepsilon_k)$, it follows that

$$\begin{aligned} f(x_{k+1}) &= f(x_k + \eta_k\alpha_k d_k) \le f(x_k) + \rho\alpha_k g_k^T d_k - \frac{(a_k + (b_k + 2\alpha_k\varepsilon_k))^2}{2(b_k + 2\alpha_k\varepsilon_k)} \\ &= f(x_k) - \left[\frac{(a_k + (b_k + 2\alpha_k\varepsilon_k))^2}{2(b_k + 2\alpha_k\varepsilon_k)} - \rho a_k\right] \le f(x_k), \end{aligned} \tag{5.16}$$

since $a_k \le 0$, (d_k is a descent direction).

Now, neglecting the contribution of ε_k in (5.16), an improvement of the function values is still obtained as

$$f(x_{k+1}) \le f(x_k) - \left[\frac{(a_k + b_k)^2}{2b_k} - \rho a_k\right] \le f(x_k). \tag{5.17}$$

In order to get the algorithm, a procedure for the computation of b_k is needed. For this, at point $z = x_k + \alpha_k d_k$,

$$f(z) = f(x_k + \alpha_k d_k) = f(x_k) + \alpha_k g_k^T d_k + \frac{1}{2}\alpha_k^2 d_k^T \nabla^2 f(\tilde{x}_k) d_k,$$

where $\tilde{x}_k$ is a point on the line segment connecting x_k and z. On the other hand, at point $x_k = z - \alpha_k d_k$,

$$f(x_k) = f(z - \alpha_k d_k) = f(z) - \alpha_k g_z^T d_k + \frac{1}{2}\alpha_k^2 d_k^T \nabla^2 f(\bar{x}_k) d_k,$$

where $g_z = \nabla f(z)$ and $\bar{x}_k$ is a point on the line segment connecting x_k and z. Having in view the local character of searching and that the distance between x_k and z is small enough, we can consider $\tilde{x}_k = \bar{x}_k = x_k$. So, adding the above equalities, the following value for b_k is obtained

$$b_k = -\alpha_k y_k^T d_k, \tag{5.18}$$

where $y_k = g_k - g_z$. Observe that the computation of b_k needs an additional evaluation of the gradient in point z. Therefore, neglecting the contribution of ε_k and considering $\eta_k = \eta_m = -a_k/b_k$ in (5.11), the following algorithm can be presented.

Algorithm 5.1 Accelerated conjugate gradient algorithm

1.	Choose a starting point $x_0 \in dom f$ and compute: $f_0 = f(x_0)$ and $g_0 = \nabla f(x_0)$. Select $\varepsilon_A > 0$ sufficiently small and positive values $0 < \rho < \sigma < 1$ used in Wolfe line search conditions. Set $d_0 = -g_0$ and $k = 0$
2.	Test a criterion for stopping the iterations. If the test is satisfied, then stop; otherwise continue with step 3
3.	Using the standard Wolfe line search conditions (5.6) and (5.7) determine the stepsize α_k
4.	Compute: $z = x_k + \alpha_k d_k$, $g_z = \nabla f(z)$ and $y_k = g_k - g_z$
5.	Compute: $a_k = \alpha_k g_k^T d_k$, and $b_k = -\alpha_k y_k^T d_k$
6.	If $\lvert b_k \rvert \geq \varepsilon_A$, then compute $\eta_k = -a_k/b_k$ and update the variables as $x_{k+1} = x_k + \eta_k \alpha_k d_k$, otherwise update the variables as $x_{k+1} = x_k + \alpha_k d_k$. Compute f_{k+1} and g_{k+1}. Compute $y_k = g_{k+1} - g_k$ and $s_k = x_{k+1} - x_k$
7.	Determine β_k according to the conjugate gradient in use
8.	Compute the search direction as: $d_{k+1} = -g_{k+1} + \beta_k d_k$
9.	Restart criterion. If $\lvert g_{k+1}^T g_k \rvert > 0.2\lVert g_{k+1} \rVert^2$ then set $d_{k+1} = -g_{k+1}$
10.	Set $k = k + 1$ and go to step 2 ♦

Usually, ε_A is epsilon machine. The unaccelerated conjugate gradient algorithm can immediately be obtained from the above algorithm by skipping steps 4 and 5 and by modifying step 6 where the variables are updated. In step 7, where the conjugate gradient parameter β_k is computed, we can consider the formula corresponding to the conjugate gradient algorithm selected for acceleration.

It is well known that, if f is bounded along the direction d_k, then there exists a stepsize α_k satisfying the Wolfe line search conditions (5.6) and (5.7) (see Proposition 1.2). In our algorithm, when the Powell restart criterion is satisfied, then the algorithm is restarted with the negative gradient $-g_{k+1}$. Under reasonable assumptions, the Wolfe conditions and the Powell restart criterion are sufficient to prove the global convergence of the algorithm.

Observe that, if $|a_k| > b_k$, then $\eta_k > 1$. In this case, $\eta_k \alpha_k > \alpha_k$ and it is also possible that $\eta_k \alpha_k \leq 1$ or $\eta_k \alpha_k > 1$. Hence, the stepsize $\eta_k \alpha_k$ can be greater than 1. On the other hand, if $|a_k| \leq b_k$, then $\eta_k \leq 1$. In this case, $\eta_k \alpha_k \leq \alpha_k$, so the stepsize $\eta_k \alpha_k$ is reduced. Therefore, if $|a_k| \neq b_k$, then $\eta_k \neq 1$ and the stepsize α_k computed by Wolfe conditions will be modified by its increasing or its reducing through factor η_k.

Neglecting ε_k in (5.14), we see that $\Psi_k(1) = 0$ and if $|a_k| \leq b_k/2$, then $\Psi_k(0) = -a_k - b_k/2 \leq 0$ and $\eta_k < 1$. Therefore, for any $\eta \in [0, 1]$, $\Psi_k(\eta) \leq 0$. As a consequence, for any $\eta \in (0, 1)$ it follows that $f(x_k + \eta \alpha_k d_k) < f(x_k)$. In this case, for any $\eta \in [0, 1]$, $\eta_k \alpha_k \leq \alpha_k$. However, in our algorithm, we selected $\eta_k = \eta_m$ as the point achieving the minimum value of $\Psi_k(\eta)$.

In the following, for strongly convex functions, let us prove the *linear convergence* of the acceleration scheme (Andrei, 2009c). Recall that a function f is strongly convex on the level set $S = \{x \in \mathbb{R}^n : f(x) \leq f(x_0)\}$ if there is a positive constant m so that

$$f(y) \geq f(x) + \nabla f(x)^T (y - x) + \frac{1}{2} m \|y - x\|^2$$

for all $x, y \in S$. For strongly convex functions, it is easy to prove that

$$\|\nabla f(x)\|^2 \geq 2m(f(x) - f(x^*)),$$

for all $x \in S$, where x^* is a local solution of (5.1) (See Appendix A).

Proposition 5.1 *Suppose that f is a strongly convex function on the level set S and d_k satisfies the sufficient descent condition $g_k^T d_k < -c_1 \|g_k\|^2$, where $c_1 > 0$, and $\|d_k\|^2 \leq c_2 \|g_k\|^2$, where $c_2 > 0$. Then, the sequence $\{x_k\}$ generated by the accelerated conjugate gradient Algorithm* 5.1 *converges linearly to x^*, solution to problem* (5.1).

Proof From (5.16), $f(x_{k+1}) \leq f(x_k)$ for all k. Since f is bounded below, it follows that

$$\lim_{k \to \infty} (f(x_k) - f(x_{k+1})) = 0.$$

Now, since f is strongly convex, there exist positive constants m and M so that $mI \leq \nabla^2 f(x) \leq MI$ on S. Suppose that $x_k + \alpha d_k \in S$ and $x_k + \eta_m \alpha d_k \in S$ for all $\alpha > 0$. Therefore,

$$f(x_k + \eta_m \alpha d_k) \leq f(x_k + \alpha d_k) - \frac{(a_k + b_k)^2}{2b_k}.$$

But, from strong convexity, the following quadratic upper bound of $f(x_k + \alpha d_k)$ is true

$$f(x_k + \alpha d_k) \leq f(x_k) + \alpha g_k^T d_k + \frac{1}{2} M \alpha^2 \|d_k\|^2.$$

Therefore,

$$\begin{aligned} f(x_k + \alpha d_k) &\leq f(x_k) - \alpha c_1 \|g_k\|^2 + \frac{1}{2} M c_2 \alpha^2 \|g_k\|^2 \\ &= f(x_k) + \left[-c_1 \alpha + \frac{1}{2} M c_2 \alpha^2 \right] \|g_k\|^2. \end{aligned}$$

Observe that for $0 \leq \alpha \leq c_1/(Mc_2)$, $-c_1\alpha + \frac{1}{2} M c_2 \alpha^2 \leq -\frac{c_1}{2}\alpha$ which follows from the convexity of $-c_1\alpha + (Mc_2/2)\alpha^2$. Using this result, it follows that

$$f(x_k + \alpha d_k) \leq f(x_k) - \frac{1}{2} c_1 \alpha \|g_k\|^2 \leq f(x_k) - \rho c_1 \alpha \|g_k\|^2,$$

since $\rho < 1/2$.

From Proposition 1.2, the Wolfe line search terminates with a value $\alpha \geq \gamma > 0$, where $\gamma > 0$ is a constant. Therefore, for $0 \leq \alpha \leq c_1/(Mc_2)$, this provides a lower bound on the decrease in function f, i.e.,

$$f(x_k + \alpha d_k) \leq f(x_k) - \rho c_1 \gamma \|g_k\|^2. \tag{5.19}$$

On the other hand,

$$\frac{(a_k + b_k)^2}{2b_k} \geq \frac{(\alpha^2 M c_2 - \alpha c_1)^2 \|g_k\|^4}{2\alpha^2 M c_2 \|g_k\|^2} \geq \frac{(\gamma M c_2 - c_1)^2}{2Mc_2} \|g_k\|^2. \tag{5.20}$$

Considering (5.19) and (5.20) together, it follows that

$$f(x_k + \eta_m \alpha d_k) \leq f(x_k) - \rho c_1 \gamma \|g_k\|^2 - \frac{(\gamma M c_2 - c_1)^2}{2Mc_2} \|g_k\|^2. \tag{5.21}$$

Therefore,

$$f(x_k) - f(x_k + \eta_m \alpha d_k) \geq \left[\rho c_1 \gamma + \frac{(\gamma M c_2 - c_1)^2}{2Mc_2}\right] \|g_k\|^2.$$

But, $f(x_k) - f(x_{k+1}) \to 0$ and consequently g_k goes to zero as well, i.e. x_k converges to x^*. Having in view that $f(x_k)$ is a nonincreasing sequence, it follows that $f(x_k)$ converges to $f(x^*)$. From (5.21), it results that

$$f(x_{k+1}) \leq f(x_k) - \left[\rho c_1 \gamma + \frac{(\gamma M c_2 - c_1)^2}{2Mc_2}\right] \|g_k\|^2. \tag{5.22}$$

Combining this with $\|g_k\|^2 \geq 2m(f(x_k) - f(x^*))$ and subtracting f^* from both sides of (5.22), the conclusion is that

$$f(x_{k+1}) - f(x^*) \leq c(f(x_k) - f(x^*)),$$

where

$$c = 1 - 2m\left[\rho c_1 \gamma + \frac{(\gamma M c_2 - c_1)^2}{2Mc_2}\right] < 1.$$

Therefore, $f(x_k)$ converges to $f(x^*)$ at least as fast as a geometric series with a factor that depends on the parameter ρ in the first Wolfe condition and on the bounds m and M, i.e., the convergence is at least linear. ♦

Remark 5.1 Basically, the acceleration scheme modifies the stepsize α_k in a multiplicative manner to improve the reduction of the function values along the iterations. In the accelerated algorithm, instead of (5.2), the new estimation of the minimum point is computed as

$$x_{k+1} = x_k + \xi_k \alpha_k d_k, \tag{5.23}$$

where the acceleration factor ξ_k is computed as

$$\xi_k = -\frac{a_k}{b_k}, \tag{5.24}$$

$a_k = \alpha_k g_k^T d_k$, $b_k = -\alpha_k (g_k - g_z)^T d_k$, $g_z = \nabla f(z)$, and $z = x_k + \alpha_k d_k$. Hence, if $|b_k| \geq \varepsilon_A$, where $\varepsilon_A > 0$ is sufficiently small, then the new estimation of the solution is computed as $x_{k+1} = x_k + \xi_k \alpha_k d_k$, otherwise $x_{k+1} = x_k + \alpha_k d_k$. ♦

Observe that

$$b_k = \alpha_k (g_z - g_k)^T d_k = \alpha_k (d_k^T \nabla^2 f(\bar{x}_k) d_k),$$

where $\bar{x}_k$ is a point on the line segment connecting x_k and z. Since $\alpha_k > 0$, it follows that for convex functions, $b_k \geq 0$.

5.3 Numerical Study

In this section, let us report some numerical results obtained with a Fortran implementation of conjugate gradient algorithms and their accelerated variants. The test functions are from the UOP collection (Andrei, 2018 g), which includes 80 functions. For each test function, ten numerical experiments with the number of variables $n = 1000, 2000, \ldots, 10000$ have been considered, thus obtaining a number of 800 problems. In the following, the numerical performance of conjugate gradient algorithms and their accelerated version corresponding to two formulae for β_k computation: PRP+ and DY are presented. The stepsize is computed using the Fortran code of the line search procedure shown in Figure 5.1. The comparisons of algorithms are given in the context of Remark 1.1.

In the first set of numerical experiments, we compare the PRP+ and DY conjugate gradient methods versus their accelerated variants. The accelerated version of PRP+ is called ACCPRP+. The accelerated version of DY is called ACCDY. In other words, Algorithm 5.1 is particularized in step 7 for the conjugate gradient methods PRP+, where $\beta_k^{PRP+} = \max\{0, g_{k+1}^T y_k / g_k^T g_k\}$ and DY, where $\beta_k^{DY} = g_{k+1}^T g_{k+1} / d_k^T y_k$. Figure 5.2 illustrates the Dolan and Moré performance profiles of ACCPRP+ versus PRP+ and of ACCDY versus DY.

From Figure 5.2, we see that the accelerated versions of the PRP+ and DY are more efficient and more robust. For example, subject to the CPU time metric, ACCPRP+ was faster in 225 problems, while PRP + was faster in 165 problems, etc.

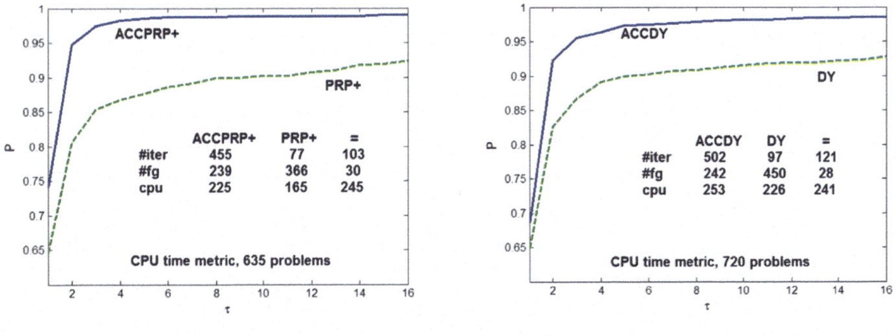

Figure 5.2 Performance profiles of ACCPRP+ versus PRP+ and of ACCDY versus DY

In the second set of numerical experiments, let us solve the applications from the MINPACK-2 collection with the accelerated conjugate gradient algorithms: ACCHS, ACCFR, ACCPRP, ACCPRP+, ACCCD, ACCLS, and ACCDY, the accelerated variants of HS, FR, PRP, PRP+, CD, LS, and DY (see Table 4.1).

Tables 5.1, 5.2 and 5.3 present the number of variables of the application (n), the performances of these accelerated algorithms subject to the number of iterations (*#iter*), the number of function and its gradient evaluations (*#fg*), and the CPU computing time (*cpu*) in seconds for solving all five applications from the

Table 5.1 Performances of ACCHS, ACCFR, and ACCPRP for solving five applications from the MINPACK-2 collection

		ACCHS			ACCFR			ACCPRP		
	n	*#iter*	*#fg*	*cpu*	*#iter*	*#fg*	*cpu*	*#iter*	*#fg*	*cpu*
A1	40,000	246	488	3.99	258	501	3.27	253	506	3.34
A2	40,000	617	1177	30.67	771	1329	13.27	632	1227	12.05
A3	40,000	937	1769	28.37	4188	7511	119.16	857	1684	26.64
A4	40,000	299	602	31.13	322	625	32.03	296	603	27.51
A5	40,000	374	679	8.44	1065	1853	22.79	360	688	8.63
Total	–	2473	4715	102.60	6604	11819	190.52	2398	4708	78.17

Table 5.2 Performances of ACCPRP+ and ACCCD for solving five applications from the MINPACK-2 collection

		ACCPRP+			ACCCD		
	n	*#iter*	*#fg*	*cpu*	*#iter*	*#fg*	*cpu*
A1	40,000	270	523	3.45	344	605	3.97
A2	40,000	617	1202	11.69	911	1531	15.15
A3	40,000	857	1684	26.64	7605	14413	269.50
A4	40,000	296	603	23.08	362	670	25.57
A5	40,000	360	688	8.44	722	1266	15.39
Total	–	2400	4700	73.30	9944	18485	329.58

Table 5.3 Performances of ACCLS and ACCDY for solving five applications from the MINPACK-2 collection

		ACCLS			ACCDY		
	n	*#iter*	*#fg*	*cpu*	*#iter*	*#fg*	*cpu*
A1	40,000	252	499	6.10	256	500	12.09
A2	40,000	602	1181	11.51	685	1246	12.34
A3	40,000	809	1608	25.42	4660	7954	128.66
A4	40,000	296	598	22.85	309	612	23.35
A5	40,000	365	691	8.48	1007	1763	21.80
Total	–	2324	4577	74.36	6917	12075	198.24

MINPACK-2 collection. For example, for solving the application A1, ACCHS needs 246 iterations, 488 evaluations of the minimizing function and its gradient and 3.99 s, etc. It is worth comparing the performances of the accelerated conjugate gradient methods presented in Tables 5.1, 5.2 and 5.3 versus their unaccelerated versions, shown in Tables 4.2, 4.3 and 4.4. For example, for solving all five applications, the HS method needs a total of 4579 iterations, 5620 evaluations of function and its gradient, and 139.65 s. In contrast, the accelerated HS method, ACCHS, needs a total of 2473 iterations, 4715 evaluations of function and its gradient, and 102.60 s. The accelerated variants of the standard conjugate gradient methods considered in this numerical study are more efficient subject to all criteria given by the number of iterations, by the number of evaluations of function and its gradient and by the CPU computing time. An explanation for this behavior is that in the conjugate gradient methods, the evolution of the stepsize is totally unpredictable. Thus, there is more room for changing the stepsize along the iterations given by (5.11).

Notes and References

The acceleration of the steepest descent method with backtracking was first given by Andrei (2006a). The acceleration of the conjugate gradient methods was presented in Andrei (2009c, 2010b). An accelerated conjugate gradient algorithm with modified secant condition for unconstrained optimization was presented in (Andrei, 2009i). The acceleration technique using the modification of the stepsize, described in this chapter, is specific to conjugate gradient methods. This is motivated by the fact that the stepsize in any conjugate gradient algorithm has an unpredictable behavior. The conjugate gradient algorithms presented in this book may include the acceleration procedure discussed in this chapter. Numerical experiments proved that the accelerated versions of the algorithms may have better performances versus the unaccelerated algorithms (see Andrei, 2009i). In Chapter 10, another acceleration of the conjugate gradient algorithms will be introduced, based on the concept of preconditioning.

Chapter 6
Hybrid and Parameterized Conjugate Gradient Methods

Numerical experiments with standard conjugate gradient methods showed that the methods FR, DY, and CD have modest numerical performances, being affected by jamming, although they have strong convergence properties. On the other hand, the computational performances of HS, PRP, and LS methods are better, even if their convergence properties are weaker.

The idea behind the hybrid conjugate gradient methods considered in this chapter is to combine the standard conjugate gradient methods in order to exploit the attractive features of each of them. The standard conjugate gradient methods may be combined in two distinct ways, thus obtaining two classes of hybrid conjugate gradient methods. The first class is based on the *projection concept*. They consider a pair of standard conjugate gradient methods and use one of them as soon as a certain criterion is satisfied. When the criterion is violated, then the other conjugate gradient method from the pair is used. The hybrid conjugate gradient methods based on the projection concept have simplicity of their algebraic expression. As soon as a conjugate gradient method enters a jamming phase, then the hybrid scheme triggers another conjugate gradient method, thus trying to improve the numerical performances of the hybrid one. The second class of hybrid conjugate gradient methods is based on the *convex combination* of the standard methods. In this case, some standard conjugate gradient methods are combined in a convex manner, thus obtaining a hybrid conjugate gradient method. In general, the hybrid methods are more efficient and more robust than the standard ones.

For solving the unconstrained optimization problem

$$\min f(x), \tag{6.1}$$

where $f:\mathbb{R}^n \to \mathbb{R}$ is a continuous differentiable function lower bounded, a *hybrid conjugate gradient* method generates the sequence $\{x_k\}$ by using the recursive scheme

N. Andrei, *Nonlinear Conjugate Gradient Methods for Unconstrained Optimization*, Springer Optimization and Its Applications 158,
https://doi.org/10.1007/978-3-030-42950-8_6

$$x_{k+1} = x_k + \alpha_k d_k, k = 0, 1, \ldots, \tag{6.2}$$

where x_0 is a known initial point. In (6.2), the scalar α_k is the stepsize, often determined by the standard Wolfe line search (1.12) and (1.13), or by the strong Wolfe line search (1.12) and (1.14). Knowing the initial search direction $d_0 = -g_0$, where $g_0 = \nabla f(x_0)$, then the search direction d_k is updated as

$$d_{k+1} = -g_{k+1} + \beta_k d_k, k = 0, 1, \ldots, \tag{6.3}$$

where β_k is a scalar parameter which defines the hybrid conjugate gradient algorithm. In hybrid conjugate gradient methods, this parameter is computed as projection or as convex combinations of the standard methods.

6.1 Hybrid Conjugate Gradient Methods Based on the Projection Concept

Table 6.1 presents some hybrid selections of the parameter β_k based on the projection of the standard conjugate gradient methods presented in Table 4.1.

As it can be seen, there is a large variety of hybrid conjugate gradient methods. Their purpose is to combine the properties of the standard ones in order to get new ones, rapid convergent to the solution. The idea is to avoid jamming. As we know, the FR method has strong convergence properties, but it may not perform well in computational experiments. On the other hand, although the PRP method and the HS method may not generally converge, they often perform better than FR. Therefore, the combination of these methods tries to exploit attractive features of each one. Thus, in the hybrid conjugate gradient method TAS, proposed by Touati-Ahmed and Storey (1990), if the iterations are affected by jamming, then the method commutes from FR to PRP. The same motivation is for the hybrid computational scheme PRP-FR, proposed by Hu and Storey (1991). Indeed, the PRP method possesses a built-in restart feature that addresses directly to jamming. When the step s_k is small, the factor y_k in the numerator of β_k^{PRP} tends to zero. Hence, in this case, the search direction d_{k+1} computed as in (6.3) with $\beta_k = \beta_k^{\mathrm{PRP}}$ is essentially the steepest descent direction $-g_{k+1}$. The DY method has even better global convergence properties than the FR method. Consequently, Dai and Yuan (2001a) combined their algorithm with the HS algorithm and proposed the hybrid scheme with $\beta_k^{\mathrm{HS-DY}}$ or $\beta_k^{h\mathrm{DY}}$ in (6.3). The HS-DY hybrid conjugate gradient method was also discussed by Andrei (2008g). Gilbert and Nocedal (1992) showed that even for strongly convex functions, it is quite possible for β_k^{PRP} to be negative. Therefore, in order to extend the number of iterations in which β_k^{PRP} is used, they suggest the hybrid method with β_k^{GN} in (6.3) to maintain the global convergence.

Table 6.1 Hybrid selection of β_k based on the projection concept

$\beta_k^{TAS} = \begin{cases} \beta_k^{PRP}, & 0 \le \beta_k^{PRP} \le \beta_k^{FR}, \\ \beta_k^{FR}, & \text{otherwise.} \end{cases}$	Proposed by Touati-Ahmed and Storey (1990)
$\beta_k^{PRP-FR} = \max\{0, \min\{\beta_k^{PRP}, \beta_k^{FR}\}\}$	Proposed by Hu and Storey (1991)
$\beta_k^{GN} = \max\{-\beta_k^{FR}, \min\{\beta_k^{PRP}, \beta_k^{FR}\}\}$	Proposed by Gilbert and Nocedal (1992)
$\beta_k^{HS-DY} = \max\{0, \min\{\beta_k^{HS}, \beta_k^{DY}\}\}$	Proposed by Dai and Yuan (2001a) and Dai and Ni (2003)
$\beta_k^{hDY} = \max\left\{-\left(\frac{1-\sigma}{1+\sigma}\right)\beta_k^{DY}, \min\{\beta_k^{HS}, \beta_k^{DY}\}\right\}$	Proposed by Dai and Yuan (2001a). σ is the parameter from the second Wolfe line search condition
$\beta_k^{DDF} = \frac{\|g_{k+1}\|^2}{\max\{-d_k^T g_k, d_k^T y_k\}}.$	Proposed by Dai (2002a)
$\beta_k^{VFR} = \frac{\mu_1 \|g_{k+1}\|^2}{\mu_2 \lvert g_{k+1}^T d_k \rvert + \mu_3 \|g_k\|^2}$	Proposed by Wei, Li, and Qi (2006b). $\mu_1 > 0$, $\mu_3 > 0$ and $\mu_2 > \mu_1$ are parameters
$\beta_k^{VPRP} = \frac{\|g_{k+1}\|^2 - \|g_{k+1}\|(g_{k+1}^T g_k)/\|g_k\|}{\|g_k\|^2}$	Proposed by Wei, Yao, and Liu (2006)
$\beta_k^{YWH} = \frac{\|g_{k+1}\|^2 - \|g_{k+1}\|(g_{k+1}^T g_k)/\|g_k\|}{d_k^T y_k}$	Proposed by Yao, Wei, and Huang (2007)
$\beta_k^{JCP} = \begin{cases} \beta_k^{DY}, & \text{if } g_{k+1}^T d_k \ge \|g_{k+1}\|^2, \\ \beta_k^{FR}, & \text{else.} \end{cases}$	Proposed by Jiao, Chen, and Pan (2007)
$\beta_k^{LS-CD} = \max\{0, \min\{\beta_k^{LS}, \beta_k^{CD}\}\}$	Proposed by Andrei (2008f)
$\beta_k^{DPRP} = \frac{\|g_{k+1}\|^2 - \|g_{k+1}\|(g_{k+1}^T g_k)/\|g_k\|}{\mu \lvert g_{k+1}^T d_k \rvert + \|g_k\|^2}$	Proposed by Dai and Wen (2012). $\mu > 1$ is a parameter
$\beta_k^{JHJ} = \frac{\|g_{k+1}\|^2 - \|g_{k+1}\|\max\{0, (g_{k+1}^T g_k)\}/\|g_k\|}{\max\{\|g_k\|^2, d_k^T y_k\}}$	Proposed by Jian, Han, and Jiang (2015)

The hybrid conjugate gradient method DDF was proposed by Dai (2002a). It employs either the DY or the FR standard conjugate gradient algorithms. He established that this hybrid scheme generates descent directions, independent of the line search.

The VFR hybrid conjugate gradient, which is a variation of the FR method proposed by Wei, Li, and Qi (2006b), satisfies the sufficient descent condition $g_k^T d_k \le -(1 - \mu_1/\mu_2)\|g_k\|^2$ independent of the line search.

The VPRP hybrid conjugate gradient method proposed by Wei, Yao, and Liu (2006) is a variant of the PRP method. This hybrid conjugate gradient method inherits the properties of PRP. Under the strong Wolfe line search with $0 < \sigma < 1/4$, Huang, Wei, and Yao (2007) showed that the search direction of the VPRP method satisfies the sufficient descent condition and the algorithm is globally convergent. An interesting property of VPRP is that it has the Property (*), first introduced by Gilbert and Nocedal (1992).

The YWH hybrid method proposed by Yao, Wei, and Huang (2007) is a variant of the HS method. They established that under the strong Wolfe line search with $\sigma \in (0, 1/3)$, the YWH method generates sufficient descent directions.

Jiao, Chen, and Pan (2007) proposed a hybrid conjugate gradient method JCP and established the global convergence under the Goldstein line search (1.11).

The hybrid conjugate gradient method LS-CD was proposed by Andrei (2008f). Its global convergence under the Wolfe line search was proved by Yang, Luo, and Dai (2013). Numerical experiments showed that the hybrid LS-CD method outperforms both the LS and the CD methods.

Dai and Wen (2012) proposed the hybrid DPRP method, where $\mu > 1$, and proved that the corresponding search direction satisfies the sufficient descent condition for any line search. They proved that the DPRP method is globally convergent under the standard Wolfe line search.

A hybridization of the HS and DY conjugate gradient methods was proposed by Babaie-Kafaki and Ghanbari (2015a). In their algorithm, the hybridization parameter is computed by solving the least-squares problem of minimizing the distance between the search direction of the hybrid method and the search direction corresponding to the three-term conjugate gradient method proposed by Zhang, Zhou, and Li (2007), which possesses the sufficient descent property.

The hybrid method JHJ proposed by Jian, Han, and Jiang (2015) generates sufficient descent directions for any line search and is globally convergent under the standard Wolfe line search.

Other hybrid conjugate gradient methods were proposed by: Zhang (2009a), Han, Zhang, and Chen (2017). Two descent hybrid conjugate gradient algorithms as projection of PRP and HS, and of HS and DY, respectively, were developed by Zhang and Zhou (2008).

The following general result proved by Dai, Han, Liu, Sun, Yin, and Yuan (1999) shows that there are a lot of possibilities to generate hybrid conjugate gradient methods, taking into consideration the FR method.

Theorem 6.1 *Suppose that the Assumption CG holds and consider any general method of the form* (6.2)–(6.3) *where the stepsize is determined by the strong Wolfe line search* (1.12) *and* (1.14) *with* $0<\rho<\sigma<1$ *and* β_k *satisfying*

$$\sigma|\beta_k| \leq \bar{\sigma}\beta_k^{FR}, \tag{6.4}$$

with $\bar{\sigma} \in (0, 1/2]$ *and*

$$\|g_{k+1}\|^2 \sum_{j=0}^{k} \prod_{i=j}^{k} \left(\frac{\beta_i}{\beta_i^{FR}} \right)^2 \leq c_2 k, \tag{6.5}$$

for some constant $c_2 > 0$. *Then,*

$$\liminf_{k\to\infty} \|g_k\| = 0. \tag{6.6}$$

Proof As known, $\beta_k^{\mathrm{FR}} = g_{k+1}^T g_{k+1}/g_k^T g_k$. Then, from (6.3), from the strong Wolfe line search $\left|g(x_k + \alpha_k d_k)^T d_k\right| \le -\sigma g_k^T d_k$ and (6.4), it results that

$$\begin{aligned}
\frac{-g_{k+1}^T d_{k+1}}{\|g_{k+1}\|^2} &= 1 - \beta_k \frac{g_{k+1}^T d_k}{\|g_{k+1}\|^2} = 1 - \left(\frac{\beta_k}{\beta_k^{FR}}\right) \frac{g_{k+1}^T d_k}{\|g_k\|^2} \\
&\le 1 + \left|\frac{\beta_k}{\beta_k^{FR}}\right| \frac{-\sigma g_k^T d_k}{\|g_k\|^2} \le 1 + \bar{\sigma}\left(\frac{-g_k^T d_k}{\|g_k\|^2}\right) \\
&\le \cdots \le \sum_{j=0}^{k} \bar{\sigma}^j + \bar{\sigma}^{k+1}\left(\frac{-g_0^T d_0}{\|g_0\|^2}\right) = \frac{1-\bar{\sigma}^{k+1}}{1-\bar{\sigma}} < \frac{1}{1-\bar{\sigma}}. \qquad (6.7)
\end{aligned}$$

Similarly, since $\bar{\sigma} \le 1/2$,

$$\frac{-g_{k+1}^T d_{k+1}}{\|g_{k+1}\|^2} \ge 1 - \bar{\sigma}\frac{1-\bar{\sigma}^k}{1-\bar{\sigma}} > 0. \qquad (6.8)$$

Therefore, d_k is a descent direction. Since $d_{k+1} + g_{k+1} = \beta_k d_k$, it follows that

$$\|d_{k+1}\|^2 = -\|g_{k+1}\|^2 - 2g_{k+1}^T d_{k+1} + \beta_k^2 \|d_k\|^2.$$

Thus,

$$\|d_{k+1}\|^2 \le -2g_{k+1}^T d_{k+1} + \beta_k^2 \|d_k\|^2. \qquad (6.9)$$

Using (6.9) recursively and noting that $d_0 = -g_0$, it follows that

$$\begin{aligned}
\|d_{k+1}\|^2 &\le -2g_{k+1}^T d_{k+1} - 2\sum_{j=0}^{k} \prod_{i=j}^{k} \beta_i^2 (g_j^T d_j) \\
&= -2g_{k+1}^T d_{k+1} - 2\|g_{k+1}\|^4 \sum_{j=0}^{k} \prod_{i=j}^{k} \left(\frac{\beta_i}{\beta_i^{\mathrm{FR}}}\right)^2 \left(\frac{g_j^T d_j}{\|g_j\|^4}\right). \qquad (6.10)
\end{aligned}$$

If the theorem is not true, then (6.5) holds and there exists a positive constant γ so that $\|g_k\| \ge \gamma$ for all k. Thus, from (6.7) and (6.10), it results that

$$\frac{\|d_{k+1}\|^2}{\|g_{k+1}\|^2} \le \frac{2}{1-\bar{\sigma}}\left(1 + \frac{\|g_{k+1}\|^2}{\gamma^2} \sum_{j=0}^{k} \prod_{i=j}^{k} \left(\frac{\beta_i}{\beta_i^{\mathrm{FR}}}\right)^2\right). \qquad (6.11)$$

From (6.11) and (6.5), it follows that

$$\sum_{k=0}^{\infty} \frac{\|g_k\|^2}{\|d_k\|^2} = +\infty.$$

This, together with Corollary 3.1, imply that $\liminf_{k\to\infty} \|g_k\| = 0$. ♦

The theorem shows that for $|\beta_k| \le (\bar{\sigma}/\sigma)\beta_k^{FR}$, the algorithm (6.2)–(6.3) with $d_0 = -g_0$ is globally convergent. In other words, to use β_k^{FR} in the hybrid schemes presented in Table 6.1 is quite natural. The theorem extends the result of Gilbert and Nocedal (1992) and Hu and Storey (1991) to the case when $\bar{\sigma} = 1/2$, i.e., when $2\sigma|\beta_k| < \beta_k^{FR}$. If $\bar{\sigma} \in (0, 1/2)$, then from (6.8) it follows that the sufficient descent $-g_{k+1}^T d_{k+1} \ge c\|g_{k+1}\|^2$ holds for any positive constant c. However, if $\bar{\sigma} = 1/2$, then only

$$\frac{-g_k^T d_k}{\|g_k\|^2} \ge \frac{1}{2^k},$$

holds, which does not imply the sufficient descent condition.

As known, the DY method has slightly better global convergence properties than FR (see also Figure 4.2). Therefore, Dai and Yuan (2001a) studied the possibility to combine their scheme with other conjugate gradient methods. Under the Wolfe line search and the Lipschitz continuity, they established the global convergence of the general conjugate gradient method (6.2) and (6.3) with $\beta_k \in [-\eta\beta_k^{DY}, \beta_k^{DY}]$, where $\eta = (1-\sigma)/(1+\sigma)$. In a research effort, Dai and Yuan (2001a) proposed the hybrid schemes with β_k^{HS-DY} and β_k^{hDY}.

Numerical study. In the following, let us present the performances of some hybrid conjugate gradient methods shown in Table 6.1: TAS, PRP-FR, GN, hDY, HS-DY, and LS-CD, based on the projection concept. For this, the set of 80 unconstrained optimization problems from the UOP collection is used (Andrei, 2018 g). Each problem from this collection is tested 10 times with the number of variables increasing as $n = 1000, 2000, \ldots, 10000$. Thus, all in all, a number of 800 unconstrained optimization problems have been solved for each method. The maximum number of iterations is limited to 2000. The comparisons among the algorithms are given in the context of Remark 1.1. The comparisons include the Dolan and Moré (2002) performance profiles.

Figure 6.1 presents the Dolan and Moré performance profiles of some hybrid conjugate gradient methods subject to the CPU time metric. The tables inside the plots show the performances of the algorithms subject to the number of iterations (*#iter*), the number of function and its gradient evaluations (*#fg*) and the CPU time in seconds (*cpu*).

For example, when comparing TAS versus PRP-FR in Figure 6.1, we see that TAS was better in 150 problems (i.e., TAS achieved the minimum number of iterations in 150 problems). PRP-FR was better in 307 problems and they achieved

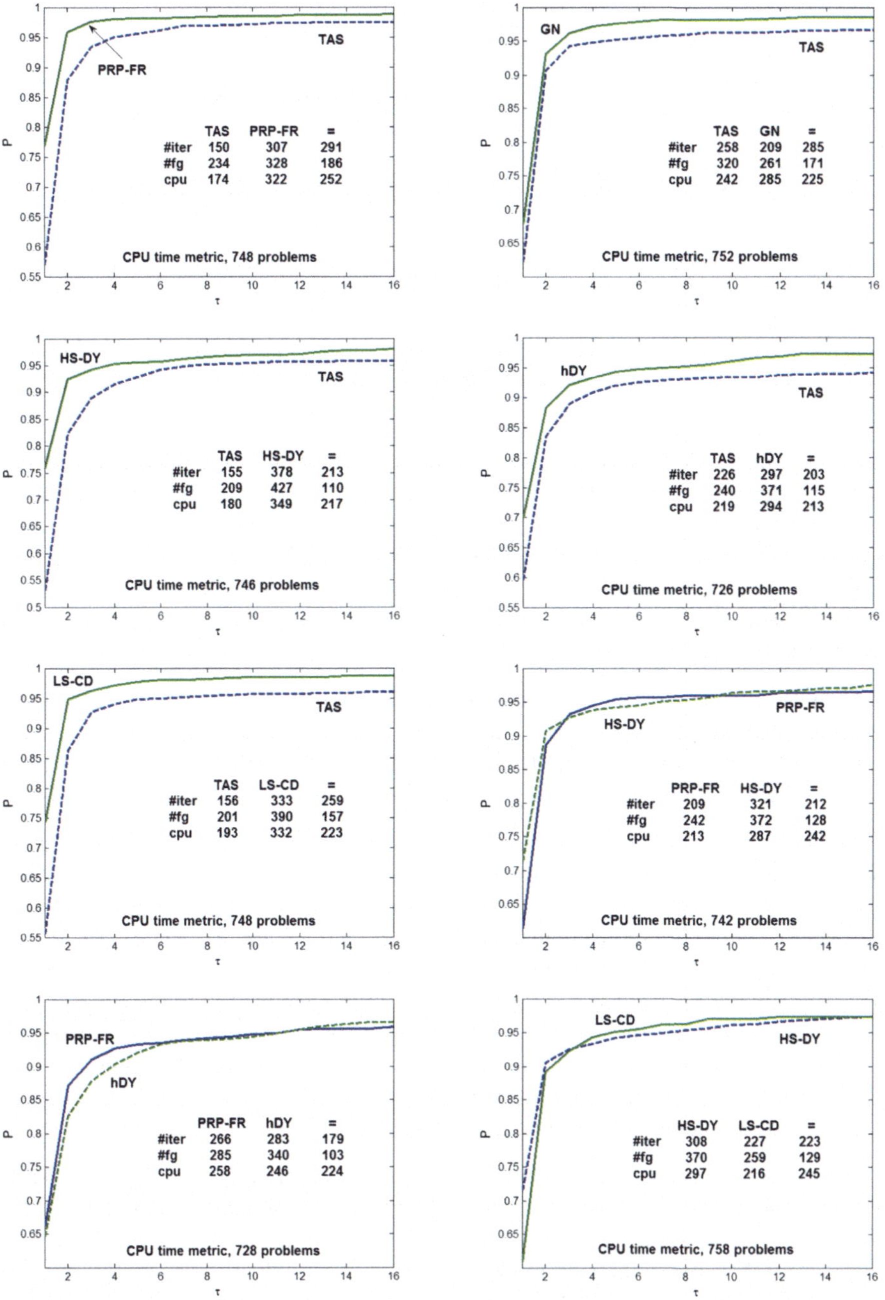

Figure 6.1 Performance profiles of some hybrid conjugate gradient methods based on the projection concept

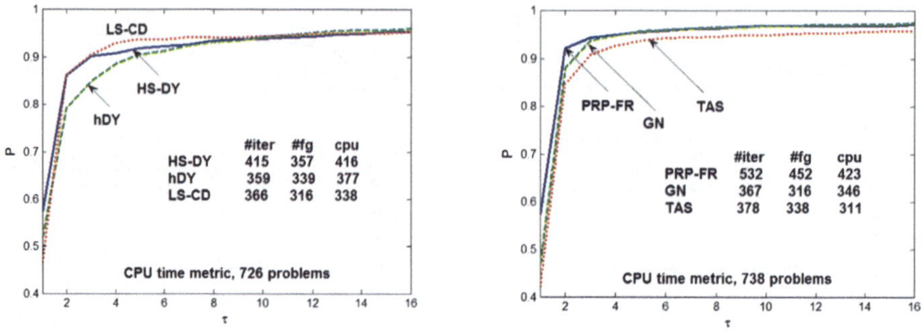

Figure 6.2 Performance profiles of the hybrid conjugate gradient methods HS-DY, hDY LS-CD, and of PRP-FR, GN, and TAS based on the projection concept

the same number of iterations in 291 problems, etc. Out of 800 problems considered in this numerical experiment, only for 748 problems does the criterion (1.118) hold.

From Figure 6.1, subject to the CPU time metric, it is clear that the TAS method is less efficient and less robust than PRP-FR, GN, HS-DY, hDY, and LS-CD hybrid conjugate gradient methods. On the other hand, the PRP-FR hybrid conjugate gradient method is faster than hDY. Notice that HS-DY is more efficient than LS-CD, etc.

Figure 6.2 separately presents the performance profiles of the hybrid conjugate gradient methods HS-DY, hDY, and LS-CD, and of PRP-FR, GN, and TAS, respectively. Obviously, subject to the CPU time metric, HS-DY is more efficient than hDY and LS-CD. Similarly, PRP-FR is more efficient than GN and TAS.

Figure 6.3 presents a global comparison among the hybrid conjugate gradient methods. Concerning their robustness, observe that, for this set of unconstrained optimization problems, subject to the CPU time metric, the LS-CD algorithm is top performer, being more robust than the hybrid conjugate gradient methods considered in this study. On the other hand, HS-DY is the most efficient. Less efficient are TAS and GN. Anyway, the performance profiles of the compared hybrid methods are grouped, one method being slightly more efficient or more robust than the other one. Out of 800 problems in this numerical experiment, only for 714 problems does the criterion (1.118) hold. The table inside Figure 6.3 shows that out of 714 problems, HS-DY was faster in 328 problems, followed by hDY, which was faster in 316 problems, etc.

Figure 6.4 presents comparisons of the hybrid conjugate gradient methods (HS-DY and PRP-FR) versus the standard conjugate gradient methods (HS, PRP, PRP+, LS). Observe that HS-DY is more efficient and slightly more robust than PRP+. HS-DY is more efficient, but less robust than LS. The hybrid HS-DY is more efficient than both HS and PRP, but it is less robust than them. On the other hand, PRP-FR is less efficient than PRP+, LS, and HS and less robust than LS, HS, and PRP. These comparisons point out that PRP-FR and HS-DY are less robust than the standard conjugate gradient methods LS, HS, and PRP.

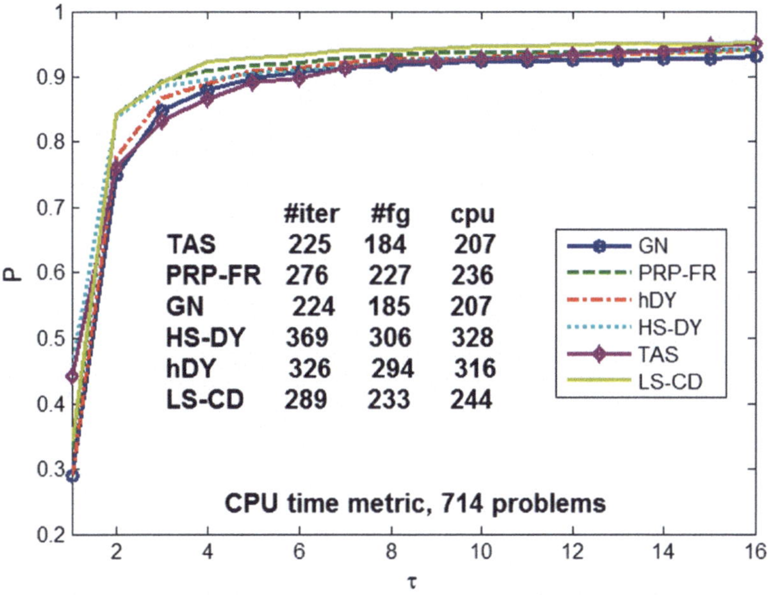

Figure 6.3 Global performance profiles of six hybrid conjugate gradient methods

The DY method is known to have better convergence properties that the FR method. Therefore, Dai and Yuan (2001a) studied the possibility of combining DY with other conjugate gradient methods in a projection way. As a result, the HS-DY and the hDY hybrid methods were proposed. Our numerical results are in agreement with those of Dai and Ni (2003), indicating that hDY is the most efficient performing more efficiently that the PRP+ , LS, and HS methods.

From these computational experiments with a large set of nonlinear unconstrained optimization problems with different structure and complexities, we see that the hybrid conjugate gradient methods do not have spectacular performances versus the standard conjugate gradient methods, subject to the CPU time metric. The first hybrid conjugate gradient method based on the projection concept was given by Touati-Ahmed and Storey (1990) and by Hu and Storey (1991). The motivation of introducing the hybrid conjugate gradient methods was to use the PRP update parameter when the iterations jam. However, the PRP-FR method is less efficient and less robust than LS, HS, and PRP+.

In the following, the performances of the hybrid conjugate gradient methods based on the projection concept for solving some applications from the MINPACK-2 applications are presented. Tables 6.2 and 6.3 show the performances of the hybrid conjugate gradient methods for solving five MINPACK-2 applications, each of them having 40,000 variables.

For this set of large-scale applications, it appears that the hybrid PRP-FR is the fastest method, followed by GN, etc. The most time consuming is HS-DY.

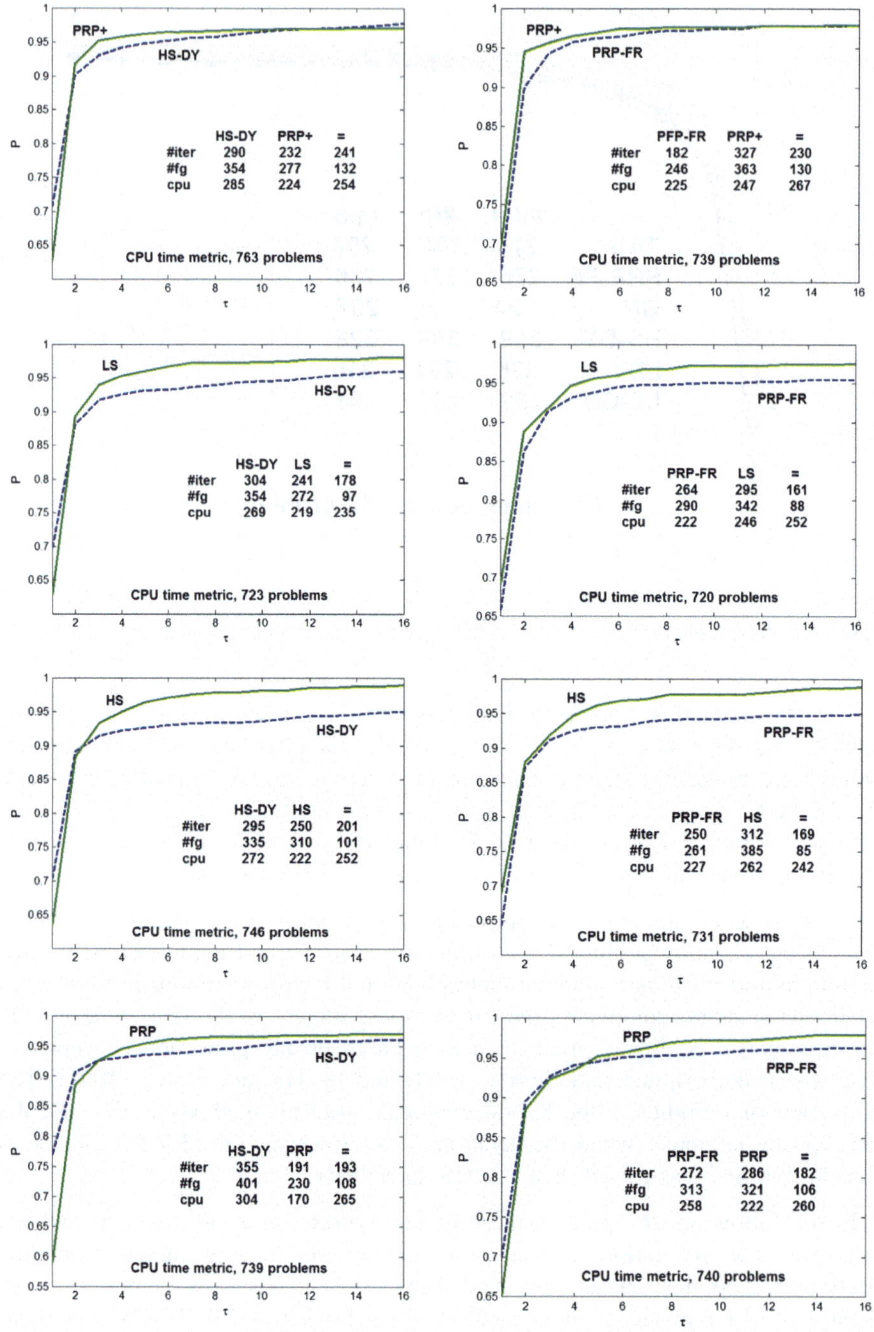

Figure 6.4 Performance profiles of the hybrid conjugate gradient methods (HS-DY, PRP-FR) versus the standard conjugate gradient methods (PRP+ , LS, HS, PRP)

Table 6.2 Performances of TAS, PRP-FR, and GN for solving five applications from the MINPACK-2 collection

		TAS			PRP-FR			GN		
	n	*#iter*	*#fg*	*cpu*	*#iter*	*#fg*	*cpu*	*#iter*	*#fg*	*cpu*
A1	40,000	402	576	6.80	385	540	6.65	460	650	7.97
A2	40,000	602	1179	15.46	602	1179	15.46	602	1179	15.45
A3	40,000	806	1615	36.76	806	1615	34.86	806	1615	34.55
A4	40,000	294	595	30.21	294	595	30.29	294	595	30.21
A5	40,000	368	701	11.45	368	701	11.42	368	701	11.44
Total	–	2472	4666	100.68	2455	4630	98.68	2530	4740	99.62

Table 6.3 Performances of HS-DY, hDY, and LS-CD for solving five applications from the MINPACK-2 collection

		HS-DY			*h*DY			LS-CD		
	n	*#iter*	*#fg*	*cpu*	*#iter*	*#fg*	*cpu*	*#iter*	*#fg*	*cpu*
A1	40,000	533	684	6.97	591	753	6.89	402	563	6.44
A2	40,000	935	1771	23.67	935	1771	23.73	649	1242	16.36
A3	40,000	1034	1939	42.53	1034	1939	41.90	1159	2073	45.28
A4	40,000	577	1148	58.39	577	1148	58.32	294	594	30.30
A5	40,000	400	760	12.62	400	760	12.64	375	702	11.53
Total	–	3479	6302	144.18	3537	6371	143.48	2879	5174	109.91

A comparison of the performances of the limited-memory BFGS method, L-BFGS $(m = 5)$, presented in Table 1.2, shows that for solving all five applications, L-BFGS $(m = 5)$ needs 102.92 s. Similarly, the performances of the truncated Newton method, TN, presented in Table 1.3, show that for solving all five applications, TN needs 104.57 s. On the other hand, PRP-FR needs only 98.68 s, as seen in Table 6.2. Although the differences are not spectacular, the hybrid PRP-FR method is faster than both L-BFGS $(m = 5)$ and TN. In the same realm of comparisons, GN is also faster versus L-BFGS $(m = 5)$ and TN.

However, the differences are more dramatic when the performances of the hybrid conjugate gradient methods are compared versus the performances of the standard conjugate gradient methods. Table 4.3 shows that for solving all five applications, CD needs only 137.09 s, the minimum time among all the methods considered in the numerical study. Clearly, the hybrid methods are top performers versus the standard methods. But the things are not as they seem to be. Table 5.2 points out that the accelerated method ACCPRP+ needs only 73.30 s.

6.2 Hybrid Conjugate Gradient Methods as Convex Combinations of the Standard Conjugate Gradient Methods

The hybrid conjugate gradient methods based on the convex combination of the standard conjugate gradient methods are defined by (6.2) and (6.3), where the conjugate gradient parameter β_k is computed as

$$\beta_k = (1-\theta_k)\beta_k^{M1} + \theta_k \beta_k^{M2}, \tag{6.12}$$

where β_k^{M1} and β_k^{M2} are the conjugate gradient parameters of the standard conjugate gradient methods which we want to hybridize (see Table 4.1) and $0 \le \theta_k \le 1$ is a parameter. The idea is to combine in a convex way the standard algorithms in order to get algorithms with better performances. The parameter θ_k in the convex combination (6.12) can be determined by means of two procedures. The first one is given by the conjugacy condition $y_k^T d_{k+1} = 0$, where

$$d_{k+1} = -g_{k+1} + ((1-\theta_k)\beta_k^{M1} + \theta_k \beta_k^{M2})d_k. \tag{6.13}$$

From the equality $y_k^T d_{k+1} = 0$, where d_{k+1} is given by (6.13), the following value for the parameter θ_k is obtained

$$\theta_k = \frac{y_k^T g_{k+1} - \beta_k^{M1}(y_k^T d_k)}{(\beta_k^{M2} - \beta_k^{M1})(y_k^T d_k)}. \tag{6.14}$$

Obviously, instead of the standard conjugacy condition $y_k^T d_{k+1} = 0$, it is quite possible to use the Dai and Liao conjugacy condition

$$y_k^T d_{k+1} = -t s_k^T g_{k+1}, \tag{6.15}$$

where $t \ge 0$ is a scalar parameter. In this case, from (6.15), where d_{k+1} is given by (6.13), it results that

$$\theta_k = \frac{y_k^T g_{k+1} - t s_k^T g_{k+1} - \beta_k^{M1}(y_k^T d_k)}{(\beta_k^{M2} - \beta_k^{M1})(y_k^T d_k)}. \tag{6.16}$$

This is another value for the parameter θ_k obtained from the Dai and Liao conjugacy condition (6.15). In this case, however, a value for t must be chosen, which is quite difficult [see Andrei (2011b)].

On the other hand, if the point x_{k+1} is close enough to a local minimizer x^*, then a good direction to follow is the one given by the Newton direction, that is, $d_{k+1} = -\nabla^2 f(x_{k+1})^{-1} g_{k+1}$. Therefore, the second procedure to determine θ_k in

(6.12) considers the formal equality between the search direction of the hybrid algorithm and the Newton direction, i.e.,

$$-g_{k+1} + ((1-\theta_k)\beta_k^{M1} + \theta_k\beta_k^{M2})d_k = -\nabla^2 f(x_{k+1})^{-1} g_{k+1}. \tag{6.17}$$

Observe that (6.17) is only a technical argument to get a value for θ_k. With some simple algebraic manipulations, from (6.17),

$$\theta_k = \frac{s_k^T \nabla^2 f(x_{k+1}) g_{k+1} - s_k^T g_{k+1} - \beta_k^{M1}(s_k^T \nabla^2 f(x_{k+1}) d_k)}{(\beta_k^{M2} - \beta_k^{M1})(s_k^T \nabla^2 f(x_{k+1}) d_k)}. \tag{6.18}$$

Both these procedures for θ_k computation are operational and can be used to generate hybrid conjugate gradient methods. The salient point in (6.18) for θ_k computation is the presence of the Hessian $\nabla^2 f(x_{k+1})$. For large-scale problems, choices for the update parameters that do not require the evaluation of the Hessian are preferred in practice over the methods that require the Hessian at each iteration. In order to avoid the exact computation of $\nabla^2 f(x_{k+1})$, a solution is to use the secant equation. In quasi-Newton methods, the secant equation plays an essential role for the approximation of the Hessian of the objective function at each iteration. Therefore, in the second procedure for θ_k computation, the search direction d_k can be calculated as solution of the following algebraic linear system $B_k d_k = -g_k$, where B_k is an approximation of $\nabla^2 f(x_k)$. In the quasi-Newton methods, the matrix B_k, which is symmetric and positive definite, is effectively updated to obtain a new matrix, B_{k+1}, also symmetric and positive definite, as an approximation of $\nabla^2 f(x_{k+1})$. The matrix B_{k+1} needs to satisfy some suitable equations, namely the secant equations, which include the second-order information. The most popular is the standard secant equation

$$B_{k+1} s_k = y_k. \tag{6.19}$$

By using the standard secant equation in (6.18), the following value for θ_k could be obtained

$$\theta_k = \frac{y_k^T g_{k+1} - s_k^T g_{k+1} - \beta_k^{M1}(y_k^T d_k)}{(\beta_k^{M2} - \beta_k^{M1})(y_k^T d_k)}. \tag{6.20}$$

The parameter θ_k, computed as in (6.14), (6.16), (6.18), or (6.20) may be outside the interval [0,1]. To get a convex combination in (6.12), a simple procedure is followed: if $\theta_k \le 0$, then in (6.12) $\theta_k = 0$, that is, $\beta_k = \beta_k^{M1}$; if $\theta_k \ge 1$, then $\theta_k = 1$ in (6.12), that is, $\beta_k = \beta_k^{M2}$.

Hence, the general hybrid conjugate gradient method based on the convex combination of the standard conjugate gradient methods is as follows.

Algorithm 6.1 General hybrid conjugate gradient algorithm by using the convex combination of standard schemes

1.	Choose an initial point $x_0 \in \mathbb{R}^n$, $\varepsilon \geq 0$ sufficiently small and $\varepsilon_b \geq 0$. Compute $f(x_0)$ and g_0. Set $d_0 = -g_0$ and the initial guess $\alpha_0 = 1/\|g_0\|$. Set $k = 0$
2.	Test a criterion for stopping the iterations. For example if $\|g_k\|_\infty \leq \varepsilon$, then stop, otherwise continue with step 3
3.	Compute the stepsize α_k satisfying the Wolfe line search conditions
4.	Set $x_{k+1} = x_k + \alpha_k d_k$. Compute $f(x_{k+1})$ and g_{k+1}. Compute $s_k = x_{k+1} - x_k$ and $y_k = g_{k+1} - g_k$
5.	If $\left|(\beta_k^{M2} - \beta_k^{M1})(y_k^T d_k)\right| \leq \varepsilon_b$ then set $\theta_k = 0$. Otherwise, compute θ_k by (6.14) or by (6.16) or by (6.20) according to the procedure we selected
6.	If $0 < \theta_k < 1$, then compute β_k by (6.12). If $\theta_k \leq 0$, then set $\beta_k = \beta_k^{M1}$. If $\theta_k \geq 1$, then set $\beta_k = \beta_k^{M2}$
7.	Compute $d = -g_{k+1} + \beta_k d_k$
8.	If the Powell restart criterion $\left|g_{k+1}^T g_k\right| \geq 0.2\|g_{k+1}\|^2$ is satisfied, then set $d_{k+1} = -g_{k+1}$, otherwise set $d_{k+1} = d$.
9.	Compute the initial guess $\alpha_k = \alpha_{k-1}\|d_{k-1}\|/\|d_k\|$
10.	Set $k = k + 1$ and go to step 2 ♦

Algorithm 6.1 is general. In step 5, it can be particularized by combining in a convex manner different standard conjugate gradient methods, where the parameter θ_k in the convex combination (6.12) may be selected by means of the conjugacy or the Newton direction procedures.

Remark 6.1 The standard secant equation uses only the gradient values available at the previous and at the current iterations, completely ignoring the function values. Therefore, in the second procedure for θ_k computation, the efforts have been directed to modify the standard secant equation in order to use more available information at the current iteration, including the function values and hence providing a better approximation to the Hessian. ♦

As it can be seen, there is a large variety of possibilities to combine in a convex way the standard conjugate gradient methods by using the conjugacy conditions or the standard or the modified secant equations. In the following, let us present some hybrid conjugate gradient methods based on the convex combination of the standard conjugate gradient algorithms.

Hybrid convex combination of LS and DY

This hybrid conjugate gradient method is based on a convex combination of LS and DY (Liu and Li, 2014). In it,

$$\beta_k^{M1} = \beta_k^{\mathrm{LS}} = -\frac{y_k^T g_{k+1}}{d_k^T g_k} \text{ and } \beta_k^{M2} = \beta_k^{\mathrm{DY}} = \frac{g_{k+1}^T g_{k+1}}{d_k^T y_k}. \tag{6.21}$$

Using the Dai and Liao conjugacy condition (6.15), from (6.16), where β_k^{M1} and β_k^{M2} are given as in (6.21), the following value for the parameter θ_k is obtained

$$\theta_k^{\mathrm{DL}} = \frac{(y_k^T g_{k+1})(d_k^T g_{k+1}) - t(s_k^T g_{k+1})(d_k^T g_k)}{\|g_{k+1}\|^2 (d_k^T g_k) + (y_k^T g_{k+1})(d_k^T y_k)}. \tag{6.22}$$

It is interesting to see the value of θ_k obtained from the second procedure based on the equality of the Newton direction and the direction corresponding to the hybrid conjugate gradient algorithm. Indeed, from (6.20), where β_k^{M1} and β_k^{M2} are given as in (6.21), after some algebraic manipulations it results that

$$\theta_k^{\mathrm{NT}} = \frac{(y_k^T g_{k+1})(d_k^T g_{k+1}) - (s_k^T g_{k+1})(d_k^T g_k)}{\|g_{k+1}\|^2 (d_k^T g_k) + (y_k^T g_{k+1})(d_k^T y_k)}. \tag{6.23}$$

Observe that $\theta_k^{\mathrm{DL}} = \theta_k^{\mathrm{NT}}$ when $t = 1$. Therefore, an interesting property of this hybrid conjugate gradient algorithm is that the search direction d_{k+1} not only satisfies the Dai and Liao conjugacy condition, but it is also the Newton direction when $t = 1$. Hence, in this hybrid algorithm, $t = 1$ is considered.

The algorithm corresponding to this convex combination of LS and DY, which we call NDLSDY, is a particularization of the Algorithm 6.1, where step 5 is modified as: "If $\left| \|g_{k+1}\|^2 (d_k^T g_k) + (y_k^T g_{k+1})(d_k^T y_k) \right| \le \varepsilon_b$, then set $\theta_k = 0$. Otherwise, compute θ_k by (6.23)", and in step 6, set $\beta_k^{M1} = \beta_k^{\mathrm{LS}}$ and $\beta_k^{M2} = \beta_k^{\mathrm{DY}}$.

The following theorem shows that the search direction d_k generated by the algorithm NDLSDY satisfies the sufficient descent condition.

Theorem 6.2 *Let $\{g_k\}$ and $\{d_k\}$ be the sequences generated by the algorithm NDLSDY with strong Wolfe line search. Then, the search direction d_k satisfies the sufficient descent condition*

$$d_k^T g_k \le -c\|g_k\|^2 \tag{6.24}$$

for any $k \ge 0$, where $c = (1 - 1.2\sigma)/(1 - \sigma)$, $\sigma < 0.5$.

Proof Assume that the Powell restart criterion does not hold, i.e.,

$$\left|g_{k+1}^T g_k\right| < 0.2\|g_{k+1}\|^2. \tag{6.25}$$

The proof is given by induction. For $k = 0$, $g_0^T d_0 = -\|g_0\|^2$. Since $c < 1$, it follows that (6.24) is satisfied. Now, assume that (6.24) holds for some $k \ge 1$.

From the second strong Wolfe condition $\left|\nabla f(x_k + \alpha_k d_k)^T d_k\right| \le -\sigma d_k^T g_k$,

$$d_k^T y_k = d_k^T g_{k+1} - d_k^T g_k \ge -(1-\sigma) d_k^T g_k \ge 0. \tag{6.26}$$

From (6.3),

$$d_{k+1}^T g_{k+1} = -\|g_{k+1}\|^2 + \beta_k d_k^T g_{k+1}. \tag{6.27}$$

Now, when $\theta_k \ge 1$, as seen in step 6 of the algorithm NDLSDY, it follows that $\beta_k = \beta_k^{\mathrm{DY}}$. Therefore, from the above relations (6.26) and (6.27), it results that

$$d_{k+1}^T g_{k+1} \le -\|g_{k+1}\|^2 + \frac{\|g_{k+1}\|^2}{d_k^T y_k}\left|d_k^T g_{k+1}\right| \le -\frac{1-2\sigma}{1-\sigma}\|g_{k+1}\|^2. \tag{6.28}$$

From step 6 of the algorithm NDLSDY, when $\theta_k \le 0$, $\beta_k = \beta_k^{\mathrm{LS}}$. Therefore, from the second strong Wolfe line search and (6.25), it results that

$$d_{k+1}^T g_{k+1} \le -\|g_{k+1}\|^2 + \frac{\left|g_{k+1}^T y_k\right|}{\left|d_k^T g_k\right|}\left|d_k^T g_{k+1}\right| \le -(1-1.2\sigma)\|g_{k+1}\|^2. \tag{6.29}$$

Finally, when $\theta_k \in (0,1)$, as seen in step 6 of the algorithm NDLSDY, β_k is computed as in (6.12). Observe that β_k^{DY} can be written as $\beta_k^{\mathrm{DY}} = d_{k+1}^T g_{k+1}/d_k^T g_k$. Now, by using (6.25), the above relations and the definition of β_k^{LS}, it follows that

$$\begin{aligned} d_{k+1}^T g_{k+1} &\le -\|g_{k+1}\|^2 + \left|\beta_k^{\mathrm{LS}}\right|\left|d_k^T g_{k+1}\right| + \left|\beta_k^{\mathrm{DY}}\right|\left|d_k^T g_{k+1}\right| \\ &\le -\|g_{k+1}\|^2 + \sigma\left|\beta_k^{\mathrm{LS}}\right|\left|d_k^T g_k\right| + \sigma\left|\beta_k^{\mathrm{DY}}\right|\left|d_k^T g_k\right| \\ &= -\|g_{k+1}\|^2 + \sigma\left|g_{k+1}^T y_k\right| + \sigma\left|d_{k+1}^T g_{k+1}\right| \\ &\le -\|g_{k+1}\|^2 + \sigma\|g_{k+1}\|^2 + \sigma\left|g_{k+1}^T g_k\right| + \sigma\left|d_{k+1}^T g_{k+1}\right| \\ &\le -\|g_{k+1}\|^2 + 1.2\sigma\|g_{k+1}\|^2 + \sigma\left|d_{k+1}^T g_{k+1}\right|. \end{aligned} \tag{6.30}$$

But, from (6.30),

$$d_{k+1}^T g_{k+1} - \sigma\left|d_{k+1}^T g_{k+1}\right| \le -(1-1.2\sigma)\|g_{k+1}\|^2.$$

Since $\sigma < 0.5$, it follows that there is always a constant $v > 0$ so that

$$d_{k+1}^T g_{k+1} - \sigma\left|d_{k+1}^T g_{k+1}\right| = v(d_{k+1}^T g_{k+1}).$$

Therefore,

$$d_{k+1}^T g_{k+1} \leq -q\|g_{k+1}\|^2, \tag{6.31}$$

where $q = (1 - 1.2\sigma)/v$, $v = 1 + \sigma$ or $1 - \sigma$.

In conclusion, (6.28), (6.29), and (6.31) show that (6.24) holds for $k + 1$. ♦

Suppose that the Assumption CG holds. Then, the following theorem proves the global convergence of the algorithm NDLSDY.

Theorem 6.3 *Suppose that the Assumption CG holds. Let $\{g_k\}$ and $\{d_k\}$ be the sequences generated by the algorithm NDLSDY with strong Wolfe line search. Then,* $\liminf_{k\to\infty}\|g_k\| = 0$.

Proof From the Assumption CG, it follows that there exists a positive constant $\Gamma > 0$ so that $\|g(x)\| \leq \Gamma$ for all $x \in S$. Observe that from Proposition 1.2, $\alpha_k \geq \lambda$, where λ is a positive constant.

Suppose that $\liminf_{k\to\infty}\|g_k\| \neq 0$. Then, there exists a positive constant $r > 0$ so that for all k sufficiently large,

$$\|g_k\| \geq r. \tag{6.32}$$

By the second strong Wolfe condition and from (6.24),

$$d_k^T y_k = d_k^T g_{k+1} - d_k^T g_k \geq -(1-\sigma)d_k^T g_k \geq c(1-\sigma)\|g_k\|^2. \tag{6.33}$$

From the Lipschitz continuity of the gradient, it results that

$$\|y_k\| = \|g_{k+1} - g_k\| \leq L\|x_{k+1} - x_k\| \leq \mathrm{LD}, \tag{6.34}$$

where $D = \max\{\|x - y\| : x, y \in S\}$ is the diameter of the level set S.

Now, having in view the above inequalities, it follows that

$$\begin{aligned}
|\beta_k| &= \left|(1-\theta_k)\beta_k^{\mathrm{LS}} + \theta_k\beta_k^{\mathrm{DY}}\right| \\
&\leq |\beta_k^{LS}| + |\beta_k^{\mathrm{DY}}| = \frac{|g_{k+1}^T y_k|}{|d_k^T g_k|} + \frac{\|g_{k+1}\|^2}{|d_k^T y_k|} \\
&\leq \frac{\|g_{k+1}\|\|y_k\|}{c\|g_k\|^2} + \frac{\|g_{k+1}\|^2}{c(1-\sigma)\|g_k\|^2} \leq \frac{\Gamma\mathrm{LD}}{cr^2} + \frac{\Gamma^2}{c(1-\sigma)r^2} \equiv M.
\end{aligned}$$

According to the selection of the parameter β_k, in step 6 of the algorithm NDLSDY, when $\theta_k \notin (0, 1)$, it is easy to see that the above inequality also holds. Therefore,

$$\|d_{k+1}\| \le \|g_{k+1}\| + |\beta_k|\|d_k\| = \|g_{k+1}\| + \frac{|\beta_k|\|s_k\|}{\alpha_k} \le \Gamma + \frac{\mathrm{MD}}{\lambda} \equiv E,$$

which implies that

$$\sum_{k \ge 0} \frac{1}{\|d_k\|^2} = \infty. \tag{6.35}$$

On the other hand, from (6.24), (6.32) and from the Zoutendijk condition (3.34), it results that

$$c^2 r^4 \sum_{k \ge 0} \frac{1}{\|d_k\|^2} \le \sum_{k \ge 0} \frac{c^2 \|g_k\|^4}{\|d_k\|^2} \le \sum_{k \ge 0} \frac{(g_k^T d_k)^2}{\|d_k\|^2} < \infty,$$

which contradicts (6.35). Therefore, (6.32) does not hold and hence $\liminf_{k \to \infty} \|g_k\| = 0$ is proved. ♦

Numerical study. In the following, let us present some numerical results with NDLSDY for solving unconstrained optimization problems from the UOP collection, as well as some comparisons with standard conjugate gradient methods and with hybrid methods based on the projection concept.

The algorithms compared are tested for solving 80 unconstrained optimization test problems from the UOP collection (Andrei, 2018g). Each problem is tested 10 times for a gradually increasing number of variables: $n = 1000, 2000, \ldots, 10000$. The comparisons of algorithms are given in the context of Remark 1.1 and the maximum number of iterations was limited to 2000.

Figure 6.5 presents the Dolan and Moré's performance profiles of NDLSDY versus LS, DY, PRP, CD, FR, and HS conjugate gradient methods for solving the problems from the UOP collection. For this set of unconstrained optimization test problems, Figure 6.5 shows that NDLSDY, as a convex combination of LS and DY, is more efficient than DY. However, LS is slightly more efficient and more robust than NDLSDY. Figure 4.2 points out the computational evidence that LS is more efficient and more robust than DY. Therefore, in this convex combination of LS and DY, the main role is played by LS. Comparisons with PRP show that NDLSDY is more efficient, but PRP is slightly more robust. NDLSDY is seen as being more efficient and more robust than CD and FR. Even if HS is slightly more robust than NDLSDY, NDLSDY is close to HS as regards its efficiency.

Figure 6.6 presents the performance profiles of NDLSDY versus the hybrid conjugate gradient methods based on the projection concept. Notice that NDLSDY is more robust than the hybrid conjugate gradient hDY, HS-DY, and PRP-FR. Again, observe that even if NDLSDY is slightly more efficient than LS-CD, they have mainly the same robustness. It seems that the hybrid conjugate gradient methods based on the convex combinations of the standard methods are top performers versus the hybrid conjugate gradient methods using the projection concept.

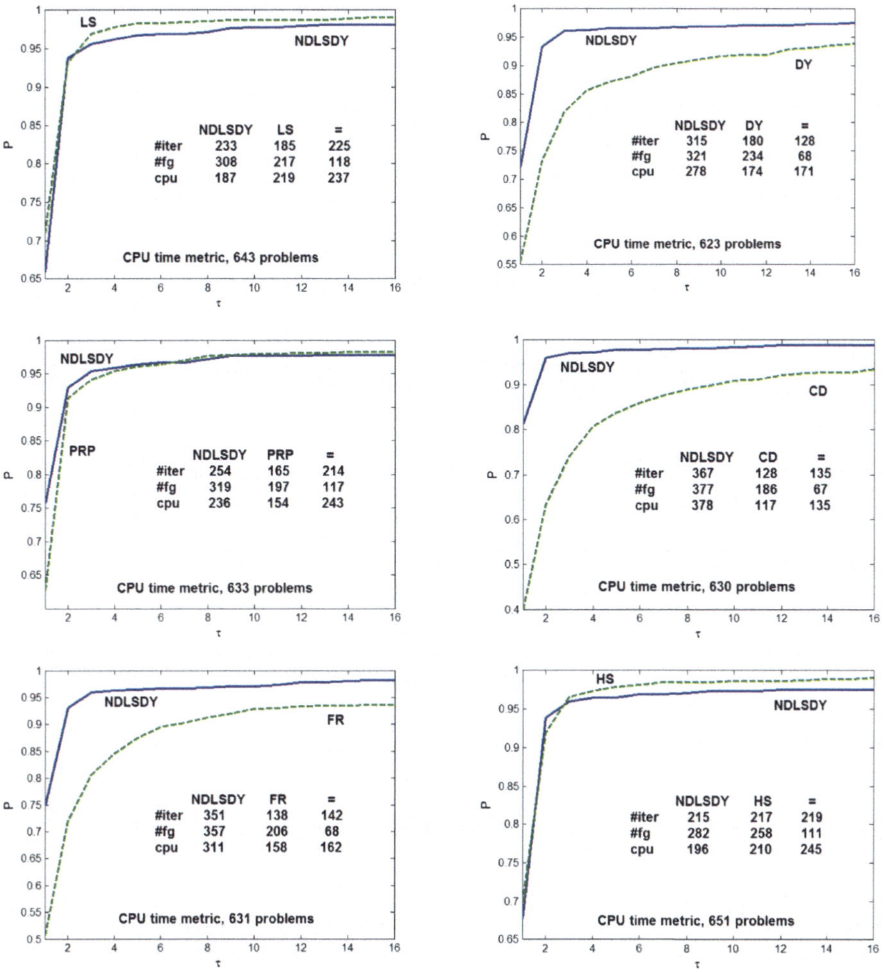

Figure 6.5 Performance profiles of NDLSDY versus the standard conjugate gradient methods LS, DY, PRP, CD, FR, and HS

Hybrid convex combination of HS and DY

Another hybrid conjugate gradient method, NDHSDY, was proposed by Andrei (2008b, 2008g). It consists of the hybridization of the standard conjugate gradient methods HS and DY. The search direction of this hybrid conjugate gradient method is as in (6.13), where $\beta_k^{M1} = \beta_k^{\mathrm{HS}}$ and $\beta_k^{M2} = \beta_k^{\mathrm{DY}}$. Using the second procedure for θ_k computation based on the standard secant equation (6.19), from (6.20), it follows that

$$\theta_k = -\frac{s_k^T g_{k+1}}{g_k^T g_{k+1}}. \tag{6.36}$$

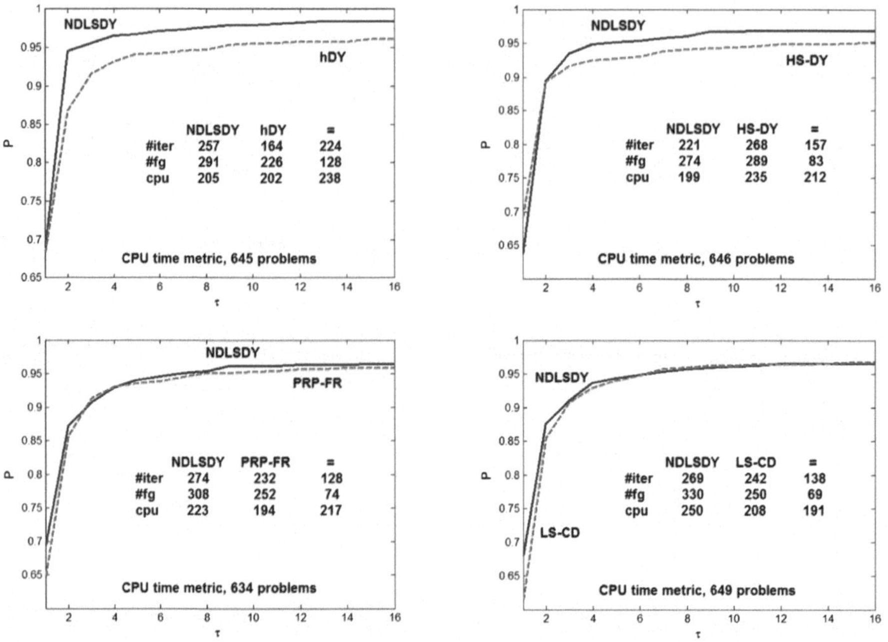

Figure 6.6 Performance profiles of NDLSDY versus the hybrid conjugate gradient methods *h*DY, HS-DY, PRP-FR, and LS-CD

The corresponding algorithm for NDHSDY is the same as Algorithm 6.1, where in step 5, $\theta_k = 0$ if $|g_k^T g_{k+1}| \leq \varepsilon_b$. Otherwise, the parameter θ_k is computed as in (6.36).

Numerical study. Figure 6.7 shows the performance profiles of NDHSDY versus NDLSDY, subject to the CPU time metric for solving the problems from the UOP collection. Obviously, NDLSDY is top performer, being more efficient and slightly more robust than NDHSDY.

Hybrid combination of PRP and DY

Other hybrid conjugate gradient algorithms CCPRPDY, NDPRPDY combine in a convex way the standard conjugate gradient algorithms PRP and DY (Andrei, 2008d, 2009a). The search direction of this hybrid conjugate gradient methods is as in (6.13), where $\beta_k^{M1} = \beta_k^{\mathrm{PRP}}$ and $\beta_k^{M2} = \beta_k^{\mathrm{DY}}$. The combination in CCPRPDY is based on the conjugacy condition $y_k^T d_{k+1} = 0$. From (6.14), where $\beta_k^{M1} = \beta_k^{\mathrm{PRP}}$ and $\beta_k^{M2} = \beta_k^{\mathrm{DY}}$, the following value for the parameter θ_k is obtained

$$\theta_k \equiv \theta_k^{\mathrm{CCPRPDY}} = \frac{(y_k^T g_{k+1})(y_k^T s_k) - (y_k^T g_{k+1})(g_k^T g_k)}{(y_k^T g_{k+1})(y_k^T s_k) - \|g_{k+1}\|^2 \|g_k\|^2}. \tag{6.37}$$

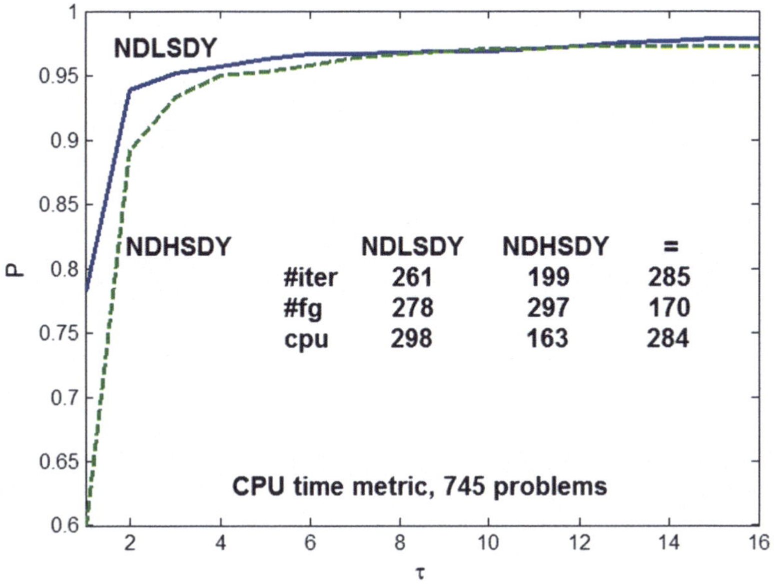

Figure 6.7 Performance profiles of NDHSDY versus NDLSDY

On the other hand, the combination in NDPRPDY considers the formal equality between the search direction of the hybrid algorithm and the Newton direction by using the standard secant condition (6.19). Therefore, using the second procedure for θ_k computation based on the standard secant equation (6.19), from (6.20), where $\beta_k^{M1} = \beta_k^{\mathrm{PRP}}$ and $\beta_k^{M2} = \beta_k^{\mathrm{DY}}$, it results that

$$\theta_k \equiv \theta_k^{\mathrm{NDPRPDY}} = \frac{(y_k^T g_{k+1} - s_k^T g_{k+1})\|g_k\|^2 - (g_{k+1}^T y_k)(y_k^T s_k)}{\|g_{k+1}\|^2\|g_k\|^2 - (g_{k+1}^T y_k)(y_k^T s_k)}. \tag{6.38}$$

The corresponding algorithms for these hybrid conjugate gradient methods are the same as Algorithm 6.1, except for step 5, where the parameter θ_k is computed as in (6.37) for the CCPRPDY algorithm if $\left|(y_k^T g_{k+1})(y_k^T s_k) - \|g_{k+1}\|^2\|g_k\|^2\right| > \varepsilon_b$, or as in (6.38) for the NDPRPDY if $\left|\|g_{k+1}\|^2\|g_k\|^2 - (g_{k+1}^T y_k)(y_k^T s_k)\right| > \varepsilon_b$. Otherwise, $\theta_k = 0$. Obviously, in step 6 of the algorithm 6.1, $\beta_k^{M1} = \beta_k^{\mathrm{PRP}}$ and $\beta_k^{M2} = \beta_k^{\mathrm{DY}}$.

Numerical study. Figure 6.8 presents the performance profiles of NDLSDY versus CCPRPDY and versus NDPRPDY as well as the performance profiles of NDHSDY versus CCPRPDY and versus NDPRPDY for solving the problems from the UOP collection. Observe that the convex combination based on the Newton direction is more efficient and more robust than the same convex combination using the conjugacy condition. This is because the hybrid convex combination based on

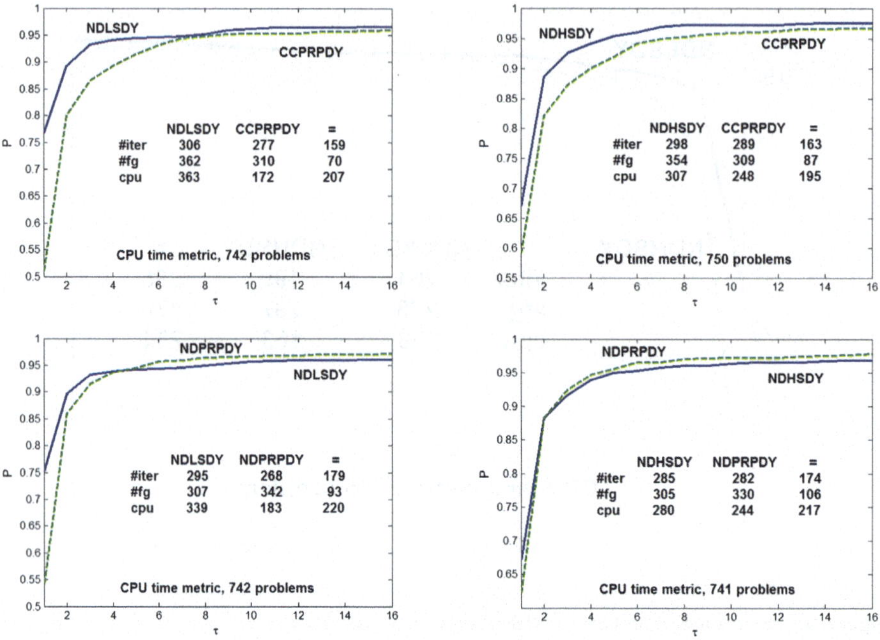

Figure 6.8 Performance profiles of NDLSDY and NDHSDY versus CCPRPDY and NDPRPDY

the Newton direction includes the second-order information by the secant equation. Similarly, in the frame of the hybrid conjugate gradient methods based on the Newton direction, NDPRPDY is top performer versus both NDLSDY and NDHSDY. Even if the convergence of the PRP method in theory is uncertain, see the example by Powell (1984a) or the example by Dai (1997), it proved in numerical experiments to be more efficient and more robust than other conjugate gradient methods. This property is also maintained in the convex combination of PRP with DY using the Newton direction.

In the following, let us present the performances of the hybrid algorithms for solving the applications from the MINPACK-2 collection. Table 6.4 presents the performances of the NDHSDY and NDLSDY hybrid conjugate gradient methods for solving five applications from the MINPACK-2 collection.

Table 6.4 shows that subject to the CPU time metric, for solving the applications from the MINPACK-2 collection, each of them having 40,000 variables, the hybrid convex combination of HS and DY based on the Newton direction is more efficient. Tables 4.2 and 4.4 show that HS with 139.65 s is top performer versus DY with 229.20 s. Table 4.4 reveals that LS with 168.03 s is again better than DY. In the above hybrid algorithms, DY with $\|g_{k+1}\|^2$ in the numerator of β_k, is combined in a convex way with HS and LS, which have $g_{k+1}^T y_k$ in the numerator of β_k. Observe that the hybrid convex combination with the Newton direction, using HS, is top performer versus the hybrid combination with LS.

Table 6.4 Performances of NDHSDY and NDLSDY for solving five applications from the MINPACK-2 collection

		NDHSDY			NDLSDY		
	n	*#iter*	*#fg*	*cpu*	*#iter*	*#fg*	*cpu*
A1	40,000	494	578	5.56	451	521	13.39
A2	40,000	1201	1465	15.61	1411	1740	23.55
A3	40,000	1517	1700	28.31	1166	1272	38.44
A4	40,000	1098	1274	49.39	1018	1172	65.59
A5	40,000	364	413	5.43	274	301	4.04
Total	–	4674	5430	104.30	4320	5006	145.01

Table 6.5 Performances of CCPRPDY and NDPRPDY for solving five applications from the MINPACK-2 collection

		CCPRPDY			NDPRPDY		
	n	*#iter*	*#fg*	*cpu*	*#iter*	*#fg*	*cpu*
A1	40,000	456	563	5.36	496	577	5.81
A2	40,000	1191	1435	15.80	1121	1283	14.01
A3	40,000	1171	1224	21.49	978	1016	17.85
A4	40,000	933	1175	47.14	874	971	39.72
A5	40,000	399	485	6.53	314	337	4.67
Total	–	4150	4882	96.32	3783	4184	82.06

Table 6.5 presents the performances of the CCPRPDY and NDPRPDY hybrid conjugate gradient methods for solving five applications from the MINPACK-2 collection.

Table 6.5 compares the hybrid convex combination of PRP and DY based on the conjugacy condition CCPRPDY versus the hybrid combination of PRP and DY based on the Newton direction NDPRPDY for solving five applications from the MINPACK-2 collection. Obviously, using the Newton direction in hybrid convex combinations is more profitable than the hybrid convex combinations based on the conjugacy condition. The value of parameter θ_k given by (6.20) at each iteration better captures the curvature of the function f than (6.14) or (6.16) based on the conjugacy condition.

The global convergence of these hybrid conjugate gradient methods presented above is proved by using the same methodology as in Theorems 6.1 and 6.2. The idea is to prove the sufficient descent condition for the search directions associated with these algorithms and to show that the norm of the search direction is bounded from below.

Chapter 5 presented an acceleration of the conjugate gradient methods by using a simple modification of the iterative scheme, as in (5.11). In the following, the results of the acceleration of the NDHSDY and NDLSDY algorithms are shown. The accelerated versions of NDHSDY and NDLSDY are called NDHSDYa and NDLSDYa, respectively. Figure 6.9 presents the performance profiles of NDHSDY

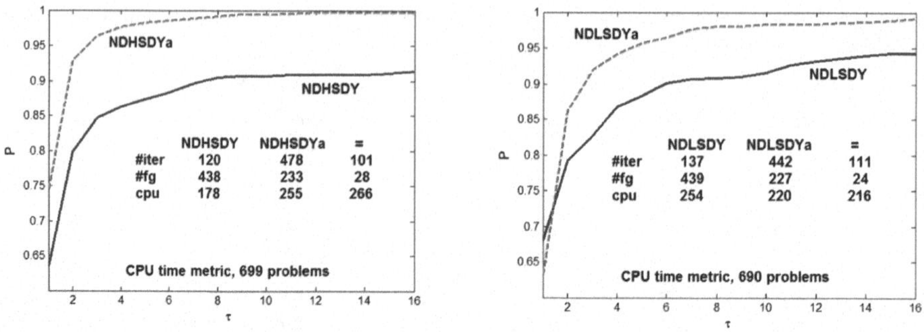

Figure 6.9 Performance profiles of NDHSDY versus NDHSDYa and of NDLSDY versus NDLSDYa

versus its accelerated version NDHSDYa and of NDLSDY versus its accelerated version NDLSDYa for solving 800 unconstrained optimization test problems from the UOP collection.

The acceleration of the conjugate gradient algorithms represents an important ingredient in increasing their performances. Figure 6.9 shows that the accelerated versions are by far more robust.

As seen in Table 6.5, the hybrid convex combination based on the Newton direction and the secant equation (6.19) is the best compared to the hybrid convex combination based on the conjugacy condition. In Remark 6.1, it was mentioned that at each iteration, the usual secant equation (6.19) employs only the gradient in two successive points, ignoring the available function values. In order to get a higher accuracy to the approximate Hessian of the minimizing function, an idea was to modify the secant equation to use both the gradient and the function values. In this context, using an interpolation of the minimizing function f, Davidon (1980) proposed a nonquadratic approximation model, namely the conic model, using both the function and the gradient values available at the current iteration. On the other hand, using a quadratic approximation for f, Yuan (1991), Li, Tang, and Wei (2007), Wei, Li, and Qi (2006a) developed modified secant equations which approximate the curvature of the objective function more accurately than the standard secant equation (6.19). The modified secant equation obtained by Li, Tang, and Wei (2007) and by Babaie-Kafaki and Mahdavi-Amiri (2013) is

$$B_{k+1}s_k = z_k, z_k = y_k + \frac{\eta_k}{\|s_k\|^2}s_k, \tag{6.39}$$

where $\eta_k = 2(f_k - f_{k+1}) + (g_k + g_{k+1})^T s_k$.

In the same avenue of research, but in the context of the BFGS quasi-Newton method, Yuan and Byrd (1995) and Wei, Yu, Yuan, and Lian (2004) proposed modified secant equations. Zhang, Deng, and Chen (1999) and Zhang and Xu (2001), using a quadratic interpolation of the gradient, developed another modified secant equation with a higher order of accuracy in approximating the curvature of the minimizing function:

$$B_{k+1}s_k = \bar{z}_k, \bar{z}_k = y_k + \frac{\mu_k}{\|s_k\|^2}s_k, \tag{6.40}$$

where $\mu_k = 6(f_k - f_{k+1}) + 3(g_k + g_{k+1})^T s_k$.

In order to adaptively switch from the standard secant equation (6.19) to the modified secant equation (6.40), some improvements were developed by Yabe, Ogasawara, and Yoshino (2007), Yabe and Takano (2004), Babaie-Kafaki, Ghanbari, and Mahdavi-Amiri (2010), and Babaie-Kafaki, Fatemi, and Mahdavi-Amiri (2011) by embedding a parameter into the modified secant equation:

$$B_{k+1}s_k = \hat{z}_k, \hat{z}_k = y_k + \rho_k\frac{\max\{0, \mu_k\}}{\|s_k\|^2}s_k \tag{6.41}$$

and

$$\rho_k = \begin{cases} 0, & \|s_k\| > 1, \\ 1, & \|s_k\| \leq 1. \end{cases}$$

A hybridization of the HS and DY conjugate gradient method, which is globally convergent for strongly convex functions, was proposed by Babaie-Kafaki (2011) and Babaie-Kafaki and Mahdavi-Amiri (2013). They introduced the hybrid secant equation

$$B_{k+1}s_k = z_k, z_k = y_k + \frac{\eta_k}{s_k^T u_k}u_k, \tag{6.42}$$

where $u_k \in \mathbb{R}^n$ is a convex combination of the vectors s_k and y_k

$$u_k = (1 - \lambda_k)y_k + \lambda_k s_k, \lambda_k \in [0, 1]. \tag{6.43}$$

An effective choice for λ_k was given by Babaie-Kafaki and Mahdavi-Amiri (2013).

In order to get a numerically efficient hybrid conjugate gradient method, Babaie-Kafaki and Ghanbari (2015b) proposed the hybridization of PRP and FR by their convex combination

$$\beta_k^{BKG} = (1 - \theta_k)\beta_k^{PRP} + \theta_k\beta_k^{FR}.$$

The computation of parameter θ_k in β_k^{BKG} is in such a way that the search direction approaches the search direction of an efficient three-term conjugate gradient method suggested by Zhang, Zhou, and Li (2007) under some suitable conditions. In other words, the parameter θ_k is computed as solution of the problem

$$\min_{\theta_k} \left\| d_{k+1}^{\mathrm{BKG}} - d_{k+1}^{\mathrm{ZZL}} \right\|^2,$$

where d_{k+1}^{ZZL} is the search direction corresponding to the Zhang, Zhou, and Li (2007) three-term conjugate gradient algorithm. The global convergence of this method is proved under the strong Wolfe line search without convexity assumption on the objective function.

Plenty of modified secant equations have been proposed in order to catch the curvature of the objective function and to use it for generating hybrid convex combination conjugate gradient methods based on the Newton direction. In the following, let us present an extension of NDHSDY by hybridizing HS and DY based on the Newton direction, this time using the modified secant equation (6.39). The search direction in this algorithm, called NDHSDYM (Andrei, 2010a), is computed as in (6.13), where $\beta_k^{M1} = \beta_k^{\mathrm{HS}}$ and $\beta_k^{M2} = \beta_k^{\mathrm{DY}}$. Using the modified secant equation (6.39), from (6.18), the corresponding parameter θ_k is computed as

$$\theta_k = \frac{z_k^T g_{k+1} - s_k^T g_{k+1} - \beta_k^{\mathrm{HS}}(z_k^T d_k)}{(\beta_k^{\mathrm{DY}} - \beta_k^{\mathrm{HS}})(z_k^T d_k)}, \tag{6.44}$$

where z_k is defined in (6.39) and includes the function values in two successive points. The corresponding algorithm is as in Algorithm 6.1, where in step 5, θ_k is computed as in (6.44) if $\left|(\beta_k^{\mathrm{DY}} - \beta_k^{\mathrm{HS}})(z_k^T d_k)\right| > \varepsilon_b$.

Both for strongly convex functions and for general nonlinear functions, the algorithm NDHSDYM with strong Wolfe line search is globally convergent.

Numerical study. Figure 6.10 shows the performance profiles of NDHSDYM versus NDHSDY for solving the unconstrained optimization problems from the UOP collection. Observe that this variant of the hybrid conjugate gradient algorithm based on the Newton direction, which uses the modified secant equation, is top performer versus the same variant which uses the standard secant equation. It seems that the modified secant equation better captures the curvature of the minimizing function.

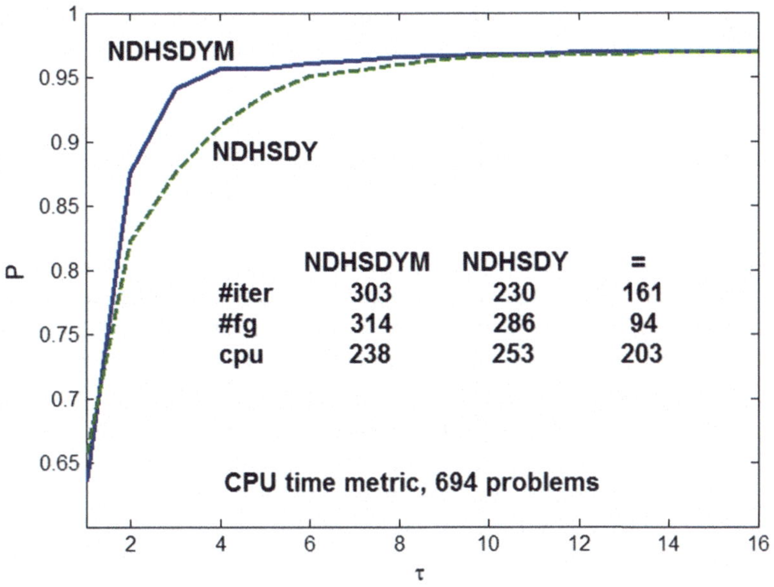

Figure 6.10 Performance profiles of NDHSDYM versus NDHSDY

6.3 Parameterized Conjugate Gradient Methods

The *parameterized conjugate gradient* methods combine the standard conjugate gradient methods by means of some parameters (Dai & Yuan, 2003) and (Nazareth, 1999). Dai and Yuan (2003) proposed a one-parameter family of conjugate gradient methods with

$$\beta_k = \frac{\|g_{k+1}\|^2}{\lambda_k \|g_k\|^2 + (1-\lambda_k) d_k^T y_k}, \tag{6.45}$$

where $\lambda_k \in [0, 1]$ is a parameter. The FR method corresponds to $\lambda_k = 1$, while for $\lambda_k = 0$, the DY method is obtained. By considering convex combinations of the numerators and denominators of β_k^{FR} and β_k^{HS}, Nazareth (1999) proposed a two-parameter family of conjugate gradient methods

$$\beta_k = \frac{\mu_k \|g_{k+1}\|^2 + (1-\mu_k) g_{k+1}^T y_k}{\lambda_k \|g_k\|^2 + (1-\lambda_k) d_k^T y_k}, \tag{6.46}$$

where $\lambda_k, \mu_k \in [0, 1]$. This two-parameter family includes the FR, DY, PRP, and HS methods. Dai and Yuan (2001b) introduced a wider family of conjugate gradient methods with three parameters

$$\beta_k = \frac{\mu_k \|g_{k+1}\|^2 + (1 - \mu_k) g_{k+1}^T y_k}{(1 - \lambda_k - \omega_k)\|g_k\|^2 + \lambda_k d_k^T y_k - \omega_k d_k^T g_k}, \tag{6.47}$$

where $\lambda_k, \mu_k \in [0, 1]$ and $\omega_k \in [0, 1 - \lambda_k]$. This three-parameter family includes the six standard conjugate gradient methods, the previous one-parameter and two-parameter families and many hybrid methods as special cases. In order to ensure that the search directions generated by this family will be descent, Powell's restart criterion is employed: set $d_k = -g_k$ if $|g_{k+1}^T g_k| > \xi \|g_{k+1}\|^2$, where $\xi > 0$ (often $\xi = 0.2$). Dai and Yuan (2001b) showed that for strong Wolfe line search where $(1 + \xi)\sigma \leq 1/2$, the search directions corresponding to these parameterized conjugate gradient methods are descent directions. The global convergence of these methods was established by Dai and Yuan (2001b).

Dai (2003b) presented two new parameterized conjugate gradient methods for which the descent property is proved.

Notes and References

Hybrid and parameterized conjugate gradient methods represent an important contribution to improving the performances of conjugate gradient methods. In this chapter, only some hybrid conjugate gradient methods based both on the projection concept and on the convex combination of the standard schemes have been presented. The hybrid methods using the Newton direction are top performers. Their advantage is that they may include the second-order information concerning the curvature of the minimizing function. There are a lot of possibilities to combine standard schemes to get new hybrid conjugate gradient methods. The hybridization of some other standard conjugate gradient algorithms follows the same methodology as presented in this chapter. Besides, another idea is to combine in a convex manner the hybrid conjugate gradient methods based on the projection concept. It is worth saying that all these hybrid algorithms may be very easy accelerated using the procedure described in Chapter 5.

Chapter 7
Conjugate Gradient Methods as Modifications of the Standard Schemes

Due to their simplicity and low memory requirements, conjugate gradient methods represent an important contribution to the class of methods for solving unconstrained optimization problems. These methods have good convergence properties and their iterations do not involve any matrices, making them extremely attractive for solving large-scale problems. In the previous chapter, the hybrid conjugate gradient methods have been presented. They combine the standard conjugate gradient methods using the projection or the convex combination concepts. In general, the hybrid methods are more efficient or more robust than the standard schemes.

In this chapter, some conjugate gradient methods obtained as modifications of the standard scheme presented in Chapter 4 are developed. The idea is to modify the formula of the standard conjugate gradient methods for computing the conjugate gradient parameter β_k in order to improve the numerical performances of the algorithm. Any standard conjugate gradient method (see Table 4.1) may be modified, but some modifications of the HS method are to be presented in this chapter. The interest is to see how the modifications of the standard conjugate gradient method HS lead to better algorithms. Firstly, a modification of the HS method based on the Dai–Liao conjugacy condition is presented. Thus, the Dai–Liao conjugate gradient algorithm, DL, is obtained. This is a very general conjugate gradient algorithm, which depends on a parameter. Different values of this parameter lead to different conjugate gradient algorithms. Another modification of the HS method is CG-DESCENT of Hager and Zhang (2005). In this chapter, CG-DESCENT is interpreted as a particular value of the parameter in the Dai–Liao conjugate gradient algorithm. Another interpretation of CG-DESCENT is a particularization of the Perry–Shanno self-scaling memoryless BFGS algorithm, presented in Chapter 8. Finally, a deeper modification of the HS method is to determine the search direction satisfying both the sufficient descent and the conjugacy conditions. Thus, the DESCON conjugate gradient algorithm is obtained, in which the stepsize is determined by a modification of the standard Wolfe line search.

As a general characteristic, any kind of modifications of the standard schemes are allowed, but the main requirement is that the modified scheme satisfies the

N. Andrei, *Nonlinear Conjugate Gradient Methods for Unconstrained Optimization*,
Springer Optimization and Its Applications 158,
https://doi.org/10.1007/978-3-030-42950-8_7

descent or the sufficient descent condition. The advanced modifications of standard schemes try to incorporate the second-order information.

For solving the unconstrained optimization problem

$$\min f(x), \tag{7.1}$$

where $f : \mathbb{R}^n \to \mathbb{R}$ is a continuous differentiable function, lower bounded, a conjugate gradient method generates the sequence $\{x_k\}$ by using the recursive scheme

$$x_{k+1} = x_k + \alpha_k d_k, k = 0, 1, \ldots, \tag{7.2}$$

where x_0 is a given initial point. Knowing the initial search direction $d_0 = -g_0$, where $g_0 = \nabla f(x_0)$, then, in conjugate gradient algorithms the search direction d_k is updated as

$$d_{k+1} = -g_{k+1} + \beta_k d_k, k = 0, 1, \ldots, \tag{7.3}$$

where β_k is a scalar parameter which defines the conjugate gradient algorithm. The scalar α_k in (7.2) is the stepsize, usually determined by the standard Wolfe line search

$$f(x_k + \alpha_k d_k) \le f(x_k) + \rho \alpha_k d_k^T g_k, \tag{7.4}$$

$$\nabla f(x_k + \alpha_k d_k)^T d_k \ge \sigma d_k^T g_k, \tag{7.5}$$

or by the strong Wolfe line search given by (7.4) and

$$\left|\nabla f(x_k + \alpha_k d_k)^T d_k\right| \le -\sigma d_k^T g_k, \tag{7.6}$$

where $0 < \rho < \sigma < 1$.

7.1 Conjugate Gradient with Dai and Liao Conjugacy Condition (DL)

For the quadratic functions, it is well known that the linear conjugate gradient methods generate a sequence of search direction d_k, $k = 1, 2, \ldots$ so that the following conjugacy condition holds: $d_i^T B d_j = 0$ for all $i \ne j$, where B is the Hessian of the objective function. For general nonlinear functions, by the mean value theorem (see Appendix A) there exists $\xi \in (0, 1)$ so that

$$d_{k+1}^T g_{k+1} = d_{k+1}^T g_k + \alpha_k d_{k+1}^T \nabla^2 f(x_k + \xi \alpha_k d_k) d_k. \tag{7.7}$$

Defining $y_k = g_{k+1} - g_k$, the following can be written

$$d_{k+1}^T y_k = \alpha_k d_{k+1}^T \nabla^2 f(x_k + \xi \alpha_k d_k) d_k. \tag{7.8}$$

Therefore, for nonlinear optimization, it is reasonable to replace the conjugacy condition from the linear case with the following one

$$d_{k+1}^T y_k = 0. \tag{7.9}$$

But, for unconstrained optimization methods, the search direction d_{k+1} can be written as $d_{k+1} = -H_{k+1} g_{k+1}$, where H_{k+1} is an approximation to the inverse of the Hessian $\nabla^2 f(x_{k+1})$, symmetric and positive definite, which satisfies the secant equation $H_{k+1} y_k = s_k$, where $s_k = x_{k+1} - x_k$. Therefore,

$$d_{k+1}^T y_k = -(H_{k+1} g_{k+1})^T y_k = -g_{k+1}^T (H_{k+1} y_k) = -g_{k+1}^T s_k.$$

Hence, the conjugacy condition $d_{k+1}^T y_k = 0$ is satisfied if the line search is exact, since in this case $g_{k+1}^T s_k = 0$. However, in practical situations, the exact line search is not used. Therefore, it is quite natural to replace the conjugacy condition $d_{k+1}^T y_k = 0$ with

$$d_{k+1}^T y_k = -t g_{k+1}^T s_k, \tag{7.10}$$

where $t \geq 0$ is a scalar.

To determine the parameter β_k in the search direction (7.3), satisfying the conjugacy condition (7.10), let us multiply (7.3) by y_k and use (7.10), thus obtaining

$$\beta_k^{DL} = \frac{g_{k+1}^T y_k - t g_{k+1}^T s_k}{y_k^T d_k}, \tag{7.11}$$

known as the Dai–Liao conjugate gradient parameter. The method where the search direction is computed as in (7.3) with β_k computed as in (7.11) is called the Dai–Liao conjugate gradient method (Dai & Liao, 2001). It is obvious that

$$\beta_k^{DL} = \beta_k^{HS} - t \frac{g_{k+1}^T s_k}{y_k^T d_k}, \tag{7.12}$$

showing that the DL method is a modification of the HS method. If the line search is exact ($g_{k+1}^T s_k = 0$), then $\beta_k^{DL} = \beta_k^{HS}$. Dai and Liao (2001) proved that for strongly convex functions the norm of the search directions generated by the DL method is bounded. Therefore, the DL method with strong Wolfe line search is globally convergent.

Similar to the PRP+ method introduced by Powell (1984a), in order to ensure the global convergence for general nonlinear functions, Dai and Liao restrict β_k^{DL} to be positive, thus suggesting β_k^{DL+}, where

$$\beta_k^{DL+} = \max\left\{\frac{g_{k+1}^T y_k}{d_k^T y_k}, 0\right\} - t\frac{g_{k+1}^T s_k}{d_k^T y_k}. \tag{7.13}$$

If the Lipschitz and boundedness assumptions hold and if d_k satisfies the sufficient descent condition $g_k^T d_k \leq -c\|g_k\|^2$, where $c > 0$ is a constant, then Dai and Liao (2001) showed that DL+ with strong Wolfe line search is globally convergent.

The HS method has the property that the conjugacy condition $y_k^T d_{k+1} = 0$ always holds independent of the line search used. Expecting the fast convergence of the method, Dai and Liao (2001) introduced their method, which is a modification of the HS method based on the secant equation of the quasi-Newton methods. Under the Assumption CG, for strongly convex functions they proved the global convergence of their method with strong Wolfe line search (7.4) and (7.6). The new feature of this method is that the high-order accuracy in approximating the second-order curvature of the minimizing function is achieved.

Now, let us consider the search directions

$$\begin{aligned} d_{k+1}^{HS} &= -g_{k+1} + \beta_k^{HS} d_k, \\ d_{k+1}^{DL} &= -g_{k+1} + \beta_k^{DL} d_k = -g_{k+1} + \beta_k^{HS} d_k - t(g_{k+1}^T s_k / y_k^T d_k) d_k, \end{aligned}$$

where d_{k+1}^{DL} is defined for any $t \geq 0$. Suppose that $g_{k+1}^T s_k > 0$ and the line search satisfies the curvature condition $y_k^T d_k > 0$, then if d_{k+1}^{HS} is descent, it follows that d_{k+1}^{DL} is also descent. Having in view the numerical example given by Powell (1984a), the DL method with exact line search may not be convergent. To ensure its convergence, Dai and Liao used the same idea specified by Powell in PRP+, by modifying the DL method to get the DL+ method. The convergence of this method follows the same methodology as the one given by Gilbert and Nocedal (1992). Observe that if $g_{k+1}^T g_k / \|g_{k+1}\|^2 \geq 1$, then the DL+ method is restarted with the direction

$$d_{k+1} = -g_{k+1} - t\frac{g_{k+1}^T s_k}{d_k^T y_k} d_k, t > 0.$$

Since this direction includes the second-order information, it is quite reasonable to expect that it is better than the negative direction of the gradient. If d_k satisfies the sufficient descent condition and ∇f is Lipschitz continuous, then the DL+ method with strong Wolfe line search is globally convergent.

Following Dai and Liao (2001), let us discuss the properties and the convergence of the DL and DL+ methods. A particularization of Corollary 3.1 under the strong

Wolfe line search, which is valid for any conjugate gradient method, is given by the following proposition.

Proposition 7.1 *Suppose that the Assumption CG holds and consider a general conjugate gradient method given by* (7.2) *and* (7.3), *where* d_k *is a descent direction and* α_k *is computed by the strong Wolfe line search* (7.4) *and* (7.6). *If*

$$\sum_{k\geq 0}\frac{1}{\|d_k\|^2}=\infty, \tag{7.14}$$

then

$$\liminf_{k\to\infty}\|g_k\|=0. \tag{7.15}$$

With this, for strongly convex functions, it is easy to prove that the norm of the search direction of the Dai–Liao method is bounded above. Therefore, using Proposition 7.1, the convergence of the DL methods may be established. As usual, define $S=\{x\in\mathbb{R}^n : f(x)\leq f(x_0)\}$ as the level set, where x_0 is the initial point.

Theorem 7.1 *Suppose that the Assumption CG holds and consider the general conjugate gradient method given by* (7.2) *and* (7.3), *where* d_{k+1} *with* $\beta_k=\beta_k^{DL}$ *is descent and* α_k *is determined by the strong Wolfe line search. If there exists a positive constant* $\mu>0$ *so that*

$$(\nabla f(x)-\nabla f(y))^T(x-y)\geq\mu\|x-y\|^2, \tag{7.16}$$

for any $x,y\in S$, *then*

$$\lim_{k\to\infty}\|g_k\|=0. \tag{7.17}$$

Proof From (7.16), it follows that f is strongly convex on S and then

$$d_k^Ty_k\geq\mu\alpha_k\|d_k\|^2.$$

Since $\beta_k^{DL}=g_{k+1}^T(y_k-ts_k)/d_k^Ty_k$, from the Lipschitz continuity of the gradient the following estimation is obtained

$$\|d_{k+1}\|\leq\|g_{k+1}\|+\frac{(L+t)\|g_{k+1}\|\|s_k\|}{\mu\alpha_k\|d_k\|^2}\|d_k\|\leq\frac{1}{\mu}(L+t+\mu)\Gamma,$$

which implies that (7.14) is true. Hence, from Proposition 7.1, it follows that (7.15) is true, which for strongly convex functions is equivalent with (7.17). ♦

The following result was proved by Dai and Liao (2001). It shows that the normalized search directions change slowly, asymptotically.

Proposition 7.2 *Suppose that the Assumption CG holds and consider the method DL+, where d_k is a descent direction and α_k is determined by the strong Wolfe line search. If there exists a constant $\gamma > 0$ so that*

$$\|g_k\| \geq \gamma \tag{7.18}$$

for all $k \geq 0$, then $d_k \neq 0$ and

$$\sum_{k \geq 1} \|u_k - u_{k-1}\|^2 < \infty, \tag{7.19}$$

where $u_k = d_k / \|d_k\|$.

Proof Observe that $d_k \neq 0$, otherwise the sufficient descent condition $g_k^T d_k < -c\|g_k\|^2$ is false. Hence, u_k is well defined. Besides, from (7.18) and Proposition 7.1 it follows that

$$\sum_{k \geq 0} \frac{1}{\|d_k\|^2} < \infty, \tag{7.20}$$

otherwise we have that (7.15) holds, thus contradicting (7.18). Now, using a standard technique, β_k^{DL+} can be expressed as a sum of two parts: $\beta_k^{DL+} = \beta_k^1 + \beta_k^2$, where

$$\beta_k^1 = \max\left\{0, \frac{g_{k+1}^T y_k}{d_k^T y_k}\right\} \text{ and } \beta_k^2 = -t\frac{g_{k+1}^T s_k}{d_k^T y_k}. \tag{7.21}$$

Let us define:

$$r_k = \frac{v_k}{\|d_k\|} \text{ and } \delta_k = \beta_{k-1}^1 \frac{\|d_{k-1}\|}{\|d_k\|}, \tag{7.22}$$

where $v_k = -g_k + \beta_{k-1}^2 d_{k-1}$.

From (7.3), it follows that $u_{k+1} = r_{k+1} + \delta_{k+1} u_k$. But $\|u_k\| = \|u_{k-1}\| = 1$. Therefore, $\|r_k\| = \|u_k - \delta_k u_{k-1}\| = \|\delta_k u_k - u_{k-1}\|$. Since $\delta_k \geq 0$, then by the triangle inequality and the above relations it follows that

$$\begin{aligned}\|u_k - u_{k-1}\| &\leq \|(1+\delta_k)u_k - (1+\delta_k)u_{k-1}\| \\ &\leq \|u_k - \delta_k u_{k-1}\| + \|\delta_k u_k - u_{k-1}\| = 2\|r_k\|.\end{aligned} \tag{7.23}$$

On the other hand, the second strong Wolfe line search (7.6) gives

$$d_k^T y_k \geq (\sigma - 1) g_k^T d_k. \tag{7.24}$$

Thus, from (7.24) and the second strong Wolfe condition (7.6),

$$\left|\frac{g_{k+1}^T d_k}{d_k^T y_k}\right| \leq \frac{\sigma}{1-\sigma}. \tag{7.25}$$

From the definition of v_k, (7.25) and having in view that $\|x_k\| \leq B$ and $\|g_k\| \leq \Gamma$ (see Assumption CG) it follows that

$$\|v_{k+1}\| \leq \|g_{k+1}\| + t\left|\frac{g_{k+1}^T d_k}{d_k^T y_k}\right| \|s_k\| \leq \Gamma + \frac{2t\sigma B}{1-\sigma}. \tag{7.26}$$

Therefore, from the definition of r_k, (7.23), (7.26), and (7.20), it follows that (7.19) holds, which completes the proof of the proposition. ♦

Following the developments given by Dai and Liao (2001), let us now present a property of β_k^{DL+} which is similar to but slightly different from Property (*) defined by Gilbert and Nocedal (1992). For this, consider that the standard Assumption CG holds, $\|g_k\| \geq \gamma$ for all $k \geq 0$ and α_k is determined by the strong Wolfe line search (7.4) and (7.6).

Property of β_k^{DL+} : *If* $g_k^T d_k < -c\|g_k\|^2$ *holds for some positive constant* $c > 0$, *then there exist the constants* $b > 1$ *and* $\lambda > 0$ *so that for all* k,

$$|\beta_k^{DL+}| \leq b \text{ and } \|s_k\| \leq \lambda \text{ implies that } |\beta_k^{DL+}| \leq \frac{1}{b}.$$

As above, from the strong Wolfe line search it follows that

$$d_k^T y_k \geq (\sigma - 1) g_k^T d_k \geq (1-\sigma)c\|g_k\|^2 \geq (1-\sigma)c\gamma^2. \tag{7.27}$$

Now, having in view that $\|x_k\| \leq B$, $\|g_k\| \leq \Gamma$, then from the Lipschitz continuity of the gradient it results that

$$|\beta_k^{DL+}| \leq \frac{(L+t)\|g_{k+1}\|\|s_k\|}{(1-\sigma)c\gamma^2} \leq \frac{2(L+t)\Gamma B}{(1-\sigma)c\gamma^2} \equiv b. \tag{7.28}$$

Observe that in (7.28), b can be defined so that $b > 1$. Therefore, it can be supposed that $b > 1$. With this, define:

$$\lambda = \frac{(1-\sigma)c\gamma^2}{b(L+t)\Gamma}. \tag{7.29}$$

Hence, from the first inequality in (7.28), it follows that if $\|s_k\| \leq \lambda$, then

$$\left|\beta_k^{DL+}\right| \le \frac{(L+t)\Gamma\lambda}{(1-\sigma)c\gamma^2} = \frac{1}{b}. \tag{7.30}$$

With b and λ defined as above in (7.28) and in (7.29), respectively, observe that $\left|\beta_k^{DL+}\right| \le b$ and $\|s_k\| \le \lambda$ imply that $\left|\beta_k^{DL+}\right| \le \frac{1}{b}$. ♦

Now, let N be the set of positive integers. For $\lambda > 0$ and a positive integer Δ, denote

$$K_{k,\Delta}^{\lambda} = \{i \in N : k \le i \le k+\Delta-1,\ \|s_{i-1}\| > \lambda\}. \tag{7.31}$$

Let $\left|K_{k,\Delta}^{\lambda}\right|$ denote the number of elements in $K_{k,\Delta}^{\lambda}$. Having in view the above property of DL+, the following result can be proved.

Proposition 7.3 *Suppose that the standard Assumption CG holds and consider the DL+ method, where d_k satisfies the sufficient descent condition $g_k^T d_k < -c\|g_k\|^2$ for some positive constant $c > 0$ and α_k is determined by the strong Wolfe line search* (7.4) *and* (7.6). *If $\|g_k\| \ge \gamma$ for all k, then there exists $\lambda > 0$ so that for any $\Delta \in N$ and any index k_0 there is an index $k > k_0$ so that*

$$\left|K_{k,\Delta}^{\lambda}\right| > \frac{\Delta}{2}. \tag{7.32}$$

Proof As in Dai and Liao (2001), let us proceed by contradiction. Suppose that for any $\lambda > 0$ there exist $\Delta \in N$ and k_0 so that

$$\left|K_{k,\Delta}^{\lambda}\right| \le \frac{\Delta}{2}, \quad \text{for all}\ k \ge k_0 \tag{7.33}$$

Now, let $b > 1$ and $\lambda > 0$ given as in (7.28) and in (7.29), respectively. For $\lambda > 0$, let us choose Δ and k_0 so that (7.33) holds. Then, from the above property of β_k^{DL+} and (7.33), it follows that

$$\prod_{k=k_0+i\Delta+1}^{k_0+(i+1)\Delta} \left|\beta_k^{DL+}\right| \le b^{\Delta/2}\left(\frac{1}{b}\right)^{\Delta/2} = 1, \quad \text{for any}\ i \ge 0. \tag{7.34}$$

If $\beta_k^{DL+} = 0$, then $d_{k+1} = -g_{k+1}$. Therefore, the method either gives the convergence $\liminf_{k\to\infty}\|g_k\| = 0$, or a certain x_k can be considered as a new initial point. Thus, without loss of generality, let us suppose that $\beta_k^{DL+} \ne 0$ for all $k \ge 0$. From (7.34) and from this assumption, it results that

$$\prod_{j=0}^{k_0+i\Delta} \frac{1}{(\beta_j^{DL+})^2} \geq \prod_{j=0}^{k_0} \frac{1}{(\beta_j^{DL+})^2} \text{ for any } i \geq 0, \tag{7.35}$$

which shows that

$$\sum_{k\geq 1} \prod_{j=0}^{k} \frac{1}{(\beta_j^{DL+})^2} = \infty. \tag{7.36}$$

But Dai (2003a) proved that for any conjugate gradient method with strong Wolfe line search, (7.36) gives the convergence relation $\liminf_{k\to\infty} \|g_k\| = 0$. [See (3.76)].

Actually, from (7.3), for any k, the DL+ method is $d_{k+1} + g_{k+1} = \beta_k^{DL+} d_k$. Hence,

$$\begin{aligned} \|d_{k+1}\|^2 &= -2g_{k+1}^T d_{k+1} - \|g_{k+1}\|^2 + (\beta_k^{DL+})^2 \|d_k\|^2 \\ &\leq \frac{(g_{k+1}^T d_{k+1})^2}{\|g_{k+1}\|^2} + (\beta_k^{DL+})^2 \|d_k\|^2. \end{aligned} \tag{7.37}$$

Therefore,

$$\begin{aligned} \|d_{k+1}\|^2 &\leq \left(1 - \frac{(g_{k+1}^T d_{k+1})^2}{\|g_{k+1}\|^2 \|d_{k+1}\|^2}\right)^{-1} (\beta_k^{DL+})^2 \|d_k\|^2 \leq \cdots \\ &\leq \prod_{j=1}^{k+1} \left(1 - \frac{(g_j^T d_j)^2}{\|g_j\|^2 \|d_j\|^2}\right)^{-1} \prod_{j=0}^{k} (\beta_j^{DL+})^2 \|d_0\|^2. \end{aligned} \tag{7.38}$$

But, any descendent method with Wolfe line search satisfies the relation

$$\sum_{j\geq 0} \frac{(g_j^T d_j)^2}{\|d_j\|^2} < \infty. \tag{7.39}$$

Since $\|g_k\| \geq \gamma$ for all k, then from (7.39), there is an integer j_0 so that

$$\prod_{j\geq j_0} \left(1 - \frac{(g_j^T d_j)^2}{\|g_j\|^2 \|d_j\|^2}\right) \geq c_1, \tag{7.40}$$

for some positive constant c_1. Thus, from (7.36), (7.38), and (7.40), it follows that (7.14) holds. Therefore, from Proposition 7.1, it follows that (7.15) holds, which is in contradiction with the assumption $\|g_k\| \geq \gamma$. Therefore, (7.32) must be true. ♦

With this result, the following theorem proves the convergence of the DL+ method under the strong Wolfe line search (Dai & Liao, 2001).

Theorem 7.2 *Suppose that the standard Assumption CG holds and consider the DL+ method, where d_k satisfies the sufficient descent condition $g_k^T d_k < -c\|g_k\|^2$ for some positive constant $c > 0$ and α_k is determined by the strong Wolfe line search* (7.4) *and* (7.6). *Then* $\liminf_{k\to\infty}\|g_k\| = 0$.

Proof The proof is given again by contradiction. Since $\liminf_{k\to\infty}\|g_k\| > 0$, then it follows that $\|g_k\| \geq \gamma$ for all $k \geq 0$. However, from the Assumption CG there exists a constant $\Gamma \geq 0$ so that $\|\nabla f(x)\| \leq \Gamma$ for all $x \in S$. Then, the conditions of Proposition 7.2 hold. Defining $u_i = d_i/\|d_i\|$, then for any two indices l and k with $l \geq k$, it follows that

$$\begin{aligned} x_l - x_{k-1} &= \sum_{i=k}^{l} \|s_{i-1}\| u_{i-1} \\ &= \sum_{i=k}^{l} \|s_{i-1}\| u_{k-1} + \sum_{i=k}^{l} \|s_{i-1}\| (u_{i-1} - u_{k-1}). \end{aligned} \tag{7.41}$$

Since $\|x\| \leq B$, for any $x \in S$ and $\|u_k\| = 1$,

$$\begin{aligned} \sum_{i=k}^{l} \|s_{i-1}\| &\leq \|x_l - x_{k-1}\| + \sum_{i=k}^{l} \|s_{i-1}\| \, \|u_{i-1} - u_{k-1}\| \\ &\leq 2B + \sum_{i=k}^{l} \|s_{i-1}\| \, \|u_{i-1} - u_{k-1}\|. \end{aligned} \tag{7.42}$$

Let $\lambda > 0$ be given as in Proposition 7.3 and define $\Delta = \lceil 8B/\lambda \rceil$ as the smallest integer not less than $8B/\lambda$. Then, by Proposition 7.2, an index $k_0 \geq 0$ can be found so that

$$\sum_{i \geq k_0} \|u_i - u_{i-1}\|^2 \leq \frac{1}{4\Delta}. \tag{7.43}$$

With this Δ and k_0, Proposition 7.3 gives an index $k \geq k_0$ for which

$$\left| K_{k,\Delta}^{\lambda} \right| > \frac{\Delta}{2}. \tag{7.44}$$

Next, for any index $i \in [k, k + \Delta - 1]$, by the Cauchy–Schwarz inequality and (7.43), it follows that

$$\|u_i - u_{k-1}\| \leq \sum_{j=k}^{i} \|u_j - u_{j-1}\| \leq (i-k+1)^{1/2}\left(\sum_{j=k}^{i} \|u_j - u_{j-1}\|^2\right)^{1/2}$$
$$\leq \Delta^{1/2}\left(\frac{1}{4\Delta}\right)^{1/2} = \frac{1}{2}.$$

From this relation, (7.44), Proposition 7.3 and taking $l = k + \Delta - 1$ in (7.42) it results that

$$2B \geq \frac{1}{2}\sum_{i=k}^{k+\Delta-1} \|s_{i-1}\| > \frac{\lambda}{2}\left|K_{k,\Delta}^{\lambda}\right| > \frac{\lambda\Delta}{4}.$$

Thus, $\Delta < 8B/\lambda$, which contradicts the definition of Δ. Hence, the theorem is true. ♦

It is obvious that the above results may be considered as a methodology for proving the convergence of conjugate gradient methods. Similar techniques may be used to prove the convergence of some other modified conjugate gradient methods (Gilbert & Nocedal, 1992).

Numerical study. In the first numerical experiment, let us present some results with the DL (t = 1) and DL+ (t = 1) methods in comparison with some standard conjugate gradient methods. For this, 80 unconstrained optimization problems from the UOP collection (Andrei, 2018g) are taken, each of them with the number of variables as $n = 1000, 2000, \ldots, 10000$. The numerical experiments are given in the context of Remark 1.1. The maximum number of iterations was limited to 2000.

Both DL and DL+ are modifications of HS based on the conjugacy condition (7.10), where $t \geq 0$ is a parameter. In these numerical experiments, t = 1. Figure 7.1 presents the Dolan and Moré performance profiles of DL (t = 1) versus DL+ (t = 1) for solving 800 unconstrained optimization problems from the UOP collection. Observe that DL+ is more efficient and more robust than DL.

Figure 7.2 presents the Dolan and Moré performance profiles of the DL (t = 1) and DL+ (t = 1) methods versus two methods with $g_{k+1}^T y_k$ at the numerator of the conjugate gradient parameter β_k (HS and PRP) and versus two methods with $\|g_{k+1}\|^2$ at the numerator of β_k (FR and DY). The performances of the methods with $g_{k+1}^T y_k$ at the numerator of β_k are known to be better than the performances of the methods with $\|g_{k+1}\|^2$ at the numerator of β_k.

In these numerical experiments presented in Figure 7.2, observe that both DL and DL+ are more efficient and more robust than FR and DY, i.e., than the methods with $\|g_{k+1}\|^2$ at the numerator of β_k and the differences are substantial. On the

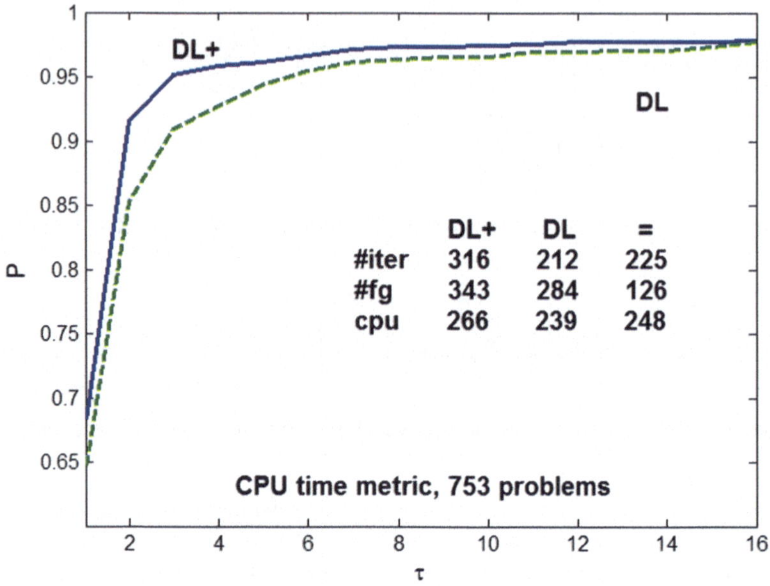

Figure 7.1 Performance profiles of DL+ ($t = 1$) versus DL ($t = 1$)

other hand, the performances of DL and DL+ are similar to those of HS and PRP, both methods DL and DL+ being slightly more robust.

In Figure 7.2, we see that the differences between DL+ and FR or DY are greater than the differences between DL and FR or DY.

In the second set of numerical experiments, Table 7.1 presents the performances of the DL ($t = 1$) and DL+ ($t = 1$) conjugate gradient methods for solving five applications from the MINPACK-2 collection, each of them with 40,000 variables.

It is worth comparing the performances of the DL and DL+ methods from Table 7.1 versus the performances of the standard conjugate gradient methods from Tables 4.2, 4.3, and 4.4. Table 4.3 shows that, subject to the CPU time metric, CD with 137.09 s is a top performer among the standard conjugate gradient methods for solving all five applications. However, DL+ with 96.53 s is much faster. Similarly, comparisons of the performances of DL and DL+ from Table 7.1 versus the performances of the hybrid conjugate gradient method NDPRPDY from Table 6.5 show that NDPRPDY with 82.06 s is faster for solving all five MINPACK-2 applications considered in this numerical study.

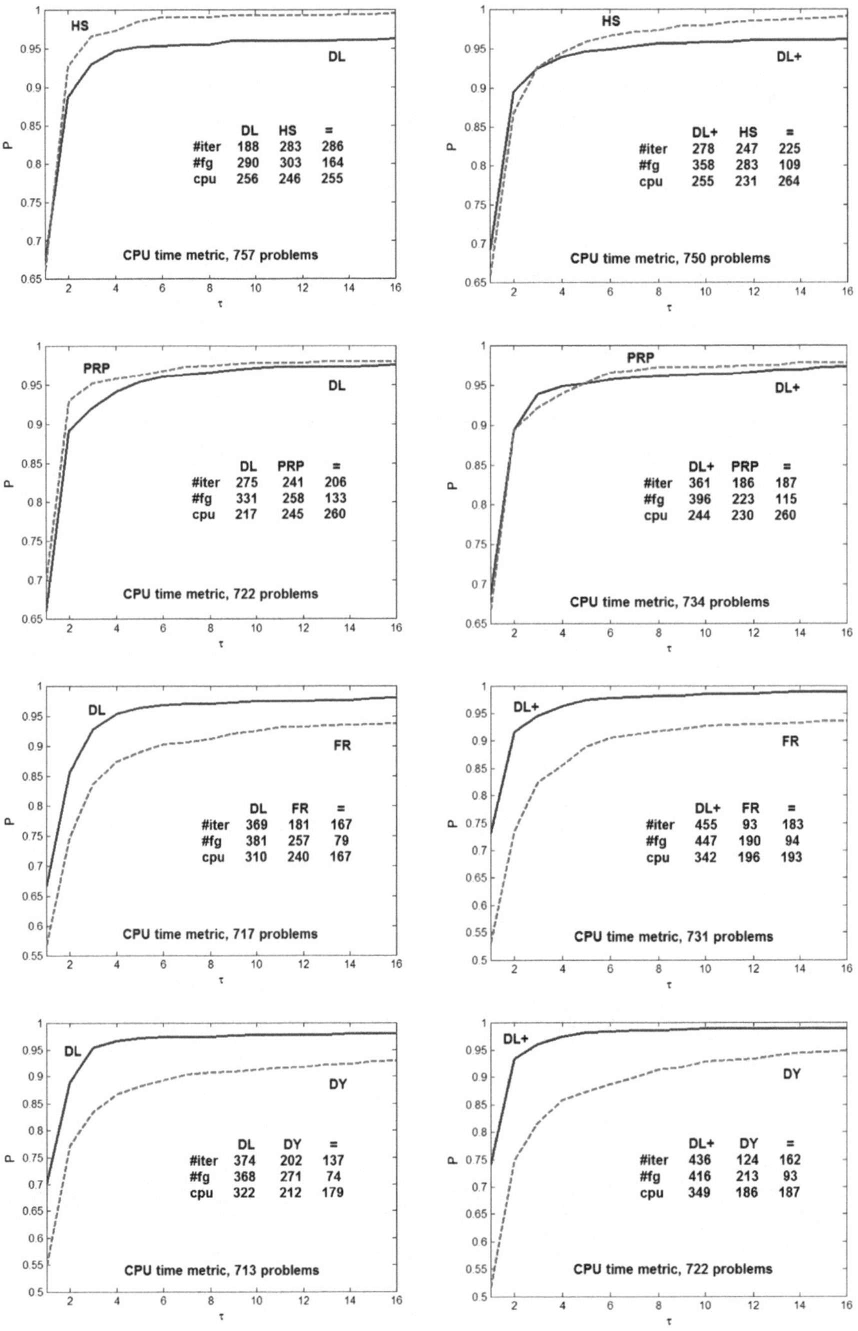

Figure 7.2 Performance profiles of DL ($t = 1$) and DL+ ($t = 1$) versus HS, PRP, FR, and DY

Table 7.1 Performances of DL (t = 1) and DL+ (t = 1) for solving five applications from the MINPACK-2 collection

		DL			DL+		
	n	*#iter*	*#fg*	*cpu*	*#iter*	*#fg*	*cpu*
A1	40,000	528	607	5.76	525	601	5.73
A2	40,000	629	1185	11.86	629	1185	11.86
A3	40,000	951	1743	47.10	906	1678	36.63
A4	40,000	299	602	30.87	299	602	30.87
A5	40,000	374	679	11.45	374	679	11.44
Total	–	2781	4816	107.04	2733	4745	96.53

7.2 Conjugate Gradient with Guaranteed Descent (CG-DESCENT)

For solving the problem (7.1), Hager and Zhang (2005, 2006a) proposed one of the most respected conjugate gradient algorithms:

$$x_{k+1} = x_k + \alpha_k d_k, \tag{7.45}$$

$$d_{k+1} = -g_{k+1} + \bar{\beta}_k^N d_k, d_0 = -g_0, \tag{7.46}$$

$$\bar{\beta}_k^N = max\{\beta_k^N, \eta_k\}, \tag{7.47}$$

$$\eta_k = \frac{-1}{\|d_k\| min\{\eta, \|g_k\|\}}, \tag{7.48}$$

$$\beta_k^N = \frac{1}{y_k^T d_k}\left(y_k - 2\frac{\|y_k\|^2}{y_k^T d_k} d_k\right)^T g_{k+1}, \tag{7.49}$$

where $\eta > 0$ is a constant ($\eta = 0.01$). Similar to the PRP+ method, in order to achieve the global convergence for general nonlinear functions, the truncation (7.47) is introduced. Observe that in the restricted scheme (7.47), the lower bound on $\bar{\beta}_k^N$ is dynamically adjusted in order to make the lower bound smaller as iterates converge. Observe that (7.49) is a modification of the HS method.

Hager and Zhang obtained this scheme by deleting a term from the direction of the memoryless quasi-Newton method of Perry (1977) and Shanno (1978b). Indeed, the search direction of Hager and Zhang may be written as

$$d_{k+1}^{HZ} = -g_{k+1} - \frac{y_k^T y_k}{d_k^T y_k}\left[2\frac{d_k^T g_{k+1}}{d_k^T y_k} - \frac{y_k^T g_{k+1}}{y_k^T y_k}\right] d_k. \tag{7.50}$$

On the other hand, the search direction of Perry–Shanno d_k^{PS} is

$$d_{k+1}^{PS} = -\frac{y_k^T s_k}{y_k^T y_k} g_{k+1} - \left[2\frac{s_k^T g_{k+1}}{y_k^T s_k} - \frac{y_k^T g_{k+1}}{y_k^T y_k} \right] s_k + \frac{s_k^T g_{k+1}}{y_k^T y_k} y_k. \tag{7.51}$$

(In Chapter 8, we are presenting the Perry–Shanno direction.) Observe that the relationship between the search direction of Perry–Shanno and that of Hager–Zhang is as follows

$$d_{k+1}^{PS} = \frac{s_k^T y_k}{y_k^T y_k} \left(d_{k+1}^{HZ} + \frac{d_k^T g_{k+1}}{y_k^T d_k} y_k \right), \tag{7.52}$$

or

$$d_{k+1}^{HZ} = \frac{y_k^T y_k}{y_k^T s_k} d_{k+1}^{PS} - \frac{d_k^T g_{k+1}}{y_k^T d_k} y_k. \tag{7.53}$$

Obviously, d_{k+1}^{HZ} may be written as

$$d_{k+1}^{HZ} = -\left[I - \frac{d_k y_k^T}{y_k^T d_k} + 2\frac{y_k^T y_k}{(y_k^T d_k)^2} d_k d_k^T \right] g_{k+1} \equiv -Q_{k+1}^{HZ} g_{k+1}. \tag{7.54}$$

Observe that Q_{k+1}^{HZ} is not symmetric and does not satisfy the quasi-Newton equation, properties satisfied by d_{k+1}^{PS}. If in a canonical manner Q_{k+1}^{HZ} is symmetrized and imposed to satisfy the quasi-Newton equation, then the Perry–Shanno direction (7.51) is obtained. From (7.52), when the angle between d_k and g_{k+1} is sufficiently small and f is strongly convex, we see that the term d_{k+1}^{HZ} dominates the y_k term. In this case, the directions d_{k+1}^{HZ} are approximately multiples of d_{k+1}^{PS}. The Perry–Shanno scheme, analyzed by Shanno and Phua (1980) and by Shanno (1985), has global convergence for convex functions and an inexact line search (Shanno, 1978b), but in general it does not necessarily converge, even when the line search is exact (Powell, 1984a). However, the Perry–Shanno scheme is convergent if the restarts are employed, but in this case the speed of convergence can decrease. Han, Liu, and Yin (1997) proved that if the standard Wolfe line search is employed, then the convergence to a stationary point of minimizing function f is achieved when $\lim_{k\to\infty} \|y_k\| = 0$ and the gradient of f is Lipschitz continuous.

The following theorem shows that if $y_k^T d_k \neq 0$, a condition which is satisfied when f is strongly convex or the line search satisfies the Wolfe conditions, then the computational method given by (7.45)–(7.49) always generates descent directions (Hager & Zhang, 2005).

Theorem 7.3 *If* $y_k^T d_k \neq 0$ *and*

$$d_{k+1} = -g_{k+1} + \tau d_k, d_0 = -g_0, \tag{7.55}$$

for any $\tau \in [\beta_k^N, max\{0, \beta_k^N\}]$, *then*

$$g_k^T d_k \leq -\frac{7}{8}\|g_k\|^2. \tag{7.56}$$

Proof Since $d_0 = -g_0$, it follows that $g_0^T d_0 = -\|g_0\|^2$, which satisfies (7.56). Suppose that $\tau = \beta_k^N$. Multiplying (7.55) by g_{k+1}^T it results that

$$\begin{aligned} g_{k+1}^T d_{k+1} &= -\|g_{k+1}\|^2 + \beta_k^N g_{k+1}^T d_k \\ &= -\|g_{k+1}\|^2 + \left(\frac{y_k^T g_{k+1}}{y_k^T d_k} - 2\frac{\|y_k\|^2 d_k^T g_{k+1}}{(y_k^T d_k)^2}\right) g_{k+1}^T d_k \\ &= \frac{(y_k^T g_{k+1})(y_k^T d_k)(g_{k+1}^T d_k) - \|g_{k+1}\|^2 (y_k^T d_k)^2 - 2\|y_k\|^2 (g_{k+1}^T d_k)^2}{(y_k^T d_k)^2}. \end{aligned}$$

Now, for the first term $(y_k^T g_{k+1})(y_k^T d_k)(g_{k+1}^T d_k)$ from the above equality let us apply the classical inequality $|u^T v| \leq \frac{1}{2}\left(\|u\|^2 + \|v\|^2\right)$, where $u = \frac{1}{2}(y_k^T d_k) g_{k+1}$ and $v = 2(g_{k+1}^T d_k) y_k$, to get (7.56). On the other hand, if $\tau \neq \beta_k^N$, then $\beta_k^N \leq \tau \leq 0$. After multiplying (7.55) with g_{k+1}^T, it follows that

$$g_{k+1}^T d_{k+1} = -\|g_{k+1}\|^2 + \tau g_{k+1}^T d_k.$$

If $g_{k+1}^T d_k \geq 0$, then (7.56) immediately follows since $\tau \leq 0$. If $g_{k+1}^T d_k < 0$, then

$$g_{k+1}^T d_{k+1} = -\|g_{k+1}\|^2 + \tau g_{k+1}^T d_k \leq -\|g_{k+1}\|^2 + \beta_k^N g_{k+1}^T d_k$$

since $\beta_k^N \leq \tau \leq 0$. Therefore, (7.56) is true by the above analysis. ♦

In the following, let us prove the convergence of the method (7.45)–(7.49) for strongly convex functions.

Theorem 7.4 *Suppose that f is Lipschitz continuous and strongly convex on the level set S. That is, there exist constants L and* $\mu > 0$ *so that*

$$\|\nabla f(x) - \nabla f(y)\| \leq L\|x - y\|, \tag{7.57}$$

$$(\nabla f(x) - \nabla f(y))^T (x - y) \geq \mu\|x - y\|^2, \tag{7.58}$$

for all $x, y \in S$. *If the conjugate gradient method*

$$d_{k+1} = -g_{k+1} + \beta_k^N d_k, d_0 = -g_0, \tag{7.59}$$

where β_k^N *given by* (7.49) *is implemented with a line search that satisfies either the Wolfe or the Goldstein conditions in each step, then either* $g_k = 0$ *for some* k, *or*

$$\lim_{k \to \infty} g_k = 0. \tag{7.60}$$

Proof Suppose that $g_k \neq 0$ for all k. Then, from the strong convexity assumption, it follows that

$$y_k^T d_k = (g_{k+1} - g_k)^T d_k \geq \mu \alpha_k \|d_k\|^2. \tag{7.61}$$

Theorem 7.3 and the assumption $g_k \neq 0$ ensure that $d_k \neq 0$. Since $\alpha_k > 0$, from (7.61) it results that $y_k^T d_k > 0$. Since f is strongly convex over S, then f is bounded from below. After summing over k the upper bound in the Goldstein condition (1.11) or in the Wolfe condition (7.4), it results that

$$\sum_{k=0}^{\infty} \alpha_k g_k^T d_k > -\infty.$$

Combination of this with the lower bound for α_k given by Proposition 1.2 and the descent property (7.56) gives

$$\sum_{k=0}^{\infty} \frac{\|g_k\|^4}{\|d_k\|^2} < \infty. \tag{7.62}$$

From the Lipschitz continuity (7.57),

$$\|y_k\| = \|g_{k+1} - g_k\| = \|\nabla f(x_k + \alpha_k d_k) - \nabla f(x_k)\| \leq L\alpha_k \|d_k\|. \tag{7.63}$$

Now, using (7.61) and (7.49), the following estimation is obtained

$$\begin{aligned}
|\beta_k^N| &= \left| \frac{y_k^T g_{k+1}}{y_k^T d_k} - 2\frac{\|y_k\|^2 (d_k^T g_{k+1})}{(y_k^T d_k)^2} \right| \leq \frac{\|y_k\| \|g_{k+1}\|}{\mu \alpha_k \|d_k\|^2} + 2\frac{\|y_k\|^2 \|d_k\| \|g_{k+1}\|}{\mu^2 \alpha_k^2 \|d_k\|^4} \\
&\leq \frac{L\alpha_k \|d_k\| \|g_{k+1}\|}{\mu \alpha_k \|d_k\|^2} + 2\frac{L^2 \alpha_k^2 \|d_k\|^3 \|g_{k+1}\|}{\mu^2 \alpha_k^2 \|d_k\|^4} \\
&\leq \left(\frac{L}{\mu} + \frac{2L^2}{\mu^2} \right) \frac{\|g_{k+1}\|}{\|d_k\|}.
\end{aligned} \tag{7.64}$$

Hence,

$$\|d_{k+1}\| \leq \|g_{k+1}\| + |\beta_k^N| \|d_k\| \leq \left(1 + \frac{L}{\mu} + \frac{2L^2}{\mu^2}\right) \|g_{k+1}\|.$$

Using this upper bound for d_k in (7.62), it follows that

$$\sum_{k=1}^{\infty} \|g_k\|^2 < \infty,$$

which completes the proof. ♦

The convergence analysis for general nonlinear functions is based on the insights developed by Gilbert and Nocedal (1992) in their analysis of the PRP+ conjugate gradient method. Firstly, Hager and Zhang (2005) proved that *if the level set* $S = \{x \in \mathbb{R}^n : f(x) \leq f(x_0)\}$ *is bounded and the Lipschitz condition* (7.57) *holds, then for the method given by* (7.45)–(7.49) *under a line search satisfying the Wolfe conditions* (7.4) *and* (7.5) *it follows that* $d_k \neq 0$ *for all k and*

$$\sum_{k=0}^{\infty} \|u_{k+1} - u_k\|^2 < \infty,$$

whenever $\inf\{\|g_k\| : k \geq 0\} > 0$, *where* $u_k = d_k / \|d_k\|$.

With this, Hager and Zhang (2005) proved that *if the level set* $S = \{x \in \mathbb{R}^n : f(x) \leq f(x_0)\}$ *is bounded and the Lipschitz condition* (7.57) *holds, then for the method given by* (7.45)–(7.49) *under the Wolfe line search* (7.4) *and* (7.5), *either* $g_k = 0$ *for some k, or* $\liminf_{k \to \infty} \|g_k\| = 0$.

Recall that Hager and Zhang obtained their conjugate gradient algorithm by deleting a term from the direction of the memoryless quasi-Newton method of Perry (1977) and Shanno (1978b). Now, it can be proved that if the cosine of the angle between d_k and g_{k+1} is small and f is strongly convex, then the second term in the brackets in (7.52), namely

$$\frac{d_k^T g_{k+1}}{y_k^T d_k} y_k,$$

can be neglected. Indeed, by (7.61) and (7.63) it follows that

$$\frac{|d_k^T g_{k+1}|}{|y_k^T d_k|} \|y_k\| = \frac{|u_k^T g_{k+1}| \|d_k\|}{|y_k^T d_k|} \|y_k\| \leq \frac{L}{\mu} |u_k^T g_{k+1}| = c_1 \varepsilon \|g_{k+1}\|, \tag{7.65}$$

where $u_k = d_k / \|d_k\|$ is the unit vector in the direction d_k, $c_1 = L/\mu$ and ε is the cosine of the angle between d_k and g_{k+1}. By definition of d_{k+1} in (7.46),

$$\|d_{k+1}\|^2 \geq \|g_{k+1}\|^2 - 2\beta_k^N (d_k^T g_{k+1}). \tag{7.66}$$

But, by the bound for β_k^N in (7.64), it follows that

$$\begin{aligned} |\beta_k^N (d_k^T g_{k+1})| &\leq |\beta_k^N| |d_k^T g_{k+1}| \leq c_2 \frac{\|g_{k+1}\|}{\|d_k\|} |d_k^T g_{k+1}| \\ &= c_2 \|g_{k+1}\| |u_k^T g_{k+1}| = c_2 \|g_{k+1}\|^2 \frac{|u_k^T g_{k+1}|}{\|g_{k+1}\|} = c_2 \varepsilon \|g_{k+1}\|^2, \end{aligned} \tag{7.67}$$

where $c_2 = L/\mu + 2L^2/(\mu^2)$.

Combining (7.66) and (7.67), it results that

$$\|d_{k+1}\| \geq \sqrt{1 - 2c_2\varepsilon} \|g_{k+1}\|. \tag{7.68}$$

Therefore, the upper bound (7.65) for the y_k term in (7.52) and the lower bound for $\|d_{k+1}\|$ in (7.68) imply that the ratio between them is bounded by $c_1\varepsilon/\sqrt{1 - 2c_2\varepsilon}$. Hence, when ε is small, the direction (7.46) is approximately a multiple of the Perry–Shanno direction (7.52). In other words, if the function f is strongly convex and the cosine of the angle between d_k and g_{k+1} is sufficiently small, then the directions generated by the computational scheme (7.45)–(7.49) approximately point along the Perry–Shanno direction (7.51).

Observe that the jamming phenomenon in the sense of Powell may also appear in this algorithm. However, the algorithm of Hager and Zhang is one of the best conjugate gradient algorithms for unconstrained optimization.

One of the main characteristics of the conjugate gradient method of Hager and Zhang is the implementation of both the standard Wolfe line search (7.4) and (7.5) and the approximate Wolfe line search (1.21). The advantage of using the approximate Wolfe conditions is that the local minimizers can be computed with accuracy on the order of the machine epsilon rather than with accuracy on the order of the square root of the machine epsilon. Therefore, it can be a speed advantage associated with the approximate Wolfe conditions. Hager and Zhang (2006a) give some details on implementing the approximate Wolfe line search (see Section 1.2), showing that the quadratic interpolant $q(\alpha)$ that matches $\varphi_k(\alpha) = f(x_k + \alpha d_k)$ at $\alpha = 0$ and $\varphi_k'(\alpha)$ at $\alpha = 0$ and $\alpha = \alpha_k$, determines that the first Wolfe condition is approximated by $(\varphi_k'(\alpha_k) + \varphi_k'(0))/2$, which is more accurate (and numerically more stable) than $(\varphi_k(\alpha_k) - \varphi_k(0))/\alpha_k$, which defines the original first Wolfe condition. The conjugate gradient method with the parameter $\bar{\beta}_k^N$ in which the line search satisfies the Wolfe line search (7.4) and (7.5) or the approximate Wolfe conditions (1.21) is called CG-DESCENT.

Numerical study. In the following, let us see the performances of CG-DESCENT (version 1.4) for solving 80 unconstrained optimization problems from the UOP collection (Andrei, 2018g), where the number of variables is $n = 1000, 2000, \ldots, 10000$. The numerical experiments are given in the context of Remark 1.1. The maximum number of iterations was limited to 2000.

We mention that in our numerical experiments we compare the conjugate gradient algorithms presented in this book versus the Fortran CG-DESCENT version 1.4 of Hager and Zhang (2005). The idea was to see the performances CG-DESCENT with $\bar{\beta}_k^N$ given by (7.47) versus some conjugate gradient algorithms without any other ingredients included in some other versions of CG-DESCENT or in the limited-memory conjugate gradient algorithm L-CG-DESCENT proposed by Hager and Zhang (2013). Our goal is not to exhaustively compare the algorithms to each other, but to see the performances of these algorithms in similar conditions.

Figure 7.3 presents the Dolan and Moré performance profiles of CG-DESCENT (version 1.4) with Wolfe line search versus the standard conjugate gradient methods HS, PRP, DY, and LS. On the other hand, Figure 7.4 presents the performance profiles of CG-DESCENT with approximate Wolfe line search (CG-DESCENTaw) versus the standard conjugate gradient methods HS, PRP, DY, and LS.

From Figures 7.3 and 7.4, it is clear that CG-DESCENT and CG-DESCENTaw are more efficient and more robust than the standard conjugate gradient methods considered in this numerical study. CG-DESCENT is a modification of the self-scaling BFGS method of Perry and Shanno. Therefore, it is able to better capture the curvature of the minimizing function. Besides, CG-DESCENTaw

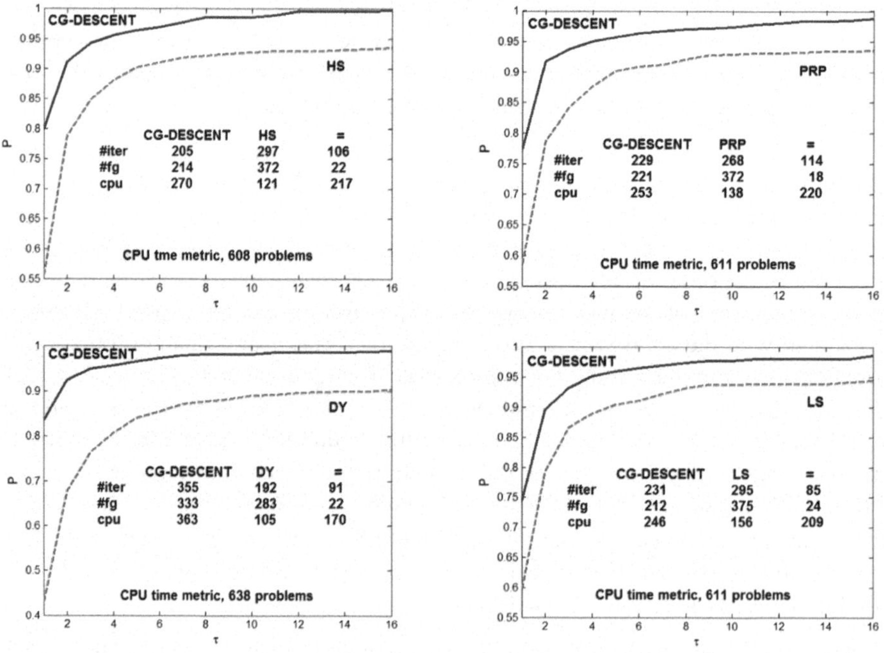

Figure 7.3 Performance profiles of CG-DESCENT versus HS, PRP, DY, and LS

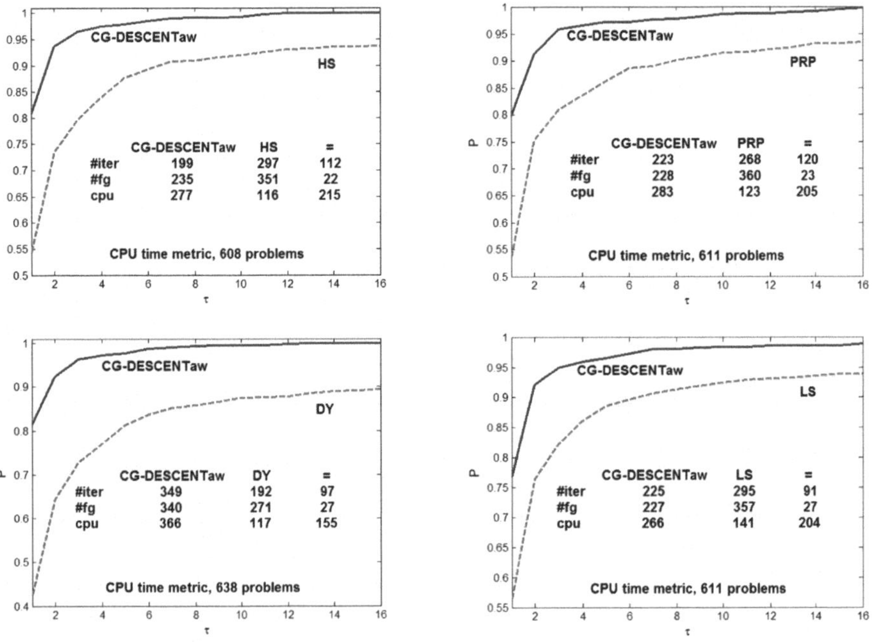

Figure 7.4 Performance profiles of CG-DESCENTaw (CG-DESCENT with approximate Wolfe conditions) versus HS, PRP, DY, and LS

implements the approximate Wolfe conditions. The difference between these two versions of CG-DESCENT is important.

In contrast to the Wolfe conditions, the approximate Wolfe conditions (1.21) are satisfied at a minimizer of $\varphi_k(\alpha) = f(x_k + \alpha d_k)$. Therefore, when trying to satisfy the approximate Wolfe conditions, we focus on minimizing φ_k; when trying to satisfy the usual Wolfe conditions, we focus on minimizing $\phi_k(\alpha) = \varphi_k(\alpha) - \varphi_k(0) - \alpha\rho\dot{\varphi}_k(0)$.

It is easy to see that since $\phi_k(0) = 0$, it is required that the local minimizer α^* satisfy $\phi_k(\alpha^*) < 0$ and $\phi'_k(\alpha^*) = 0$. But, these two relations together imply that the Wolfe conditions hold in a neighborhood of α^* when $\rho < \sigma$. Although there is no theory to guarantee the convergence of the algorithm with approximate Wolfe conditions, however, Hager and Zhang pointed out that there is a numerical advantage in using the approximate Wolfe conditions: with approximate Wolfe conditions, the local minimizers are computed with accuracy on the order of the machine epsilon rather than with accuracy on the order of the square root of the machine epsilon. We must emphasize that the conjugate gradient method has an n-step quadratic convergence property when α_k is the minimum of $\varphi_k(\alpha) = f(x_k + \alpha d_k)$.

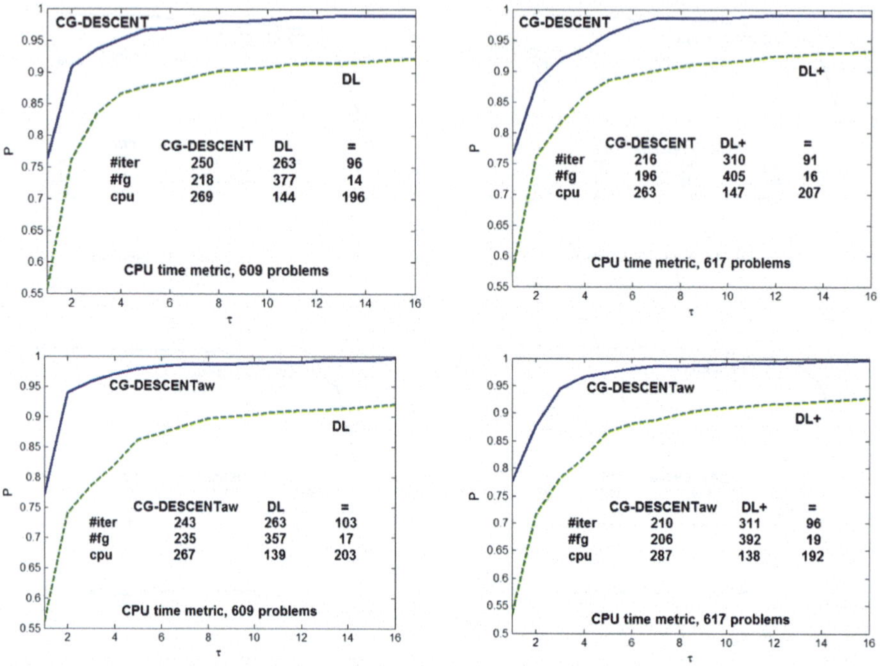

Figure 7.5 Performance profiles of CG-DESCENT and CG-DESCENTaw (CG-DESCENT with approximate Wolfe conditions) versus DL (t = 1) and DL+ (t = 1)

Table 7.2 Performances of CG-DESCENT and CG-DESCENTaw for solving five applications from the MINPACK-2 collection

		CG-DESCENT			CG-DESCENTaw		
	n	*#iter*	*#f*	*cpu*	*#iter*	*#f*	*cpu*
A1	40,000	323	647	9.67	323	647	7.77
A2	40,000	788	1577	31.35	788	1577	27.05
A3	40,000	1043	2088	64.96	1043	2088	66.03
A4	40,000	435	871	81.40	435	871	72.24
A5	40,000	286	573	9.89	286	573	13.25
Total	–	2875	5756	197.27	2875	5756	186.34

Figure 7.5 shows the performance profiles of CG-DESCENT and CG-DESCENTaw versus DL (t = 1) and DL+ (t = 1). Both CG-DESCENT and CG-DESCENTaw are more efficient and more robust than DL or DL+.

The performances of CG-DESCENT with Wolfe line search (CG-DESCENT) and of CG-DESCENT with approximate Wolfe line search (CG-DESCENTaw) for solving the MINPACK-2 applications are presented in Table 7.2. (#f represents the number of function calls.)

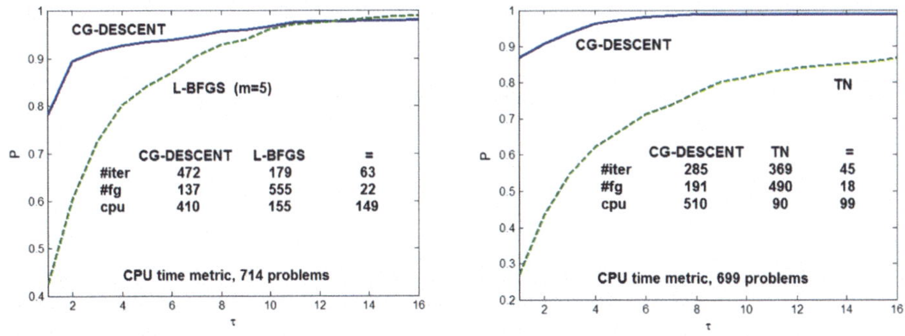

Figure 7.6 Performance profile of CG-DESCENT versus L-BFGS ($m = 5$) and versus TN

For solving the applications from the MINPACK-2, the performances of CG-DESCENTaw are similar to those of CG-DESCENT. Table 7.1 shows that, for solving these applications from the MINPACK-2 collection, both DL with 107.04 s and DL+ with 96.53 s are faster than CG-DESCENTaw with 186.34 s.

The performance profiles of CG-DESCENT versus L-BFGS ($m = 5$) and versus TN are given in Figure 7.6. Observe that CG-DESCENT is a top performer. L-BFGS uses the curvature information from only the most recent iterations (in our case 5) with which the approximation of the Hessian and the search direction are constructed. TN uses a BFGS approximation of the Hessian and the search direction is determined by an approximate solution of the Newton system. On the other hand, CG-DESCENT is indeed a modification of the HS conjugate gradient algorithm, but more importantly in another interpretation, it uses a self-scaling memoryless BFGS approximation of the Hessian, initialized with the identity matrix that better captures the curvature of the minimizing function.

7.3 Conjugate Gradient with Guaranteed Descent and Conjugacy Conditions and a Modified Wolfe Line Search (DESCON)

For solving the unconstrained optimization problem (7.1), Andrei (2013c) developed the following conjugate gradient algorithm

$$x_{k+1} = x_k + \alpha_k d_k, \tag{7.69}$$

where $\alpha_k > 0$ is obtained by a variant of the Wolfe line search discussed below and the directions d_k are generated as

$$d_{k+1} = -\theta_k g_{k+1} + \beta_k s_k, \tag{7.70}$$

$$\beta_k = \frac{y_k^T g_{k+1} - t_k s_k^T g_{k+1}}{y_k^T s_k}, \tag{7.71}$$

$d_0 = -g_0$, where θ_k and t_k are scalar parameters which are to be determined. Observe that in d_{k+1}, given by (7.70), g_{k+1} is scaled by parameter θ_k while parameter t_k in (7.71) is changed at every iteration. Algorithms of this form or variations of them were studied by many authors. For example, Andrei (2007a, 2007c) considered a preconditioned conjugate gradient algorithm where the preconditioner is a scaled memoryless BFGS matrix and the parameter scaling the gradient is selected as the spectral gradient. On the other hand, Birgin and Martínez (2001) suggested a spectral conjugate gradient method, where $\theta_k = s_k^T s_k / s_k^T y_k$. Stoer and Yuan (1995) studied the conjugate gradient algorithm on a subspace, where the search direction d_{k+1} is taken from the subspace $span\{-g_{k+1}, d_k\}$. Observe that, if for every $k \geq 1, \theta_k = 1$ and $t_k = t$, then (7.70) reduces to the Dai and Liao search direction [see (7.11)].

Suppose that the Assumption CG holds. Therefore, there exists a constant $\Gamma \geq 0$ so that $\|\nabla f(x)\| \leq \Gamma$ for all $x \in S = \{x \in \mathbb{R}^n : f(x) \leq f(x_0)\}$. Besides, it is easy to see that $\|s_k\| = \|x_{k+1} - x_k\| \leq \|x_{k+1}\| + \|x_k\| \leq 2B$.

In our algorithm, for all $k \geq 0$, the scalar parameters θ_k and t_k in (7.70) and (7.71), respectively, are determined in such a way so that both the sufficient descent and the conjugacy conditions are satisfied. Therefore, from the *sufficient descent condition*

$$g_{k+1}^T d_{k+1} \leq -w\|g_{k+1}\|^2 \tag{7.72}$$

it follows that

$$-\theta_k \|g_{k+1}\|^2 + \frac{(y_k^T g_{k+1})(s_k^T g_{k+1})}{y_k^T s_k} - t_k \frac{(s_k^T g_{k+1})^2}{y_k^T s_k} = -w\|g_{k+1}\|^2 \tag{7.73}$$

and from the Dai–Liao *conjugacy condition*

$$d_{k+1}^T y_k = -v(g_{k+1}^T s_k) \tag{7.74}$$

it results that

$$-\theta_k y_k^T g_{k+1} + y_k^T g_{k+1} - t_k s_k^T g_{k+1} = -v(s_k^T g_{k+1}), \tag{7.75}$$

where $v > 0$ and $w > 0$ are *known* scalar parameters. Observe that in (7.73) the classical sufficient descent condition (7.72) is modified with equality. It is worth saying that the main condition in any conjugate gradient algorithm is the descent condition $g_k^T d_k < 0$ or the sufficient descent condition (7.72). In (7.72) w is selected close to 1. This is quite a reasonable value. For example, Hager and Zhang (2005, 2006a) showed that in their CG-DESCENT algorithm, $w = 7/8$. On the other hand, the conjugacy condition is not so strict. In fact, very few conjugate gradient algorithms satisfy this condition. For example, the Hestenes and Stiefel algorithm has the property that the pure conjugacy condition always holds, independent of the line search.

If $v = 0$, then (7.75) is the "pure" conjugacy condition. However, in order to improve the algorithm and to incorporate the second-order information, $v > 0$ is taken.

Now, let us define

$$\bar{\Delta}_k \equiv (y_k^T g_{k+1})(s_k^T g_{k+1}) - \|g_{k+1}\|^2 (y_k^T s_k), \tag{7.76}$$

$$\Delta_k \equiv (s_k^T g_{k+1}) \bar{\Delta}_k, \tag{7.77}$$

$$a_k \equiv v(s_k^T g_{k+1}) + y_k^T g_{k+1}, \tag{7.78}$$

$$b_k \equiv w\|g_{k+1}\|^2 (y_k^T s_k) + (y_k^T g_{k+1})(s_k^T g_{k+1}). \tag{7.79}$$

Supposing that $\Delta_k \neq 0$ and $y_k^T g_{k+1} \neq 0$, then, from the linear algebraic system given by (7.73) and (7.75), t_k and θ_k can be determined as

$$t_k = \frac{b_k (y_k^T g_{k+1}) - a_k (y_k^T s_k)\|g_{k+1}\|^2}{\Delta_k}, \tag{7.80}$$

$$\theta_k = \frac{a_k - t_k (s_k^T g_{k+1})}{y_k^T g_{k+1}}, \tag{7.81}$$

with which the parameter β_k and the direction d_{k+1} can immediately be computed.

Observe that using (7.80) in (7.81), θ_k can be expressed as

$$\theta_k = \frac{a_k}{y_k^T g_{k+1}} \left[1 + \frac{(y_k^T s_k)\|g_{k+1}\|^2}{\bar{\Delta}_k} \right] - \frac{b_k}{\bar{\Delta}_k}. \tag{7.82}$$

Again, using (7.80) in (7.71), β_k can be determined as

$$\beta_k = \frac{y_k^T g_{k+1}}{y_k^T s_k} \left(1 - \frac{b_k}{\bar{\Delta}_k} \right) + a_k \frac{\|g_{k+1}\|^2}{\bar{\Delta}_k}. \tag{7.83}$$

Observe that the crucial element in our algorithm is $\bar{\Delta}_k$. Therefore, the proposed conjugate gradient algorithm with guaranteed descent and conjugacy conditions is defined by (7.69) and (7.70), where the scalar parameters θ_k and β_k are given by

(7.82) and (7.83), respectively, and α_k is computed by a variant of the Wolfe line search which is discussed in the following.

Modified Wolfe line search conditions. In order to define the algorithm, a small modification of the second Wolfe line search condition (7.5) is considered as

$$g(x_k + \alpha_k d_k)^T d_k \geq \sigma_k g_k^T d_k, \tag{7.84}$$

where σ_k is a sequence of parameters satisfying the condition $0<\rho<\sigma_k<1$ for all k. The interpretation of (7.84) is that the rate of decrease of f in the direction d_k at x_{k+1} is larger than a fraction σ_k, modified at every iteration, of the rate of decrease of f in the direction d_k at x_k. The condition $\rho<\sigma_k$ for all $k\geq 0$ guarantees that the Wolfe line search (7.4) and (7.84) can be satisfied simultaneously. Relations (7.4) and (7.84) are called *the modified Wolfe line search conditions.*

Proposition 7.4 *If*

$$\frac{1}{2}<\sigma_k \leq \frac{\|g_{k+1}\|^2}{|y_k^T g_{k+1}| + \|g_{k+1}\|^2}, \tag{7.85}$$

then, for all $k\geq 1$, $\bar{\Delta}_k<0$.

Proof Observe that

$$s_k^T g_{k+1} = s_k^T y_k + s_k^T g_k < s_k^T y_k. \tag{7.86}$$

The modified Wolfe condition (7.84) gives

$$g_{k+1}^T s_k \geq \sigma_k g_k^T s_k = -\sigma_k y_k^T s_k + \sigma_k g_{k+1}^T s_k. \tag{7.87}$$

Since $\sigma_k<1$, (7.87) can be rearranged to obtain

$$g_{k+1}^T s_k \geq \frac{-\sigma_k}{1-\sigma_k} y_k^T s_k. \tag{7.88}$$

Now, the combination of this lower bound for $g_{k+1}^T s_k$ with the upper bound (7.86), since $y_k^T s_k > 0$ (if $\|g_k\| \neq 0$), gives

$$|g_{k+1}^T s_k| \leq |y_k^T s_k| \max\left\{1, \frac{\sigma_k}{1-\sigma_k}\right\}. \tag{7.89}$$

But, since $\sigma_k > 1/2$, from (7.89) it follows that

$$|g_{k+1}^T s_k| < \frac{\sigma_k}{1-\sigma_k} |y_k^T s_k|. \tag{7.90}$$

If (7.85) is true, then

$$\frac{\sigma_k}{1-\sigma_k}\left|y_k^T g_{k+1}\right| \le \|g_{k+1}\|^2. \tag{7.91}$$

From (7.91), since $y_k^T s_k > 0$ it follows that

$$\frac{\sigma_k}{1-\sigma_k}\left|y_k^T s_k\right|\left|g_{k+1}^T y_k\right| \le \left|y_k^T s_k\right|\|g_{k+1}\|^2. \tag{7.92}$$

Now, from (7.90) and (7.92) it results that

$$\left|s_k^T g_{k+1}\right|\left|y_k^T g_{k+1}\right| < \frac{\sigma_k}{1-\sigma_k}\left|y_k^T s_k\right|\left|y_k^T g_{k+1}\right| \le \left|y_k^T s_k\right|\|g_{k+1}\|^2, \tag{7.93}$$

i.e. $\bar{\Delta}_k < 0$ for all $k \ge 1$. ♦

Therefore, in the algorithm, σ_k is computed as

$$\sigma_k = \frac{\|g_{k+1}\|^2}{\left|y_k^T g_{k+1}\right| + \|g_{k+1}\|^2}. \tag{7.94}$$

If $g_k \ne 0$ for all $k \ge 0$, then $0 < \sigma_k < 1$ for all $k \ge 0$.

Proposition 7.5 *Suppose that the Assumption CG holds and that* $\|g_k\| \ge \gamma > 0$ *for all* $k \ge 0$, *i.e., the norm of the gradient is bounded away from zero for all* $k \ge 0$. *Then the sequence* $\{\sigma_k\}$ *is uniformly bounded away from zero, independent of k.*

Proof From the above assumptions, observe that

$$\left|y_k^T g_{k+1}\right| \le \|y_k\|\|g_{k+1}\| \le L\|s_k\|\Gamma \le L\Gamma(2B).$$

Therefore,

$$|\sigma_k| = \frac{\|g_{k+1}\|^2}{\left|y_k^T g_{k+1}\right| + \|g_{k+1}\|^2} \ge \frac{\gamma^2}{2BL\Gamma + \Gamma^2} \equiv \eta > 0.$$

Since $|\sigma_k| \ge \eta$ for any $k \ge 0$, it follows that $\{\sigma_k\}$ is uniformly bounded away from zero. ♦

Observe that, since $\left|y_k^T g_{k+1}\right| \le \|g_{k+1}\|^2 + \left|g_k^T g_{k+1}\right|$, it follows that

$$\sigma_k \ge \frac{\|g_{k+1}\|^2}{2\|g_{k+1}\|^2 + \left|g_k^T g_{k+1}\right|}.$$

Obviously, in the bounded sequence $\{\sigma_k\}$ there exists a convergent subsequence.

Proposition 7.6 *Suppose that d_k satisfies the descent condition $g_k^T d_k = -w\|g_k\|^2$, where $w > 0$, and ∇f satisfies the Lipschitz condition $\|\nabla f(x) - \nabla f(x_k)\| \leq L\|x - x_k\|$ for all x on the line segment connecting x_k and x_{k+1}, where L is a positive constant. Besides, assume that $\|g_k\| \geq \gamma > 0$ for all $k \geq 0$. If the line search satisfies the modified Wolfe conditions (7.4) and (7.84), where $0 < \sigma_k < 1$ for all $k \geq 0$, then*

$$\alpha_k \geq \frac{(1-\sigma_k)}{L}\frac{w\gamma^2}{\|d_k\|^2} \equiv \omega_k. \tag{7.95}$$

Proof To prove (7.95), subtract $g_k^T d_k$ from both sides of (7.84) and, by using the Lipschitz condition, it results that

$$(\sigma_k - 1)g_k^T d_k \leq (g_{k+1} - g_k)^T d_k \leq \alpha_k L\|d_k\|^2.$$

However, d_k is a descent direction and $\sigma_k < 1$. From the descent condition, it follows that

$$\alpha_k \geq \frac{(1-\sigma_k)}{L}\frac{|g_k^T d_k|}{\|d_k\|^2} = \frac{(1-\sigma_k)}{L}\frac{w\|g_k\|^2}{\|d_k\|^2} \geq \frac{(1-\sigma_k)}{L}\frac{w\gamma^2}{\|d_k\|^2} > 0.$$

♦

Consider $\omega = \inf\{\omega_k\}$, where ω_k is defined in (7.95).

Some properties of the algorithm. In the following, let us present some properties of the elements which define the algorithm. Assume that the stepsize α_k is computed by the modified Wolfe line search conditions (7.4) and (7.84).

Proposition 7.7 *Suppose that d_k satisfies the descent condition $g_k^T d_k = -w\|g_k\|^2$, where $w > 0$, and $\nabla f(x)$ is Lipschitz continuous on the level set $S = \{x \in \mathbb{R}^n : f(x) \leq f(x_0)\}$. Besides, assume that $\|g_k\| \geq \gamma > 0$ for all $k \geq 0$. Then the sequence $\{\bar{\Delta}_k\}$ given by (7.76) is uniformly bounded away from zero, independent of k.*

Proof Since $g_k \neq 0$ for all $k \geq 0$, from (7.94) it follows that $\sigma_k < 1$ for all $k \geq 1$. Observe that with this value for σ_k, from (7.90) it follows that $\bar{\Delta}_k < 0$ for all $k \geq 1$. Now, from Proposition 7.6, from the modified Wolfe condition (7.84) and the descent condition $g_k^T d_k = -w\|g_k\|^2$, since $\sigma_k < 1$, for all $k \geq 1$, it follows that

$$\begin{aligned} y_k^T s_k &= \alpha_k y_k^T d_k = \alpha_k (g_{k+1} - g_k)^T d_k \geq \alpha_k(\sigma_k - 1) g_k^T d_k \\ &= -\alpha_k(\sigma_k - 1)w\|g_k\|^2 \geq \omega_k(1-\sigma_k)w\gamma^2 > 0. \end{aligned}$$

Therefore, $|y_k^T s_k|\|g_{k+1}\|^2 \geq \omega_k(1-\sigma_k)w\gamma^4 > 0$, for all $k \geq 1$, i.e. $(y_k^T s_k)\|g_{k+1}\|^2$ is uniformly bounded away from zero, independent of k.

On the other hand, observe that the first Wolfe condition (7.4) limits the accuracy of the algorithm to the order of the square root of the machine precision (Hager & Zhang, 2005). But d_k is a descent direction for any $k \geq 0$, therefore, even if the line search is not exact, the line search based on the modified Wolfe conditions is accurate enough to ensure that $s_k^T g_{k+1}$ tends to zero along the iterations. Therefore, since $|y_k^T g_{k+1}|$ is bounded as $|y_k^T g_{k+1}| \leq 2BL\Gamma$, it follows that $(y_k^T g_{k+1})(s_k^T g_{k+1}) \to 0$. Since $\bar{\Delta}_k < 0$ for all $k \geq 1$, it results that the sequence $\{\bar{\Delta}_k\}$ is uniformly bounded away from zero, independent of k. ♦

Proposition 7.8 *Suppose that d_k satisfies the descent condition $g_k^T d_k = -w\|g_k\|^2$, where $w > 0$ and $\|g_k\| \geq \gamma > 0$ for all $k \geq 0$. Then, the parameter θ_k defined in* (7.82) *tends to $w > 0$, i.e. $\theta_k \to w$.*

Proof From (7.70), using the descent condition $g_k^T d_k = -w\|g_k\|^2$,

$$\beta_k(s_k^T g_{k+1}) = (\theta_k - w)\|g_{k+1}\|^2 \geq (\theta_k - w)\gamma^2.$$

Since d_k is a descent direction and the stepsize α_k is computed by the modified Wolfe line search conditions, it follows that $s_k^T g_{k+1}$ tends to zero. Therefore, θ_k tends to $w > 0$ and hence $\theta_k > 0$. ♦

Observe that, since w is a real positive and finite constant and $\theta_k \to w$, there exist the real arbitrary and positive constants $0 < c_1 \leq w$ and $c_2 \geq w$, so that, for any $k \geq 1$, $c_1 \leq \theta_k \leq c_2$.

Proposition 7.9 *Suppose that d_k satisfies the descent condition $g_k^T d_k = -w\|g_k\|^2$, $\|g_k\| \geq \gamma > 0$ for all $k \geq 0$ and $w > 1$. Then, the scalar parameter b_k given by* (7.79) *is positive, i.e., $b_k > 0$.*

Proof By the second modified Wolfe condition (7.84), $y_k^T s_k = (g_{k+1} - g_k)^T s_k \geq (\sigma_k - 1) g_k^T s_k$. However, from the descent condition, it follows that $g_k^T s_k = \alpha_k g_k^T d_k = -\alpha_k w\|g_k\|^2$. From Proposition 7.6,

$$\begin{aligned} y_k^T s_k &\geq (\sigma_k - 1) g_k^T s_k = -\alpha_k(\sigma_k - 1) w\|g_k\|^2 \\ &\geq \omega_k w(1 - \sigma_k)\|g_k\|^2 > \omega_k w(1 - \sigma_k)\gamma^2 > 0. \end{aligned}$$

Therefore, by the modified second Wolfe condition (7.84), for all $k \geq 0$, $y_k^T s_k > 0$. On the other hand, since $w > 1$, from (7.93) it follows that

$$w\|g_{k+1}\|^2 (y_k^T s_k) \geq |y_k^T g_{k+1}| \, |s_k^T g_{k+1}|.$$

Since d_k is a descent direction and the stepsize α_k is computed by the modified Wolfe line search conditions, it follows that $s_k^T g_{k+1}$ tends to zero along the iterations. Therefore, from (7.79), $b_k > 0$ for all $k \geq 0$. ♦

The algorithm is equipped with an acceleration scheme presented in Chapter 5. Basically, the acceleration scheme modifies the stepsize α_k in a multiplicative way to improve the reduction of the function values along the iterations. In the accelerated algorithm, instead of (7.69), the new estimation of the minimum point is computed as

$$x_{k+1} = x_k + \xi_k \alpha_k d_k, \tag{7.96}$$

where

$$\xi_k = -\frac{\bar{a}_k}{\bar{b}_k}, \tag{7.97}$$

$\bar{a}_k = \alpha_k g_k^T d_k$, $\bar{b}_k = -\alpha_k (g_k - g_z)^T d_k$, $g_z = \nabla f(z)$, and $z = x_k + \alpha_k d_k$. Hence, if $\bar{b}_k > 0$, then the new estimation of the solution is computed as $x_{k+1} = x_k + \xi_k \alpha_k d_k$, otherwise $x_{k+1} = x_k + \alpha_k d_k$.

Observe that $\bar{b}_k = \alpha_k (g_z - g_k)^T d_k = \alpha_k (d_k^T \nabla^2 f(\bar{x}_k) d_k)$, where $\bar{x}_k$ is a point on the line segment connecting x_k and z. Since $\alpha_k > 0$, it follows that for convex functions, $\bar{b}_k \geq 0$. Hence, for convex functions, from the sufficient descent condition $g_k^T d_k = -w\|g_k\|^2$ it results that

$$\xi_k = -\frac{\bar{a}_k}{\bar{b}_k} = \frac{-\alpha_k (g_k^T d_k)}{\alpha_k (d_k^T \nabla^2 f(\bar{x}_k) d_k)} = \frac{w\|g_k\|^2}{d_k^T \nabla^2 f(\bar{x}_k) d_k} \geq 0. \tag{7.98}$$

For convex functions, there exist the constants $m > 0$ and $M < \infty$ so that

$$m\|u\|^2 \leq u^T \nabla^2 f(x) u \leq M\|u\|^2,$$

for any $u \neq 0$. Suppose that $\|g_k\| \geq \gamma > 0$ for all $k \geq 0$, otherwise a stationary point is obtained. From (7.98), it follows that

$$0 \leq \frac{w\gamma^2}{M\|d_k\|^2} \leq \xi_k \leq \frac{w\Gamma^2}{m\|d_k\|^2}.$$

Therefore, in (7.96), the stepsize α_k is modified by a finite and positive value ξ_k. With this modification of the stepsize, by Proposition 7.7 the sequence $\{\bar{\Delta}_k\}$ continues to be uniformly bounded away from zero, independent of k.

Observe that since ρ in the first Wolfe condition (7.4) is small enough (usually $\rho = 0.0001$), the Wolfe line search leads to very small reductions in the function values along the iterations. The acceleration scheme (7.96) emphasizes the reduction of function values, since often, in conjugate gradient algorithms, $\alpha_k > 1$ along the iterations (Andrei, 2009c).

Taking into consideration the above acceleration scheme (7.96) (see also Remark 5.1), where the acceleration factor ξ_k is computed as in (7.97), according to the value of the parameter "*acceleration*" (true or false), the following algorithms DESCON and DESCONa can be presented. DESCONa is the accelerated version of DESCON.

Algorithm 7.1 Guaranteed descent and conjugacy conditions with a modified Wolfe line search: DESCON/DESCONa

1.	Select a starting point $x_0 \in domf$ and compute: $f_0 = f(x_0)$ and $g_0 = \nabla f(x_0)$. Select some positive values for ρ and σ_0, and for v and w. Set $d_0 = -g_0$ and $k = 0$. Select the small positive constants ε_A and ε_m
2.	Test a criterion for stopping the iterations. If the test is satisfied, then stop; otherwise continue with step 3
3.	Determine the stepsize α_k by the modified Wolfe line search conditions (7.4) and (7.84). Update the variables as $x_{k+1} = x_k + \alpha_k d_k$. Compute f_{k+1} and g_{k+1}. Compute $y_k = g_{k+1} - g_k$ and $s_k = x_{k+1} - x_k$
4.	If *acceleration* equal true, then a) Compute: $z = x_k + \alpha_k d_k$, $g_z = \nabla f(z)$ and $y_k = g_k - g_z$ b) Compute: $\bar{a}_k = \alpha_k g_k^T d_k$, and $\bar{b}_k = -\alpha_k y_k^T d_k$ c) If $\lvert\bar{b}_k\rvert \geq \varepsilon_A$, then compute $\xi_k = -\bar{a}_k/\bar{b}_k$ and update the variables as $x_{k+1} = x_k + \xi_k \alpha_k d_k$. Compute f_{k+1} and g_{k+1}. Compute $y_k = g_{k+1} - g_k$ and $s_k = x_{k+1} - x_k$
5.	Compute $\bar{\Delta}_k$ as in (7.76)
6.	If $\lvert\bar{\Delta}_k\rvert \geq \varepsilon_m$, then determine θ_k and β_k as in (7.82) and (7.83), respectively, else set $\theta_k = 1$ and $\beta_k = 0$
7.	Compute the search direction as: $d_{k+1} = -\theta_k g_{k+1} + \beta_k s_k$
8.	Compute $\sigma_k = \lVert g_{k+1}\rVert^2 / \left(\lvert y_k^T g_{k+1}\rvert + \lVert g_{k+1}\rVert^2\right)$
9.	Restart criterion. If $\lvert g_{k+1}^T g_k\rvert > 0.2\lVert g_{k+1}\rVert^2$ then set $d_{k+1} = -g_{k+1}$
10.	Consider $k = k+1$ and go to step 2 ♦

If f is bounded along the direction d_k, then there exists a stepsize α_k satisfying the modified Wolfe line search conditions (7.4) and (7.84). When the Powell restart condition in step 9 is satisfied, then the algorithm is restarted with the negative gradient: $-g_{k+1}$. Under reasonable assumptions, the modified Wolfe line search conditions and the Powell restart criterion are sufficient to prove the global convergence of the algorithm. The first trial of the stepsize crucially affects the practical behavior of the algorithm. At every iteration $k \geq 1$, the starting guess for the step α_k in the line search is computed as $\alpha_{k-1}\lVert d_{k-1}\rVert / \lVert d_k\rVert$. Observe that in the line search procedure (step 3 of the algorithm), the stepsize α_k is computed using the updated value of the parameter σ_k computed as in step 8. For strongly convex functions, the linear convergence of the acceleration scheme is proved as in Andrei (2009c).

The DESCON algorithm can be implemented in some other variants. For example, in step 6, when $|\bar{\Delta}_k| \geq \varepsilon_m$ is not satisfied, $\theta_k = 1$ can be set and compute β_k as in standard conjugate gradient algorithms like HS, DY, PRP, etc.

Convergence analysis. In order to prove the global convergence of DESCON, assume that the stepsize α_k satisfies the strong Wolfe line search conditions (7.4) and

$$|g(x_k + \alpha_k d_k)^T d_k| \leq \sigma_k g_k^T d_k, \tag{7.99}$$

where ρ and σ_k are arbitrary positive constants so that $0 < \rho < \sigma_k < 1$. Observe that, since ρ in (7.3) is small enough, the parameter σ_k in (7.99) can be selected at each iteration as in (7.94), thus satisfying the above condition $0 < \rho < \sigma_k < 1$.

Proposition 7.10 *Suppose that the Assumption CG holds. Consider the conjugate gradient algorithm* (7.70), *where* θ_k *and* β_k *are given by* (7.82) *and* (7.83), *the descent condition* $g_k^T d_k < 0$ *is satisfied for any* $k \geq 0$ *and* α_k *is obtained by the modified Wolfe line search conditions* (7.4) *and* (7.84), *where* $1/2 \leq \sigma_k < 1$. *Then,*

$$\sum_{k=0}^{\infty} \frac{(g_k^T d_k)^2}{\|d_k\|^2} < +\infty. \tag{7.100}$$

Proof From (7.84) and from the Assumption CG,

$$(\sigma_k - 1) g_k^T d_k \leq (g_{k+1} - g_k)^T d_k \leq L\alpha_k \|d_k\|^2.$$

Since $1/2 \leq \sigma_k < 1$,it follows that

$$\alpha_k \geq \frac{-(1-\sigma_k)}{L} \frac{g_k^T d_k}{\|d_k\|^2} \geq -\frac{1}{2L} \frac{g_k^T d_k}{\|d_k\|^2}.$$

By the descent condition $g_k^T d_k < 0$, it follows that $f_{k+1} - f_k \leq \rho\alpha_k (g_k^T d_k) \leq 0$, i.e., $\{f(x_k)\}$ is a decreasing sequence. Therefore, the basic assumptions imply that there exists a constant f^* so that $\lim_{k\to\infty} f_k = f^*$. With this, it results that

$$\sum_{k=1}^{\infty} (f_k - f_{k+1}) = \lim_{n\to\infty} \sum_{k=1}^{n} (f_k - f_{k+1}) = \lim_{n\to\infty} (f_1 - f_{n+1}) = f_1 - f^* < \infty.$$

Hence,

$$\sum_{k=1}^{\infty} -\alpha_k (g_k^T d_k) < \infty.$$

With this,

$$\sum_{k=1}^{\infty}\frac{(g_k^T d_k)^2}{\|d_k\|^2}\leq 2L\sum_{k=1}^{\infty}(-\alpha_k g_k^T d_k),$$

which implies that (7.100) holds. ♦

Proposition 7.11 *Suppose that the Assumption CG holds. Consider the conjugate gradient algorithm* (7.70), *where* θ_k *and* β_k *are given by* (7.82) *and* (7.83), *for all* $k\geq 1$ d_k *is a descent direction satisfying* $d_{k+1}^T g_{k+1} = -w\|g_{k+1}\|^2 < 0$, *where* $w > 0$, *and* α_k *is obtained by the strong Wolfe line search* (7.4) *and* (7.99), *where* $0 < \sigma_k < 1$. *Then either*

$$\liminf_{k\to\infty}\|g_k\| = 0, \tag{7.101}$$

or

$$\sum_{k=0}^{\infty}\frac{\|g_k\|^4}{\|d_k\|^2}<\infty. \tag{7.102}$$

Proof Observe that in Proposition 7.8 it is proved that $\theta_k > 0$ and $\theta_k \to w$. Now, squaring both terms of $d_{k+1} + \theta_k g_{k+1} = \beta_k s_k$ it follows that

$$\|d_{k+1}\|^2 + \theta_k^2\|g_{k+1}\|^2 + 2\theta_k d_{k+1}^T g_{k+1} = \beta_k^2\|s_k\|^2. \tag{7.103}$$

However, $d_{k+1}^T g_{k+1} = -w\|g_{k+1}\|^2$. Therefore,

$$\|d_{k+1}\|^2 = -(\theta_k^2 - 2\theta_k w)\|g_{k+1}\|^2 + \beta_k^2\|s_k\|^2. \tag{7.104}$$

Using Proposition 7.8, observe that for $\theta_k \in (0, 2w]$, $\theta_k^2 - 2\theta_k w \leq 0$ is bounded below by $-w^2$. On the other hand, from (7.70), it results that $g_{k+1}^T d_{k+1} - \beta_k g_{k+1}^T s_k = -\theta_k\|g_{k+1}\|^2$. Now, using the strong Wolfe line search it follows that

$$\left|g_{k+1}^T d_{k+1}\right| + \sigma_k|\beta_k|\left|g_k^T s_k\right| \geq \theta_k\|g_{k+1}\|^2. \tag{7.105}$$

Now, let us apply the following inequality $(a+\sigma b)^2 \leq (1+\sigma^2)(a^2+b^2)$, true for all $a, b, \sigma \geq 0$, with $a = \left|g_{k+1}^T d_{k+1}\right|$ and $b = |\beta_k|\left|g_k^T s_k\right|$. After some algebraic manipulations, the following estimation is obtained

$$(g_{k+1}^T d_{k+1})^2 + \beta_k^2 (g_k^T s_k)^2 \geq \frac{\theta_k^2}{1+\sigma_k^2} \|g_{k+1}\|^4. \tag{7.106}$$

However, from Proposition 7.8, $\theta_k \geq c_1$. Besides, $0 < \sigma_k < 1$. Therefore, $\theta_k^2/(1+\sigma_k^2) \geq c_1^2/2$. Hence,

$$(g_{k+1}^T d_{k+1})^2 + \beta_k^2 (g_k^T s_k)^2 \geq e \|g_{k+1}\|^4, \tag{7.107}$$

where $e = c_1^2/2$ is a positive constant.

Using (7.104) and (7.107), the following can be written

$$\begin{aligned}
\frac{(g_{k+1}^T d_{k+1})^2}{\|d_{k+1}\|^2} + \frac{(g_k^T s_k)^2}{\|s_k\|^2} &= \frac{1}{\|d_{k+1}\|^2}\left[(g_{k+1}^T d_{k+1})^2 + \frac{\|d_{k+1}\|^2}{\|s_k\|^2}(g_k^T s_k)^2\right] \\
&= \frac{1}{\|d_{k+1}\|^2}\left[(g_{k+1}^T d_{k+1})^2 + \frac{(g_k^T s_k)^2}{\|s_k\|^2}\left(-(\theta_k^2 - 2\theta_k w)\|g_{k+1}\|^2 + \beta_k^2 \|s_k\|^2\right)\right] \\
&\geq \frac{1}{\|d_{k+1}\|^2}\left[e\|g_{k+1}\|^4 - (\theta_k^2 - 2\theta_k w)\frac{(g_k^T s_k)^2}{\|s_k\|^2}\|g_{k+1}\|^2\right] \\
&= \frac{\|g_{k+1}\|^4}{\|d_{k+1}\|^2}\left[e - (\theta_k^2 - 2\theta_k w)\frac{(g_k^T s_k)^2}{\|s_k\|^2}\frac{1}{\|g_{k+1}\|^2}\right].
\end{aligned} \tag{7.108}$$

From Proposition 7.10 observe that the left side of (7.108) is finite. Now, from Proposition 7.10, it is known that

$$\lim_{k\to\infty} \frac{(g_k^T s_k)^2}{\|s_k\|^2} = 0.$$

On the other hand, for $\theta_k \in (0, 2w]$, $\theta_k^2 - 2\theta_k w$ is finite. Therefore, if (7.101) is not true, then

$$\lim_{k\to\infty} \frac{(g_k^T s_k)^2}{\|s_k\|^2} \frac{(\theta_k^2 - 2\theta_k w)}{\|g_{k+1}\|^2} = 0. \tag{7.109}$$

Hence, from (7.108), it follows that

$$\frac{(g_{k+1}^T d_{k+1})^2}{\|d_{k+1}\|^2} + \frac{(g_k^T s_k)^2}{\|s_k\|^2} \geq e \frac{\|g_{k+1}\|^4}{\|d_{k+1}\|^2} \tag{7.110}$$

holds for all sufficiently large k. Therefore, by Proposition 7.10, it follows that (7.102) is true. ♦

Using Proposition 7.11, the following proposition can be proved, with a crucial role in showing the convergence of the algorithm DESCON.

Proposition 7.12 *Suppose that the Assumption CG holds. Consider the conjugate gradient algorithm* (7.70), *where* θ_k *and* β_k *are given by* (7.82) *and* (7.83), *and* α_k *is obtained by the strong Wolfe line search* (7.4) *and* (7.99), *where* $0<\sigma_k<1$. *If*

$$\sum_{k\geq 1}\frac{1}{\|d_k\|^2}=\infty, \tag{7.111}$$

then

$$\liminf_{k\to\infty}\|g_k\|=0. \tag{7.112}$$

Proof Suppose by contradiction that there is a positive constant γ so that $\|g_k\|\geq\gamma>0$ for all $k\geq 1$. Then, from Proposition 7.11 it follows that

$$\sum_{k\geq 1}\frac{1}{\|d_k\|^2}\leq\frac{1}{\gamma^4}\sum_{k\geq 1}\frac{\|g_k\|^4}{\|d_k\|^2}<\infty,$$

which is in contradiction with (7.111). ♦

Therefore, the iteration can fail in the sense that $\|g_k\|\geq\gamma>0$ for all k only if $\|d_k\|\to\infty$ sufficiently rapidly.

Convergence for strongly convex functions. For strongly convex functions, it can be proved that the norm of the direction d_k generated by (7.70), where θ_k and β_k are given by (7.82) and (7.83), respectively, is bounded. By using the Proposition 7.12, the following result can be shown.

Theorem 7.5 *Suppose that the Assumption CG holds. Consider the method* (7.69)–(7.71) *and* (7.76)–(7.81), *where* α_k *is obtained by the strong Wolfe line search* (7.4) *and* (7.99), *with* $1/2\leq\sigma_k<1$. *If there exists a constant* $\mu>0$ *so that*

$$(\nabla f(x)-\nabla f(y))^T(x-y)\geq\mu\|x-y\|^2 \tag{7.113}$$

for all $x,y\in S=\{x\in\mathbb{R}^n : f(x)\leq f(x_0)\}$ *then*

$$\lim_{k\to\infty}g_k=0. \tag{7.114}$$

Proof From (7.113), it follows that f is a strongly convex function on the level set S and therefore

$$y_k^T s_k \geq \mu \|s_k\|^2. \tag{7.115}$$

Again, by Lipschitz continuity, $\|y_k\| \leq L\|s_k\|$. Using (7.78) and (7.79) in (7.80), it follows that

$$t_k = \frac{(w-1)(y_k^T s_k)\|g_{k+1}\|^2 (y_k^T g_{k+1})}{(s_k^T g_{k+1})\bar{\Delta}_k} + \frac{(y_k^T g_{k+1})^2 - v(y_k^T s_k)\|g_{k+1}\|^2}{\bar{\Delta}_k}.$$

But, from (7.88),

$$t_k \leq \frac{(1-w)(1-\sigma_k)\|g_{k+1}\|^2 (y_k^T g_{k+1})}{\sigma_k \bar{\Delta}_k} + \frac{(y_k^T g_{k+1})^2 - v(y_k^T s_k)\|g_{k+1}\|^2}{\bar{\Delta}_k}.$$

However, $1/2 \leq \sigma_k < 1$, therefore

$$t_k \leq \frac{(1-w)\|g_{k+1}\|^2 (y_k^T g_{k+1})}{\bar{\Delta}_k} + \frac{(y_k^T g_{k+1})^2 - v(y_k^T s_k)\|g_{k+1}\|^2}{\bar{\Delta}_k}.$$

Observe that since $\{\bar{\Delta}_k\}$ is uniformly bounded away from zero independent of k and $\bar{\Delta}_k < 0$ for all $k \geq 1$, there exists a positive constant c_3 so that $|\bar{\Delta}_k| > c_3$. Hence,

$$|t_k| \leq \frac{|1-w|\|g_{k+1}\|^2 |y_k^T g_{k+1}| + |y_k^T g_{k+1}|^2 + v|y_k^T s_k|\|g_{k+1}\|^2}{c_3}.$$

From the Assumption CG, observe that $|y_k^T g_{k+1}| \leq \|y_k\|\|g_{k+1}\| \leq L\|s_k\| \Gamma \leq L\Gamma(2B)$ and $|y_k^T s_k| \leq \|y_k\|\|s_k\| \leq L\|s_k\|^2 \leq L(2B)^2$. With this, it follows that

$$|t_k| \leq \frac{2BL\Gamma^2[|1-w|\Gamma + 2B(L+v)]}{c_3} \equiv t,$$

where $t > 0$ is a constant. Now, from (7.71), by using the Lipschitz continuity, the following estimation is obtained

$$\begin{aligned} |\beta_k| &= \left| \frac{y_k^T g_{k+1}}{y_k^T s_k} - t_k \frac{s_k^T g_{k+1}}{y_k^T s_k} \right| \leq \frac{\|y_k\|\|g_{k+1}\|}{\mu\|s_k\|^2} + |t_k| \frac{\|s_k\|\|g_{k+1}\|}{\mu\|s_k\|^2} \\ &\leq \frac{L\|s_k\|\|g_{k+1}\|}{\mu\|s_k\|^2} + t\frac{\|s_k\|\|g_{k+1}\|}{\mu\|s_k\|^2} = \frac{L+t}{\mu}\frac{\Gamma}{\|s_k\|}. \end{aligned} \tag{7.116}$$

Hence, from (7.70) and from Proposition 7.8, it follows that

$$\|d_{k+1}\| \leq c_2\Gamma + \frac{L+t}{\mu}\frac{\Gamma}{\|s_k\|}\|s_k\| = \left(c_2 + \frac{L+t}{\mu}\right)\Gamma, \tag{7.117}$$

which implies that (7.111) is true. Therefore, by Proposition 7.12, it follows that (7.112) is true, which for strongly convex functions is equivalent to (7.114). ♦

Convergence for general nonlinear functions. Firstly, let us prove that in very mild conditions the direction d_k generated by (7.70), where θ_k and β_k are given by (7.82) and (7.83), respectively, is bounded. Again, by Proposition 7.12, the following result can be proved.

Theorem 7.6 *Suppose that the Assumption CG holds and* $\|g_k\| \geq \gamma > 0$ *for all* $k \geq 0$. *Consider the conjugate gradient algorithm* (7.69), *where the direction* d_{k+1} *given by* (7.70) *and* (7.71) *satisfies the descent condition* $g_k^T d_k = -w\|g_k\|^2$, *where* $w > 1$ *and the stepsize* α_k *is obtained by the strong Wolfe line search* (7.4) *and* (7.99), *where* $1/2 \leq \sigma_k < 1$. *Then* $\liminf_{k\to\infty}\|g_k\| = 0$.

Proof From (7.71), using (7.80), after some algebraic manipulations,

$$\beta_k = \frac{y_k^T g_{k+1}}{y_k^T s_k}\left(1 - \frac{b_k}{\bar{\Delta}_k}\right) + a_k \frac{\|g_{k+1}\|^2}{\bar{\Delta}_k}. \tag{7.118}$$

From: Proposition 7.6, the definition of ω_k, the modified Wolfe condition (7.84) and the descent condition $g_k^T d_k = -w\|g_k\|^2$, since $\|g_k\| \geq \gamma > 0$ and $\sigma_k < 1$, for all $k \geq 0$, it follows that

$$y_k^T s_k \geq w\omega_k(1-\sigma_k)\gamma^2 > w\omega(1-\sigma_k)\gamma^2 > 0.$$

However, from the Assumption CG,

$$\left|y_k^T g_{k+1}\right|\|s_k\| \leq \|y_k\|\|g_{k+1}\|\|s_k\| \leq L\|s_k\|^2\Gamma \leq L\Gamma(2B)^2.$$

Therefore,

$$\frac{\left|y_k^T g_{k+1}\right|}{\left|y_k^T s_k\right|} \leq \frac{L\Gamma(2B)^2}{w\omega(1-\sigma_k)\gamma^2}\frac{1}{\|s_k\|} = \frac{\bar{c}}{\|s_k\|}, \tag{7.119}$$

where

$$\bar{c} = \frac{L\Gamma(2B)^2}{w\omega(1-\sigma_k)\gamma^2}.$$

Now, observe that, since for all $k \geq 0$, $\bar{\Delta}_k < 0$ (by Proposition 7.4) and $b_k > 0$ (by Proposition 7.9), it follows that $-b_k/\bar{\Delta}_k > 0$. Besides, from (7.76) and (7.79), it follows that

$$-\frac{b_k}{\bar{\Delta}_k} = w + (1+w)\frac{(y_k^T g_{k+1})(s_k^T g_{k+1})}{-\bar{\Delta}_k}. \tag{7.120}$$

Since $-\bar{\Delta}_k > 0$ and $s_k^T g_{k+1}$ tends to zero along the iterations, it follows that $-b_k/\bar{\Delta}_k$ tends to $w > 0$. Hence, $1 - b_k/\bar{\Delta}_k$ tends to $1 + w$. Therefore, there exists a positive constant $c_4 > 1$ so that $1 < 1 - b_k/\bar{\Delta}_k \leq c_4$.

Again, from the Assumption CG,

$$\left|y_k^T s_k\right| \|s_k\| \leq \|y_k\| \|s_k\|^2 \leq L\|s_k\|^3 \leq L(2B)^3.$$

Therefore, $\left|y_k^T s_k\right| \leq L(2B)^3/\|s_k\|$. Now, from (7.78) and (7.89), the following estimation is obtained

$$\begin{aligned}
|a_k| &= \left|v(s_k^T g_{k+1}) + (y_k^T g_{k+1})\right| \leq v\left|s_k^T g_{k+1}\right| + \left|y_k^T g_{k+1}\right| \\
&\leq v\left|y_k^T s_k\right| \max\left\{1, \frac{\sigma_k}{1-\sigma_k}\right\} + \left|y_k^T g_{k+1}\right| \\
&\leq v\frac{L(2B)^3}{\|s_k\|} \max\left\{1, \frac{\sigma_k}{1-\sigma_k}\right\} + \frac{L\Gamma(2B)^2}{\|s_k\|}.
\end{aligned} \tag{7.121}$$

Since $1/2 \leq \sigma_k < 1$, there exists a positive constant $c_5 > 0$ so that $\max\{1, \sigma_k/(1-\sigma_k)\} \leq c_5$. Hence,

$$|a_k| \leq \left(vLc_5(2B)^3 + L\Gamma(2B)^2\right)\frac{1}{\|s_k\|} = \frac{\hat{c}}{\|s_k\|}, \tag{7.122}$$

where $\hat{c} = vLc_5(2B)^3 + L\Gamma(2B)^2$. With this, from (7.118), the following estimation is obtained

$$\begin{aligned}
|\beta_k| &\leq \left|\frac{y_k^T g_{k+1}}{y_k^T s_k}\right| \left|1 - \frac{b_k}{\bar{\Delta}_k}\right| + |a_k| \frac{\|g_{k+1}\|^2}{|\bar{\Delta}_k|} \leq \frac{\bar{c}c_4}{\|s_k\|} + \frac{\hat{c}\Gamma^2}{c_3}\frac{1}{\|s_k\|} \\
&= \left[\bar{c}c_4 + \frac{\hat{c}\Gamma^2}{c_3}\right]\frac{1}{\|s_k\|}.
\end{aligned} \tag{7.123}$$

Therefore, from (7.70), it follows that

$$\begin{aligned}
\|d_{k+1}\| &\leq |\theta_k| \|g_{k+1}\| + |\beta_k| \|s_k\| \\
&\leq c_2\Gamma + \left[\bar{c}c_4 + \frac{\hat{c}\Gamma^2}{c_3}\right]\frac{1}{\|s_k\|}\|s_k\| \equiv E,
\end{aligned} \tag{7.124}$$

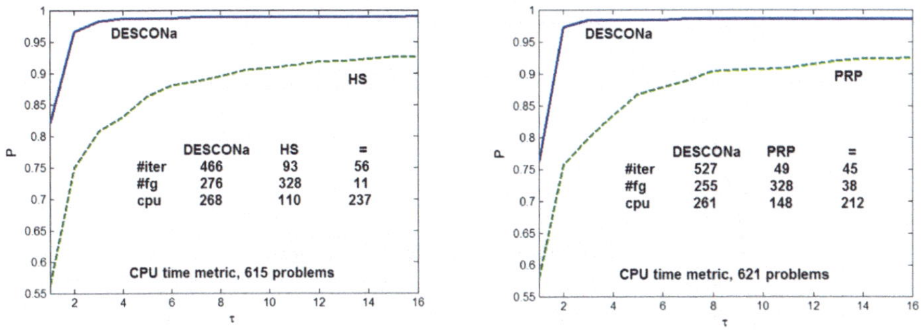

Figure 7.7 Performance profile of DESCONa versus HS and versus PRP

where E is a positive constant. Hence, for all $k \geq 0$, $\|d_k\| \leq E$, which implies (7.111). By Proposition 7.12, since d_k is a descent direction, it follows that $\liminf_{k\to\infty}\|g_k\| = 0$. ♦

Numerical study. Let us consider the set of 80 unconstrained optimization problems from the UOP collection, where for each problem 10 numerical experiments are run with the number of variables $n = 1000, 2000, \ldots, 10000$. The DESCON algorithm implements the Wolfe line search conditions with $\rho = 0.0001$, $\sigma = \|g_{k+1}\|^2/(|y_k^T g_{k+1}| + \|g_{k+1}\|^2)$. In DESCON, $w = 7/8$ and $v = 0.05$. The numerical experiments are given in the context of Remark 1.1. Figure 7.7 shows the performance profiles of the accelerated variant of DESCON (DESCONa) versus the standard conjugate gradient algorithms HS and PRP.

Figure 7.8 shows the performance profiles of DESCONa versus DL (t = 1) and versus CG-DESCENT (version 1.4) with Wolfe line search.

The performance profiles of accelerated DESCON (DESCONa) versus CG-DESCENT with approximate Wolfe line search (CG-DESCENTaw) are illustrated in Figure 7.9.

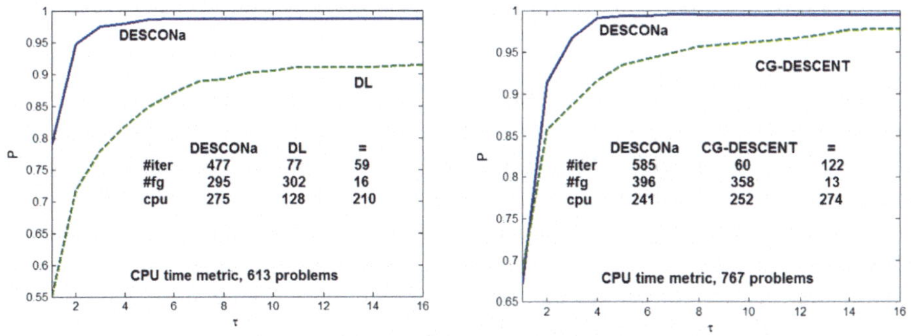

Figure 7.8 Performance profile of DESCONa versus DL (t = 1) and versus CG-DESCENT

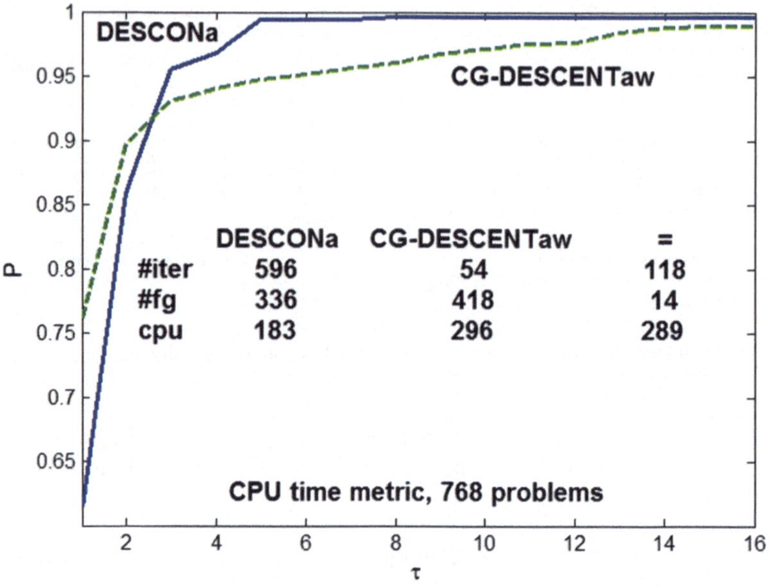

Figure 7.9 Performances of DESCONa versus CG-DESCENTaw

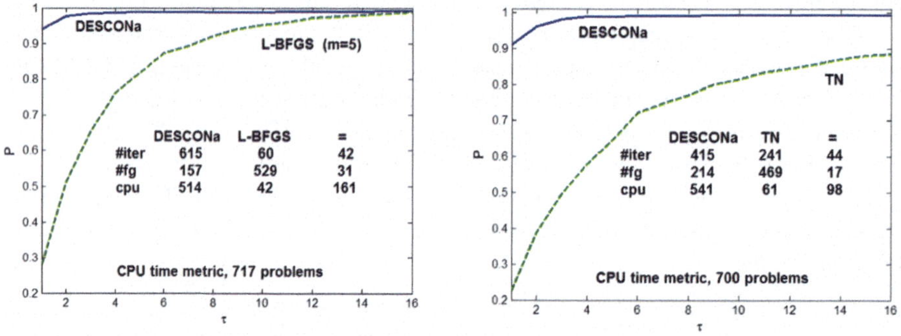

Figure 7.10 Performance profile of DESCONa versus L-BFGS (m = 5) and versus TN

Figure 7.9 points out that DESCONa is top performer in comparison with CG-DESCENTaw. However, CG-DESCENT with approximate Wolfe line search is more efficient than DESCONa. However, Figure 7.8 shows that both DESCONa and CG-DESCENT with Wolfe line search practically have the efficiency. From the table inside Figure 7.8 notice that DESCONa was faster in 241 problems, while CG-DESCENT in 252 problems. This illustrates once again the importance of the line search in conjugate gradient algorithms. An accurate line search increases the performances of the algorithm.

Table 7.3 Performances of DESCONa for solving five applications from the MINPACK-2 collection

		DESCONa		
	n	*#iter*	*#fg*	*cpu*
A1	40,000	241	510	4.12
A2	40,000	631	1297	12.11
A3	40,000	1036	2100	32.25
A4	40,000	299	632	23.80
A5	40,000	278	576	6.71
Total	–	2485	5115	78.99

Table 7.4 Total performances of L-BFGS ($m = 5$), TN, DL ($t = 1$), DL+ ($t = 1$), CG-DESCENT, CG-DESCENTaw, and DESCONa for solving five applications from the MINPACK-2 collection with 40,000 variables

Algorithms	*#iter*	*#fg*	*cpu*
L-BFGS ($m = 5$)	4842	4987	102.92
TN	153	3714	104.57
DL ($t = 1$)	2781	4816	107.04
DL+ ($t = 1$)	2733	4745	96.53
CG-DESCENT	2875	5756	197.27
CG-DESCENTaw	2875	5756	186.34
DESCONa	2485	5115	78.99

The performance profiles of DESCONa versus L-BFGS ($m = 5$) and versus TN are presented in Figure 7.10.

Table 7.3 presents the performances of DESCONa for solving the applications from the MINPACK-2 collection, where $nx = 200$ and $ny = 200$.

The total performances of L-BFGS ($m = 5$) from Table 1.2, of TN from Table 1.3, of DL ($t = 1$) and DL+ ($t = 1$) from Table 7.1, of CG-DESCENT from Table 7.2, and of DESCONa from Table 7.3 for solving the applications from the MINPACK-2 collection with 40,000 variables each are assembled in Table 7.4.

From Table 7.4, subject to the CPU time, DESCONa with 78.99 s is top performer for solving all five applications from the MINPACK-2 collection, each with 40,000 variables.

Notes and References

The algorithms described in this chapter (DL, CG-DESCENT, and DESCON) consist of different modifications of the conjugate gradient parameter β_k^{HS}. They belong to the Dai–Liao family of conjugate gradient algorithms. In CG-DESCENT, the parameter t in Dai–Liao is selected in order to ensure the descent character of the search direction. On the other hand, in DESCON the parameter t is chosen in such a way as to ensure for the search direction both the sufficient descent and the conjugacy conditions. It is known that conjugate gradient algorithms that satisfy both the sufficient descent and the conjugacy conditions are not necessarily the best algorithms. In DESCON, two additional ingredients are implemented. The first one consists of the multiplication of the negative gradient in the search direction with a positive, bounded scalar. The second one is a modification of the parameter σ from the second Wolfe line search condition for the stepsize computation.

Another interpretation of CG-DESCENT as a particularization of the Perry–Shanno self-scaling memoryless BFGS algorithm, not as a modification of the HS method, is presented in Chapter 8. A few words about this subject matter are welcomed. Perry (1977) introduced the self-scaling memoryless BFGS algorithm as a means of developing a nonlinear conjugate gradient algorithm with memory. Later on Shanno (1978b) reinterpreted Perry's algorithm as memoryless BFGS and improved it into the final self-scaling memoryless BFGS algorithm with the introduction of a scaling parameter τ_k, so that

$$d_{k+1} = -H_{k+1}g_{k+1}, \tag{7.125}$$

where

$$H_{k+1} = \left(I - \frac{s_k y_k^T}{y_k^T s_k}\right)\tau_k I\left(I - \frac{y_k s_k^T}{y_k^T s_k}\right) + \frac{s_k s_k^T}{y_k^T s_k}. \tag{7.126}$$

It easy to see that (7.126) is a limited-memory BFGS approximation to the inverse Hessian with only one update and with the initial Hessian defined as $H_0 = \tau_k I$. The resulting algorithm is known as the Perry–Shanno scheme.

Now, by substituting $s_k = \alpha_k d_k$, a search direction with the following structure is obtained

$$d_{k+1} = \tau_k(-g_{k+1} + \beta_k^{PS} d_k + \gamma_k^{PS} y_k), \tag{7.127}$$

where

$$\beta_k^{PS} = \frac{1}{y_k^T d_k}\left[g_{k+1}^T y_k - \left(\frac{\|y_k\|^2}{y_k^T d_k} + \frac{\alpha_k}{\tau_k}\right) g_{k+1}^T d_k\right], \gamma_k^{PS} = \frac{g_{k+1}^T d_k}{y_k^T d_k}.$$

To guarantee the global convergence of the algorithm, the self-scaling memoryless BFGS scheme requires that the Hessian approximation H_{k+1} be periodically reset to the identity matrix. Of course, this can decrease the rate of convergence. Therefore, Hager and Zhang (2005) improved the algorithm by deleting the $\gamma_k^{PS} y_k$ term in (7.127) and adopting the Oren and Spedicato (1976) scaling parameter

$$\tau_k = \frac{y_k^T s_k}{y_k^T y_k} = \frac{\alpha_k y_k^T d_k}{y_k^T y_k}, \tag{7.128}$$

thus obtaining the search direction (7.46) with (7.47)–(7.49).

Similar to Powel's modification to the PRP method, the parameter $\bar{\beta}_k^N$ in (7.47) is restricted with a lower bound to guarantee global convergence. Observe that the truncation in (7.47) dynamically adjusts this lower bound so that $\eta_k \to -\infty$ in (7.48) as $\|g_k\| \to 0$ in order to accelerate the convergence. More details on this method are given in Chapters 8 and 10.

Other standard conjugate gradient methods may also be modified. For example, a modification of the DY method was given by Andrei (2008c, 2010c), a modification of the PRP method was given by Zhang, Zhou, and Li (2006a) and Andrei (2011a), a modification of the FR method was given by Zhang, Zhou, and Li (2006b). A modification of the HS method based on the secant condition was given by Zhang (2009b). Another modification of the HS method with clustering the eigenvalues of the iteration matrix was given by Andrei (2016). Based on the modified secant equation, Li, Tang, and Wei (2007) obtained a modified Dai–Liao computational scheme.

Another class of modifications of the standard conjugate gradient methods refers to including in the definition of the parameter β_k a modification of the secant equation (Yabe & Sakaiwa, 2005), (Zhang, 2009b), (Babaie-Kafaki, Ghanbari, & Mahdavi-Amiri, 2010), etc.

Chapter 8
Conjugate Gradient Methods Memoryless BFGS Preconditioned

Conjugate gradient methods are widely acknowledged to be among the most efficient and robust methods for solving the large-scale unconstrained nonlinear optimization problems

$$\min f(x), \tag{8.1}$$

where $f:\mathbb{R}^n \to \mathbb{R}$ is a continuous differentiable function, lower bounded. Under mild assumptions, they have very well-established convergence results and are characterized by very modest memory requirements. Therefore, these methods are of a paramount importance for solving large-scale applications.

For a given initial point x_0, a conjugate gradient method generates the sequence $\{x_k\}$ defined by

$$x_{k+1} = x_k + \alpha_k d_k, k = 0, 1, \ldots, \tag{8.2}$$

where

$$d_{k+1} = -g_{k+1} + \beta_k d_k, k = 0, 1, \ldots, \tag{8.3}$$

where $d_0 = -g_0 = -\nabla f(x_0)$. The conjugate gradient parameter β_k is a scalar which defines the conjugate gradient algorithm.

The purpose of this chapter is to see how the second-order information of the minimizing function may be used in the formula for the search direction computation. Thus, new conjugate gradient algorithms with better convergence properties are obtained. In fact, these methods include a special preconditioning of conjugate gradient algorithms by using the BFGS updating. The general theory of preconditioning and its virtues are presented in Chapter 10. The idea of preconditioning based on BFGS update is to accelerate the convergence of the algorithms by including the second-order information in their formula for search direction. Chapter 11 will present another type of preconditioning, based on the

N. Andrei, *Nonlinear Conjugate Gradient Methods for Unconstrained Optimization*, Springer Optimization and Its Applications 158,
https://doi.org/10.1007/978-3-030-42950-8_8

limited-memory BFGS update. Memoryless quasi-Newton methods were first introduced by Perry (1977) and Shanno (1978a). They can be considered as the quasi-Newton methods for which the approximation to the inverse of the Hessian is taken as the identity matrix at every iteration.

Four such BFGS preconditioned conjugate gradient methods are presented in this chapter. The first, developed by Shanno (1978a, 1978b, 1983), is based on the idea that the conjugate gradient methods are exactly the BFGS quasi-Newton method, where the approximation to the inverse Hessian of the minimizing function is restarted as the identity matrix at every iteration. The second one, developed by Andrei (2007a, 2007b, 2007c, 2010b), is using scaling in the frame of the memoryless BFGS method. The third one chooses the search direction in one-dimensional manifold closest to the Perry–Shanno self-scaling memoryless BFGS method, developed by Dai and Kou (2013). Finally, using the trace and the determinant, or a combination of these operators of the self-scaling memoryless BFGS iteration matrix, new conjugate gradient methods are developed, (Andrei, 2019b, 2020).

Finally, let us mention that the scalar α_k in (8.2) is the stepsize. Even if there are lots of possibilities for α_k computation, the standard Wolfe conditions

$$f(x_k + \alpha_k d_k) \leq f(x_k) + \rho \alpha_k d_k^{\mathrm{T}} g_k, \tag{8.4}$$

$$\nabla f(x_k + \alpha_k d_k)^{\mathrm{T}} d_k \geq \sigma d_k^{\mathrm{T}} g_k, \tag{8.5}$$

where $0 < \rho < \sigma < 1$, are often used for its determination.

8.1 Conjugate Gradient Memoryless BFGS Preconditioned (CONMIN)

It is common knowledge that in the HS method, β_k is defined by

$$\beta_k = \frac{g_{k+1}^{\mathrm{T}} y_k}{d_k^{\mathrm{T}} y_k}, \tag{8.6}$$

where $g_k = \nabla f(x_k)$ and $y_k = g_{k+1} - g_k$. For the quadratic functions of the form

$$f(x) = \frac{1}{2} x^T A x + bx \tag{8.7}$$

with A a symmetric and positive definite matrix and b an arbitrary vector, the choice of β_k as in (8.6) guarantees that, if α_k is chosen to minimize $f(x)$ along the direction d_k at each iteration, then the directions d_k satisfy $d_k^{\mathrm{T}} A d_j = 0$ for $j \neq k$. This is the conjugacy condition which guarantees the convergence of the algorithm (8.2) and (8.3) to the minimizer of f in at most $n + 1$ iterations.

On the other hand, if α_k is chosen to minimize $f(x_k + \alpha d_k)$, then the first order condition for this minimum is

$$g_{k+1}^{\mathrm{T}} d_k = 0. \tag{8.8}$$

Observe that this reduces the denominator of (8.6) to $g_k^{\mathrm{T}} g_k$. Further, if $f(x)$ is quadratic, then $g_k^{\mathrm{T}} g_{k+1} = 0$ and in this case the FR choice of β_k is obtained

$$\beta_k = \frac{g_{k+1}^{\mathrm{T}} g_{k+1}}{g_k^{\mathrm{T}} g_k}. \tag{8.9}$$

If α_k is chosen to minimize $f(x_k + \alpha d_k)$ and the condition that $f(x)$ is quadratic is relaxed, then the PRP formula

$$\beta_k = \frac{g_{k+1}^{\mathrm{T}} y_k}{g_k^{\mathrm{T}} g_k} \tag{8.10}$$

is obtained. This is the relation among the known standard conjugate gradient methods HS, FR and PRP (see Chapter 4).

In the following, a short technical result is presented.

Proposition 8.1 *If* $x, y, z \in \mathbb{R}^n$, *then*

$$(x^T y)z = (zy^{\mathrm{T}})x \text{ and } x^T(yz^{\mathrm{T}}) = (x^T y)z^{\mathrm{T}}.$$

Proof The above equalities are obtained by direct calculation. ♦

By Proposition 8.1, Perry (1976) noted that the search direction (8.3) where β_k is given by (8.6) can be written as

$$d_{k+1} = -\left[I - \frac{d_k y_k^{\mathrm{T}}}{y_k^{\mathrm{T}} d_k}\right] g_{k+1}. \tag{8.11}$$

Defining $s_k = \alpha_k d_k = x_{k+1} - x_k$ and since $d_k y_k^{\mathrm{T}} / y_k^{\mathrm{T}} d_k = s_k y_k^{\mathrm{T}} / y_k^{\mathrm{T}} s_k$, (8.11) can be rewritten as

$$d_{k+1} = -\left[I - \frac{s_k y_k^{\mathrm{T}}}{y_k^{\mathrm{T}} s_k}\right] g_{k+1}. \tag{8.12}$$

Perry relaxed the assumption that α_k is chosen to minimize $f(x_k + \alpha d_k)$ and added the correction term $s_k s_k^{\mathrm{T}} / y_k^{\mathrm{T}} s_k$ to the matrix from (8.12) and proposed the modified search direction

$$d_{k+1} = -\left[I - \frac{s_k y_k^T}{y_k^T s_k} + \frac{s_k s_k^T}{y_k^T s_k}\right] g_{k+1} \equiv -Q_{k+1} g_{k+1}, \tag{8.13}$$

where Q_{k+1} defined in (8.13) satisfies the equation

$$y_k^T Q_{k+1} = s_k^T, \tag{8.14}$$

which is similar, but not identical to the secant equation

$$H_{k+1} y_k = s_k, \tag{8.15}$$

where H_{k+1} is an approximation to the inverse Hessian. If the line search is exact, then $s_k^T g_{k+1} = 0$ and (8.13) is identical to the HS method.

The major difficulty with this approach, not corrected by Perry, is that the matrix Q_{k+1} is not symmetric and therefore it is not positive definite. Thus, the search directions d_{k+1} given by (8.13) are not necessarily descent directions and hence numerical instabilities can result, ruining the convergence of the algorithm. To overcome this difficulty, Shanno (1978a, b) investigated the relationship between the conjugate gradient and the BFGS methods.

As known, the best quasi-Newton method is BFGS, where the updating to the inverse Hessian of the minimizing function is given by

$$H_{k+1} = H_k - \frac{s_k y_k^T H_k + H_k y_k s_k^T}{y_k^T s_k} + \left(1 + \frac{y_k^T H_k y_k}{y_k^T s_k}\right) \frac{s_k s_k^T}{y_k^T s_k}. \tag{8.16}$$

The major difference between the conjugate gradient and the quasi-Newton methods is the presence of the matrix H_k in the updating formula for the search direction d_{k+1}. For problems with a large number of variables, it is impossible to store an approximation to the inverse Hessian, and therefore, the conjugate gradient methods are preferred.

Following Shanno (1978a), to see the relationship between the conjugate gradient and the BFGS methods, the matrix Q_{k+1} defined by (8.13) is first symmetrized to get

$$Q_{k+1} = I - \frac{s_k y_k^T}{y_k^T s_k} - \frac{y_k s_k^T}{y_k^T s_k} + \frac{s_k s_k^T}{y_k^T s_k}. \tag{8.17}$$

Afterward, let us force Q_{k+1} defined by (8.17) to satisfy the quasi-Newton Equation (8.15), yielding the symmetric update

$$Q_{k+1}^* = I - \frac{s_k y_k^T + y_k s_k^T}{y_k^T s_k} + \left(1 + \frac{y_k^T y_k}{y_k^T s_k}\right) \frac{s_k s_k^T}{y_k^T s_k}. \tag{8.18}$$

Observe that, if in (8.16) $H_k = I$ then it results precisely Q^*_{k+1} from (8.18). Therefore, the important result noticed for the first time by Shanno was that *the conjugate gradient methods are precisely the BFGS quasi-Newton methods, where the approximation to the inverse Hessian is restarted as the identity matrix at every iteration*. Hence, the conjugate gradient methods are often called *memoryless quasi-Newton methods.*

It is worth seeing that, by using Proposition 8.1, a conjugate gradient method in which the search directions are computed as

$$d_{k+1} = -Q^*_{k+1} g_{k+1} \tag{8.19}$$

do not explicitly require the construction of Q^*_{k+1}. Indeed, from (8.19) and (8.18) it is easy to get

$$d_{k+1} = -g_{k+1} - \left[\left(1 + \frac{y_k^T y_k}{y_k^T s_k}\right)\frac{s_k^T g_{k+1}}{y_k^T s_k} - \frac{y_k^T g_{k+1}}{y_k^T s_k}\right] s_k + \frac{s_k^T g_{k+1}}{y_k^T s_k} y_k, \tag{8.20}$$

which shows that no additional information is needed more than the one required by known standard conjugate gradient methods.

For the Newton method, close to the minimum point the stepsize can be considered as $\alpha_k = 1$, thus avoiding the line search along the search direction d_k. The quasi-Newton methods, particularly BFGS, inherit this property since better and better approximations to the inverse Hessian are generated along the iterations. The conjugate gradient methods do not have this property. An explanation for this behavior is that at every iteration, the current estimation of the Hessian is given by the identity matrix, so that no information on the stepsize is accumulated. On the other hand, concerning the stability of the computations, observe that (8.20) always defines a descent direction if $y_k^T s_k > 0$, which is always true when the stepsize is determined by the Wolfe line search (8.4) and (8.5).

To get a superlinear convergence of a conjugate gradient method in which the search direction is computed as in (8.3) with β_k given by the HS method (8.6), Beale (1972) suggested a modification of the search direction as

$$d_{t+1} = -g_{t+1} + \beta_t d_t, \tag{8.21}$$

$$d_{k+1} = -g_{k+1} + \beta_k d_k + \mu_k d_t, \tag{8.22}$$

where

$$\mu_k = \frac{y_t^T g_{k+1}}{y_t^T d_t}, \quad \text{with} \quad t < k < t + n. \tag{8.23}$$

In particular, starting with $t = 0$, at every n iterations a new pair of vectors d_t and y_t, known as *the restart vectors of Beale*, are stored and the generated search directions are conjugate subject to these restart vectors as well as to the previous ones. After n iterations, the restart direction is replaced by the current one.

Powell (1977) made a modification of restarting the conjugate gradient methods. Powell's criterion is to use (8.22) whenever

(a) k (or $k-t$) is a multiple of n, or if
(b) $|g_{k+1}^T g_k| \geq 0.2\|g_{k+1}\|^2$.

Obviously, (8.22) may be written as

$$d_{k+1} = -\left[I - \frac{d_k y_k^T}{y_k^T d_k} - \frac{d_t y_t^T}{y_t^T d_t}\right] g_{k+1} \equiv -P_k g_{k+1}, \tag{8.24}$$

where P_k which modifies the gradient g_{k+1} is an update using the information from two prior points. It is easy to show that if $f(x)$ is quadratic and the line search is exact, then P_k is a projection matrix of rank $n-2$ rather than rank $n-1$ of the matrix from (8.11).

Using this double update with information in two points and keeping the philosophy behind the matrix Q_{k+1}^* from (8.18), based on the fact that the conjugate gradient method is exactly the memoryless BFGS method, Shanno (1978a) suggested the following search direction

$$d_{k+1} = -H_{k+1} g_{k+1}, \tag{8.25}$$

where

$$H_{k+1} = \hat{H}_k - \frac{\hat{H}_k y_k s_k^T + s_k y_k^T \hat{H}_k}{y_k^T s_k} + \left(1 + \frac{y_k^T \hat{H}_k y_k}{y_k^T s_k}\right) \frac{s_k s_k^T}{y_k^T s_k}, \tag{8.26}$$

with

$$\hat{H}_k = I - \frac{y_t s_t^T + s_t y_t^T}{y_t^T s_t} + \left(1 + \frac{y_t^T y_t}{y_t^T s_t}\right) \frac{s_t s_t^T}{y_t^T s_t}. \tag{8.27}$$

Observe that, if in (8.26) $\hat{H}_k = I$ is set, then quite the matrix Q_{k+1}^* is obtained. Besides, observe that the conjugate gradient method (8.25)–(8.27) does not imply additional memory requirements. From Proposition 8.1, it is clear that the search direction may be written as

$$\begin{aligned} d_{k+1} = &-\hat{H}_k g_{k+1} + \frac{s_k^T g_{k+1}}{y_k^T s_k} \hat{H}_k y_k \\ &- \left[\left(1 + \frac{y_k^T \hat{H}_k y_k}{y_k^T s_k}\right) \frac{s_k^T g_{k+1}}{y_k^T s_k} - \frac{y_k^T \hat{H}_k g_{k+1}}{y_k^T s_k}\right] s_k, \end{aligned} \tag{8.28}$$

where the vectors $\hat{H}_k g_{k+1}$ and $\hat{H}_k y_k$ are computed as

$$\hat{H}_k g_{k+1} = g_{k+1} - \frac{s_t^T g_{k+1}}{y_t^T s_t} y_t + \left[\left(1 + \frac{y_t^T y_t}{y_t^T s_t} \right) \frac{s_t^T g_{k+1}}{y_t^T s_t} - \frac{y_t^T g_{k+1}}{y_t^T s_t} \right] s_t, \tag{8.29}$$

$$\hat{H}_k y_k = y_k - \frac{s_t^T y_k}{y_t^T s_t} y_t + \left[\left(1 + \frac{y_t^T y_t}{y_t^T s_t} \right) \frac{s_t^T y_k}{y_t^T s_t} - \frac{y_t^T y_k}{y_t^T s_t} \right] s_t. \tag{8.30}$$

It should be emphasized that the implementation of this method requires only seven vectors to store: $x_k, x_{k+1}, g_k, g_{k+1}, d_k, d_t$ and y_t. At the time the update has been accomplished, the information in x_k is no longer required, and so $\hat{H}_k y_k$ from (8.30) may be stored in x_k. Once $\hat{H}_k y_k$ and the scalars $y_k^T s_k$ and $y_k^T (\hat{H}_k y_k)$ have been computed, g_k is no longer needed, so $\hat{H}_k g_{k+1}$ from (8.29) can be explicitly computed and stored in g_k. This double update scheme was first proposed by Perry (1976, 1977), but with y_t and s_t replaced by y_{k-1} and s_{k-1}, respectively.

Using a modification of the Broyden class of the quasi-Newton methods given by Oren (1974), Oren and Luenberger (1974) and Oren and Spedicato (1976), which makes these methods invariant under the multiplication of the objective function by a constant, a new conjugate gradient algorithm, defined by Shanno (1978a, b) and Shanno and Phua, (1976, 1980) was obtained. Indeed, the modification of the Broyden class of quasi-Newton methods proposed by Oren *et co* is as follows

$$H_{k+1} = \left[H_k - \frac{H_k y_k y_k^T H_k}{y_k^T H_k y_k} + \Phi v_k v_k^T \right] \gamma + \frac{s_k s_k^T}{y_k^T s_k}, \tag{8.31}$$

where

$$v_k = \sqrt{y_k^T H_k y_k} \left[\frac{s_k}{y_k^T s_k} - \frac{H_k y_k}{y_k^T H_k y_k} \right], \tag{8.32}$$

Φ is a scalar and the parameter γ is introduced to make the sequence of the matrices $\{H_k\}$ invariant under the multiplication of the objective function by a constant.

To get a value for Φ, Oren and Spedicato (1976) considered the minimization of the condition number of $H_k^{-1} H_{k+1}$ and derived the value

$$\Phi = \frac{b(c - b\gamma)}{\gamma(ac - b^2)}, \tag{8.33}$$

where $a = y_k^T H_k y_k$, $b = y_k^T s_k$, and $c = s_k^T H_k s_k$.

Shanno and Phua (1978), after intensive computational experiments with general nonlinear functions, found that using γ in (8.31) is harmful, as it introduced both the truncation error and the approximation error in the estimate of the inverse Hessian. However, the use of γ at the *initial step* was critical, especially for larger problems, in order to eliminate the truncation error which results from using the identity matrix as an initial approximation to the inverse Hessian (Shanno, 1978a).

Now, substituting $\Phi = 1$ in (8.33) yields

$$\gamma = \frac{y_k^T s_k}{y_k^T H_k y_k}. \tag{8.34}$$

Therefore, using this philosophy where the approximation to the inverse Hessian is restarted as the identity matrix at every step, i.e., setting $H_k = I$, it seems reasonable to always scale the sequence of updating matrices by substituting $\Phi = 1$ and introducing γ from (8.34) in (8.31). Thus, another modified conjugate gradient method is obtained as

$$d_{k+1} = -\gamma g_{k+1} - \left[\left(1 + \gamma \frac{y_k^T y_k}{y_k^T s_k}\right)\frac{s_k^T g_{k+1}}{y_k^T s_k} - \gamma \frac{y_k^T g_{k+1}}{y_k^T s_k}\right] s_k + \gamma \frac{s_k^T g_{k+1}}{y_k^T s_k} y_k, \tag{8.35}$$

where

$$\gamma = \frac{y_k^T s_k}{y_k^T y_k}. \tag{8.36}$$

Now, substituting (8.36) in (8.35), after some simple algebraic manipulations it results that

$$d_{k+1} = -\frac{y_k^T s_k}{y_k^T y_k} g_{k+1} - \left[2\frac{s_k^T g_{k+1}}{y_k^T s_k} - \frac{y_k^T g_{k+1}}{y_k^T y_k}\right] s_k + \frac{s_k^T g_{k+1}}{y_k^T y_k} y_k, \tag{8.37}$$

which is the Perry/Shanno (d_{k+1}^{PS}) search direction. Again, observe that if $s_k^T g_{k+1} = 0$, that is if the line search is exact, then

$$d_{k+1} = -\frac{y_k^T s_k}{y_k^T y_k} g_{k+1} + \frac{y_k^T g_{k+1}}{y_k^T y_k} s_k = \gamma\left[-g_{k+1} + \frac{y_k^T g_{k+1}}{y_k^T s_k} s_k\right], \tag{8.38}$$

so the effect is one of multiplying the search direction (8.12) with a scalar, thus the n- step convergence to the minimum of a quadratic function being maintained.

Therefore, using this scaling technique in (8.27), it follows that

$$\hat{H}_k = \gamma_t \left[I - \frac{s_t y_t^T + y_t s_t^T}{y_t^T s_t} + \frac{y_t^T y_t}{y_t^T s_t} \frac{s_t s_t^T}{y_t^T s_t}\right] + \frac{s_t s_t^T}{y_t^T s_t}, \tag{8.39}$$

where

$$\gamma_t = \frac{y_t^{\mathrm{T}} s_t}{y_t^{\mathrm{T}} y_t}.$$

With these developments, *the conjugate gradient memoryless BFGS preconditioned algorithm* is defined by (8.28), where this time, the vectors $\hat{H}_k g_{k+1}$ and $\hat{H}_k y_k$ are computed as

$$\hat{H}_k g_{k+1} = \frac{y_t^{\mathrm{T}} s_t}{y_t^{\mathrm{T}} y_t} g_{k+1} - \frac{s_t^{\mathrm{T}} g_{k+1}}{y_t^{\mathrm{T}} y_t} y_t + \left[2 \frac{s_t^{\mathrm{T}} g_{k+1}}{y_t^{\mathrm{T}} s_t} - \frac{y_t^{\mathrm{T}} g_{k+1}}{y_t^{\mathrm{T}} y_t} \right] s_t, \tag{8.40}$$

$$\hat{H}_k y_k = \frac{y_t^{\mathrm{T}} s_t}{y_t^{\mathrm{T}} y_t} y_k - \frac{s_t^{\mathrm{T}} y_k}{y_t^{\mathrm{T}} y_t} y_t + \left[2 \frac{s_t^{\mathrm{T}} y_k}{y_t^{\mathrm{T}} s_t} - \frac{y_t^{\mathrm{T}} y_k}{y_t^{\mathrm{T}} y_t} \right] s_t. \tag{8.41}$$

Before presenting the conjugate gradient memoryless BFGS preconditioned algorithm defined by (8.28), (8.40) and (8.41) on steps, let us discuss some of its properties. Firstly, observe that $y_k^{\mathrm{T}} s_k > 0$ is a sufficient condition to ensure the descent character of d_{k+1} defined by (8.28). Of course, this may be achieved by using the Wolfe line search (8.4) and (8.5). Secondly, the following theorem can be proved.

Theorem 8.1 *For quadratic functions*

$$f(x) = \frac{1}{2} x^T Q x - b^{\mathrm{T}} x$$

with exact line search, the conjugate gradient memoryless BFGS preconditioned algorithm defined by (8.28)*,* (8.40)*, and* (8.41) *is reduced exactly to Beale's method defined by* (8.22)*.*

Proof From (8.28) with exact line search, that is considering $s_k^{\mathrm{T}} g_{k+1} = 0$, it follows that

$$d_{k+1} = -\hat{H}_k g_{k+1} + \frac{y_k^{\mathrm{T}} \hat{H}_k g_{k+1}}{y_k^{\mathrm{T}} s_k} s_k. \tag{8.42}$$

Now, for quadratic functions, $\nabla f(x) = g = Qx - b$ and therefore,

$$g_{k+1} = Qx_{k+1} - b = Q\left(x_{t+1} + \sum_{i=t+1}^{k} s_i \right) - b = g_{t+1} + \sum_{i=t+1}^{k} Q s_i.$$

Hence,

$$s_t^{\mathrm{T}} g_{k+1} = s_t^{\mathrm{T}} g_{t+1} + \sum_{i=t+1}^{k} s_t^{\mathrm{T}} Q s_i.$$

But, for quadratic functions, $s_t^{\mathrm{T}} Q s_i = 0$, $i = t+1, \ldots k$, and having in view that the line search is exact, it follows that $s_t^{\mathrm{T}} g_{t+1} = 0$. Now, from (8.40),

$$\hat{H}_k g_{k+1} = \frac{y_t^{\mathrm{T}} s_t}{y_t^{\mathrm{T}} y_t} g_{k+1} - \frac{y_t^{\mathrm{T}} g_{k+1}}{y_t^{\mathrm{T}} y_t} s_t. \tag{8.43}$$

Substituting (8.43) in (8.42) yields

$$d_{k+1} = -\frac{y_t^{\mathrm{T}} s_t}{y_t^{\mathrm{T}} y_t} g_{k+1} + \frac{y_t^{\mathrm{T}} g_{k+1}}{y_t^{\mathrm{T}} y_t} s_t + \frac{y_k^{\mathrm{T}} g_{k+1}}{y_k^{\mathrm{T}} s_k} \frac{y_t^{\mathrm{T}} s_t}{y_t^{\mathrm{T}} y_t} s_k. \tag{8.44}$$

But (8.44) may be rewritten as

$$\begin{aligned} d_{k+1} &= \frac{y_t^{\mathrm{T}} s_t}{y_t^{\mathrm{T}} y_t} \left[-g_{k+1} + \frac{y_k^{\mathrm{T}} g_{k+1}}{y_k^{\mathrm{T}} s_k} s_k + \frac{y_t^{\mathrm{T}} g_{k+1}}{y_t^{\mathrm{T}} s_t} s_t \right] \\ &= \gamma_t \left[-g_{k+1} + \frac{y_k^{\mathrm{T}} g_{k+1}}{y_k^{\mathrm{T}} d_k} d_k + \frac{y_t^{\mathrm{T}} g_{k+1}}{y_t^{\mathrm{T}} d_t} d_t \right]. \end{aligned} \tag{8.45}$$

Thus, the search direction is identical to Beale's direction (8.22) and the stepsize is scaled by γ_t, proving the theorem. ♦

As a final remark of these developments, the fact that under the exact line search, conjugate gradient methods are simple projections of the gradient, they make their initial length a poor approximation to the desired stepsize. Fletcher (1987) proposed scaling d_{k+1} by

$$\hat{d}_{k+1} = \frac{2(f(x_{k+1}) - f(x_k))}{d_{k+1}^{\mathrm{T}} g_{k+1}} d_{k+1}.$$

Observe that this scaling of the search direction includes the function values in two successive steps.

Algorithm 8.1 Conjugate gradient memoryless BFGS preconditioned: CONMIN

1.	Choose an initial point $x_0 \in \mathbb{R}^n$ and $\varepsilon > 0$ sufficiently small. Set $k = 0$. Compute $f(x_k)$, $g_k = \nabla f(x_k)$ and set $d_k = -g_k$
2.	Compute the stepsize $\alpha_k > 0$ satisfying the Wolfe line search conditions (8.4) and (8.5)
3.	Compute $x_{k+1} = x_k + \alpha_k d_k$, $f(x_{k+1})$, g_{k+1}, $y_k = g_{k+1} - g_k$ and $s_k = x_{k+1} - x_k$
4.	Test a criterion for stopping the iterations. If this criterion is satisfied, then stop; otherwise, continue with step 5

(continued)

Algorithm 8.1 (continued)

5.	Test the criterion for restarting the iterations. If: (a) iteration k is a multiple of n, or (b) $\lvert g_{k+1}^T g_k \rvert \geq 0.2 \lVert g_{k+1} \rVert^2$, then compute the search direction d_{k+1} as in (8.35). Set $s_t = d_k$, $y_t = y_k$, $k = k+1$ and continue with step 2. Otherwise, continue with step 6
6.	Compute the search direction d_{k+1} as in (8.28), where the vectors $\hat{H}_k g_{k+1}$ and $\hat{H}_k y_k$ are computed as in (8.40) and (8.41), respectively
7.	Scale the search direction d_{k+1} as $d_{k+1} = [2(f(x_{k+1}) - f(x_k))/d_{k+1}^T g_{k+1}] d_{k+1}$ set $k = k+1$ and continue with step 2 ♦

For convex functions, using the exact line search, Shanno (1978b) proved the convergence of Algorithm 8.1 when the Hessian of the minimizing function is strictly bounded. However, since the search direction is computed using the BFGS updating strategy and since the line search is based on the Wolfe conditions and the Beale restart is implemented, the convergence of the algorithm is ensured. For general nonlinear functions bounded below with the level set bounded, Shanno showed that the algorithm is not possible to converge to a point where the gradient is bounded away from zero. Cycling in the form defined by Powell is the only way in which the conjugate gradient memoryless BFGS preconditioned can possibly fail to converge on general nonlinear functions.

Numerical study. CONMIN is one of the oldest and most respectable conjugate gradient methods implemented by Shanno (1983). In the following, let us present some results with this method and its comparisons versus some other methods for solving the unconstrained optimization test problems from the UOP collection (Andrei, 2018g). For each problem, ten numerical experiments were executed with the number of variables n = 1000, 2000, …, 10000. Therefore, in each numerical experiment a number of 800 unconstrained optimization test problems have been solved. All the numerical experiments were run in the context of Remark 1.1. The maximum number of iterations was limited to 2000.

In the first set of numerical experiments, CONMIN is compared versus the standard conjugate gradient methods HS, PRP, DY, and LS. Figure 8.1 shows the Dolan and Moré performance profiles of these methods. Figure 8.1 shows that CONMIN is top performer being more robust than all these methods. The standard conjugate gradient methods have a very simple structure and, as a characteristic, they do not try to capture the curvature of the minimizing function. The most important condition in these methods is that the conjugate gradient parameter β_k must satisfy the descent or the sufficient descent condition. On the other hand, CONMIN is a conjugate gradient memoryless BFGS preconditioned. The search direction in CONMIN uses the memoryless BFGS updating strategy which better captures the curvature of the minimizing function.

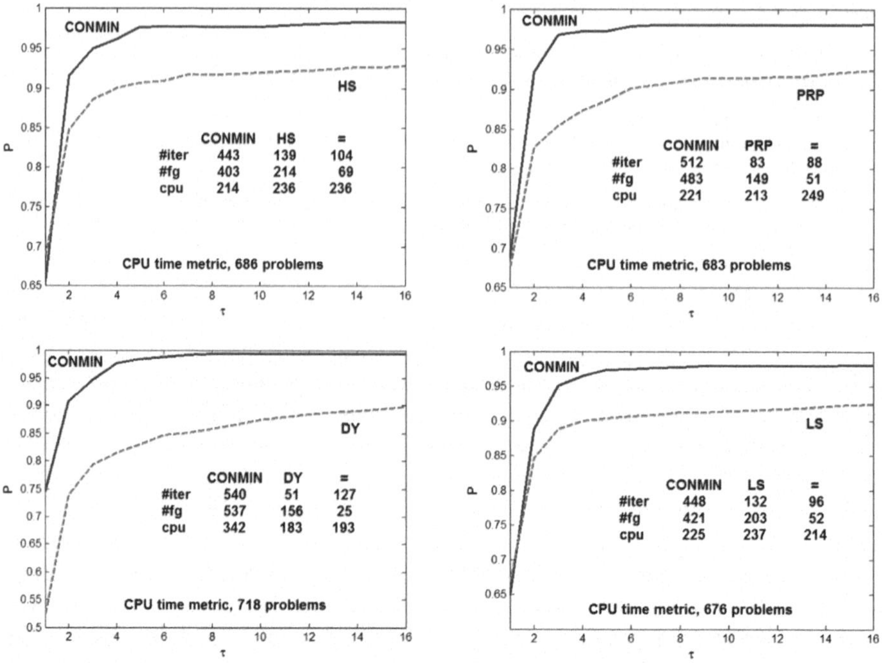

Figure 8.1 Performance profiles of CONMIN versus HS, PRP, DY, and LS

Figure 8.2 presents the performances of CONMIN versus the hybrid conjugate gradient methods hDY, HS-DY, GN, and LS-CD. Hybrid methods try to exploit the attractive features of the standard methods. However, as it can be seen in Figure 8.2, CONMIN is faster than all the methods considered in this set of numerical experiments. Hence, CONMIN appears to generate the best search direction in comparison with these hybrid methods.

Comparisons of CONMIN versus the modified conjugate gradient methods DL ($t = 1$), DL+ ($t = 1$), CG-DESCENT (version 1.4), and DESCONa are presented in Figure 8.3. Observe that CONMIN is more robust than DL and than DL+. However, both CG-DESCENT and DESCONa are more efficient and slightly more robust than CONMIN. CG-DESCENT and CONMIN are based on the memoryless BFGS quasi-Newton method. The search direction in CG-DESCENT is obtained from the Perry/Shanno search direction by deleting a term. On the other hand, DESCON is a modification of the HS method with guaranteed sufficient descent and conjugacy conditions and a modified Wolfe line search.

Now, let us compare CONMIN versus L-BFGS ($m = 5$) and versus TN for solving the same set of problems from the UOP collection. Figure 8.4 presents the performance profiles of these methods. Clearly, CONMIN is top performer versus these methods.

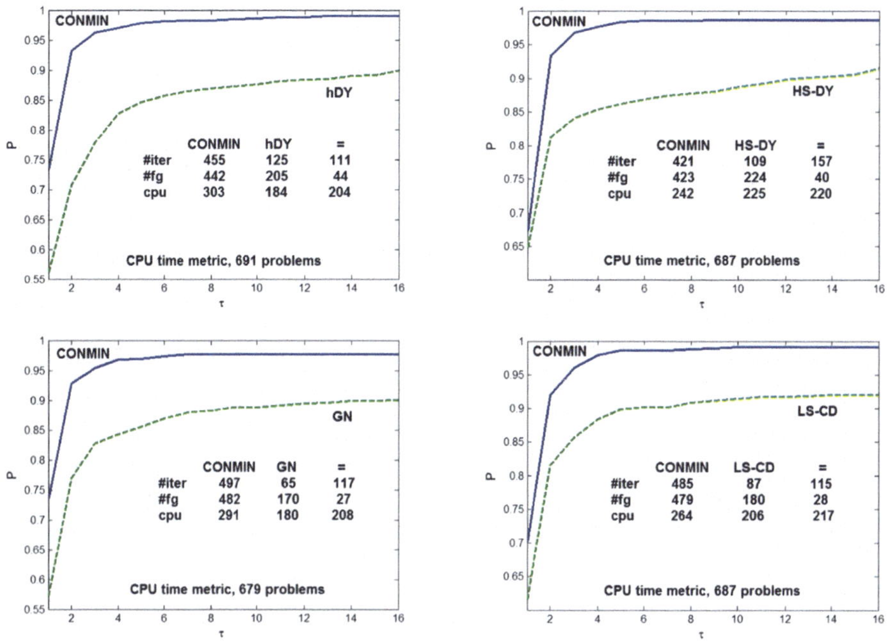

Figure 8.2 Performance profiles of CONMIN versus hDY, HS-DY, GN, and LS-CD

In the second set of numerical experiments, let us present the performances of CONMIN for solving the applications from the MINPACK-2 collection, each of them with 40,000 variables. Table 8.1 shows the performances of CONMIN for solving these applications.

8.2 Scaling Conjugate Gradient Memoryless BFGS Preconditioned (SCALCG)

For solving the unconstrained optimization problem (8.1), in the following let us develop a method based on a combination of the scaled memoryless BFGS method and the preconditioning technique (Andrei, 2007a, 2007b, 2007c, 2008a). Chapter 10 will point out that for general nonlinear functions, a good preconditioner is any matrix that approximates $\nabla^2 f(x^*)^{-1}$, where x^* is a local optimal solution of (8.1). In the algorithm presented in this chapter, the preconditioner is a scaled memoryless BFGS matrix which is reset when the Powell restart criterion holds. The scaling factor in the preconditioner is selected as spectral gradient (Raydan, 1997) or as a scalar, computed by using the information in two successive steps (see Andrei, 2004; Dai, & Yuan, 2002).

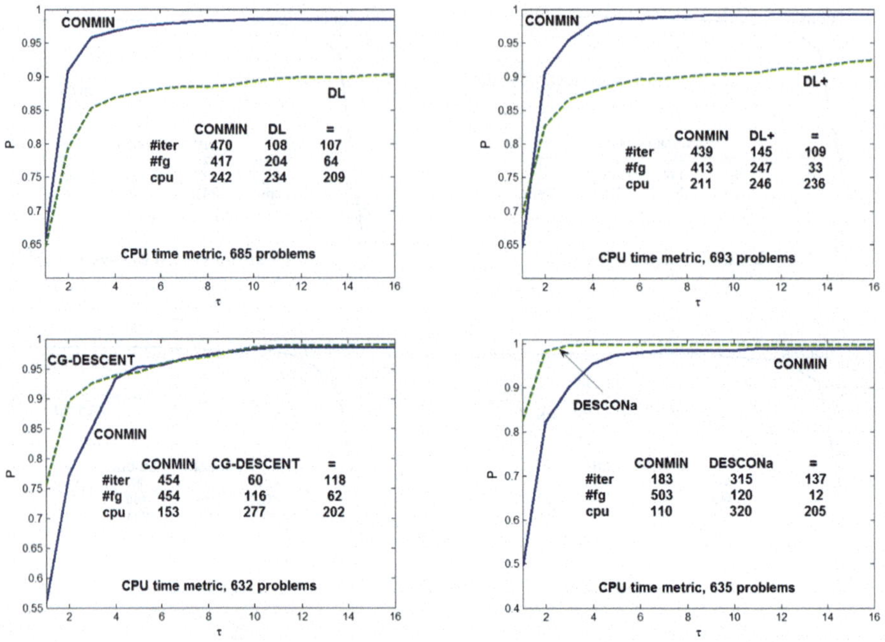

Figure 8.3 Performance profiles of CONMIN versus DL ($t = 1$), DL+ ($t = 1$). CG-DESCENT and DESCONa

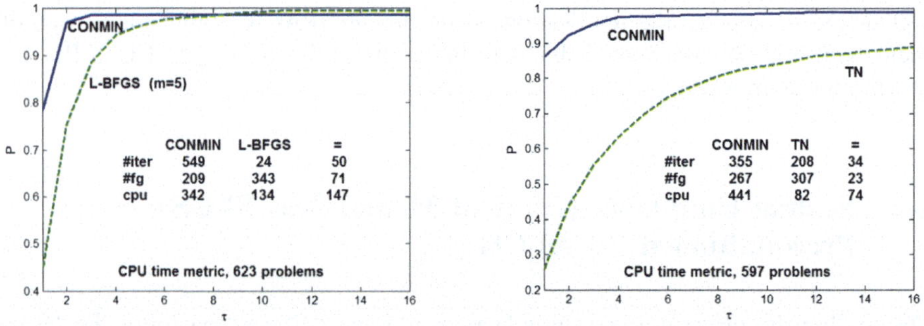

Figure 8.4 Performance profiles of CONMIN versus L-BFGS ($m = 5$) and versus TN

The algorithm generates a sequence $\{x_k\}$ of the approximations to the minimum x^* of f, in which

$$x_{k+1} = x_k + \alpha_k d_k, \tag{8.46}$$

$$d_{k+1} = -\theta_{k+1} g_{k+1} + \beta_k s_k, \tag{8.47}$$

Table 8.1 Performances of CONMIN for solving five applications from the MINPACK-2 collection

	CONMIN			
	n	*#iter*	*#fg*	*cpu*
A1	40,000	241	484	5.67
A2	40,000	827	1674	21.25
A3	40,000	1094	2217	41.65
A4	40,000	486	985	40.21
A5	40,000	374	757	11.39
Total	–	3022	6117	120.17

where β_k is a scalar parameter, $s_k = x_{k+1} - x_k$ and θ_{k+1} is a scalar parameter or a matrix to be determined. The iterative process is initialized with an initial point x_0 and $d_0 = -g_0$.

Observe that if $\theta_{k+1} = 1$, then according to the value of the scalar parameter β_k, the standard conjugate gradient method is obtained. On the other hand, if $\beta_k = 0$, then another class of methods is obtained, according to the selection of the parameter θ_{k+1}. Considering $\beta_k = 0$, there are two possibilities for θ_{k+1} : a positive scalar or a positive definite matrix. If $\theta_{k+1} = 1$, then the steepest descent method is obtained. If $\theta_{k+1} = \nabla^2 f(x_{k+1})^{-1}$, or an approximation of it, then the Newton or the quasi-Newton methods are obtained, respectively. Therefore, in the general case, when $\theta_{k+1} \neq 0$ is selected in a quasi-Newton manner and $\beta_k \neq 0$, (8.47) represents a combination between the quasi-Newton and the conjugate gradient methods. However, if θ_{k+1} is a matrix containing some useful information about the inverse Hessian of function f, it is better to use $d_{k+1} = -\theta_{k+1} g_{k+1}$, since the addition of the term $\beta_k s_k$ in (8.47) may prevent the direction d_{k+1} from being a descent one unless the line search is sufficiently accurate. Therefore, in the following, θ_{k+1} is considered as a positive scalar which contains some useful information about the inverse Hessian of function f.

As it is known, when the initial point x_0 is close enough to a local minimum point x^*, then the best search direction to be followed in the current point x_{k+1} is the Newton direction: $-\nabla^2 f(x_{k+1})^{-1} g_{k+1}$. Therefore, there is a strong motivation to choose the parameter β_k in (8.47) so that for every $k \geq 1$ the direction d_{k+1} given by (8.47) can be the best direction known, i.e., the Newton direction. Hence, using the Newton direction from the equality

$$-\nabla^2 f(x_{k+1})^{-1} g_{k+1} = -\theta_{k+1} g_{k+1} + \beta_k s_k, \tag{8.48}$$

it follows that

$$\beta_k = \frac{s_k^T \nabla^2 f(x_{k+1}) \theta_{k+1} g_{k+1} - s_k^T g_{k+1}}{s_k^T \nabla^2 f(x_{k+1}) s_k}. \tag{8.49}$$

Observe that the Newton direction is being used here only as a motivation for formula (8.49). The salient point with this formula for the β_k computation is the presence of the Hessian. If the line search is exact, the scaled Daniel (1967) method is obtained. Obviously, for large-scale problems, choices for the update parameter that do not require the evaluation of the Hessian matrix are often preferred in practice to the methods that require the Hessian.

Now, for quasi-Newton methods, an approximation matrix B_k to the Hessian $\nabla^2 f(x_k)$ is used and updated so that the new matrix B_{k+1} satisfies the secant equation $B_{k+1}s_k = y_k$, where $y_k = g_{k+1} - g_k$. Therefore, in order to have a method for solving large-scale problems, it is assumed that the pair (s_k, y_k) satisfies the secant equation. In this case, Zhang, Deng, and Chen (1999) proved that if $\|s_k\|$ is sufficiently small, then $s_k^T \nabla^2 f(x_{k+1}) s_k - s_k^T y_k = O(\|s_k\|^3)$. Therefore, using this assumption, from (8.49) for β_k the following formula is obtained

$$\beta_k = \frac{(\theta_{k+1} y_k - s_k)^T g_{k+1}}{y_k^T s_k}. \tag{8.50}$$

Birgin and Martínez (2001) arrived at the same formula for β_k by using a geometric interpretation of the quadratic function minimization. The direction corresponding to β_k given in (8.50) is as follows

$$d_{k+1} = -\theta_{k+1} g_{k+1} + \frac{(\theta_{k+1} y_k - s_k)^T g_{k+1}}{y_k^T s_k} s_k. \tag{8.51}$$

The following particularizations are obvious. If $\theta_{k+1} = 1$, then (8.51) is the direction considered by Perry (1977). At the same time, observe that (8.51) is the direction given by Dai and Liao (2001) for $t = 1$, obtained this time by an interpretation of the conjugacy condition. Additionally, if $s_j^T g_{j+1} = 0$, $j = 0, 1, \ldots, k$, then from (8.51) it follows that

$$d_{k+1} = -\theta_{k+1} g_{k+1} + \frac{\theta_{k+1} y_k^T g_{k+1}}{\alpha_k \theta_k g_k^T g_k} s_k, \tag{8.52}$$

which is the direction corresponding to a *generalization of the Polak, Ribière, and Polyak* formula. Of course, if $\theta_{k+1} = \theta_k = 1$ in (8.52), the *standard Polak, Ribière, and Polyak* (PRP) formula is obtained. If $s_j^T g_{j+1} = 0$, $j = 0, 1, \ldots, k$, and additionally the successive gradients are orthogonal, then from (8.51) it results that

$$d_{k+1} = -\theta_{k+1} g_{k+1} + \frac{\theta_{k+1} g_{k+1}^T g_{k+1}}{\alpha_k \theta_k g_k^T g_k} s_k, \tag{8.53}$$

which is the direction corresponding to a *generalization of the Fletcher and Reeves* formula. Therefore, (8.51) is a general formula for the direction computation in a conjugate gradient way, including the standard FR and PRP formulae.

Shanno (1978a, b) proved that the conjugate gradient methods are exactly the BFGS quasi-Newton method, where at every step, the approximation to the inverse Hessian is restarted as the identity matrix. *Now, this result is extended to the scaled conjugate gradient.* The direction given by (8.51) can be written as

$$d_{k+1} = -\left[\theta_{k+1} I - \theta_{k+1} \frac{s_k y_k^T}{y_k^T s_k} + \frac{s_k s_k^T}{y_k^T s_k}\right] g_{k+1} \equiv -Q_{k+1} g_{k+1}, \tag{8.54}$$

where

$$Q_{k+1} = \theta_{k+1} I - \theta_{k+1} \frac{s_k y_k^T}{y_k^T s_k} + \frac{s_k s_k^T}{y_k^T s_k}. \tag{8.55}$$

If $\theta_{k+1} = 1$, it follows that

$$d_{k+1} = -\left[I - \frac{s_k y_k^T}{y_k^T s_k} + \frac{s_k s_k^T}{y_k^T s_k}\right] g_{k+1}, \tag{8.56}$$

which is exactly the Perry formula. By direct computation, the following proposition can be proved.

Proposition 8.2

$$y_k^T Q_{k+1} = s_k^T. \tag{8.57}$$

Observe that (8.57) is similar but not identical to the quasi-Newton equation, which requires an update to the approximate of the inverse Hessian H_{k+1} to satisfy

$$H_{k+1} y_k = s_k. \tag{8.58}$$

A major difficulty with (8.54) is that the matrix Q_{k+1} defined by (8.55) is not symmetric and therefore not positive definite. Thus, the direction d_{k+1} from (8.54) is not necessarily a descent one and so numerical instability can appear. Besides, another difficulty arising from this lack of symmetry is that the true quasi-Newton Equation (8.58) is not satisfied.

In order to overcome this difficulty and to get a true quasi-Newton updating, the matrix Q_{k+1} from (8.55) is first symmetrized as

$$Q_{k+1} = \theta_{k+1} I - \theta_{k+1} \frac{s_k y_k^T + y_k s_k^T}{y_k^T s_k} + \frac{s_k s_k^T}{y_k^T s_k}. \tag{8.59}$$

Secondly, Q_{k+1} is forced to satisfy the quasi-Newton Equation (8.58) yielding the following symmetric update

$$Q_{k+1}^{*} = \theta_{k+1}I - \theta_{k+1}\frac{y_k s_k^{\mathrm{T}} + s_k y_k^{\mathrm{T}}}{y_k^{\mathrm{T}} s_k} + \left[1 + \theta_{k+1}\frac{y_k^{\mathrm{T}} y_k}{y_k^{\mathrm{T}} s_k}\right]\frac{s_k s_k^{\mathrm{T}}}{y_k^{\mathrm{T}} s_k}. \tag{8.60}$$

By direct computation, it is very easy to prove that Q_{k+1}^{*} satisfies the quasi-Newton equation, i.e.,

Proposition 8.3

$$Q_{k+1}^{*} y_k = s_k. \tag{8.61}$$

Notice that

$$d_{k+1} = -Q_{k+1}^{*} g_{k+1} \tag{8.62}$$

does not actually require the matrix Q_{k+1}^{*}, i.e., the direction d_{k+1} can be easily computed as

$$\begin{aligned} d_{k+1} = & -\theta_{k+1} g_{k+1} + \theta_{k+1}\left(\frac{g_{k+1}^{\mathrm{T}} s_k}{y_k^{\mathrm{T}} s_k}\right) y_k \\ & - \left[\left(1 + \theta_{k+1}\frac{y_k^{\mathrm{T}} y_k}{y_k^{\mathrm{T}} s_k}\right)\frac{g_{k+1}^{\mathrm{T}} s_k}{y_k^{\mathrm{T}} s_k} - \theta_{k+1}\frac{g_{k+1}^{\mathrm{T}} y_k}{y_k^{\mathrm{T}} s_k}\right] s_k. \end{aligned} \tag{8.63}$$

Again, observe that if $g_{k+1}^{\mathrm{T}} s_k = 0$, then (8.63) is reduced to

$$d_{k+1} = -\theta_{k+1} g_{k+1} + \theta_{k+1}\frac{g_{k+1}^{\mathrm{T}} y_k}{y_k^{\mathrm{T}} s_k} s_k. \tag{8.64}$$

Thus, the effect is simply one of multiplying the Hestenes and Stiefel search direction by a positive scalar.

The BFGS update to the inverse Hessian, which is currently the best update of the Broyden class, is defined by

$$H_{k+1} = H_k - \frac{H_k y_k s_k^{\mathrm{T}} + s_k y_k^{\mathrm{T}} H_k}{y_k^{\mathrm{T}} s_k} + \left[1 + \frac{y_k^{\mathrm{T}} H_k y_k}{y_k^{\mathrm{T}} s_k}\right]\frac{s_k s_k^{\mathrm{T}}}{y_k^{\mathrm{T}} s_k}. \tag{8.65}$$

Therefore, the conjugate gradient method (8.62), where Q_{k+1}^{*} is given by (8.60), is exactly the BFGS quasi-Newton method, where at every step the approximation of the inverse Hessian is restarted as the identity matrix multiplied by the scalar θ_{k+1}.

In order to ensure the convergence of the algorithm (8.46) with d_{k+1} given by (8.63), the choice of α_k must be constrained. Therefore, let us consider that α_k satisfies the Wolfe line search conditions (8.4) and (8.5).

Theorem 8.2 *Suppose that α_k in (8.46) satisfies the Wolfe conditions (8.4) and (8.5), then the direction d_{k+1} given by (8.63) is a descent direction.*

Proof Since $d_0 = -g_0$, it follows that $g_0^T d_0 = -\|g_0\|^2 \le 0$. Multiplying (8.63) by g_{k+1}^T, gives

$$g_{k+1}^T d_{k+1} = \frac{1}{(y_k^T s_k)^2}\Big[-\theta_{k+1}\|g_{k+1}\|^2 (y_k^T s_k)^2 + 2\theta_{k+1}(g_{k+1}^T y_k)(g_{k+1}^T s_k)(y_k^T s_k) - (g_{k+1}^T s_k)^2 (y_k^T s_k) - \theta_{k+1}(y_k^T y_k)(g_{k+1}^T s_k)^2\Big].$$

Applying the inequality $u^T v \le \frac{1}{2}(\|u\|^2 + \|v\|^2)$ to the second term of the right-hand side of the above equality, with $u = (s_k^T y_k) g_{k+1}$ and $v = (g_{k+1}^T s_k) y_k$, the following estimation is obtained

$$g_{k+1}^T d_{k+1} \le -\frac{(g_{k+1}^T s_k)^2}{y_k^T s_k}. \tag{8.66}$$

But, by the Wolfe condition (8.5), $y_k^T s_k > 0$. Therefore, $g_{k+1}^T d_{k+1} < 0$ for every $k = 0, 1, \ldots$ ♦

Observe that the second Wolfe condition (8.5) is crucial for the descent character of the direction (8.63). Moreover, the estimation (8.66) is independent of the parameter θ_{k+1}.

Usually, all conjugate gradient algorithms are periodically restarted. The standard restarting point occurs when the number of iterations is equal to the number of variables, but some other restarting methods can be considered as well. The Powell (1977) restarting procedure is to test if there is very little orthogonality left between the current gradient and the previous one. At step r, when

$$|g_{r+1}^T g_r| \ge 0.2\|g_{r+1}\|^2, \tag{8.67}$$

the algorithm is restarted using the direction given by (8.63). The convergence analysis with this restart criterion can be found in Dai and Yuan (1998). Another restarting procedure, considered by Birgin and Martínez (2001), consists of testing if the angle between the current direction and $-g_{k+1}$ is not very acute. Therefore, at step r, when

$$d_r^T g_{r+1} > -10^{-3}\|d_r\|_2 \|g_{r+1}\|_2, \tag{8.68}$$

the algorithm is restarted using the direction given by (8.63).

At step r, when one of the two criteria (8.67) or (8.68) is satisfied, the direction is computed as in (8.63). For $k \ge r + 1$ the same philosophy is used to get (8.60), i.e., that of modifying the gradient g_{k+1} with a positive definite matrix which best estimates the inverse Hessian without any additional storage requirements. Therefore, the direction d_{k+1} for $k \ge r + 1$ is computed using a double update scheme as

$$d_{k+1} = -H_{k+1}g_{k+1}, \tag{8.69}$$

where

$$H_{k+1} = H_{r+1} - \frac{H_{r+1}y_k s_k^T + s_k y_k^T H_{r+1}}{y_k^T s_k} + \left[1 + \frac{y_k^T H_{r+1} y_k}{y_k^T s_k}\right] \frac{s_k s_k^T}{y_k^T s_k} \tag{8.70}$$

and

$$H_{r+1} = \theta_{r+1} I - \theta_{r+1} \frac{y_r s_r^T + s_r y_r^T}{y_r^T s_r} + \left(1 + \theta_{r+1} \frac{y_r^T y_r}{y_r^T s_r}\right) \frac{s_r s_r^T}{y_r^T s_r}. \tag{8.71}$$

As above, observe that this computational scheme does not involve any matrix. Indeed, $H_{r+1}g_{k+1}$ and $H_{r+1}y_k$ can be computed as

$$\begin{aligned} v \equiv H_{r+1}g_{k+1} &= \theta_{r+1}g_{k+1} - \theta_{r+1}\left(\frac{g_{k+1}^T s_r}{y_r^T s_r}\right) y_r \\ &+ \left[\left(1 + \theta_{r+1}\frac{y_r^T y_r}{y_r^T s_r}\right)\frac{g_{k+1}^T s_r}{y_r^T s_r} - \theta_{r+1}\frac{g_{k+1}^T y_r}{y_r^T s_r}\right] s_r \end{aligned} \tag{8.72}$$

and

$$\begin{aligned} w \equiv H_{r+1}y_k &= \theta_{r+1}y_k - \theta_{r+1}\left(\frac{y_k^T s_r}{y_r^T s_r}\right) y_r \\ &+ \left[\left(1 + \theta_{r+1}\frac{y_r^T y_r}{y_r^T s_r}\right)\frac{y_k^T s_r}{y_r^T s_r} - \theta_{r+1}\frac{y_k^T y_r}{y_r^T s_r}\right] s_r. \end{aligned} \tag{8.73}$$

With this, the direction (8.69) at any nonrestart step can be computed as

$$d_{k+1} = -v + \frac{(g_{k+1}^T s_k)w + (g_{k+1}^T w)s_k}{y_k^T s_k} - \left(1 + \frac{y_k^T w}{y_k^T s_k}\right)\frac{g_{k+1}^T s_k}{y_k^T s_k} s_k. \tag{8.74}$$

Observe that d_{k+1} from (8.74) is defined as a *double quasi-Newton update scheme*. It is useful to note that $y_k^T s_k > 0$ is sufficient to ensure that the direction d_{k+1} given by (8.69) is well defined and it is always a descent one.

In the following, the computation of θ_{k+1} is discussed. As it has already been seen, θ_{k+1} is defined as a scalar approximation to the inverse Hessian. According to the procedures for a scalar estimation of the inverse Hessian, a family of scaling conjugate gradient algorithms is obtained. The following procedures can be used.

θ_{k+1} *spectral.* Motivated by the efficiency of the spectral gradient method introduced by Raydan (1997) and used by Birgin and Martínez (2001) in their spectral conjugate gradient method for unconstrained optimization, θ_{k+1} is defined as a scalar approximation to the inverse Hessian. This is given as the inverse of the Rayleigh quotient

$$s_k^T \left[\int_0^1 \nabla^2 f(x_k + ts_k) dt \right] s_k / s_k^T s_k,$$

i.e.,

$$\theta_{k+1} = \frac{s_k^T s_k}{y_k^T s_k}. \tag{8.75}$$

The inverse of the Rayleigh quotient lies between the smallest and the largest eigenvalue of the Hessian average $\int_0^1 \nabla^2 f(x_k + ts_k) dt$. Again, observe that $y_k^T s_k > 0$ is sufficient to ensure that θ_{k+1} in (8.75) is well defined.

θ_{k+1} *anticipative.* Using the information in two successive points of the iterative process, Andrei (2004) proposed another scalar approximation to the Hessian of function f, thus obtaining a new algorithm which favorably compares with the one of Barzilai and Borwein. This is only a half step of the spectral procedure. Indeed, in point $x_{k+1} = x_k + \alpha_k d_k$,

$$f(x_{k+1}) = f(x_k) + \alpha_k g_k^T d_k + \frac{1}{2} \alpha_k^2 d_k^T \nabla^2 f(z) d_k, \tag{8.76}$$

where z is on the line segment connecting x_k and x_{k+1}. Having in view the local character of the searching procedure and that the distance between x_k and x_{k+1} is small enough, $z = x_{k+1}$ can be chosen and $\gamma_{k+1} \in \mathbb{R}$ considered as a scalar approximation of $\nabla^2 f(x_{k+1})$. This is an anticipative viewpoint in which a scalar approximation of the Hessian at point x_{k+1} is computed using only the local information from two successive points x_k and x_{k+1}. Therefore,

$$\gamma_{k+1} = \frac{2}{d_k^T d_k} \frac{1}{\alpha_k^2} \left[f(x_{k+1}) - f(x_k) - \alpha_k g_k^T d_k \right]. \tag{8.77}$$

This formula can also be found in Dai and Yuan (2002). Observe that $\gamma_{k+1} > 0$ for convex functions. If $f(x_{k+1}) - f(x_k) - \alpha_k g_k^T d_k < 0$, then the reduction $f(x_{k+1}) - f(x_k)$ in the function value is smaller than $\alpha_k g_k^T d_k$. In these cases the idea is to reduce a little stepsize α_k as $\alpha_k - \eta_k$, maintaining the other quantities at their values in such a way so that γ_{k+1} is positive. To get a value for η_k let us select a real $\delta > 0$, "small enough" but comparable with the value of the function, and take

$$\eta_k = \frac{1}{g_k^T d_k} \left[f(x_k) - f(x_{k+1}) + \alpha_k g_k^T d_k + \delta \right], \tag{8.78}$$

with which a new value for γ_{k+1} can be computed as

$$\gamma_{k+1} = \frac{2}{d_k^T d_k} \frac{1}{(\alpha_k - \eta_k)^2} \left[f(x_{k+1}) - f(x_k) - (\alpha_k - \eta_k) g_k^T d_k \right]. \tag{8.79}$$

With this, the value for parameter θ_{k+1} is selected as

$$\theta_{k+1} = \frac{1}{\gamma_{k+1}}, \tag{8.80}$$

where γ_{k+1} is given by (8.77) or (8.79).

Proposition 8.4 *Assume that $f(x)$ is continuously differentiable and $\nabla f(x)$ is Lipschitz continuous, with a positive constant L. Then, at point x_{k+1},*

$$\gamma_{k+1} \leq 2L. \tag{8.81}$$

Proof From (8.77) it follows that

$$\gamma_{k+1} = \frac{2\left[f(x_k) + \alpha_k \nabla f(\xi_k)^T d_k - f(x_k) - \alpha_k \nabla f(x_k)^T d_k \right]}{\|d_k\|^2 \alpha_k^2},$$

where ξ_k is on the line segment connecting x_k and x_{k+1}. Therefore,

$$\gamma_{k+1} = \frac{2(\nabla f(\xi_k) - \nabla f(x_k))^T d_k}{\|d_k\|^2 \alpha_k}.$$

Using the inequality of Cauchy and the Lipschitz continuity, it follows that

$$\gamma_{k+1} \leq \frac{2\|\nabla f(\xi_k) - \nabla f(x_k)\|}{\|d_k\| \alpha_k} \leq \frac{2L\|\xi_k - x_k\|}{\|d_k\| \alpha_k} \leq \frac{2L\|x_{k+1} - x_k\|}{\|d_k\| \alpha_k} = 2L.$$

♦

Therefore, from (8.80) a lower bound for θ_{k+1} is obtained as

$$\theta_{k+1} \geq \frac{1}{2L},$$

i.e., it is bounded away from zero.

The scaling conjugate gradient memoryless BFGS preconditioned algorithm is equipped with the acceleration scheme developed in Chapter 5. Taking into account the acceleration scheme shown in Remark 5.1, according to the value of the parameter "*acceleration*" (true or false) the following algorithms SCALCG and SCALCGa can be presented. SCALCGa is the accelerated version of SCALCG.

Algorithm 8.2 Scaling memoryless BFGS preconditioned: SCALCG/SCALCGa

1.	Select $x_0 \in \mathbb{R}^n$, $\varepsilon_A > 0$ sufficiently small and the values $0 < \rho \le \sigma < 1$ used in Wolfe line search. Compute $f(x_0)$ and $g_0 = \nabla f(x_0)$. Set $d_0 = -g_0$ and $\alpha_0 = 1/\lVert g_0 \rVert$. Set $k = 0$
2.	*Line search*. Compute α_k satisfying the Wolfe conditions (8.4) and (8.5). Update the variables $x_{k+1} = x_k + \alpha_k d_k$. Compute f_{k+1}, g_{k+1} and $s_k = x_{k+1} - x_k$, $y_k = g_{k+1} - g_k$
3.	*Test for the continuation of iterations*. If this test is satisfied, then the iterations are stopped, else set $k = k+1$ and continue with step 4
4.	*Scaling factor computation*. Compute θ_k using a spectral (8.75) or an anticipative (8.80) approach
5.	*Restart direction*. Compute the (restart) direction d_k as in (8.63)
6.	*Line search*. Compute the initial guess of the step length as $\alpha_k = \alpha_{k-1} \lVert d_{k-1} \rVert_2 / \lVert d_k \rVert_2$ With this initialization, compute α_k satisfying the Wolfe conditions (8.4) and (8.5). Update the variables $x_{k+1} = x_k + \alpha_k d_k$. Compute f_{k+1}, g_{k+1} and $s_k = x_{k+1} - x_k$, $y_k = g_{k+1} - g_k$
7.	If *acceleration* equal true, then (a) Compute: $z = x_k + \alpha_k d_k$, $g_z = \nabla f(z)$ and $y_k = g_k - g_z$ (b) Compute: $\bar{a}_k = \alpha_k g_k^T d_k$ and $\bar{b}_k = -\alpha_k y_k^T d_k$ (c) If $\lvert \bar{b}_k \rvert \ge \varepsilon_A$, then compute $\xi_k = -\bar{a}_k / \bar{b}_k$ and update the variables as $x_{k+1} = x_k + \xi_k \alpha_k d_k$. Compute f_{k+1} and g_{k+1}. Compute $s_k = x_{k+1} - x_k$ and $y_k = g_{k+1} - g_k$
8.	*Store*: $\theta = \theta_k$, $s = s_k$ and $y = y_k$
9.	*Test for the continuation of iterations*. If this test is satisfied, then the iterations are stopped, else set $k = k+1$
10.	*Decision on the way*. If either the Powell restart criterion (8.67) or the angle restart criterion (8.68) is satisfied, then go to step 4 (a restart step); otherwise continue with step 11 (a standard step)
11.	*Standard direction*. Compute: $v = \theta g_k - \theta\left(\frac{g_k^T s}{y^T s}\right) y + \left[\left(1 + \theta \frac{y^T y}{y^T s}\right) \frac{g_k^T s}{y^T s} - \theta \frac{g_k^T y}{y^T s}\right] s,$ $w = \theta y_k - \theta\left(\frac{y_{k-1}^T s}{y^T s}\right) y + \left[\left(1 + \theta \frac{y^T y}{y^T s}\right) \frac{y_{k-1}^T s}{y^T s} - \theta \frac{y_{k-1}^T y}{y^T s}\right] s,$ and $d_k = -v + \frac{(g_k^T s_{k-1}) w + (g_k^T w) s_{k-1}}{y_{k-1}^T s_{k-1}} - \left(1 + \frac{y_{k-1}^T w}{y_{k-1}^T s_{k-1}}\right) \frac{g_k^T s_{k-1}}{y_{k-1}^T s_{k-1}} s_{k-1}$
12.	*Line search*. Compute the initial guess of the step length as $\alpha_k = \alpha_{k-1} \lVert d_{k-1} \rVert_2 / \lVert d_k \rVert_2$ With this initialization, compute α_k as to satisfy the Wolfe conditions (8.4) and (8.5). Update the variables $x_{k+1} = x_k + \alpha_k d_k$. Compute f_{k+1}, g_{k+1} and $s_k = x_{k+1} - x_k$, $y_k = g_{k+1} - g_k$
13.	If *acceleration* equal true, then (a) Compute: $z = x_k + \alpha_k d_k$, $g_z = \nabla f(z)$ and $y_k = g_k - g_z$ (b) Compute: $\bar{a}_k = \alpha_k g_k^T d_k$ and $\bar{b}_k = -\alpha_k y_k^T d_k$ (c) If $\lvert \bar{b}_k \rvert \ge \varepsilon_A$, then compute $\xi_k = -\bar{a}_k / \bar{b}_k$ and update the variables as $x_{k+1} = x_k + \xi_k \alpha_k d_k$. Compute f_{k+1} and g_{k+1}. Compute $s_k = x_{k+1} - x_k$ and $y_k = g_{k+1} - g_k$
14.	*Test for the continuation of iterations*. If this test is satisfied, then the iterations are stopped, else set $k = k+1$ and go to step 10 ♦

Observe that in SCALCGa the acceleration scheme is implemented in two places. Therefore, at every iteration, two additional evaluations of the gradient are needed. Clearly, the acceleration scheme is profitable when the evaluation of the gradient does not require too much time.

It is well known that if f is bounded below along the direction d_k, then there is a step length α_k satisfying the Wolfe conditions. The initial selection of the stepsize crucially affects the practical behavior of the algorithm. At every iteration $k \geq 1$, the starting guess for step α_k in the line search is computed as $\alpha_{k-1}\|d_{k-1}\|_2/\|d_k\|_2$. Some other procedures for the initial selection of the stepsize are discussed in Chapter 12.

Convergence analysis for strongly convex functions. Assume that f is strongly convex and its gradient is Lipschitz continuous on the level set $S = \{x \in \mathbb{R}^n : f(x) \leq f(x_0)\}$. That is, there exist constants $\mu > 0$ and L so that

$$(\nabla f(x) - \nabla f(y))^{\mathrm{T}}(x - y) \geq \mu\|x - y\|^2 \tag{8.82}$$

and

$$\|\nabla f(x) - \nabla f(y)\| \leq L\|x - y\|, \tag{8.83}$$

for all x and y from S.

Proposition 8.5 *Assume that f is strongly convex and ∇f is Lipschitz continuous on S. If θ_{k+1} is selected by the spectral gradient, then the direction d_{k+1} given by (8.63) satisfies*

$$\|d_{k+1}\| \leq \left(\frac{2}{\mu} + \frac{2L}{\mu^2} + \frac{L^2}{\mu^3}\right)\|g_{k+1}\|. \tag{8.84}$$

Proof By the Lipschitz continuity (8.83), it follows that

$$\begin{aligned} \|y_k\| &= \|g_{k+1} - g_k\| \\ &= \|\nabla f(x_k + \alpha_k d_k) - \nabla f(x_k)\| \leq L\alpha_k\|d_k\| = L\|s_k\|. \end{aligned} \tag{8.85}$$

On the other hand, by the strong convexity (8.82), it results that

$$y_k^{\mathrm{T}} s_k \geq \mu\|s_k\|^2. \tag{8.86}$$

Selecting θ_{k+1} as in (8.75), it follows that

$$\theta_{k+1} = \frac{s_k^{\mathrm{T}} s_k}{y_k^{\mathrm{T}} s_k} \leq \frac{\|s_k\|^2}{\mu\|s_k\|^2} = \frac{1}{\mu}. \tag{8.87}$$

Now, using the triangle inequality and the above estimates (8.85)–(8.87), after some algebra on $\|d_{k+1}\|$, where d_{k+1} is given by (8.63), (8.84) is obtained. ♦

Proposition 8.6 *Assume that f is strongly convex and ∇f is Lipschitz continuous on S. Assume that $\nabla^2 f(x) \geq mI$ for all $x \in S$. If θ_{k+1} is selected by the anticipative procedure, then the direction d_{k+1} given by (8.63) satisfies*

$$\|d_{k+1}\| \leq \left(\frac{1}{m} + \frac{2L}{m\mu} + \frac{1}{\mu} + \frac{L^2}{m\mu^2}\right)\|g_{k+1}\|. \tag{8.88}$$

Proof By strong convexity on S, there exists the constant $m > 0$ so that $\nabla^2 f(x) \geq mI$ for all $x \in S$ (see Appendix A). Therefore, $\gamma_{k+1} \geq m$ for every k. Now, from (8.80) it is clear that for all k,

$$\theta_{k+1} \leq \frac{1}{m}. \tag{8.89}$$

With this, like in Proposition 8.5, (8.88) is obtained. ♦

Convergence of the algorithm for strongly convex functions. From the second Wolfe line search (8.5), it follows that

$$s_k^T y_k = s_k^T g_{k+1} - s_k^T g_k \geq -(1-\sigma)s_k^T g_k > 0, \tag{8.90}$$

i.e., the inequality (8.66) ensures that the search directions (8.62) are descent directions (see also Theorem 8.2). Although the descent condition is adequate (Dai, Han, Liu, Sun, Yin, and Yuan, 1999), the sufficient descent condition is crucial in the convergence analysis of conjugate gradient methods. Therefore, in the following let us prove that the search direction (8.62) satisfies the sufficient descent condition

$$g_{k+1}^T d_{k+1} \leq -c\|g_{k+1}\|^2 \tag{8.91}$$

for any $k \geq 0$, where c is a positive constant.

Since from (8.90) the Wolfe line search conditions ensure that $s_k^T y_k > 0$, it follows that the matrix Q_{k+1}^* is positive definite and consequently it is nonsingular. Based on the relationship between the BFGS and DFP updates (Nocedal and Wright, 2006), it can be shown that the matrix $W_{k+1}^* \in \mathbb{R}^{n \times n}$ defined by

$$W_{k+1}^* = \frac{1}{\theta_{k+1}} I - \frac{1}{\theta_{k+1}} \frac{s_k s_k^T}{s_k^T s_k} + \frac{y_k y_k^T}{y_k^T s_k} \tag{8.92}$$

is the inverse of Q^*_{k+1} given by (8.60). Therefore, W^*_{k+1} is also a positive definite matrix.

Theorem 8.3 *Suppose that the Assumption CG holds. If the function f is strongly convex on S and the stepsize α_k is determined by the Wolfe line search (8.4) and (8.5), then the search direction (8.62) with θ_{k+1} spectral satisfies the sufficient descent condition (8.91).*

Proof Since Q^*_{k+1} and W^*_{k+1} are $n \times n$ symmetric and positive definite matrices, they have n positive eigenvalues. To prove the theorem, let us first show that for all $k \geq 0$ the eigenvalues of Q^*_{k+1} are bounded below by a positive constant.

From (8.90), it results that $s_k^T y_k > 0$ and consequently $s_k \neq 0$ and $y_k \neq 0$. So, there exists a set of mutually orthogonal unit vectors $\{u_k^i\}_{i=1}^{n-2}$ so that

$$s_k^T u_k^i = y_k^T u_k^i = 0, \quad i = 1, \ldots, n-2,$$

which leads to

$$Q^*_{k+1} u_k^i = \theta_{k+1} u_k^i, \quad i = 1, \ldots, n-2.$$

Thus, the vectors u_k^i, $i = 1, \ldots, n-2$, are the eigenvectors of Q^*_{k+1} corresponding to the eigenvalue θ_{k+1}. Now, let $\lambda_k^{(n-1)}$ and $\lambda_k^{(n)}$ be the two remaining eigenvalues of Q^*_{k+1}. Since the trace of a matrix is exactly the sum of its eigenvalues, from (8.60) and (8.92) it follows that

$$\begin{aligned} \operatorname{tr}(Q^*_{k+1}) &= (n-2)\theta_{k+1} + \frac{s_k^T s_k}{y_k^T s_k}\left(1 + \theta_{k+1}\frac{y_k^T y_k}{y_k^T s_k}\right) \\ &= (n-2)\theta_{k+1} + \lambda_k^{(n-1)} + \lambda_k^{(n)}. \end{aligned} \tag{8.93}$$

Similarly,

$$\begin{aligned} \operatorname{tr}(W^*_{k+1}) &= \frac{n-1}{\theta_{k+1}} + \frac{y_k^T y_k}{y_k^T s_k} \\ &= \frac{n-2}{\theta_{k+1}} + \frac{1}{\lambda_k^{(n-1)}} + \frac{1}{\lambda_k^{(n)}}. \end{aligned} \tag{8.94}$$

Therefore, from (8.75), (8.93) and (8.94), it results that

$$\lambda_k^{(n-1)} + \lambda_k^{(n)} = \theta_{k+1}\left(1 + \theta_{k+1}\frac{y_k^T y_k}{y_k^T s_k}\right), \tag{8.95}$$

$$\frac{1}{\lambda_k^{(n-1)}} + \frac{1}{\lambda_k^{(n)}} = \frac{1}{\theta_{k+1}} \left(1 + \theta_{k+1} \frac{y_k^T y_k}{y_k^T s_k} \right). \tag{8.96}$$

From the above relations, it follows that

$$\lambda_k^{(n-1)} \lambda_k^{(n)} = \theta_{k+1}^2. \tag{8.97}$$

Assume that $\lambda_k^{(n)} \leq \lambda_k^{(n-1)}$. Now, from (8.75), from the Lipschitz continuity (8.83) and strong convexity (8.86), the following estimation is obtained

$$\frac{1}{L} \leq \theta_{k+1} \leq \frac{1}{\mu}, \tag{8.98}$$

which together with (8.83), (8.86), (8.95) and (8.98) yield

$$\begin{aligned} \lambda_k^{(n)} &= \theta_{k+1}^2 \frac{1}{\lambda_k^{(n-1)}} \geq \theta_{k+1}^2 \frac{1}{\lambda_k^{(n-1)} + \lambda_k^{(n)}} \\ &= \frac{s_k^T s_k}{y_k^T s_k + \theta_{k+1} y_k^T y_k} \geq \frac{\mu}{L^2 + L\mu}. \end{aligned} \tag{8.99}$$

Now, from (8.62) and (8.99), for all $k \geq 0$ it follows that

$$g_{k+1}^T d_{k+1} = -g_{k+1}^T Q_{k+1}^* g_{k+1} \leq -\lambda_k^{(n)} \|g_{k+1}\|^2 \leq -\frac{\mu}{L^2 + L\mu} \|g_{k+1}\|^2,$$

which proves the theorem with $c = \mu/(L^2 + L\mu)$. ♦

For general nonlinear functions, the convergence of the algorithm comes from Theorem 8.2 and the restart procedure. Therefore, for convex functions and under inexact line search, it is global convergent. If restarts are used, then the algorithm is convergent, but the speed of convergence can decrease. To a great extent, however, SCALCG algorithm is very close to the Perry/Shanno computational scheme. In fact, SCALCG is a scaled memoryless BFGS preconditioned algorithm, where the scaling factor is the inverse of a scalar approximation of the Hessian. For general functions that are bounded from below with bounded level sets and bounded second partial derivatives, the convergence of the proposed algorithm can be established using exactly the same analysis given by Shanno (1978b).

Numerical study. In the first set of numerical experiments, let us compare SCALCG (θ_k spectral) versus its accelerated version SCALCGa. Figure 8.5 presents the performance profiles of these algorithms for solving 80 unconstrained optimization

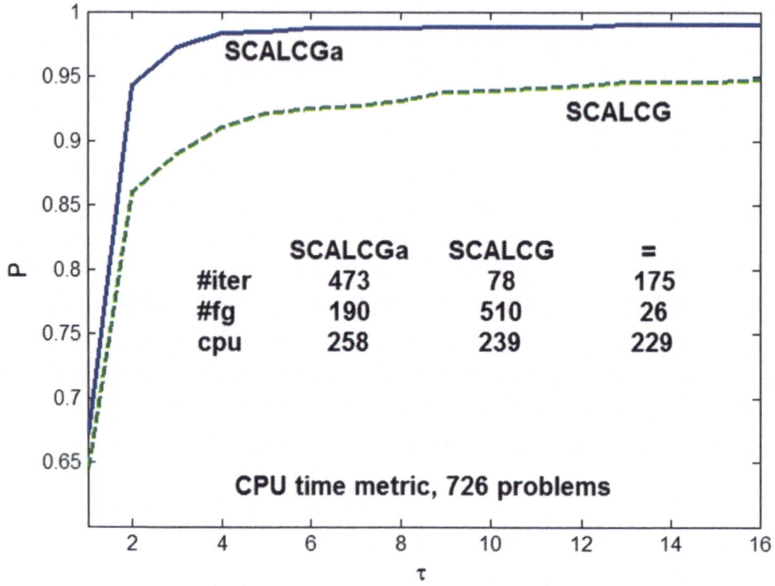

Figure 8.5 Performance profiles of SCALCG (spectral) versus SCALCGa (spectral)

test problems from the UOP collection (Andrei, 2018g), where for each problem ten numerical experiments are run with the number of variables $n = 1000, \ldots, 10000$. The numerical experiments are given in the context of Remark 1.1.

Observe that the acceleration scheme represents an improvement in the behavior of the algorithm. SCALCGa is both more efficient and far more robust than SCALCG.

Figure 8.6 contains the performance profiles of SCALCG (θ_k spectral) versus DL ($t = 1$), CG-DESCENT (version 1.4), DESCON, and CONMIN for solving 80 problems from the UOP collection. Figure 8.7 shows the performance profiles of SCALCGa (θ_k spectral) versus DL ($t = 1$), CG-DESCENT, DESCONa, and CONMIN.

From Figures 8.6 and 8.7, it is obvious that both SCALCG and SCALCGa are more robust than DL ($t = 1$). On the other hand, CG-DESCENT, DESCONa, and CONMIN are top performers in comparison both with SCALCG and SCALCGa. In these comparisons, DESCONa remains the most efficient and the most robust versus SCALCG. The accelerated version improves the performances of SCALCG, but in the end it is inferior to the compared algorithms.

Table 8.2 contains the performances of SCALCG (θ_k—spectral) and of SCALCG (θ_k—anticipative) for solving the applications from the MINPACK-2 collection with 40,000 variables ($nx = 200$ and $ny = 200$).

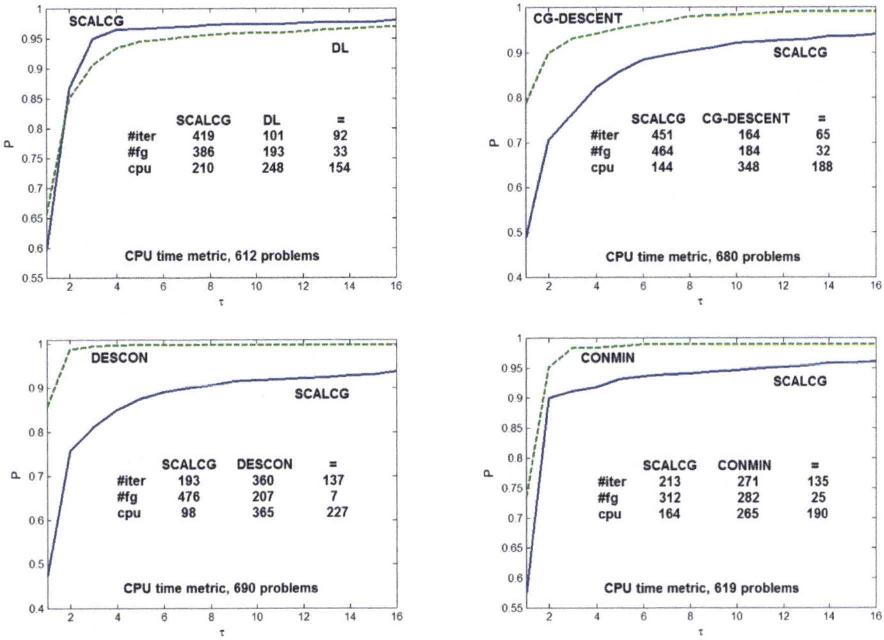

Figure 8.6 Performance profiles of SCALCG (spectral) versus DL ($t = 1$), CG-DESCENT, DESCON, and CONMIN

Observe that both variants of SCALCG (with θ_k spectral, or with θ_k anticipative) have similar performances. The variant of SCALCG with θ_k anticipative takes into account the function values in two successive points, but without any considerable advantage.

SCALCG is a scalar scaling conjugate gradient memoryless BFGS preconditioned algorithm derived from the Newton direction, using a double quasi-Newton updating scheme. The scaling parameter θ_{k+1} in (8.47) is a very crude scalar approximation to the inverse Hessian. This explains its modest performances versus some other conjugate gradient algorithms. The conclusion is that if θ_{k+1} (a scalar or a matrix) contains useful information about the Hessian of the minimizing function, it is better to use the search direction $d_{k+1} = -\theta_{k+1}g_{k+1}$ since $\beta_k s_k$, the addition of the last term in (8.47), may prevent d_{k+1} from being a descent direction unless the line search is highly accurate.

Observe that SCALCG is not a standard BFGS preconditioned conjugate gradient algorithm. The parameter θ_{k+1} is a scalar approximation to the Hessian far away from a real preconditioner matrix P_k which satisfies $P_k\nabla^2 f(x_k) \approx I$ (see Chapter 10).

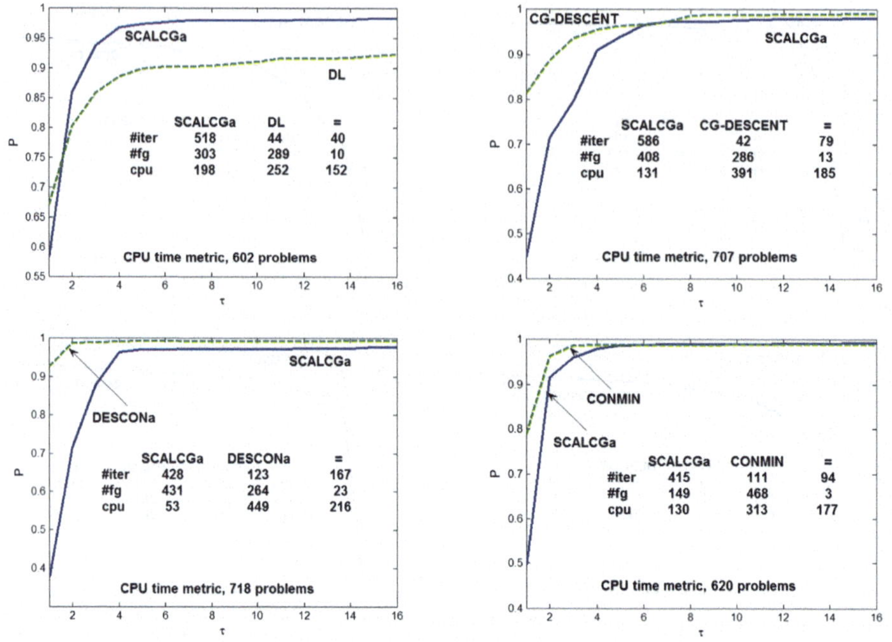

Figure 8.7 Performance profiles of SCALCGa (SCALCG accelerated) versus DL ($t = 1$). CG-DESCENT, DESCONa and CONMIN

Table 8.2 Performances of SCALCG (spectral) and SCALCG (anticipative) for solving five applications from the MINPACK-2 collection

		SCALCG (spectral)			SCALCG (anticipative)		
	n	*#iter*	*#fg*	*cpu*	*#iter*	*#fg*	*cpu*
A1	40,000	354	460	7.61	433	555	6.78
A2	40,000	902	1185	17.06	908	1176	15.67
A3	40,000	1404	1731	33.96	1503	1899	37.40
A4	40,000	703	907	37.63	690	874	36.21
A5	40,000	458	582	9.21	556	708	11.08
Total	–	3821	4865	105.47	4090	5212	107.14

8.3 Conjugate Gradient Method Closest to Scaled Memoryless BFGS Search Direction (DK/CGOPT)

The search directions in the quasi-Newton methods are computed as

$$d_k = -\bar{H}_k g_k, \tag{8.100}$$

where $\bar{H}_k \in \mathbb{R}^{n\times n}$ is an approximation to the inverse Hessian of the minimizing function. At the iteration k, the approximation $\bar{H}_k$ to the inverse Hessian is updated to achieve $\bar{H}_{k+1}$ as a new approximation to the inverse Hessian in such a way so that $\bar{H}_{k+1}$ satisfies a particular equation, namely the secant equation, which includes the second-order information. The most used is the standard secant equation

$$\bar{H}_{k+1} y_k = s_k, \tag{8.101}$$

where $s_k = x_{k+1} - x_k$ and $y_k = g_{k+1} - g_k$.

Given the initial approximation $\bar{H}_0$ to the inverse Hessian as an arbitrary symmetric and positive definite matrix, the most known quasi-Newton updating formula is the BFGS update

$$\bar{H}_{k+1} = \bar{H}_k - \frac{s_k y_k^T \bar{H}_k + \bar{H}_k y_k s_k^T}{y_k^T s_k} + \left(1 + \frac{y_k^T \bar{H}_k y_k}{y_k^T s_k}\right) \frac{s_k s_k^T}{y_k^T s_k}. \tag{8.102}$$

The self-scaling memoryless BFGS method of Perry (1977) and Shanno (1978a) is obtained by updating the scaled identity matrix $(1/\tau_k)I$ by the BFGS updating formula (8.102), i.e., by considering $\bar{H}_k = (1/\tau_k)I$ in (8.102), where I is the $n \times n$ identity matrix and τ_k is the scaling parameter. Therefore, the search direction in the self-scaling memoryless BFGS method is computed as

$$d_{k+1} = -H_{k+1} g_{k+1}, \tag{8.103}$$

where

$$H_{k+1} = \frac{1}{\tau_k}\left(I - \frac{s_k y_k^T + y_k s_k^T}{y_k^T s_k}\right) + \left(1 + \frac{1}{\tau_k}\frac{\|y_k\|^2}{y_k^T s_k}\right)\frac{s_k s_k^T}{y_k^T s_k}, \tag{8.104}$$

and τ_k is the scaling parameter, known as the *SSML-BFGS updating*. Now, substituting (8.104) in (8.103), the SSML-BFGS search direction, i.e., *the self-scaling memoryless BFGS search direction of Perry and Shanno* (with a multiplier difference) is obtained as

$$d_{k+1}^{PS} = -g_{k+1} + \left[\frac{g_{k+1}^T y_k}{y_k^T s_k} - \left(\tau_k + \frac{\|y_k\|^2}{y_k^T s_k}\right)\frac{g_{k+1}^T s_k}{y_k^T s_k}\right] s_k + \frac{g_{k+1}^T s_k}{y_k^T s_k} y_k. \tag{8.105}$$

(Observe that the Perry/Shanno search direction (8.37) is a scaling of (8.105) with a particular value for the scaling parameter τ_k.) Thus, subject to the parameter τ_k, a family of the Perry–Shanno self-scaling memoryless BFGS quasi-Newton methods is obtained. Now, the following particularizations of the search direction d_{k+1}^{PS} may be considered.

(1) Having in view that $s_k = \alpha_k d_k$, by deletion of the last term in (8.105) the following search direction is obtained:

$$d_{k+1} = -g_{k+1} + \left[\frac{y_k^T g_{k+1}}{y_k^T d_k} - \left(\tau_k + \frac{\|y_k\|^2}{y_k^T s_k} \right) \frac{s_k^T g_{k+1}}{y_k^T d_k} \right] d_k. \tag{8.106}$$

As suggested by Oren and Spedicato (1976), if τ_k in (8.106) is chosen as

$$\tau_k^{OS} = \frac{\|y_k\|^2}{y_k^T s_k}, \tag{8.107}$$

then (8.106) reduces to the well-known conjugate gradient algorithm CG-DESCENT proposed by Hager and Zhang (2005)

$$d_{k+1} = -g_{k+1} + \beta_k^{\text{CG-DESCENT}} d_k, \tag{8.108}$$

where

$$\beta_k^{\text{CG-DESCENT}} = \frac{g_{k+1}^T y_k}{y_k^T d_k} - 2\frac{\|y_k\|^2}{y_k^T s_k} \frac{g_{k+1}^T s_k}{y_k^T d_k}. \tag{8.109}$$

To establish the global convergence for general nonlinear functions, the conjugate gradient parameter (8.109) is truncated as

$$\beta_k^{\text{CG-DESCENT}+} = \max\left\{ \beta_k^{\text{CG-DESCENT}}, \frac{-1}{\|d_k\| \min\{\eta, \|g_k\|\}} \right\}, \tag{8.110}$$

where $\eta > 0$ is a constant ($\eta = 0.01$). The numerical experiments showed that CG-DESCENT is more efficient and more robust than the self-scaling memoryless BFGS method given by (8.105) (Dai and Kou, 2013).

Other values for the scaling parameter τ_k in (8.106) were proposed as follows. Oren (1974) and Oren and Luenberger (1974) proposed for τ_k the value $y_k^T s_k / (s_k^T B_k s_k)$ with $B_k = H_k^{-1}$. If H_k is the identity matrix, then this value reduces to

$$\tau_k^{OL} = \frac{y_k^T s_k}{\|s_k\|^2}. \tag{8.111}$$

Al-Baali (1998) suggested the following two choices:

$$\bar{\tau}_k^H = \min\left\{ 1, \frac{\|y_k\|^2}{y_k^T s_k} \right\} \quad \text{and} \quad \bar{\tau}_k^B = \min\left\{ 1, \frac{y_k^T s_k}{\|s_k\|^2} \right\}. \tag{8.112}$$

For a general nonlinear convex objective function, Nocedal and Yuan (1993) proved the global convergence of the self-scaling BFGS method with τ_k given by (8.111) and with Wolfe line search. They also presented numerical results indicating that the unscaled BFGS method is in general superior to the self-scaling BFGS with τ_k given by (8.111). Also, the proposals (8.112) for τ_k analyzed by Al-Baali ensure global convergence of the self-scaling BFGS method which is competitive with the unscaled BFGS method.

(2) Observe that the self-scaling memoryless BFGS search direction of Perry and Shanno (8.105) is a three-term conjugate gradient algorithm. A more reasonable way to deal with the last term in (8.105) was suggested by Dai and Kou (2013), who proposed to seek the search direction as a vector on the manifold $S_{k+1} = \{-g_{k+1} + \beta d_k : \beta \in \mathbb{R}\}$ that is closest to d_{k+1}^{PS}. This approach is not new. For example, Andrei (2017b) presented accelerated adaptive Perry conjugate gradient algorithms based on the minimization of the Frobenius norm of the difference between the symmetrical scaled Perry conjugate gradient direction and the self-scaling memoryless BFGS update. Also, Livieris, Tampakas, and Pintelas (2018) developed a convex hybridization of the conjugate gradient algorithms DY and HS, in which the hybridization parameter is computed by minimizing the distance between the hybrid direction and the self-scaling memoryless BFGS direction.

The search direction in S_{k+1} closest to d_{k+1}^{PS} is obtained as solution of the following least-squares problem

$$d_{k+1} = \arg\min\left\{ \|d - d_{k+1}^{\mathrm{PS}}\|_2 : d \in S_{k+1} \right\}, \tag{8.113}$$

which is

$$d_{k+1}^{\mathrm{DK}} = -g_{k+1} + \beta_k^{\mathrm{DK}}(\tau_k) s_k, \tag{8.114}$$

where

$$\beta_k^{\mathrm{DK}}(\tau_k) = \frac{g_{k+1}^{\mathrm{T}} y_k}{y_k^{\mathrm{T}} s_k} - \left(\tau_k + \frac{\|y_k\|^2}{y_k^{\mathrm{T}} s_k} - \frac{s_k^{\mathrm{T}} y_k}{\|s_k\|^2} \right) \frac{g_{k+1}^{\mathrm{T}} s_k}{y_k^{\mathrm{T}} s_k}. \tag{8.115}$$

If the line search is exact, $s_k^{\mathrm{T}} g_{k+1} = 0$, then the second term in (8.115) is missing and the search direction reduces to the HS formula.

In order to avoid the nonconvergence of the algorithm, similarly to Gilbert and Nocedal (1992) who proved the global convergence of the PRP methods for general nonlinear functions by restricting $\beta_k \geq 0$, (8.115) is truncated, being replaced by

$$\beta_k^{DK+}(\tau_k) = \max\left\{\beta_k^{DK}(\tau_k), \eta\frac{g_{k+1}^T d_k}{\|d_k\|^2}\right\}, \tag{8.116}$$

where $\eta \in [0, 1)$ is a parameter. ($\eta = 0.5$).

Hence, the Dai–Kou conjugate gradient algorithm belongs to the same family of conjugate gradient methods obtained from the memoryless BFGS method by Perry and Shanno. Under the Assumption CG, the family of conjugate gradient methods (8.114) and (8.115) generate sufficient descent directions. Numerical experiments with this family of conjugate gradient algorithms showed that the most efficient is the one where the parameter τ_k is given by (8.111). Therefore, substituting this choice of τ_k in (8.115) and (8.116), it results that

$$\beta_k^{DK} = \frac{y_k^T g_{k+1}}{d_k^T y_k} - \frac{\|y_k\|^2}{d_k^T y_k}\frac{d_k^T g_{k+1}}{d_k^T y_k} \tag{8.117}$$

and

$$\beta_k^{DK+} = \max\left\{\beta_k^{DK}, \eta\frac{d_k^T g_{k+1}}{\|d_k\|^2}\right\}, \tag{8.118}$$

where $\eta \in [0, 1)$. Observe that (8.117) is exactly the Dai–Liao conjugate gradient algorithm with $t = \|y_k\|^2/(s_k^T y_k)$. Again, note that (8.117) differs from the Hager and Zhang algorithm only in a constant coefficient in the second term. Thus, the family of Dai–Kou self-scaling memoryless BFGS quasi-Newton methods is obtained. The following result shows that the search direction given by (8.114) and (8.115) satisfies the Dai and Liao conjugacy condition.

Proposition 8.7 *The search direction* (8.114)*, where the parameter* $\beta_k(\tau_k)$ *is determined by* (8.115) *satisfies the Dai and Liao conjugacy condition* $d_{k+1}^T y_k = -t_k(g_{k+1}^T s_k)$, *where*

$$t_k = \tau_k + \frac{\|y_k\|^2}{y_k^T s_k} - \frac{s_k^T y_k}{\|s_k\|^2}$$

for any $k \geq 0$.

Proof From (8.114) and (8.115), by direct computation it follows that

$$d_{k+1}^T y_k = -\left(\tau_k + \frac{\|y_k\|^2}{y_k^T s_k} - \frac{s_k^T y_k}{\|s_k\|^2}\right)(g_{k+1}^T s_k) \equiv -t_k(g_{k+1}^T s_k).$$

Since $\|y_k\|^2/(y_k^T s_k) - (s_k^T y_k)/\|s_k\|^2 \geq 0$, it results that $t_k \geq 0$. ♦

Proposition 8.7 is a generalization of the result of Shanno (1978a), who proved that the conjugate gradient methods are precisely the BFGS quasi-Newton method, where the approximation to the inverse Hessian is restarted as the identity matrix at every step. In this case, the approximation to the inverse Hessian is restarted as a scaled identity matrix at every step. Therefore, (8.2), (8.115), or (8.116) is a conjugate gradient algorithm.

It is worth mentioning that if τ_k in (8.115) is selected as τ_k^{OL}, then the CGOPT conjugate gradient algorithm of Dai and Kou (2013) is obtained, where the search direction is computed as

$$d_{k+1}^{\mathrm{CGOPT}} = -g_{k+1} + \beta_k^{\mathrm{CGOPT}} d_k, \tag{8.119}$$

where

$$\beta_k^{\mathrm{CGOPT}} = \frac{g_{k+1}^{\mathrm{T}} y_k}{y_k^{\mathrm{T}} d_k} - \frac{\|y_k\|^2}{y_k^{\mathrm{T}} s_k} \frac{g_{k+1}^{\mathrm{T}} s_k}{y_k^{\mathrm{T}} d_k}, \tag{8.120}$$

which is identical to β_k^{DK} given by (1.117). Observe that the difference between the conjugate gradient parameters of CG-DESCENT given by (8.109) and of the CGOPT given by (8.120) is the absence of the constant factor 2 in the second term of the parameter from (8.120). Again, the numerical experiments showed that CGOPT performs more efficiently than the self-scaling memoryless BFGS method given by (8.105) (Dai and Kou, 2013). If the line search is exact, i.e., $g_{k+1}^{\mathrm{T}} s_k = 0$, then the second term in (8.115), (or in (8.109), or in (8.120)) is missing and the search direction reduces to that of the HS algorithm.

Dai and Kou (2013) (see Lemma 2.1) proved that if $y_k^{\mathrm{T}} s_k > 0$, than the search direction given by (8.114) and (8.115) satisfies

$$g_{k+1}^{\mathrm{T}} d_{k+1} \leq -\min\left\{ \tau_k \frac{\|s_k\|^2}{y_k^{\mathrm{T}} s_k}, \frac{3}{4} \right\} \|g_{k+1}\|^2. \tag{8.121}$$

More generally, if function f is continuously differentiable and bounded below and its gradient g is Lipschitz continuous, then Dai and Kou (2013) (see Lemma 2.2) proved that the search direction (8.114), where τ_k in (8.115) is chosen to be any of τ_k^{OS}, τ_k^{OL}, $\bar{\tau}_k^{H}$ or $\bar{\tau}_k^{B}$ and $y_k^{\mathrm{T}} s_k > 0$ satisfies $g_{k+1}^{\mathrm{T}} d_{k+1} \leq -c\|g_{k+1}\|^2$ for some positive constant $c > 0$.

Dai and Kou (2013) implemented the algorithm (8.2) with (8.119) and (8.120), which they called CGOPT, endowed with two ingredients which improve its performances. The first ingredient is an improved Wolfe line search, which avoids the numerical drawback of the first Wolfe line search condition and guarantees the global convergence of the algorithm. The second one is an adaptive restart of the algorithm along the negative gradient based on how the minimizing function is close to some quadratic function. The numerical experiments with this family of self-scaling memoryless BFGS methods, where the parameter τ_k in (8.115) is chosen as τ_k^{OS}, τ_k^{OL}, $\bar{\tau}_k^{H}$, or $\bar{\tau}_k^{B}$, proved that the selection τ_k^{OL} of τ_k is the most efficient one. With this selection of the parameter τ_k, Dai and Kou (2013) showed that the algorithm (8.2) with (8.114) and (8.115), with improved Wolfe line search is more efficient and more robust than CG-DESCENT.

Numerical strudy. In the following, let us present the performances of the conjugate gradient algorithm given by (8.2), where the search direction is computed as

$$d_{k+1} = -g_{k+1} + \beta_k^{\text{DK+}} s_k \tag{8.122}$$

and $\beta_k^{\text{DK}+}$ is given by (8.118) *with standard Wolfe line search* (8.4) and (8.5), denoted as DK+w, for solving all the problems from the UOP collection. Even if $\beta_k^{\text{DK}+} \equiv \beta_k^{\text{CGOPT}+}$, where $\beta_k^{\text{CGOPT}+} = \max\{\beta_k^{\text{CGOPT}}, \eta(s_k^{\text{T}} g_{k+1})/\|s_k\|^2\}$, the algorithm given by (8.122) and (8.118) with standard Wolfe line search is called DK+w since in this algorithm we implemented neither the dynamic restart strategy used in CGOPT, nor any other ingredients specific to CGOPT. For each test problem from the UOP collection (Andrei, 2018g), ten numerical experiments were run with the number of variables $n = 1000, \ldots, 10000$. The initial value for the stepsize is computed as $\alpha_k^0 = \alpha_{k-1}\|d_{k-1}\|/\|d_k\|$. All the numerical experiments were run in the context of Remark 1.1.

Figure 8.8 presents the performance profiles of DK+w with standard Wolfe line search versus CONMIN, SCALCG, CG-DESCENT (version 1.4), and DESCONa. Figure 8.8 shows that among these algorithms, DK+w is top performer versus CONMIN and SCALCG. CG-DESCENT is more efficient than DK+w, but DK+w is slightly more robust. DESCONa is clearly more efficient and more robust than DK+w. Observe that DK+w and CONMIN are quite close to each other, DK+w being slightly more efficient and more robust. This is because both these algorithms derive from the Perry–Shanno self-scaling BFGS preconditioned conjugate gradient algorithm. The difference between CONMIN and DK+w is that in CONMIN the search direction given by (8.28) is obtained as a double update with information in two points, while in DK+w the search direction is closest to the Perry–Shanno direction expressed as in (8.122) with (8.118). Besides, in CONMIN, the restart direction is exactly the Perry–Shanno search direction while in DK+w, the restart direction is simply the negative gradient.

To compare the algorithms, in (Andrei, 2019a) we manufactured a new code implementing (8.2), (8.122) with (8.118), where the stepsize is determined by the *approximate Wolfe conditions*

$$\sigma d_k^{\text{T}} g_k \leq d_k^{\text{T}} g_{k+1} \leq (2\rho - 1) d_k^{\text{T}} g_k, \tag{8.123}$$

where $0<\rho<1/2$ and $\rho<\sigma<1$, given by Hager and Zhang (2005). The algorithm (8.2), (8.122), (8.118) and (8.123) is called DK+aw (DK+ with approximate Wolfe line search). Now, as in Chapter 7, let us denote CG-DESCENTaw as the

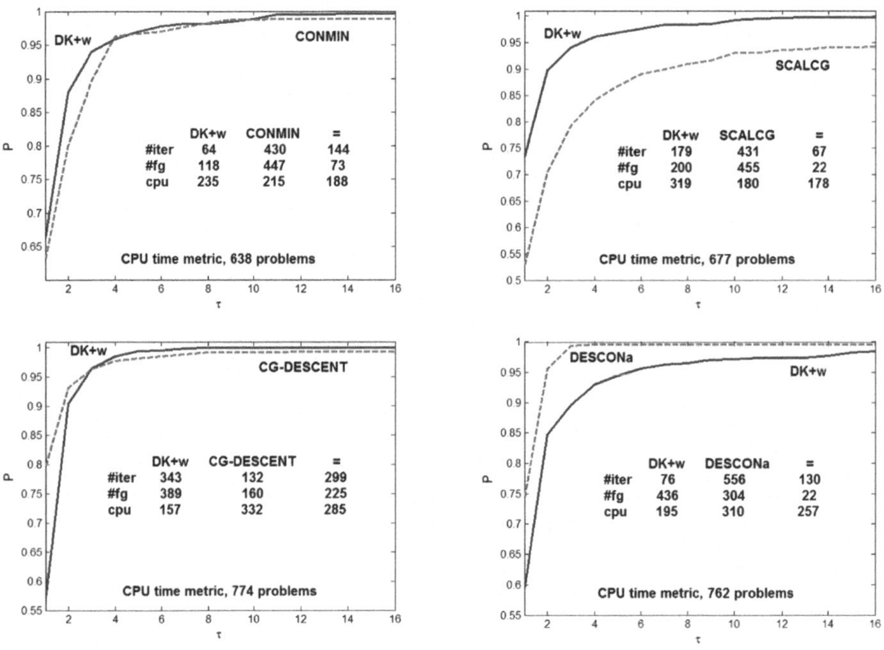

Figure 8.8 Performance profiles of DK+w versus CONMIN, SCALCG (spectral), CG-DESCENT and DESCONa

CG-DESCENT (version 1.4) algorithm with approximate Wolfe line search. Figure 8.9 shows the performance profiles of DK+aw versus CONMIN, SCALCG (spectral), CG-DESCENTaw and DESCONa.

Figure 8.9 shows that DK+aw is top performer versus CONMIN and SCALCG. CG-DESCENTaw is faster than DK+aw. Observe that DESCONa is top performer versus DK+aw.

Observe that the difference between the performances of DK+aw (DK+ with approximate Wolfe line search) versus CONMIN, SCALCG, CG-DESCENT, and DESCONa (see Figure 8.9) is greater than the difference between the performances of DK+w (DK+ with standard Wolfe line search) and the algorithms mentioned above (see Figure 8.8). These numerical experiments emphasize the importance of the line search in conjugate gradient methods. The conjugate gradient methods are very sensitive to the procedures for stepsize computation. This is in sharp contrast to the Newton or quasi-Newton methods, for which, for the vast majority of iterations, the stepsize is equal to 1.

It is worth saying that the CGOPT algorithm given by (8.119) and (8.120) is not obtained from a principle. Instead, it is derived from (8.115) using a particular value for the parameter τ_k obtained from trial and error. Besides, it is remarkable that the DK algorithm (8.117) is a simple modification with a constant coefficient of the second term of the Hager and Zhang algorithm. Another point which must be

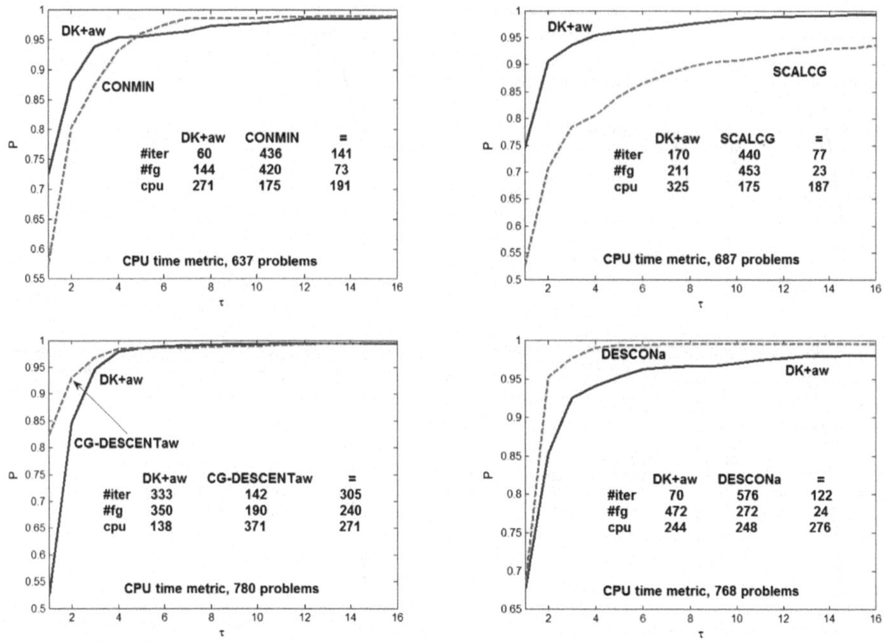

Figure 8.9 Performance profiles of DK+aw versus CONMIN, SCALCG (spectral). CG-DESCENTaw and DESCONa

emphasized is that the truncation procedure of β_k improves the performances of the algorithm.

As already mentioned, in their algorithm CGOPT, Dai and Kou (2013) introduced two important ingredients: the *dynamic restart strategy* and an *improved Wolfe line search*.

The dynamic restart strategy is based on measuring how the minimizing function is close to some quadratic function. The deviation from quadratic can be measured by using an approximation of the size of the third derivative, as given by Nash and Nocedal (1991). Instead, similarly to the ratio used for adjusting the trust-region radius, Dai and Kou (2013) evaluated the following quantity

$$r_k = \frac{2(f_{k+1} - f_k)}{\alpha_k (g_k^T d_k + g_{k+1}^T d_k)},$$

$k = 0, \ldots$. If r_k is close to 1, then it can be supposed that $\phi_k(\alpha) = f(x_k + \alpha d_k)$ is close to some quadratic function. If there are continuously a number of iterations so that r_k is close to 1, then the algorithm is restarted with the steepest descent direction.

On the other hand, the *improved Wolfe line search* consists in the modified first Wolfe condition as

$$\phi_k(\alpha) \leq \phi_k(0) + \min\{\varepsilon|\phi_k'(0)|, \rho\alpha\phi_k'(0) + \eta_k\}, \tag{8.124}$$

where $\phi_k(\alpha) = f(x_k + \alpha d_k)$, $\varepsilon > 0$ is a given constant and $\{\eta_k\}$ is a positive sequence satisfying $\sum_{k\geq 1}\eta_k < \infty$, as well as in the second Wolfe line search

$$\phi_k'(\alpha) \geq \sigma\phi_k'(0), \tag{8.125}$$

where $0 < \rho < \sigma < 1$.

Dai and Kou claimed that these two ingredients are important arguments for the efficiency and robustness of the conjugate gradient algorithms (8.117) or (8.118). They reported that the algorithm given by (8.117) or (8.118) with dynamic restart strategy and improved Wolfe line search for solving 105 problems from the CUTEr collection was more efficient and more robust than CG-DESCENT (version 5.3).

To see the performances of the DK+ algorithm with the improved Wolfe line search, in (Andrei, 2019a) we manufactured a new code implementing (8.122) with (8.118), where this time the stepsize is determined by the improved Wolfe line search (8.124) and (8.125), (without the dynamic restart strategy). The algorithm using (8.122), (8.118), (8.124) and (8.125) is called DK+iw (DK+ with improved Wolfe line search).

Figure 8.10 shows the performance profiles of DK+iw versus DK+w (DK+ with standard Wolfe line search) and versus DK+aw (DK+ with approximate Wolfe line search). The improved Wolfe line search (8.124) is implemented with $\varepsilon = 10^{-6}$ and $\eta_k = 1/(k^2)$, where k is the iteration number. Observe that DK+iw is more efficient and more robust than DK+w. In other words, in the context of the conjugate gradient algorithms, the improved Wolfe line search (8.124) represents an improvement versus the standard Wolfe line search given by (8.4). From Figure 8.10, we have computational evidence that the performances of DK+iw are better than those of DK+aw. DK+iw is more efficient and more robust than DK+aw.

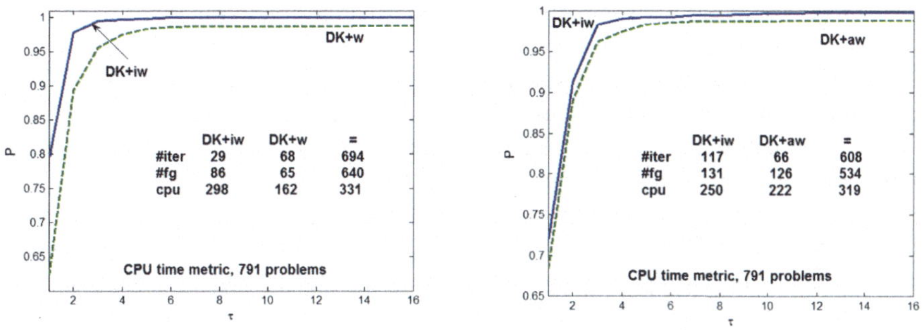

Figure 8.10 Performance profiles of DK+iw versus DK+w and versus DK+aw

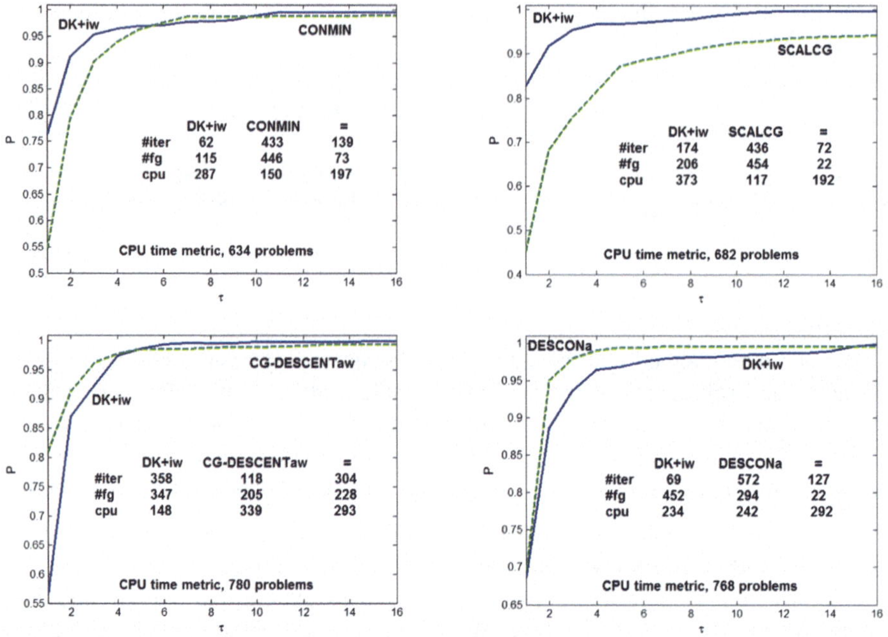

Figure 8.11 Performance profiles of DK+iw versus CONMIN, SCALCG (spectral). CG-DESCENTaw, and DESCONa

In Figure 8.11, the performance profiles of DK+iw versus CONMIN, SCALCG, CG-DESCENTaw, and DESCONa are presented. Observe that this variant of DK+ with improved Wolfe line search, without the dynamic restart strategy, is less efficient than CG-DESCENTaw (CG-DESCENT with approximate Wolfe line search). However, DK+iw is slightly more robust than CG-DESCENTaw. However, DESCONa, which is a conjugate gradient algorithm with guaranteed descent and conjugacy conditions and with a modified Wolfe line search, is top performer.

From Figures 8.9 and 8.11, observe that DK+aw and DK+iw have similar performances in comparison with CONMIN, SCALCG, CG-DESCENT, and DESCONa. Both the approximate Wolfe line search condition (8.123) and the improved Wolfe line search condition (8.124), (8.125) were introduced to avoid the numerical drawback of the first standard Wolfe condition responsible for the sufficient reduction of the objective function values. The second inequality in (8.123) is an approximation to the first Wolfe condition. When the iterates are near a local minimum, this approximation can be evaluated with greater accuracy than the original condition, since the approximate Wolfe conditions are expressed in terms of the derivative of function $\phi_k(\alpha) = f(x_k + \alpha d_k)$, not as the difference of function values which numerically may be rather unstable. On the other hand, since the improved Wolfe line search (8.124) is expressed as the difference of the function values $\phi_k(\alpha) - \phi_k(0)$, the numerical instabilities are not avoided. The extra positive

Table 8.3 Performances of DK+w and DK+aw for solving five applications from the MINPACK-2 collection

		DK+w			DK+aw		
	n	*#iter*	*#fg*	*cpu*	*#iter*	*#fg*	*cpu*
A1	40,000	406	637	7.92	323	647	7.17
A2	40,000	940	1508	20.54	791	1583	32.64
A3	40,000	4001	6260	182.63	987	1976	65.87
A4	40,000	670	1065	84.22	435	871	80.42
A5	40,000	417	654	8.80	289	580	9.54
Total	–	6434	10124	304.11	2825	5657	195.64

term η_k in (8.124), which allows a slight increase in the function value, compensates this deficiency. Observe that the improved Wolfe line search conditions (8.124) and (8.125) are weaker than the standard Wolfe line search (8.4) and (8.5). Anyway, all in all, DK+iw is faster than DK+aw.

The conclusion of these numerical experiments is that the performances of the conjugate gradient methods are dependent on the computation formula of the conjugate gradient parameter β_k, but they are strongly dependent on the procedure for the stepsize computation and possibly on some extra ingredients like: initial estimation of the stepsize, restart strategy, truncation of the parameter β_k, stopping criteria, etc. Anyway, the critical (weak) point in conjugate gradient methods is the stepsize computation.

In the following, let us present the performances of DK+w and DK+aw for solving the applications from the MINPACK-2 collection as in Table 8.3.

Table 8.3 points out that DK+ with approximate Wolfe line search (DK+aw) is top performer versus DK+ with Wolfe line search (DK+w).

Table 8.4 contains the total performances of L-BFGS ($m = 5$) (See Table 1.2), of TN (see Table 1.3), of CONMIN (see Table 8.1), of SCALCG (θ_k—anticipative), of SCALCG (θ_k—spectral) (see Table 8.2), of DK+w and DK+aw (see Table 8.3) for solving all five applications from the MINPACK-2 collection, each of them with 40,000 variables.

Table 8.4 The total performances of L-BFGS ($m = 5$), TN, CONMIN, SCALCG, DK+w and DK+aw for solving five applications from the MINPACK-2 collection with 40,000 variables

Algorithms	*#iter*	*#fg*	*cpu*
L-BFGS ($m = 5$)	4842	4987	102.92
TN	153	3714	104.57
CONMIN	3022	6117	120.17
SCALCG (spectral)	3821	4865	105.47
SCALCG (anticipative)	4090	5212	107.14
DK+w	6434	10124	304.11
DK+aw	2825	5657	195.64

From Table 8.4, by comparing these algorithms, the performances of SCALCG (θ_k—spectral) with 105.47 s and of SCALCG (θ_k—anticipative) with 107.14 s are comparable with those of L-BFGS and TN. Observe that DK+aw is 1.55 times faster than DK+w. Even if SCALCG proved to be less efficient than the compared algorithms for solving the problems from the UOP collection (see Figures 8.6 and 8.7), it proved to be the fastest for solving real large-scale applications. Chapter 12 shows that SCALCG it top performer versus DK+ and DK+aw for solving these applications, each of them with 250,000 variables.

8.4 New Conjugate Gradient Algorithms Based on Self-Scaling Memoryless BFGS Updating

The Dai–Kou family of the self-scaling memoryless BFGS quasi-Newton methods given by (8.114) with (8.115) depends on the scaling parameter τ_k. In this section, we present *three* different ways to choose the scaling parameter τ_k in (8.115). From the very beginning observe that H_{k+1} given by (8.104) satisfies the secant equation $H_{k+1}y_k = s_k$. Besides it is symmetric, positive definite and therefore it has n positive eigenvalues. If $y_k^T s_k > 0$, which is always satisfied when the stepsize is determined by the Wolfe line search, then there exists a set of mutually orthogonal unit vectors $\{u_k^i\}_{i=1}^{i=n-2}$ so that

$$s_k^T u_k^i = y_k^T u_k^i = 0, \quad i = 1,\ldots,n-2,$$

which leads to

$$H_{k+1}u_k^i = \frac{1}{\tau_k}u_k^i, \quad i = 1,\ldots,n-2.$$

Thus, the vectors u_k^i, $i = 1,\ldots,n-2$, are the eigenvectors of H_{k+1} corresponding to the eigenvalues $1/\tau_k$. Therefore, H_{k+1} given by (8.104) has $n-2$ eigenvalues all equal to $1/\tau_k$. Let $\lambda_k^{(n-1)}$ and $\lambda_k^{(n)}$ be the two remaining eigenvalues of H_{k+1}.

In a small neighborhood of the current point, the nonlinear objective function f from the unconstrained optimization problem (8.1) behaves like a quadratic one for which the results from the linear conjugate gradient can be applied. For faster convergence of linear conjugate gradient algorithms, some approaches can be considered as follows: the presence of isolated smallest and/or largest eigenvalues of the matrix H_{k+1} as well as gaps inside the eigenvalues spectrum (Axelsson and Lindskog, 1986), clustering of the eigenvalues about one point (Winther, 1980) or about several points (Kratzer, Parter, and Steuerwalt, 1983), or preconditioning (Kaporin, 1994). If the matrix has a number of certain distinct eigenvalues contained in m disjoint intervals of very small length, then the linear conjugate gradient method will produce a very small residual after m iterations (Luenberger, 1984). This is an important property of the linear conjugate gradient method and we try to extend it to the nonlinear case. Therefore, let us consider the extension of the method of clustering the eigenvalues of the matrix defining the search direction from the linear conjugate gradient algorithm to the nonlinear case. For this, we need to compute the determinant and the trace of the self-scaling memoryless BFGS matrix H_{k+1} given by (8.104). After some simple algebraic manipulation (see Appendix A) from (8.104), the following is obtained:

$$\det(H_{k+1}) = \frac{1}{(\tau_k)^{n-1}} \frac{\|s_k\|^2}{y_k^T s_k} \tag{8.126}$$

and

$$\operatorname{tr}(H_{k+1}) = \frac{n-2}{\tau_k} + \left(1 + \frac{1}{\tau_k} \frac{\|y_k\|^2}{y_k^T s_k}\right) \frac{\|s_k\|^2}{y_k^T s_k}. \tag{8.127}$$

The extension of the clustering of the eigenvalues from the linear case to the nonlinear one can be achieved in the following ways.

(1) The first extension of the clustering of the eigenvalues from the linear case to the nonlinear one is based on the determinant of the self-scaling memoryless BFGS matrix H_{k+1} given by (8.104). The idea of this variant of the self-scaling memoryless BFGS algorithm is to determine τ_k by clustering the eigenvalues of H_{k+1} in a point. Since H_{k+1} has $n-2$ eigenvalues all equal to $1/\tau_k$, then, imposing that the remaining eigenvalues have the same value, i.e., $\lambda_k^{(n-1)} = \lambda_k^{(n)} = 1/\tau_k$, from the equality

$$\det(H_{k+1}) = \frac{1}{(\tau_k)^n},$$

where $\det(H_{k+1})$ is given by (8.126), it results that

$$\tau_k = \frac{y_k^T s_k}{\|s_k\|^2}. \tag{8.128}$$

From (8.120) observe that this choice of τ_k, based on the determinant, is exactly the choice proposed by Dai and Kou in their CGOPT algorithm, using the Oren and Luenberger choice: $\frac{y_k^T s_k}{s_k^T B_k s_k}$, with $B_k = H_k^{-1}$ for the BFGS method, where H_k is the identity matrix. Substituting (8.128) into (8.115) leads to the conjugate gradient parameter

$$\beta_k^{DE} = \frac{g_{k+1}^T y_k}{y_k^T d_k} - \frac{\|y_k\|^2}{y_k^T s_k} \frac{g_{k+1}^T s_k}{y_k^T d_k} \tag{8.129}$$

and from (8.116), the truncated form is obtained as

$$\beta_k^{DE+} = \max\left\{ \beta_k^{DE}, \eta \frac{g_{k+1}^T d_k}{\|d_k\|^2} \right\}, \tag{8.130}$$

based on the determinant. *Observe that* $\beta_k^{DE} = \beta_k^{CGOPT}$, *i.e., the algorithm obtained by clustering the eigenvalues of the iteration matrix* H_{k+1} *based on the determinant of* H_{k+1} *is exactly the CGOPT algorithm of Dai and Kou*. The formula (8.129) differs from (8.109) only in a constant coefficient in the second term of the Hager and Zhang method.

In the following, let us prove that for strongly convex functions f the search directions (8.103) and (8.104), where τ_k is computed as in (8.128), satisfy the sufficient descent condition $g_{k+1}^T d_{k+1} \leq -c\|g_{k+1}\|^2$ for any $k \geq 0$, where c is a positive constant.

Theorem 8.4 *Suppose that the Assumption CG holds. For the method (8.2), (8.103) and (8.104), if f is a strongly convex function on the level set $S = \{x \in \mathbb{R}^n : f(x) \leq f(x_0)\}$ and the stepsize α_k is determined by the Wolfe line search (8.4) and (8.5), then the search directions (8.103) and (8.104), where the parameter τ_k is computed as in (8.128), satisfy the sufficient descent condition $g_{k+1}^T d_{k+1} \leq -c\|g_{k+1}\|^2$ for any $k \geq 0$, where c is a positive constant.*

Proof The matrix H_{k+1} given by (8.104) has $n-2$ eigenvalues all equal to $1/\tau_k$, as well as $\lambda_k^{(n-1)}$ and $\lambda_k^{(n)}$. Since $\mathrm{tr}(H_{k+1})$ is equal to the summation of the eigenvalues of H_{k+1} and $\det(H_{k+1})$ is equal to the product of them (see Appendix A), from (8.127) and (8.126) it is easy to see that

$$\lambda_k^{(n-1)} + \lambda_k^{(n)} = \left(1 + \frac{1}{\tau_k} \frac{\|y_k\|^2}{y_k^T s_k}\right) \frac{\|s_k\|^2}{y_k^T s_k} \tag{8.131}$$

and

$$\lambda_k^{(n-1)} \lambda_k^{(n)} = \frac{1}{\tau_k} \frac{\|s_k\|^2}{y_k^T s_k}. \tag{8.132}$$

Now, from the Assumption CG, $\|y_k\| \leq L\|s_k\|$. On the other hand, from the strong convexity of function f on S it follows that $y_k^T s_k \geq \mu\|s_k\|^2$. Assume that $\lambda_k^{(n)} \leq \lambda_k^{(n-1)}$. With this, from (8.132) and (8.131) it results that

$$\lambda_k^{(n)} = \frac{1}{\tau_k} \frac{\|s_k\|^2}{y_k^T s_k} \frac{1}{\lambda_k^{(n-1)}} \geq \frac{1}{\tau_k} \frac{\|s_k\|^2}{y_k^T s_k} \frac{1}{(\lambda_k^{(n-1)} + \lambda_k^{(n)})} = \frac{y_k^T s_k}{\tau_k (y_k^T s_k) + \|y_k\|^2}. \tag{8.133}$$

But, from (8.128)

$$\tau_k (y_k^T s_k) = \frac{(y_k^T s_k)^2}{\|s_k\|^2} \leq L^2 \|s_k\|^2.$$

Therefore,

$$\lambda_k^{(n)} \geq \frac{\mu \|s_k\|^2}{L^2 \|s_k\|^2 + L^2 \|s_k\|^2} = \frac{\mu}{2L^2}. \tag{8.134}$$

Now, from (8.103) and (8.134), for all $k \geq 0$, it results that

$$d_{k+1}^T g_{k+1} = -g_{k+1}^T H_{k+1} g_{k+1} \leq -\lambda_k^{(n)} \|g_{k+1}\|^2 \leq -\frac{\mu}{2L^2} \|g_{k+1}\|^2,$$

i.e., the search direction (8.103), where τ_k is determined as in (8.128), satisfies the sufficient descent condition $g_{k+1}^T d_{k+1} \leq -c \|g_{k+1}\|^2$ with $c = \mu/(2L^2)$. ♦

(2) The second extension of the clustering of the eigenvalues from the linear case to the nonlinear one is based on the trace of the self-scaling memoryless BFGS matrix H_{k+1}. Again, the idea of this variant of the self-scaling memoryless BFGS algorithm is to determine τ_k by clustering the eigenvalues of H_{k+1} in a point, but this time using the trace of H_{k+1}. Since H_{k+1} has $n-2$ eigenvalues all equal to $1/\tau_k$, then, imposing that the remaining eigenvalues have the same value, i.e., $\lambda_k^{(n-1)} = \lambda_k^{(n)} = 1/\tau_k$, from the equality

$$\mathrm{tr}(H_{k+1}) = \frac{n}{\tau_k}$$

where $\mathrm{tr}(H_{k+1})$ is given by (8.127), it results that

$$\tau_k = \left(2 - \frac{\|y_k\|^2 \|s_k\|^2}{(y_k^T s_k)^2} \right) \frac{y_k^T s_k}{\|s_k\|^2}. \tag{8.135}$$

Now, substituting (8.135) into (8.115) leads to the conjugate gradient parameter

$$\beta_k^{TR} = \frac{g_{k+1}^T y_k}{y_k^T d_k} - \frac{y_k^T s_k}{\|s_k\|^2} \frac{g_{k+1}^T s_k}{y_k^T d_k}, \tag{8.136}$$

and from (8.116), the truncated form is obtained as

$$\beta_k^{TR+} = \max\left\{\beta_k^{TR}, \eta \frac{g_{k+1}^T d_k}{\|d_k\|^2}\right\}, \tag{8.137}$$

based on the trace of H_{k+1}.

Theorem 8.5 *Suppose that the Assumption CG holds. For the method (8.2), (8.103) and (8.104), if f is a strongly convex function on the level set $S = \{x \in \mathbb{R}^n : f(x) \le f(x_0)\}$ and the stepsize α_k is determined by the Wolfe line search (8.4) and (8.5), then the search directions (8.103) and (8.104), where the parameter τ_k is computed as in (8.135), satisfy the sufficient descent condition $g_{k+1}^T d_{k+1} \le -c\|g_{k+1}\|^2$ for any $k \ge 0$, where c is a positive constant.*

Proof The scaling parameter τ_k from (8.135) can be written as

$$\tau_k = 2\frac{y_k^T s_k}{\|s_k\|^2} - \frac{\|y_k\|^2}{y_k^T s_k}.$$

From (8.133) it is easy to see that

$$\lambda_k^{(n)} \ge \frac{\mu\|s_k\|^2}{3L^2\|s_k\|^2 + L^2\|s_k\|^2} = \frac{\mu}{4L^2}. \tag{8.138}$$

Now, from (8.103) and (8.138), for all $k \ge 0$, it follows that

$$d_{k+1}^T g_{k+1} = -g_{k+1}^T H_{k+1} g_{k+1} \le -\lambda_k^{(n)}\|g_{k+1}\|^2 \le -\frac{\mu}{4L^2}\|g_{k+1}\|^2,$$

i.e., the search directions (8.103), where τ_k is determined as in (8.135), satisfy the sufficient descent condition $g_{k+1}^T d_{k+1} \le -c\|g_{k+1}\|^2$ with $c = \mu/(4L^2)$. ♦

(3) Another possibility to determine a value for the scaling parameter τ_k in the self-scaling memoryless BFGS method considered in this section is to minimize a combination of the determinant and the trace of the iteration matrix H_{k+1} given by (8.104). Byrd and Nocedal (1989) introduced such a combination of $\det(H_{k+1})$ and $\operatorname{tr}(H_{k+1})$ as the function

$$\varphi(H_{k+1}) = \mathrm{tr}(H_{k+1}) - \ln(\det(H_{k+1})), \tag{8.139}$$

where $\ln(.)$ denotes the natural logarithm, known as the measure function. Since H_{k+1} is positive definite, it follows that $\varphi(H_{k+1})$ is well defined. Fletcher (1991) observed that the BFGS formula can be derived by a variational argument using function φ. This is an elegant and efficient tool for analyzing the global behavior of quasi-Newton methods and now we intend to use it to generate new and efficient algorithms for the unconstrained optimization. Observe that function φ works simultaneously with trace and determinant, thus simplifying the analysis of the quasi-Newton methods. In fact, this function is a measure of matrices involving all the eigenvalues of the iteration matrix, not only the smallest and the largest as it is traditionally used in the analysis of the quasi-Newton methods based on the condition number of matrices (see for example: Andrei (2015a, 2016, 2017a), Babaie-Kafaki (2015, 2016)). Observe that this function is strictly convex on the set of symmetric and positive definite matrices and it is minimized by $H_{k+1} = I$. Besides, it becomes unbounded as H_{k+1} becomes singular or infinite, and therefore, it works as a barrier function that keeps H_{k+1} positive definite.

Therefore, the idea of this variant of the self-scaling memoryless BFGS algorithm is to determine τ_k by minimizing the measure function $\varphi(H_{k+1})$ of Byrd and Nocedal, defined in (8.139). From (8.126) and (8.127),

$$\varphi(H_{k+1}) = \frac{n-2}{\tau_k} + \frac{\|s_k\|^2}{y_k^T s_k} + \frac{1}{\tau_k}\frac{\|y_k\|^2\|s_k\|^2}{(y_k^T s_k)^2} + (n-1)\ln(\tau_k) - \ln\left(\frac{\|s_k\|^2}{y_k^T s_k}\right). \tag{8.140}$$

Obviously, the solution of the problem $\min\varphi(H_{k+1})$, where $\varphi(H_{k+1})$ is given by (8.140), is the solution of the equation

$$\frac{d\varphi(H_{k+1})}{d\tau_k} = -\frac{n-2}{\tau_k^2} - \frac{1}{\tau_k^2}\frac{\|y_k\|^2\|s_k\|^2}{(y_k^T s_k)^2} + (n-1)\frac{1}{\tau_k} = 0,$$

which is

$$\tau_k = \frac{n-2}{n-1} + \frac{1}{n-1}\frac{\|y_k\|^2\|s_k\|^2}{(y_k^T s_k)^2} > 0. \tag{8.141}$$

The measure function $\varphi(H_{k+1})$ of Byrd and Nocedal is a special combination of the determinant and of the trace of the iteration matrix H_{k+1}. Observe that the determinant in $\varphi(H_{k+1})$ is under the natural logarithm. Since H_{k+1} is positive definite, it follows that $\mathrm{tr}(H_{k+1}) > 0$. It is quite possible that along the iterations, $\ln(\det(H_{k+1})) < 0$. This is more harmful for the minimization of $\varphi(H_{k+1})$. Therefore, a variant of the algorithm based on minimizing the measure function

$\varphi(H_{k+1})$ is to update the value of the parameter τ_k by using (8.141) only when $\det(H_{k+1}) > 1$, otherwise τ_k remains to be updated, for example, by (8.135).

Hence, to minimize the measure function $\varphi(H_{k+1})$, the following procedure is considered. From (8.126), compute $\det(H_{k+1})$ with τ_k given by (8.135). With this value of $\det(H_{k+1})$, compute

$$\bar{\tau}_k = \begin{cases} \dfrac{y_k^T s_k}{\|s_k\|^2}, & \text{if} \quad \det(H_{k+1}) \le 1, \\ \dfrac{n-2}{n-1} + \dfrac{1}{n-1} \dfrac{\|y_k\|^2 \|s_k\|^2}{(y_k^T s_k)^2} & \text{if} \quad \det(H_{k+1}) > 1. \end{cases} \tag{8.142}$$

Now, considering $\tau_k = \bar{\tau}_k$ into (8.115), it results that

$$\beta_k^{FI} = \frac{g_{k+1}^T y_k}{y_k^T d_k} - \left(\bar{\tau}_k + \frac{\|y_k\|^2}{y_k^T s_k} - \frac{s_k^T y_k}{\|s_k\|^2} \right) \frac{g_{k+1}^T s_k}{y_k^T d_k} \tag{8.143}$$

and its truncated value

$$\beta_k^{FI+} = \max\left\{ \beta_k^{FI}, \eta \frac{g_{k+1}^T d_k}{\|d_k\|^2} \right\}, \tag{8.144}$$

based on minimizing the measure function of Byrd and Nocedal. Besides, taking into account the insights gained from the example given by Powell (1984a), β_k^{FI+} parameter is further constrained to be positive, i.e.,

$$\beta_k^{FI+} = \max\{\beta_k^{FI+}, 0\}. \tag{8.145}$$

Theorem 8.6 *Suppose that the Assumption CG holds. For the method* (8.2), (8.103) *and* (8.104), *if f is a strongly convex function on the level set $S = \{x \in \mathbb{R}^n : f(x) \le f(x_0)\}$ and the stepsize α_k is determined by the Wolfe line search* (8.4) *and* (8.5), *then the search directions* (8.103) *and* (8.104), *where the parameter $\tau_k = \bar{\tau}_k$ is computed as in* (8.142), *satisfy the sufficient descent condition $g_{k+1}^T d_{k+1} \le -c\|g_{k+1}\|^2$ for any $k \ge 0$, where c is a positive constant.*

Proof Having in view that $\|y_k\| \le L\|s_k\|$ and $y_k^T s_k \ge \mu\|s_k\|^2$, following the same procedure as in the previous theorems, from (8.141) the quantity $\tau_k(y_k^T s_k)$ in (8.133) can be estimated as

$$\begin{aligned} \tau_k(y_k^T s_k) &= \frac{n-2}{n-1}(y_k^T s_k) + \frac{1}{n-1} \frac{\|y_k\|^2 \|s_k\|^2}{y_k^T s_k} \le y_k^T s_k + \frac{\|y_k\|^2 \|s_k\|^2}{y_k^T s_k} \\ &\le \|y_k\| \|s_k\| + \frac{\|y_k\|^2 \|s_k\|^2}{\mu\|s_k\|^2} \le \left(L + \frac{L^2}{\mu} \right) \|s_k\|^2. \end{aligned} \tag{8.146}$$

Therefore, from (8.133), using (8.146) it follows that

$$\lambda_k^{(n)} \geq \frac{y_k^T s_k}{\tau_k (y_k^T s_k) + \|y_k\|^2} \geq \frac{\mu^2}{L^2 + \mu(L + L^2)}. \tag{8.147}$$

Now, from (8.103) and (8.147), for all $k \geq 0$,

$$d_{k+1}^T g_{k+1} = -g_{k+1}^T H_{k+1} g_{k+1} \leq -\lambda_k^{(n)} \|g_{k+1}\|^2 \leq -\frac{\mu^2}{L^2 + \mu(L + L^2)} \|g_{k+1}\|^2,$$

i.e., the search directions (8.103) where $\tau_k = \bar{\tau}_k$ is determined as in (8.142) satisfy the sufficient descent condition $g_{k+1}^T d_{k+1} \leq -c\|g_{k+1}\|^2$ with $c = \mu^2/[L^2 + \mu(L + L^2)]$. ♦

For *general nonlinear functions* to establish the sufficient descent condition for the family of conjugate gradient methods (8.2), where the search direction is given by (8.114) and (8.115), let us define:

$$p_k = \frac{\|d_k\|^2 \|y_k\|^2}{(d_k^T y_k)^2} \text{ and } \gamma_k = \tau_k \frac{\|s_k\|^2}{y_k^T s_k},$$

where τ_k is given as in (8.128), (8.135), or (8.141). Observe that $p_k \geq 1$. Then, for general nonlinear functions like in (Dai and Kou 2013) (see Lemma 2.1), we can prove that if $d_k^T y_k \neq 0$, then the search direction computed as in (8.114) with (8.115), satisfies the sufficient descent condition, i.e.,

$$-d_{k+1}^T g_{k+1} \geq \min\left\{\gamma_k, \frac{3}{4}\right\} \|g_{k+1}\|^2. \tag{8.148}$$

Theorem 8.7 *Suppose that the Assumption CG holds. Consider the family of conjugate gradient methods given by (8.2), where the search direction is computed as in (8.114) with (8.115). Suppose that for any $k \geq 0$, $d_k^T y_k \neq 0$. If τ_k in (8.115) is selected to be computed as in (8.128), or as in (8.135) with $p_k \leq 2$, or as in (8.141), then there is a positive constant c such that*

$$-d_{k+1}^T g_{k+1} \geq c\|g_{k+1}\|^2. \tag{8.149}$$

Proof If τ_k is chosen as in (8.128), then $\gamma_k = 1$. Therefore, $\min\{\gamma_k, 3/4\} = 3/4$, i.e., $c = 3/4$ in (8.149). If τ_k is chosen as in (8.135), then $\gamma_k = 2 - p_k$, where $1 \leq p_k \leq 2$. Obviously, for $1.25 \leq p_k \leq 2$, $\min\{\gamma_k, 3/4\} = \gamma_k$, where $0 \leq \gamma_k \leq 3/4$. On the other hand, for $1 \leq p_k \leq 1.25$, $\min\{\gamma_k, 3/4\} = 3/4$. Therefore, there is a positive constant c such that (8.149) holds. If τ_k is chosen as in (8.141), then since $p_k \geq 1$ then it is easy to see that

$$\gamma_k = \frac{n-2}{n-1}\frac{\|s_k\|^2}{y_k^T s_k} + \frac{1}{n-1}p_k\frac{\|s_k\|^2}{y_k^T s_k} \geq \frac{1}{n-1}\frac{\|s_k\|^2}{y_k^T s_k}(1+p_k)$$
$$\geq \frac{1}{n}\frac{\|s_k\|^2}{\|y_k\|\|s_k\|}(1+p_k) \geq \frac{2}{nL},$$

i.e., $\min\{\gamma_k, 3/4\} \geq \min\{2/(nL), 3/4\}$, where $L > 0$ is the Lipschitz constant of the gradient and we supposed that the dimension of the problem $n > 2$. Therefore, there is a positive constant c such that (8.149) holds. ♦

With these developments the following general self-scaling memoryless BFGS quasi-Newton algorithm may be presented.

Algorithm 8.3 CGSSML—conjugate gradient self-scaling memoryless BFGS

1.	Initialization. Choose an initial point $x_0 \in \mathbb{R}^n$. Choose the constants σ, ρ with $0<\rho<\sigma<1$ and $\varepsilon > 0$ sufficiently small. Compute $g_0 = \nabla f(x_0)$. Set $d_0 = -g_0$ and $k = 0$
2.	Test a criterion for stopping the iterations. If this test is satisfied, then stop the iterations
3.	Compute the stepsize $\alpha_k > 0$ using the Wolfe line search conditions, or some variants of them (approximate or improved)
4.	Update the variables $x_{k+1} = x_k + \alpha_k d_k$ and compute f_{k+1} and g_{k+1}
5.	Compute the scaling parameter τ_k using clustering the eigenvalues of the iteration matrix, or by minimizing the measure function of Byrd and Nocedal
6.	Compute the parameter β_k according the values of parameter τ_k
7.	Update the search direction $d_{k+1} = -g_{k+1} + \beta_k d_k$
8.	Restart criterion. If $\lvert g_{k+1}^T g_k \rvert > 0.2\|g_{k+1}\|^2$ then set $d_{k+1} = -g_{k+1}$
9.	Set $k = k+1$ and go to step 2 ♦

For computing the stepsize α_k in step 3 of the algorithm, the Wolfe line search (8.4) and (8.5) or the approximate Wolfe line search (8.123) of Hager and Zhang (2005, 2006a), or the improved Wolfe line search (8.124) and (8.125) of Dai and Kou (2013) may be implemented. Observe that in step 5 the parameter τ_k may be computed using the clustering of the eigenvalues of H_{k+1} by the determinant of H_{k+1} (8.128) or by the trace of H_{k+1} (8.135) or by minimizing the measure function of Byrd and Nocedal (8.142). In our algorithm, when the Powell restart condition is satisfied (step 8), then the algorithm is restarted with the negative gradient $-g_{k+1}$. Some other restarting procedures may be implemented in CGSSML, like $d_{k+1}^T g_{k+1} \geq 10^{-3}\|d_{k+1}\|\|g_{k+1}\|$ of Birgin and Martínez (2001) or the adaptive restarting strategy of Dai and Kou (2013), but we are interested in seeing the performances of CGSSML implementing the Powell restarting technique. Of course, the acceleration scheme of the conjugate gradient algorithms may be introduced after step 3 of CGSSML.

Convergence of CGSSML for strongly convex functions. For strongly convex functions, the convergence of CGSSML follows from the Assumption CG.

Proposition 8.8 *Suppose that the Assumption CG holds. Then for τ_k chosen as in (8.128) we have $\tau_k \le L$.*

Proof From the Lipschitz continuity of the gradient it follows that $\|y_k\| \le L\|s_k\|$. Therefore, using the Cauchy–Schwarz inequality in (8.128) we have

$$|\tau_k| = \frac{|y_k^T s_k|}{\|s_k\|^2} \le \frac{L\|s_k\|^2}{\|s_k\|^2} = L. \tag{8.150}$$

♦

Proposition 8.9 *Suppose that the Assumption CG holds. Then for τ_k chosen as in (8.135) we have $\tau_k \le L^3/(\mu^2)$.*

Proof Notice that $(y_k^T s_k)^2 \le \|y_k\|^2\|s_k\|^2$. Again from the strong convexity of function f, from Lipschitz continuity of the gradient and using the Cauchy–Schwarz inequality in (8.135) we have

$$|\tau_k| = \left|2 - \frac{\|y_k\|^2\|s_k\|^2}{(y_k^T s_k)^2}\right| \frac{|y_k^T s_k|}{\|s_k\|^2} \le \frac{\|y_k\|^2\|s_k\|^2}{(y_k^T s_k)^2} L \le \frac{\|y_k\|^2\|s_k\|^2}{\mu^2\|s_k\|^4} L \le \frac{L^3}{\mu^2}. \tag{8.151}$$

♦

Proposition 8.10 *Suppose that the Assumption CG holds. Then for τ_k chosen as in (1.141) we have $\tau_k \le 1 + L^2/(\mu^2)$.*

Proof As above we have

$$|\tau_k| = \left|\frac{n-2}{n-1} + \frac{1}{n-1}\frac{\|y_k\|^2\|s_k\|^2}{(y_k^T s_k)^2}\right| \le 1 + \frac{\|y_k\|^2\|s_k\|^2}{(y_k^T s_k)^2} \le 1 + \frac{L^2}{\mu^2}. \tag{8.152}$$

♦

For strongly convex functions, the following theorem prove the global convergence of the algorithm (8.2), (8.114) and (8.115), where the scaling parameter τ_k is chosen as in (8.128), (8.135) or (8.141), under the Wolfe line search.

Theorem 8.8 *Suppose that the Assumption CG holds. Consider the algorithm (8.2) in which the search direction is defined by (8.114) and (8.115), where τ_k is chosen to be as in (8.128), (8.135) or (8.141) and the stepsize α_k is determined by the Wolfe line search (8.4) and (8.5). If the function f is strongly convex, then the algorithm CGSSML is global convergent, i.e., $\lim_{k\to\infty}\|g_k\| = 0$.*

Proof From the Assumption CG and the strong convexity of function f it follows that $\|y_k\| \le L\|s_k\|$ and $y_k^T s_k \ge \mu\|s_k\|^2$. Therefore, from Propositions 8.8–8.10, for any τ_k given by (8.128), (8.135) or (8.141), there exists a positive constant c_τ such that $|\tau_k| \le c_\tau$. Now, from (8.114) and (8.115) it follows that

$$\begin{aligned}\|d_{k+1}\| &\le \|g_{k+1}\| + \frac{|g_{k+1}^T y_k|}{|y_k^T d_k|}\|d_k\| + \left|\tau_k + \frac{\|y_k\|^2}{y_k^T s_k} + \frac{s_k^T y_k}{\|s_k\|^2}\right| \frac{|g_{k+1}^T s_k|}{|y_k^T d_k|}\|d_k\| \\ &\le \|g_{k+1}\| + \frac{\|g_{k+1}\|L\|s_k\|}{\mu\|s_k\|^2}\|s_k\| + \left(c_\tau + \frac{L^2\|s_k\|^2}{\mu\|s_k\|^2} + \frac{L\|s_k\|^2}{\|s_k\|^2}\right)\frac{\|g_{k+1}\|\|s_k\|}{\mu\|s_k\|^2}\|s_k\| \\ &\le \left(1 + \frac{L^2 + 2\mu L + \mu c_\tau}{\mu^2}\right)\|g_{k+1}\|.\end{aligned} \tag{8.153}$$

On the other hand, since from the Theorems 8.4–8.6 for any τ_k given by (8.128), (8.135) or (8.141) the search direction (8.114) and (8.115) satisfies the sufficient descent condition, it follows that

$$\sum_{k\ge 1} \frac{\|g_k\|^4}{\|d_k\|^2} < \infty. \tag{8.154}$$

From (8.153) we see that the sequence $\{\|d_k\|/\|g_k\|\}$ is bounded. Hence by (8.154) we get

$$\sum_{k\ge 1} \|g_k\|^2 < \infty,$$

which implies that $\lim_{k\to\infty}\|g_k\| = 0$. ♦

Convergence of CGSSML for general nonlinear functions.

For general nonlinear functions, the global convergence of the algorithm (8.2) with (8.114) and (8.115), where the scaling parameter τ_k is chosen as in (8.128), (8.135) or (8.141) under the Wolfe line search follows the methodology given by Dai and Kou (2013) and by Gilbert and Nocedal (1992).

Proposition 8.11 *Suppose that the Assumption CG holds. Consider the family of conjugate gradient algorithms given by* (8.2) *in which the search direction d_{k+1} is computed as in* (8.114) *and* (8.116) *and the stepsize α_k is determined by the Wolfe line search* (8.4) *and* (8.5). *If $\|g_k\| \ge \gamma$ for any $k \ge 1$, then $d_k \ne 0$ and*

$$\sum_{k\ge 2} \|u_k - u_{k-1}\|^2 < \infty, \tag{8.155}$$

where $u_k = d_k/\|d_k\|$.

Proof Observe that $d_k \neq 0$, since otherwise the sufficient descent condition $g_k^T d_k \leq -c\|g_k\|^2$ would imply $g_k = 0$. Hence, u_k is well defined. Formula (8.116) for β_k^{DK+} can be expressed as $\beta_k^{DK+} = \beta_k^{(1)} + \beta_k^{(2)}$, where

$$\beta_k^{(1)} = \max\left\{\frac{g_{k+1}^T y_k}{d_k^T y_k} - \left(1 + \tau_k \frac{y_k^T s_k}{\|y_k\|^2}\right)\frac{\|y_k\|^2}{d_k^T y_k}\frac{g_{k+1}^T d_k}{d_k^T y_k} + (1-\eta)\frac{g_{k+1}^T d_k}{\|d_k\|^2}, 0\right\}, \tag{8.156}$$

$$\beta_k^{(2)} = \eta \frac{g_{k+1}^T d_k}{\|d_k\|^2}. \tag{8.157}$$

Now, let us define:

$$w_k = \frac{-g_{k+1} + \beta_k^{(2)} d_k}{\|d_{k+1}\|} \text{ and } \delta_k = \frac{\beta_k^{(1)}\|d_k\|}{\|d_{k+1}\|}. \tag{8.158}$$

Since $d_{k+1} = -g_{k+1} + \beta_k d_k$, for $k \geq 1$, it follows that

$$u_{k+1} = w_k + \delta_k u_k. \tag{8.159}$$

But, using the identity $\|u_k\| = \|u_{k+1}\| = 1$, we get that

$$\|w_k\| = \|u_{k+1} - \delta_k u_k\| = \|\delta_k u_{k+1} - u_k\|. \tag{8.160}$$

Using the triangle inequality and since $\delta_k \geq 0$ from (8.160), we have

$$\begin{aligned} \|u_{k+1} - u_k\| &\leq \|(1+\delta_k)u_{k+1} - (1+\delta_k)u_k\| \\ &\leq \|u_{k+1} - \delta_k u_k\| + \|\delta_k u_{k+1} - u_k\| \\ &= 2\|w_k\|. \end{aligned} \tag{8.161}$$

But

$$\left\|-g_{k+1} + \beta_k^{(2)} d_k\right\| \leq \|g_{k+1}\| + \left|\beta_k^{(2)}\right|\|d_k\| \leq (1+\eta)\|g_{k+1}\|. \tag{8.162}$$

Therefore, from definition of $\beta_k^{(2)}$, the bound of the numerator of w_k given by (8.162), using (8.161) it follows that

$$\|u_{k+1} - u_k\| \leq 2\|w_k\| \leq 2(1+\eta)\frac{\|g_{k+1}\|}{\|d_{k+1}\|}. \tag{8.163}$$

Since $\|g_{k+1}\| \geq \gamma$, the sufficient descent condition $-(g_{k+1}^T d_{k+1}) \geq c\|g_{k+1}\|^2$, where c is a constant and the Zoutendijk condition (3.34) it follows that

$$\sum_{k\geq 0}\frac{\|g_{k+1}\|^2}{\|d_{k+1}\|^2}\leq\frac{1}{\gamma^2}\sum_{k\geq 0}\frac{\|g_{k+1}\|^4}{\|d_{k+1}\|^2}\leq\frac{1}{\gamma^2 c^2}\sum_{k\geq 0}\frac{(g_{k+1}^T d_{k+1})^2}{\|d_{k+1}\|^2}<+\infty. \tag{8.164}$$

Therefore, (8.155) follows from (8.163) and (8.164). ♦

This result, similar to Lemma 4.1 in (Gilbert and Nocedal, 1992), is used for proving the global convergence of the CGSSML algorithm with Wolfe line search. For this in the following proposition we prove that $\beta_k(\tau_k)$ in (8.115) has Property (*) defined by Gilbert and Nocedal (see also (Dai, 2010)).

Proposition 8.12 *Suppose that the Assumption CG holds. Consider the family of conjugate gradient algorithms given by* (8.2) *in which the search direction d_{k+1} is computed as in* (8.114) *and* (8.115) *and the stepsize α_k is determined by the Wolfe line search* (8.4) *and* (8.5)*. If the sequence $\{x_k\}$ generated by the algorithm CGSSML is bounded and if τ_k is chosen as in* (8.128)*,* (8.135) *or* (8.141)*, then $\beta_k(\tau_k)$ in* (8.115) *has Property (*).*

Proof The proof follows by contradiction, i.e., suppose that $\|g_k\|\geq\gamma$ for any $k\geq 1$. From continuity of the gradient and the boundedness of $\{x_k\}$ it follows that there exists a positive constant $\bar{\gamma}$ such that

$$\|x_k\|\leq\bar{\gamma},\ \|g_k\|\leq\bar{\gamma}, \text{for any } k\geq 1. \tag{8.165}$$

From (8.5) it follows that

$$g_{k+1}^T d_k\geq\sigma g_k^T d_k. \tag{8.166}$$

From Theorems 8.4–8.6 it follows that for any values of τ_k given by (8.128), or (8.135), or (8.141) we have $g_k^T d_k\leq -c\|g_k\|^2$, where c is a positive constant. Therefore, from (8.166) we get

$$d_k^T y_k = d_k^T g_{k+1} - d_k^T g_k\geq -(1-\sigma)d_k^T g_k\geq c(1-\sigma)\gamma^2. \tag{8.167}$$

Now, from (8.166) and since $g_k^T d_k<0$, it follows that

$$\frac{\sigma}{\sigma-1}\leq\frac{g_{k+1}^T d_k}{d_k^T y_k}\leq 1. \tag{8.168}$$

Since the sequence $\{x_k\}$ generated by the algorithm is bounded, it is easy to see that any values of τ_k given by (8.128), or (1.135), or (8.141) are bounded by a constant c_τ. Therefore,

$$\left|\tau_k(y_k^T s_k)\right|\leq c_\tau\|s_k\|^2, \text{ for any } k\geq 1. \tag{8.169}$$

Observe that $\beta_k(\tau_k)$ from (8.115) can be written as

$$\beta_k(\tau_k) = \frac{g_{k+1}^T y_k}{d_k^T y_k} - \left(1 - \frac{(d_k^T y_k)^2}{\|d_k\|^2 \|y_k\|^2}\right) \frac{\|y_k\|^2}{d_k^T y_k} \frac{g_{k+1}^T d_k}{d_k^T y_k} - \frac{\tau_k (y_k^T s_k)}{d_k^T y_k} \frac{g_{k+1}^T d_k}{d_k^T y_k}. \quad (8.170)$$

Observe that $\|y_k\| \leq L\|s_k\|$ and $0 \leq (d_k^T y_k)^2 \leq \|d_k\|^2 \|y_k\|^2$, for any $k \geq 1$. Since by (8.165), $\|s_k\| = \|x_{k+1} - x_k\| \leq \|x_{k+1}\| + \|x_k\| \leq 2\bar{\gamma}$, using (8.167), (8.169), and (8.170), we get that there exists a constant $c_\beta > 0$ such that for any $k \geq 1$,

$$|\beta_k(\tau_k)| \leq c_\beta \|s_k\|. \quad (8.171)$$

Now, like in (Gilbert and Nocedal, 1992) define $b = 2c_\beta \bar{\gamma}$ and $\lambda = 1/(2c_\beta^2 \bar{\gamma})$. From (8.171) and (8.165), it follows that for all $k \geq 1$ we have that

$$|\beta_k(\tau_k)| \leq b, \quad (8.172)$$

and

$$\|s_k\| \leq \lambda \Rightarrow |\beta_k(\tau_k)| \leq \frac{1}{b}. \quad (8.173)$$

Therefore, (8.172) and (8.173) show that $\beta_k(\tau_k)$ defined by (8.115) has Property (*). ♦

Theorem 8.9 *Suppose that the Assumption CG holds. Consider the algorithm (8.2) in which the search direction is defined by (8.114) and (8.115), where τ_k is chosen to be as in (8.128), (8.135) or (8.141) and the stepsize α_k is determined by the Wolfe line search (8.4) and (8.5). If the sequence $\{x_k\}$ generated by the algorithm CGSSML is bounded, then the algorithm is global convergent, i.e.,* $\liminf_{k\to\infty} \|g_k\| = 0$.

Proof By contradiction suppose that $\|g_k\| \geq \gamma$ for any $k \geq 1$. Since $-g_k^T d_k \geq c\|g_k\|^2$ for some positive constant $c > 0$ and for any $k \geq 1$, from Zoutendijk condition (3.34) it follows that

$$\|d_k\| \to +\infty. \quad (8.174)$$

From the continuity of the gradient, it follows that there exists a positive constant $\bar{\gamma}$ such that $\|g_k\| \leq \bar{\gamma}$, for any $k \geq 1$. By (8.116), (8.174) means that $\beta_k^{DK+}(\tau_k)$ can only be less than $\eta \frac{g_{k+1}^T d_k}{\|d_k\|^2}$ for finite times, since otherwise, we have that

$$\|d_{k+1}\| = \left\| -g_{k+1} + \eta \frac{g_{k+1}^T d_k}{\|d_k\|^2} d_k \right\| \leq (1+\eta)\|g_{k+1}\| \leq (1+\eta)\bar{\gamma}$$

for infinite k's, and therefore, we get a contradiction with (8.174). Hence, we can suppose that along the iterations $\beta_k^{DK+}(\tau_k) = \beta_k^{DK}(\tau_k)$ for sufficiently large k.

In this case, using Property (*) proved in Proposition 8.12 and the fact that $\|d_k\|$ is increasing at most linearly, similarly to Lemma 4.2 in (Gilbert and Nocedal 1992) we can prove that for any positive integers Δ and k_0, there exists an integer $k \geq k_0$ such that the size of $K = \{i : k \leq i \leq k + \Delta - 1, \|s_{i-1}\| > \lambda\}$ is greater that $\Delta/2$. With this, from (8.155) proved in Proposition 8.11, Lemma 4.2 in (Gilbert and Nocedal 1992) and the boundedness of the sequence $\{x_k\}$, we get a contradiction similarly to the proof of Theorem 4.3 in (Gilbert and Nocedal 1992). This contradiction shows that $\liminf_{k\to\infty}\|g_k\| = 0$. ♦

Numerical study. In the following, let us report some numerical results of the CGSSML algorithm for solving large-scale unconstrained optimization problems from the UOP collection (Andrei, 2018g). The algorithm CGSSML was implemented by modifying the CG-DESCENT code (Fortran version 1.4) of Hager and Zhang (2005) in order to incorporate the self-scaling memoryless BFGS algorithms. The conjugate gradient parameter β_k in the search direction is computed by clustering of the eigenvalues of the iteration matrix H_{k+1} or by minimizing the measure function of Byrd and Nocedal. The stepsize is computed with the standard Wolfe (or with the approximate Wolfe or with the improved Wolfe line searches).

It is worth emphasizing that in our numerical experiments, we compare algorithms included in CGSSML versus the CG-DESCENT version 1.4 of Hager and Zhang (2005). The aim was to see the performances of the algorithms using β_k^{DE+} given by (8.130), β_k^{TR+} given by (8.137), β_k^{FI+} given by (8.145) and $\beta_k^{CG-DESCENT+}$ given by (8.110) without any other ingredients included, for example, in some other versions of CG-DESCENT, or in the limited-memory conjugate gradient algorithm proposed by Hager and Zhang (2013), or in the CGOPT by Dai and Kou (2013). Our interest was to see the power of the conjugate gradient parameters β_k^{DE+}, β_k^{TR+}, β_k^{FI+}, and $\beta_k^{CG-DESCENT+}$ with different line searches for solving large-scale unconstrained optimization problems in similar conditions.

The algorithms compared in this section are as follows: DESW, i.e., CGSSML algorithm with β_k^{DE+} given by (8.130) and with the standard Wolfe line search (8.4) and (8.5), respectively. TRSW, i.e., CGSSML algorithm with β_k^{TR+} given by (8.137) and with the standard Wolfe line search (8.4) and (8.5). FISW, i.e., CGSSML algorithm with β_k^{FI+} given by (8.145) and with the standard Wolfe line search (8.4) and (8.5).

The parameters in the standard Wolfe line searches are $\rho = 0.0001$ and $\sigma = 0.8$. All the algorithms use the same stopping criterion $\|g_k\|_\infty \leq 10^{-6}$, where $\|.\|_\infty$ is the maximum absolute component of a vector, or when the number of iterations exceeds 2000 iterations. The rest of parameters are the same defined in CG-DESCENT by Hager and Zhang (2005). In all algorithms, the Powell restart

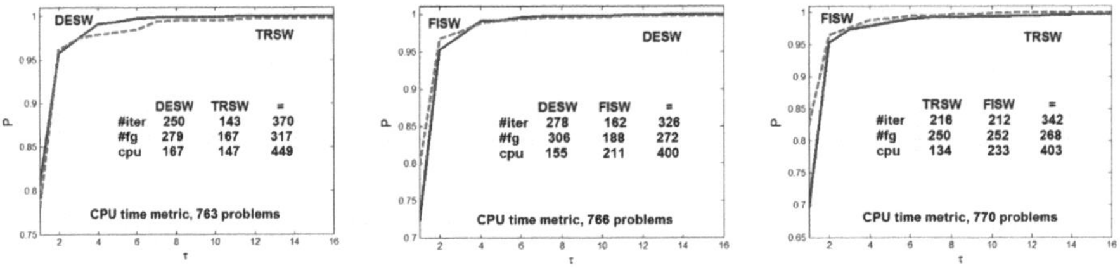

Figure 8.12 Performance profiles of DESW versus TRSW, of DESW versus FISW, and of TRSW versus FISW

criterion, described in step 8 of the CGSSML algorithm, is used. All the algorithms are not accelerated in the sense of the Remark 5.1.

In the first set of numerical experiments, let us compare the performance of CGSSML algorithms with standard Wolfe line search, namely DESW versus TRSW, DESW versus FISW, and TRSW versus FISW for solving the set of problems considered in this numerical study. Figure 8.12 shows the Dolan and Moré CPU performance profiles of these algorithms. When comparing DESW versus TRSW subject to the CPU time metric, we see that DESW is top performer. Comparing DESW versus TRSW (see Figure 8.12), subject to the number of iterations, we see that DESW was better in 250 problems (i.e., it achieved the minimum number of iterations in 250 problems). TRSW was better in 143 problems and they achieved the same number of iterations in 370 problems, etc. Out of 800 problems, only for 763 problems does the criterion (1.118) holds. Observe that both DESW and FISW are slightly more efficient and more robust than TRSW. It is because, as we have proved in the Theorem 8.7, for the algorithm TRSW based on trace of the iteration matrix not all the iterations satisfy the sufficient descent condition. However, from the viewpoint of clustering of the eigenvalues of H_{k+1} using the determinant or the trace of the iteration matrix or minimizing the measure function leads to algorithms with similar performances. From Figure 8.12, we see that FISW is top performer versus DESW and versus TRSW. This is because the FISW algorithm is based on an ad hoc procedure for minimizing a special combination of the determinant and of the trace of the iteration matrix H_{k+1}.

In the second set of numerical experiments, let us compare DESW, TRSW, and FISW versus CG-DESCENT (version 1.4) with truncated conjugate gradient parameter and with standard Wolfe line search (8.4) and (8.5). The idea was to see the performances of the algorithms using $\beta_k^{\mathrm{DE}+}$ given by (8.130), $\beta_k^{\mathrm{TR}+}$ given by (8.137), $\beta_k^{\mathrm{FI}+}$ given by (8.145) versus CG-DESCENT where the conjugate gradient parameter is $\beta_k^{\mathrm{CG-DESCENT}+}$ is given by (8.110) with $\eta = 0.0001$, without any other ingredients included in the CG-DESCENT version 5.3, or in the limited-memory conjugate gradient algorithm proposed by Hager and Zhang (2013). CG-DESCENT was devised in order to ensure sufficient descent, independent of the accuracy of the line search. In CG-DESCENT, the search direction (8.108), where the conjugate gradient parameter is computed as in (8.109), satisfies the sufficient descent

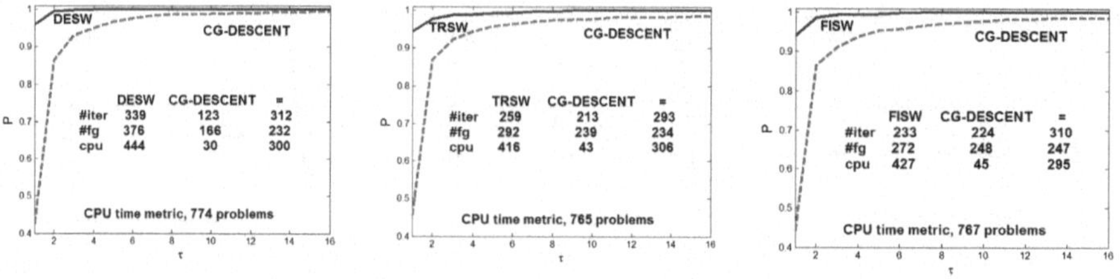

Figure 8.13 Performance profiles of DESW, TRSW, and FISW versus CG-DESCENT

condition $g_k^T d_k \leq -(7/8)\|g_k\|^2$, provided that $y_k^T d_k \neq 0$. The search directions in CG-DESCENT do not satisfy the conjugacy condition. When iterates jam the expression $\|y_k\|^2 (g_{k+1}^T s_k)/(y_k^T s_k)^2$ in (8.109) becomes negligible. If the minimizing function f is quadratic and the line search is exact, then CG-DESCENT reduces to the Hestenes and Stiefel algorithm (1952). Figure 8.13 presents the Dolan and Moré performance profiles of these algorithms. Form Figure 8.13, we see that DESW, TRSW, and FISW are top performers versus CG-DESCENT and the differences are significant. Since all these algorithms use the same line search based on Wolfe conditions (8.4) and (8.5), it follows that DESW, TRSW, and FISW generate a better search direction. Notice that the difference between DESW and CG-DESCENT is only in a constant coefficient of the second term of the Hager and Zhang method. Besides, the truncation mechanisms in these algorithms are different and this explains the differences between these algorithms.

In the third set of numerical experiments, we compare DESW, TRSW, and FISW versus DESCONa (Andrei, 2013c). DESCONa is an accelerated conjugate gradient algorithm with guaranteed descent and conjugacy conditions and a modified Wolfe line search. Figure 8.14 presents the performance profiles of these algorithms.

Observe that DESW, TRSW, and FISW are more efficient versus DESCONa. However, DESCONa is more robust versus these algorithms.

In the fourth set of numerical experiments, we compare DESW, TRSW, and FISW versus the self-scaling memoryless BFGS algorithms SBFGS-OS and SBFG-OL (Babaie-Kafaki, 2015). Babaie-Kafaki has shown that the scaling parameter proposed by Oren and Spedicato (8.107) is the unique minimizer of the

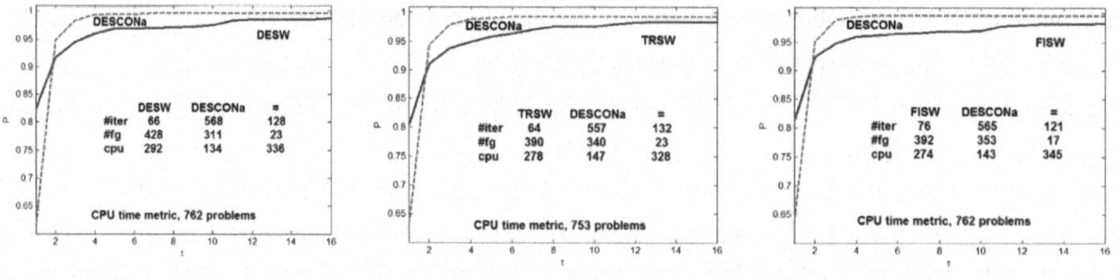

Figure 8.14 Performance profiles of DESW, TRSW, and FISW versus DESCONa

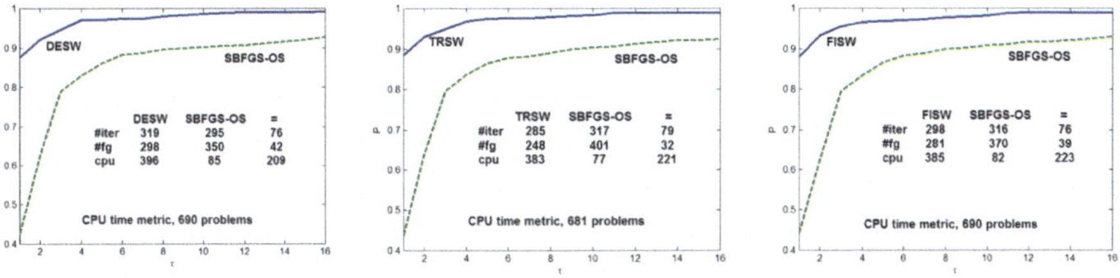

Figure 8.15 Performance profiles of DESW, TRSW, and FISW versus SBFGS-OS

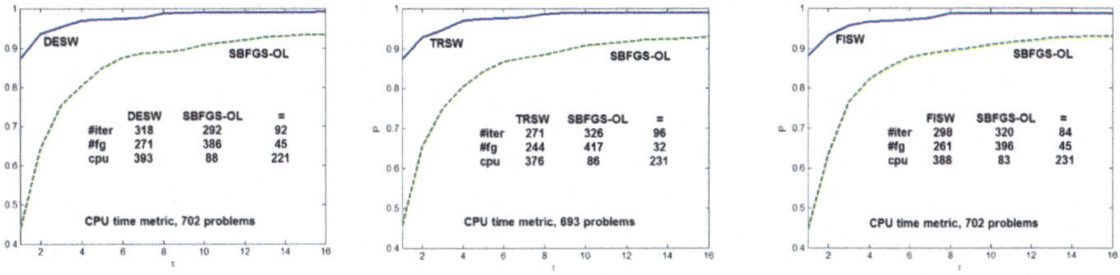

Figure 8.16 Performance profiles of DESW, TRSW, and FISW versus SBFGS-OL

given upper bound for the condition number of the scaled memoryless BFGS update. At the same time, Babaie-Kafaki proved that the scaling parameter proposed by Oren and Luenberger (8.111) is the unique minimizer of the given upper bound for the condition number of scaled memoryless DFP update. Figure 8.15 shows the performance profiles of DESW, TRSW, and FISW versus SBFGS-OS. Figure 8.16 presents the performances of the same algorithms versus SBFGS-OL.

From Figures 8.15 and 8.16, we see that the self-scaling memoryless BFGS algorithms based on clustering the eigenvalues of the iteration matrix using the determinant, the trace or minimizing the measure function of Byrd and Nocedal are more efficient than SBFGS-OS and than SBFGS-OL. Observe that DESW, TRSW, and FISW are more robust than SBFGS-OS. This is in agreement with the numerical results obtained by Babaie-Kafaki (2015) who showed that SBFGS-OS is more efficient and more robust that SBFGS-OL. Observe that the algorithms DESW, TRSW, and FISW are based on clustering the eigenvalues of the iteration matrix in a point. On the other hand, SBFGS-OS and SBFGS-OL are based on minimizing *an upper bound* of the condition number of the same iteration matrix. However, both these approaches are two different ways to basically pursue similar ideas based on eigenvalues or on minimizing (an upper bound of) the condition number of the iteration matrix.

In the next set of numerical experiments, we compare DESW, TRSW, and FISW versus L-BFGS ($m = 5$), where m is the number of the vector pairs $\{s_i, y_i\}$ used in the updating formulae of the L-BFGS (Liu & Nocedal, 1989). Figure 8.17 presents the performance profiles of these algorithms.

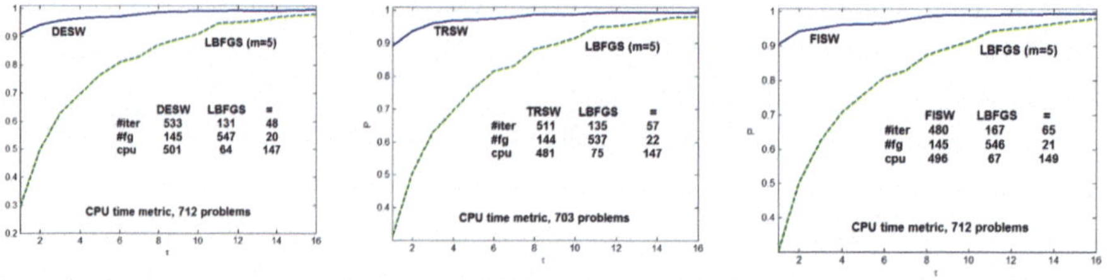

Figure 8.17 Performance profiles of DESW, TRSW, and FISW versus LBFGS

Figure 8.17 shows that the self-scaling memoryless BFGS algorithms based on clustering the eigenvalues of the iteration matrix using the determinant, the trace or minimizing the measure function of Byrd and Nocedal are more efficient and more robust that L-BFGS. L-BFGS uses a fixed, low-cost formula requiring no extra derivative information, being very effective for solving highly nonlinear unconstrained optimization problems. More than this, L-BFGS is not sensitive to the eigenvalues of the Hessian. In contrast, DESW, TRSW, and FISW are based on eigenvalues clustering, thus being able to better capture the curvature of the minimizing function in the current iteration.

In (Andrei, 2019b), we presented the numerical comparisons of CGSSML implemented with the approximate Wolfe line search (8.123) or with the improved Wolfe line search (8.124) and (8.125) versus CG-DESCENT. The numerical experiments showed that:

(1) Both the approximate Wolfe line search and the improved Wolfe line search are important ingredients for the efficiency and robustness of the self-scaling memoryless BFGS algorithms with clustering of the eigenvalues. The performances of the CGSSML algorithms with approximate or improved line searches are slightly better than the performances of the same algorithms with standard Wolfe line search.
(2) No matter how the stepsize is computed, by using the standard, the approximate or the improved Wolfe line search, the performances of the CGSSML algorithms based on the determinant or on the trace of the iteration matrix H_{k+1}, or based on minimizing the measure function $\varphi(H_{k+1})$ defined by Byrd and Nocedal are better that those of CG-DESCENT with Wolfe or with the approximate Wolfe line search.

Notes and References

The idea of the methods described in this chapter is to include the approximations to the Hessian of the minimizing function into the formula for computation the conjugate gradient parameter β_k. This was first considered by Perry (1976, 1977). In fact, the foundation of the self-scaling memoryless BFGS algorithm was first presented by Perry as a technique for developing a nonlinear conjugate gradient algorithm with memory, i.e., with stored information from the previous iterations, as an alternative to the quasi-Newton methods for large-scale problems, where it is impractical to store and handle the Hessian matrix. This method was the first effort for solving large-scale problems, preceding the introduction by Nocedal (1980) of the limited-memory BFGS method. Shanno (1978a) reinterpreted Perry's algorithm and showed that the conjugate gradient methods are exactly the BFGS quasi-Newton method where the approximation to the inverse Hessian is restarted as the identity matrix at every step. He introduced a scaling term, thus improving the final form of the self-scaling memoryless BFGS method, i.e., the SSML-BFGS method. A modification of the self-scaling memoryless BFGS method was given by Kou and Dai (2015). They multiplied the third term in (8.105) by some nonnegative parameter, thus obtaining a new self-scaling BFGS algorithm with better convergence properties.

The SSML-BFGS method provided a very good understanding on the relationship between nonlinear conjugate gradient methods and quasi-Newton methods. For convex quadratic functions if the line search is exact and the identity matrix is used as the initial approximation to the Hessian, then both BFGS and SSML-BFGS methods generate the same iterations as the conjugate gradient method. This was the starting point for the conjugate gradient methods memoryless BFGS preconditioned.

Using this approach, Shanno and Phua (1976, 1980) and Shanno (1983) developed the CONMIN algorithm, one of the most respectable algorithms and codes. Using a double quasi-Newton update scheme, Andrei (2007a, 2008a) elaborated the SCALCG algorithm. In both these algorithms, the stepsize is computed by means of the standard Wolfe line search conditions (see Figure 5.1). Hager and Zhang (2005) presented the CG-DESCENT, one of the best conjugate gradient algorithms. In CG-DESCENT, Hager and Zhang introduced an approximate Wolfe line search. Later on, Dai and Kou (2013) proposed the CGOPT algorithm where the search direction is closest to the direction of the scaled memoryless BFGS method. Similar to Hager and Zhang, Dai, and Kou developed an improved Wolfe line search. In this way, a family of conjugate gradient algorithms was obtained, where the stepsize was computed by an improved Wolfe line search. Further on, Andrei (2019b), by using the determinant, the trace or a combination of these operators, known as the measure function of Byrd and Nocedal, developed new efficient self-scaling memoryless BFGS conjugate gradient methods.

Chapter 9
Three-Term Conjugate Gradient Methods

This chapter is dedicated to presenting three-term conjugate gradient methods. For solving the nonlinear unconstrained optimization problem

$$\min\{f(x) : x \in \mathbb{R}^n\}, \tag{9.1}$$

where $f : \mathbb{R}^n \to \mathbb{R}$ is a continuously differentiable function bounded from below, starting from an initial guess $x_0 \in \mathbb{R}^n$, a three-term nonlinear conjugate gradient method generates a sequence $\{x_k\}$ as

$$x_{k+1} = x_k + \alpha_k d_k, \tag{9.2}$$

where the stepsize $\alpha_k > 0$ is obtained by line search (usually the Wolfe line search), while the directions d_k include three terms.

One of the first general three-term conjugate gradient methods was proposed by Beale (1972) as

$$d_{k+1} = -g_{k+1} + \beta_k d_k + \gamma_k d_t, \tag{9.3}$$

where $\beta_k = \beta_k^{HS}$ (or $\beta_k^{FR}, \beta_k^{DY}$ etc.),

$$\gamma_k = \begin{cases} 0, & k = t+1, \\ \frac{g_{k+1}^T y_t}{d_t^T y_t}, & k > t+1 \end{cases} \tag{9.4}$$

and d_t is a restart direction. McGuire and Wolfe (1973) and Powell (1984a) made further research into the Beale three-term conjugate gradient algorithm and established efficient restart strategies, obtaining good numerical results. Mainly, to make d_{k+1} satisfy the sufficient descent condition and two consecutive gradients not far from orthogonal, the following conditions should be imposed:

N. Andrei, *Nonlinear Conjugate Gradient Methods for Unconstrained Optimization*, Springer Optimization and Its Applications 158,
https://doi.org/10.1007/978-3-030-42950-8_9

$-g_{k+1}^T d_{k+1} \geq \xi \|g_{k+1}\| \|d_{k+1}\|$, where ξ is a small positive constant, and the Powell-Beale restart criterion $|g_k^T g_{k+1}| < 0.2\|g_{k+1}\|^2$.

It is interesting to see how Beale arrived at the three-term conjugate gradient algorithms. Powell (1977) pointed out that the restart of the conjugate gradient algorithms with negative gradient has two main drawbacks: a restart along $-g_k$ abandons the second derivative information that is found by the search along d_{k-1} and the immediate reduction in the values of the objective function is usually less than it would be without restart. Therefore, it seems more advantageous to use $-g_k + \beta_k d_{k-1}$ as a restarting direction. Beale (1972) studied this restart strategy which uses $-g_k + \beta_k d_{k-1}$ as the restart direction and extended the nonrestart direction from two terms to three terms, so that all search directions are conjugate to one another if f is convex quadratic and if the line search is exact. However, McGuire and Wolfe (1973) evaluated this algorithm and reported disappointing numerical results. By introducing a new restart criterion, namely, $|g_{k+1}^T g_k| > 0.2\|g_{k+1}\|^2$, Powell (1977) overcame the difficulties that McGuire and Wolfe encountered and obtained satisfactory numerical results. Therefore, the introduction of the three-term conjugate gradient algorithms was suggested by Beale as a procedure for restarting the conjugate gradient algorithms.

Deng and Li (1995) and Dai and Yuan (1999) studied the general three-term conjugate gradient method

$$d_{k+1} = -g_{k+1} + \beta_k d_k + \gamma_k d_{t(p)}, \tag{9.5}$$

where $t(p)$ is the number of the pth restart iteration satisfying $t(p) < k \leq t(p+1) \leq +\infty$, showing that under some mild conditions the algorithm is global convergent.

Nazareth (1977) proposed a conjugate gradient algorithm by using a three-term recurrence formula

$$d_{k+1} = -y_k + \frac{y_k^T y_k}{y_k^T d_k} d_k + \frac{y_{k-1}^T y_k}{y_{k-1}^T d_{k-1}} d_{k-1}, \tag{9.6}$$

with $d_{-1} = 0, d_0 = 0$. If f is a convex quadratic function, then for any stepsize α_k the search directions generated by (9.6) are conjugate subject to the Hessian of f, even without exact line search.

In the same context, Zhang, Zhou, and Li (2006a) proposed a descent modified PRP conjugate gradient algorithm with three terms as

$$d_{k+1} = -g_{k+1} + \frac{g_{k+1}^T y_k}{g_k^T g_k} d_k - \frac{g_{k+1}^T d_k}{g_k^T g_k} y_k \tag{9.7}$$

and a descent modified HS conjugate gradient algorithm with three terms (Zhang, Zhou, and Li, 2007) as

$$d_{k+1} = -g_{k+1} + \frac{g_{k+1}^T y_k}{s_k^T y_k} s_k - \frac{g_{k+1}^T s_k}{s_k^T y_k} y_k, \tag{9.8}$$

where $d_0 = -g_0$. A remarkable property of these methods is that they produce descent directions, i.e., $g_k^T d_k = -\|g_k\|$ for any $k \geq 1$. The convergence properties of (9.8) for convex optimization were given by Zhang and Zhou (2012).

Motivated by this nice descent property, Zhang, Xiao, and Wei (2009) introduced another three-term conjugate gradient method based on the Dai–Liao method as

$$d_{k+1} = -g_{k+1} + \frac{g_{k+1}^T (y_k - t s_k)}{y_k^T s_k} s_k - \frac{g_{k+1}^T s_k}{y_k^T s_k} (y_k - t s_k), \tag{9.9}$$

where $d_0 = -g_0$ and $t \geq 0$. Again, it is easy to see that the sufficient descent condition also holds, independent of the line search, i.e., for this method, $g_k^T d_k = -\|g_k\|^2$ for all k. A specialization of this three-term conjugate gradient given by (9.9) was developed by Al-Bayati and Sharif (2010), where the search direction is computed as

$$d_{k+1} = -g_{k+1} + \beta_k^{DL+} s_k - \frac{g_{k+1}^T s_k}{y_k^T s_k} (y_k - t s_k), \tag{9.10}$$

where $\beta_k^{DL+} = \max\left\{\frac{y_k^T g_{k+1}}{y_k^T s_k}, 0\right\} - t\frac{s_k^T g_{k+1}}{y_k^T s_k}$ and $t = 2\frac{\|y_k\|^2}{y_k^T s_k}$. It is easy to see that (9.10) satisfies the sufficient descent condition independent of the line search used.

In an effort to improve the performances of conjugate gradient algorithms for the large-scale unconstrained optimization, Andrei (2007a) developed a scaled conjugate gradient algorithm based on the quasi-Newton BFGS updating formula, in which the search direction has three terms as

$$\begin{aligned} d_{k+1} = & -\theta_{k+1} g_{k+1} + \theta_{k+1} \frac{s_k^T g_{k+1}}{y_k^T s_k} y_k \\ & - \left[\left(1 + \theta_{k+1} \frac{\|y_k\|^2}{y_k^T s_k} \right) \frac{s_k^T g_{k+1}}{y_k^T s_k} - \theta_{k+1} \frac{y_k^T g_{k+1}}{y_k^T s_k} \right] s_k \end{aligned} \tag{9.11}$$

where θ_{k+1} is a parameter defined as a scalar approximation of the inverse Hessian. In the same paper (Andrei, 2007a), using a double quasi-Newton update scheme in a restart environment, (9.11) is further modified to get another more complex three-term conjugate gradient algorithm.

Cheng (2007) gave another three-term conjugate gradient algorithm based on a modification of the Polak–Ribiére–Polyak method

$$d_{k+1} = -g_{k+1} + \frac{g_{k+1}^T y_k}{\|g_k\|^2}\left(I - \frac{g_{k+1}g_{k+1}^T}{\|g_{k+1}\|^2}\right)d_k, \tag{9.12}$$

showing its global convergence under an appropriate line search.

Another three-term conjugate gradient algorithm was given by Narushima, Yabe, and Ford (2011), where the searching direction is computed as

$$d_{k+1} = \begin{cases} -g_{k+1} & \text{if } k = 0 \text{ or } g_{k+1}^T p_k = 0, \\ -g_{k+1} + \beta_k d_k - \beta_k \frac{g_{k+1}^T d_k}{g_{k+1}^T p_k} p_k & \text{otherwise,} \end{cases} \tag{9.13}$$

where $\beta_k \in \mathbb{R}$ is a parameter and $p_k \in \mathbb{R}^n$ is any vector. This is a general three-term conjugate gradient method for which a sufficient descent condition for its global convergence is proved. In the same paper, Narushima, Yabe, and Ford (2011) proposed a specific three-term conjugate gradient algorithm based on the multi-step quasi-Newton method, for which the global convergence property is proved. The numerical experiments showed that the CG-DESCENT algorithm by Hager and Zhang (2005) performs better than these three-term conjugate gradient algorithms.

Recently, Andrei (2011a) suggested another three-term conjugate gradient algorithm

$$d_{k+1} = -\frac{y_k^T s_k}{\|g_k\|^2} g_{k+1} + \frac{y_k^T g_{k+1}}{\|g_k\|^2} s_k - \frac{s_k^T g_{k+1}}{\|g_k\|^2} y_k, \tag{9.14}$$

which is a modification of the Polak–Ribière–Polyak conjugate gradient algorithm for which, independent of the line search at each iteration, both the sufficient descent condition and the conjugacy condition are satisfied. Intensive numerical experiments show that the algorithm given by (9.14) is top performer versus PRP, DY, and versus the three-term conjugate gradient algorithm given by (9.7).

Another three-term conjugate gradient method was given by Andrei (2013a), where the search direction is computed as

$$d_{k+1} = -g_{k+1} - \delta_k s_k - \eta_k y_k \tag{9.15}$$

and

$$\delta_k = \left(1 + \frac{\|y_k\|^2}{y_k^T s_k}\right)\frac{s_k^T g_{k+1}}{y_k^T s_k} - \frac{y_k^T g_{k+1}}{y_k^T s_k}, \quad \eta_k = \frac{s_k^T g_{k+1}}{y_k^T s_k}. \tag{9.16}$$

The search direction (9.15) is descent and satisfies the Dai and Liao conjugacy condition. The numerical experiments proved that this three-term conjugate gradient method substantially outperforms the well-known CG-DESCENT (version 1.4), as well as some other three-term conjugate gradient methods by Zhang, Zhou, and Li (2006a, 2007), Zhang, Xiao, and Wei (2009), Cheng (2007), Andrei (2011a), Baluch, Salleh, and Alhawarat (2018).

Another family of three-term conjugate gradient methods with sufficient descent property for unconstrained optimization was presented by Al-Baali, Narushima, and Yabe (2015).

It is worth seeing that the SSML-BFGS search direction given by (8.105) also is a family of three-term conjugate gradient direction depending on the scaling parameter τ_k. Based on the SSML-BFGS updating (8.104), some efficient conjugate gradient algorithms, called CG-DESCENT, CGOPT, and CGSSML, have been developed. The numerical experiments with these algorithms show that under the Wolfe line search CG-DESCENT (Hager & Zhang, 2005), CGOPT (Dai & Kou, 2013), and CGSSML (Andrei, 2019b) perform more efficiently than SSML-BFGS (see Kou & Dai, 2015).

A close analysis of the three-term conjugate gradient algorithms described above shows that the search direction d_{k+1} is obtained as a linear combination of $-g_{k+1}$, d_k and y_k, where the coefficients in these linear combinations are computed using the same elements as $\|y_k\|^2$, $\|g_k\|^2$, $\|g_{k+1}\|^2$, $s_k^T y_k$, $s_k^T g_{k+1}$, and $y_k^T g_{k+1}$ in similar computational formulae, in order to satisfy the descent property, the most important property in the conjugate gradient class of algorithms.

Using these ingredients, plenty of three-term conjugate gradient algorithms can be generated and therefore the following *project* may be suggested (Andrei, 2013b). *Develop three-term conjugate gradient algorithms which generate a sequence* $\{x_k\}$ *as*

$$x_{k+1} = x_k + \alpha_k d_k, \tag{9.17}$$

where $\alpha_k > 0$ *is obtained by line search (Wolfe conditions) and the search direction is computed like*

$$d_{k+1} = -g_{k+1} - a_k s_k - b_k y_k, \tag{9.18}$$

as modifications of the conjugate gradient algorithms HS, FR, PRP, LS, DY, DL, CD, CG-DESCENT, etc., where the scalar parameters a_k *and* b_k *are determined in such a way so that the descent condition* $g_k^T d_k \leq 0$ *and the conjugacy condition* $y_k^T d_{k+1} = -t(s_k^T g_{k+1})$, *where* $t > 0$, *are simultaneously satisfied. The line search is based on the standard Wolfe conditions*

$$f(x_k + \alpha_k d_k) - f(x_k) \leq \rho \alpha_k g_k^T d_k, \tag{9.19}$$

$$g_{k+1}^T d_k \geq \sigma g_k^T d_k, \tag{9.20}$$

or on the strong Wolfe line search conditions given by (9.19) and

$$\left|g_{k+1}^T d_k\right| \leq -\sigma g_k^T d_k, \tag{9.21}$$

where d_k is a descent direction and $0 < \rho \leq \sigma < 1$.

Of course, some other conditions on d_{k+1} may be introduced. For example, it can be considered that the direction (9.18) is just the Newton direction $d_{k+1}^N = -\nabla^2 f(x_{k+1})^{-1} g_{k+1}$. This formal equality combined with the secant equation (or the modified secant equation), together with the sufficient descent condition can be used to determine the parameters a_k and b_k in (9.18), thus defining a three-term conjugate gradient algorithm. Considering this project, a lot of three-term conjugate gradient algorithms may be developed. In the following, some three-term conjugate gradient methods generated in the frame of this project will be presented.

9.1 A Three-Term Conjugate Gradient Method with Descent and Conjugacy Conditions (TTCG)

This section develops a three-term conjugate gradient algorithm, which is a modification of the Hestenes and Stiefel (1952) or a modification of the CG-DESCENT by Hager and Zhang (2005) updating formulae, for which both the *descent condition* and the *conjugacy condition* are simultaneously satisfied (Andrei, 2013b). The algorithm is given by (9.17), where the direction d_{k+1} is computed as

$$d_{k+1} = -g_{k+1} - \delta_k s_k - \eta_k y_k, \tag{9.22}$$

where

$$\delta_k = \left(1 + 2\frac{\|y_k\|^2}{y_k^T s_k}\right)\frac{s_k^T g_{k+1}}{y_k^T s_k} - \frac{y_k^T g_{k+1}}{y_k^T s_k}, \tag{9.23}$$

$$\eta_k = \frac{s_k^T g_{k+1}}{y_k^T s_k}. \tag{9.24}$$

Obviously, using (9.22)–(9.24), the direction d_{k+1} can be written as

$$d_{k+1} = \underbrace{-g_{k+1} + \frac{y_k^T g_{k+1}}{y_k^T s_k} s_k}_{d_{k+1}^{HS}} - \left(1 + 2\frac{\|y_k\|^2}{y_k^T s_k}\right) \frac{s_k^T g_{k+1}}{y_k^T s_k} s_k - \frac{s_k^T g_{k+1}}{y_k^T s_k} y_k, \tag{9.25}$$

or as

$$d_{k+1} = \underbrace{-g_{k+1} + \left(y_k - 2\frac{\|y_k\|^2}{y_k^T s_k} s_k\right)^T \frac{g_{k+1}}{y_k^T s_k} s_k}_{d_{k+1}^{HZ}} - (s_k + y_k) \frac{s_k^T g_{k+1}}{y_k^T s_k}. \tag{9.26}$$

Observe that the direction d_{k+1} from (9.22)–(9.24) can be written as

$$d_{k+1} = -Q_k g_{k+1}, \tag{9.27}$$

where the matrix Q_k is given by

$$Q_k = I - \frac{s_k y_k^T - y_k s_k^T}{y_k^T s_k} + \left(1 + 2\frac{\|y_k\|^2}{y_k^T s_k}\right) \frac{s_k s_k^T}{y_k^T s_k}. \tag{9.28}$$

As it is known, the BFGS updating of the inverse approximation to the Hessian of function f is

$$H_{k+1} = H_k - \frac{s_k y_k^T H_k + H_k y_k s_k^T}{y_k^T s_k} + \left(1 + \frac{y_k^T H_k y_k}{y_k^T s_k}\right) \frac{s_k s_k^T}{y_k^T s_k}. \tag{9.29}$$

Obviously, the matrix Q_k in (9.28) is a modification of the BFGS updating (9.29) in the sense that it is restarted with the identity matrix at every step ($H_k = I$), i.e., it is a modification of the memoryless BFGS quasi-Newton updating, and more importantly, the sign in front of $y_k s_k^T$ in the second term of (9.28) is modified to get the descent property as it is proved in the following proposition. It is worth saying that for strongly convex functions and a relatively accurate line search, the search directions using factor 2 which multiplies $\|y_k\|^2 / y_k^T s_k$ in (9.28) are approximately multiples of the search directions generated by the memoryless quasi-Newton method of Shanno (1978b).

Proposition 9.1 *Suppose that the line search satisfies the Wolfe conditions (9.19) and (9.20). Then, d_{k+1} given by (9.22) with (9.23) and (9.24) is a descent direction.*

Proof Since the line search satisfies the Wolfe conditions, it follows that $y_k^T s_k > 0$. Now, by direct computation, it results that

$$g_{k+1}^T d_{k+1} = -\|g_{k+1}\|^2 - \left(1 + 2\frac{\|y_k\|^2}{y_k^T s_k}\right)\frac{(s_k^T g_{k+1})^2}{y_k^T s_k} \le 0.$$

♦

Day and Liao (2001) extended in a very natural way the classical conjugate condition $y_k^T d_{k+1} = 0$, suggesting the following one $y_k^T d_{k+1} = -t(s_k^T g_{k+1})$, where $t \ge 0$ is a given scalar. The proposition below proves that the direction d_{k+1} given by (9.22) with (9.23) and (9.24) satisfies the Dai–Liao conjugacy condition.

Proposition 9.2 *Suppose that the line search satisfies the Wolfe conditions (9.19) and (9.20). Then, d_{k+1} given by (9.22) with (9.23) and (9.24) satisfies the Dai–Liao conjugacy condition $y_k^T d_{k+1} = -t_k(s_k^T g_{k+1})$, where $t_k > 0$ for all k.*

Proof By direct computation,

$$y_k^T d_{k+1} = -\left(1 + 3\frac{\|y_k\|^2}{y_k^T s_k}\right)(s_k^T g_{k+1}) \equiv -t_k(s_k^T g_{k+1}), \tag{9.30}$$

where since $y_k^T s_k > 0$,

$$t_k = \left(1 + 3\frac{\|y_k\|^2}{y_k^T s_k}\right) > 0.$$

♦

Now, if f is strongly convex or the line search satisfies the Wolfe conditions (9.19) and (9.20), then $y_k^T s_k > 0$ and therefore the above computational scheme yields descent. Besides, the direction d_{k+1} satisfies the Dai–Liao conjugacy condition (9.30), where $t_k > 0$ at every iteration. Observe that if the line search is exact, i.e., $s_k^T g_{k+1} = 0$, then (9.22) reduces to the HS method.

Therefore, taking into consideration, the acceleration scheme from Remark 5.1, where the acceleration factor ξ_k is computed as in (5.24), according to the value of the parameter "*acceleration*" (true or false), the following algorithms TTCG and TTCGa can be presented. TTCGa is the accelerated version of TTG.

Algorithm 9.1 Three-term descent and conjugacy conditions: TTCG/TTCGa

1.	Select a starting point $x_0 \in domf$ and compute $f_0 = f(x_0)$ and $g_0 = \nabla f(x_0)$. Select $\varepsilon_A > 0$ sufficiently small and some positive values $0 < \rho < \sigma < 1$ used in Wolfe line search. Set $d_0 = -g_0$ and $k = 0$
2.	Test a criterion for stopping the iterations. If the test is satisfied, then stop; otherwise continue with step 3
3.	Determine the stepsize α_k by using the Wolfe line search conditions (9.19) and (9.20). Update the variables $x_{k+1} = x_k + \alpha_k d_k$. Compute f_{k+1}, g_{k+1} and $s_k = x_{k+1} - x_k, y_k = g_{k+1} - g_k$

(continued)

Algorithm 9.1 (continued)

4.	If the parameter *acceleration* is true, then (a) Compute: $z = x_k + \alpha_k d_k$, $g_z = \nabla f(z)$ and $y_k = g_k - g_z$ (b) Compute: $a_k = \alpha_k g_k^T d_k$, and $b_k = -\alpha_k y_k^T d_k$ (c) If $\lvert b_k \rvert \geq \varepsilon_A$, then compute $\xi_k = -a_k / b_k$ and update the variables as $x_{k+1} = x_k + \xi_k \alpha_k d_k$. Compute f_{k+1} and g_{k+1}. Compute $s_k = x_{k+1} - x_k$ and $y_k = g_{k+1} - g_k$
5.	Determine δ_k and η_k as in (9.23) and (9.24), respectively
6.	Compute the search direction as: $d_{k+1} = -g_{k+1} - \delta_k s_k - \eta_k y_k$
7.	Powell restart criterion. If $\lvert g_{k+1}^T g_k \rvert > 0.2 \lVert g_{k+1} \rVert^2$ then set $d_{k+1} = -g_{k+1}$
8.	Consider $k = k + 1$ and go to step 2 ♦

If f is bounded along the direction d_k, then there exists a stepsize α_k satisfying the Wolfe line search conditions (9.19) and (9.20). When the Powell restart condition is satisfied (step 7), then the algorithm is restarted with the negative gradient $-g_{k+1}$. More sophisticated reasons for restarting the algorithms have been proposed in the literature, but we are interested in the performance of a conjugate gradient algorithm that uses this restart criterion associated to a direction satisfying both the descent and the conjugacy conditions. Under reasonable assumptions, the Wolfe conditions and the Powell restart criterion are sufficient to prove the global convergence of the algorithm. At every iteration $k \geq 1$, the starting guess for step α_k in the line search is computed as $\alpha_{k-1} \lVert d_{k-1} \rVert / \lVert d_k \rVert$.

Convergence analysis. To prove the global convergence of nonlinear conjugate gradient algorithms, the Zoutendijk condition is often used. The analysis is given under the Assumption CG. Under this Assumption on f, there exists a constant $\Gamma \geq 0$ so that $\lVert \nabla f(x) \rVert \leq \Gamma$ for all $x \in S$. Besides, it is easy to see that $\lVert s_k \rVert = \lVert x_{k+1} - x_k \rVert \leq \lVert x_{k+1} \rVert + \lVert x_k \rVert \leq 2B$. The following proposition proves that in the above three-term conjugate gradient method, the Zoutendijk condition holds under the standard Wolfe line search (9.19) and (9.20).

Proposition 9.3 *Suppose that the Assumption CG holds. Consider the algorithm* (9.17) *with* (9.22)–(9.24)*, where d_k is a descent direction and α_k is computed by the standard Wolfe line search* (9.19) *and* (9.20)*. Then,*

$$\sum_{k=0}^{\infty} \frac{(g_k^T d_k)^2}{\lVert d_k \rVert^2} < +\infty. \tag{9.31}$$

Proof From (9.19) and from Proposition 1.2 it follows that

$$f_k - f_{k+1} \geq -\rho \alpha_k g_k^T d_k \geq \rho \frac{(1-\sigma)(g_k^T d_k)^2}{L \lVert d_k \rVert^2}.$$

Therefore, from the Assumption CG the Zoutendijk condition, (9.31) is obtained ♦

The conjugate gradient algorithms can fail, in the sense that $||g_k|| \geq \gamma > 0$ for all k, only if $||d_k|| \to \infty$ fast enough. More exactly, the sequence of gradient norms $||g_k||$ can be bounded away from zero only if $\sum_{k \geq 0} 1/||d_k|| < \infty$. For any conjugate gradient method with strong Wolfe line search (9.19) and (9.21), the following general result holds (see Nocedal, 1996).

Proposition 9.4 *Suppose that the Assumption CG holds and consider any conjugate gradient algorithm (9.17), where d_k is a descent direction and α_k is obtained by the strong Wolfe line search (9.19) and (9.21). If*

$$\sum_{k \geq 1} \frac{1}{||d_k||^2} = \infty, \tag{9.32}$$

then

$$\liminf_{k \to \infty} ||g_k|| = 0. \tag{9.33}$$

For strongly convex functions, let us prove that the norm of the direction d_{k+1} generated by (9.22)–(9.24) is bounded above. Therefore, by Proposition 9.4 the following result can be proved.

Theorem 9.1 *Suppose that the Assumption CG holds and consider the algorithm (9.17) where the search direction d_{k+1} given by (9.22) with (9.23) and (9.24) is a descent direction and α_k is computed by the strong Wolfe line search (9.19) and (9.21). Suppose that f is a strongly convex function on the level set S, i.e., there exists a constant $\mu > 0$ so that*

$$(\nabla f(x) - \nabla f(y))^T (x - y) \geq \mu ||x - y||^2 \tag{9.34}$$

for all $x, y \in N \subset S$, then

$$\lim_{k \to \infty} ||g_k|| = 0. \tag{9.35}$$

Proof From the Lipschitz continuity it follows that $||y_k|| \leq L||s_k||$. On the other hand, from the strong convexity, $y_k^T s_k \geq \mu ||s_k||^2$. Using the Cauchy inequality, the Assumption CG and the above inequalities δ_k can be estimated as

$$|\delta_k| \leq \frac{|s_k^T g_{k+1}|}{|y_k^T s_k|} + 2\frac{||y_k||^2 |s_k^T g_{k+1}|}{|y_k^T s_k|^2} + \frac{|y_k^T g_{k+1}|}{|y_k^T s_k|} \leq \frac{\Gamma}{\mu ||s_k||} + 2\frac{L^2 \Gamma}{\mu^2 ||s_k||} + \frac{L\Gamma}{\mu ||s_k||}$$
$$= \frac{\Gamma}{\mu}\left(1 + L + 2\frac{L^2}{\mu}\right)\frac{1}{||s_k||}. \tag{9.36}$$

At the same time,

$$|\eta_k| = \frac{|s_k^T g_{k+1}|}{|y_k^T s_k|} \le \frac{\|s_k\| \|g_{k+1}\|}{\mu \|s_k\|^2} \le \frac{\Gamma}{\mu \|s_k\|}. \tag{9.37}$$

Therefore, using (9.36) and (9.37) in (9.22), it follows that

$$\|d_{k+1}\| \le \|g_{k+1}\| + |\delta_k| \|s_k\| + |\eta_k| \|y_k\| \le \Gamma + \frac{\Gamma}{\mu}\left(1 + 2L + 2\frac{L^2}{\mu}\right), \tag{9.38}$$

showing that (9.32) is true. By Proposition 9.4 it follows that (9.33) is true, which for strongly convex functions is equivalent to (9.35). ♦

Convergence analysis for general nonlinear functions exploits the Assumption CG as well as the fact that by Wolfe line search, $y_k^T s_k > 0$ and therefore it can be bounded from below by a positive constant, i.e., there exists $\tau > 0$ so that $y_k^T s_k \ge \tau$.

Theorem 9.2 *Suppose that the Assumption CG holds and consider the algorithm (9.17) where the search direction d_{k+1} given by (9.22) with (9.23) and (9.24) is a descent direction, α_k is computed by the Wolfe line search (9.19)–(9.20) and there exists a constant $\tau > 0$ so that $y_k^T s_k \ge \tau$ for any $k \ge 1$. Then,*

$$\liminf_{k \to \infty} \|g_k\| = 0. \tag{9.39}$$

Proof Since $g_k^T s_k < 0$ for any k, it follows that $s_k^T g_{k+1} = y_k^T s_k + g_k^T s_k < y_k^T s_k$. By the Assumption CG it follows that

$$\|y_k\| = \|g_{k+1} - g_k\| = \|\nabla f(x_k + \alpha_k d_k) - \nabla f(x_k)\| \le L\|s_k\| \le 2BL.$$

Suppose that $g_k \ne 0$ for all $k \ge 1$, otherwise a stationary point is obtained. Now, from (9.23), using the Assumption CG, the following estimation is obtained

$$\begin{aligned} |\delta_k| &\le \frac{|s_k^T g_{k+1}|}{|y_k^T s_k|} + 2\frac{\|y_k\|^2 |s_k^T g_{k+1}|}{|y_k^T s_k|^2} + \frac{|y_k^T g_{k+1}|}{|y_k^T s_k|} \\ &\le 1 + 2\frac{\|y_k\|^2}{|y_k^T s_k|} + \frac{|y_k^T g_{k+1}|}{|y_k^T s_k|} \le 1 + 2\frac{L^2 \|s_k\|^2}{|y_k^T s_k|} + \frac{L\Gamma \|s_k\|}{|y_k^T s_k|} \\ &\le 1 + 2\frac{4B^2 L^2}{|y_k^T s_k|} + \frac{2BL\Gamma}{|y_k^T s_k|} \le 1 + \frac{8B^2 L^2 + 2BL\Gamma}{\tau} \equiv M_1. \end{aligned} \tag{9.40}$$

On the other hand, from (9.24),

$$|\eta_k| \le \frac{|s_k^T g_{k+1}|}{|y_k^T s_k|} \le \frac{\|s_k\| \|g_{k+1}\|}{\tau} \le \frac{2B\Gamma}{\tau} \equiv M_2. \tag{9.41}$$

Therefore, from (9.22),

$$\|d_{k+1}\| \leq \|g_{k+1}\| + |\delta_k|\|s_k\| + |\eta_k|\|y_k\| \leq \Gamma + 2BM_1 + 2BLM_2. \tag{9.42}$$

Now, from Proposition 9.4, it follows that (9.39) is true. ♦

Numerical study. In the first set of numerical experiments, the performances of the TTCG method and its accelerated variant TTCGa are presented. For this, the set of 80 unconstrained optimization test problems from the UOP collection is used, where the number of variables is $n = 1000, \ldots, 10000$. Figure 9.1 shows the Dolan and Moré performance profiles of TTCG versus TTCGa.

Figure 9.1 shows that TTCGa is more robust than TTCG. Since TTCG is a modification of HS (see (9.25)) or a modification of CG-DESCENT (see (9.26)), in the same set of numerical experiments. Figure 9.2 presents a comparison of TTCG versus HS and versus CG-DESCENT (version 1.4). CG-DESCENT is far more efficient and more robust than TTCG. The search direction in the TTCG method is given by (9.27), where the matrix Q_k given by (9.28) is a severe modification of the inverse BFGS approximation to the Hessian (9.29). Clearly, Q_k does not satisfy the quasi-Newton equation, and therefore the curvature of the minimizing function is captured in a modest way.

Figure 9.3 contains the performance profiles of TTCG versus DL ($t = 1$) and versus DESCONa. As it is known, DL is a simple modification of HS based on the Dai and Liao conjugacy condition (7.10). TTCG is a more elaborated three-term

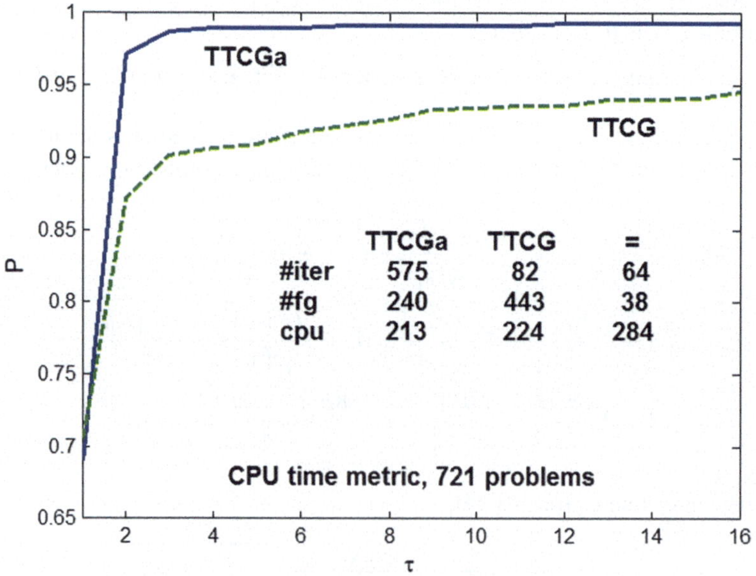

Figure 9.1 Performance profiles of TTCG versus TTCGa

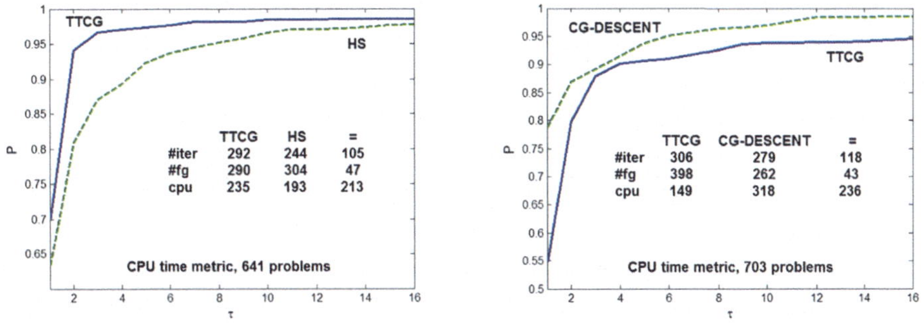

Figure 9.2 Performance profiles of TTCG versus HS and versus CG-DESCENT

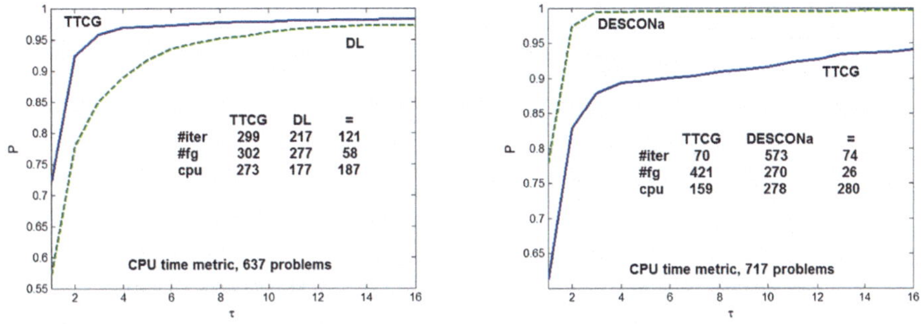

Figure 9.3 Performance profiles of TTCG versus DL ($t = 1$) and versus DESCONa

conjugate gradient method based on a special modification of the memoryless BFGS approximation to the inverse Hessian. Clearly, TTCG is top performer versus DL.

On the other hand, DESCONa is a more elaborated conjugate gradient method, satisfying both the sufficient descent and the Dai and Liao conjugacy conditions by using a modified Wolfe line search. Figure 9.3 shows that DESCONa is much more efficient and more robust than TTCG.

In Figure 9.4, the performance profiles of TTCG versus CONMIN and versus SCALCG (θ_k- spectral) are presented on the same set of unconstrained optimization problems from the UOP collection.

Figure 9.5 contains the performance profiles of TTCG versus L-BFGS ($m = 5$) and versus TN for solving the problems from the UOP collection. TTCG is more efficient than both L-BFGS ($m = 5$) and TN. Both L-BFGS and TN are highly elaborated, implemented in sophisticated software, using in one way or another BFGS approximation to the Hessian. L-BFGS captures the curvature of the minimizing function by using only a certain number of vectors pairs $\{s_i, y_i\}$ to update

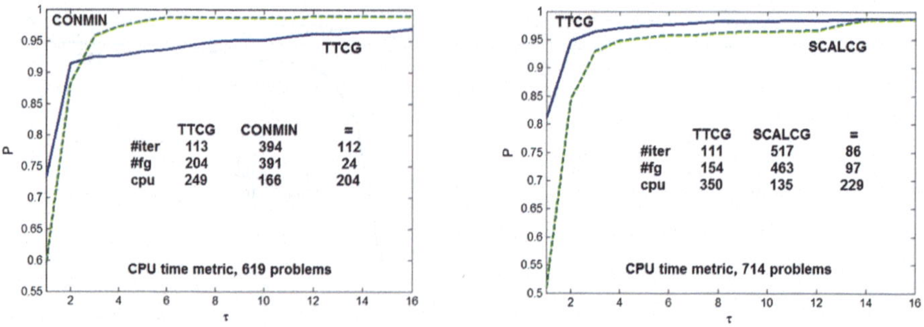

Figure 9.4 Performance profiles of TTCG versus CONMIN and versus SCALCG

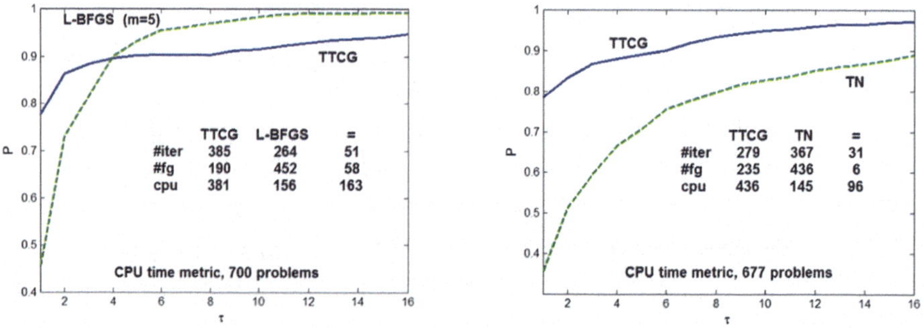

Figure 9.5 Performance profiles of TTCG versus L-BFGS ($m = 5$) and versus TN

the BFGS approximation to the Hessian. L-BFGS is more robust than TTCG. On the other hand, TN uses a different strategy. The search direction is determined by an approximate solution to the Newton system. Compared to TN, TTCG is top performer, being much more efficient and more robust.

9.2 A Three-Term Conjugate Gradient Method with Subspace Minimization (TTS)

Stoer and Yuan (1995) presented an algorithm for computing the search direction by minimizing the approximate quadratic model of function f in the two-dimensional subspace spanned by the negative current gradient and the previous search direction. Their method reduces to the conjugate gradient method when the line searches are exact and the objective function is strictly convex and quadratic. In another effort for solving large-scale unconstrained optimization problems, in (Conn, Gould, Sartenaer, & Toint, 1996), the so-called iterated-subspace minimization (ISM) method was introduced. At each iteration of this method a low-dimensional manifold, the iterate subspace, is constructed and an approximate minimizer of the objective function f in this manifold is determined. This method proves to be advantageous in some cases, but in general, it cannot be trusted and a number of important aspects remain for future investigation.

In this section, let us introduce a simple algorithm for solving large-scale unconstrained optimization problems, in which the directions are computed by minimizing the quadratic approximation of the minimizing function f in a subspace spanned by the vectors $-g_{k+1}, s_k$ and y_k, (Andrei, 2014). Consider that at the kth iteration an inexact Wolfe line search is executed, that is, the stepsize α_k satisfying (9.19) and (9.20) is computed. With this, the following elements $s_k = x_{k+1} - x_k$ and $y_k = g_{k+1} - g_k$ can immediately be determined. Let us now consider the quadratic approximate of function f in x_{k+1} as

$$\Phi_{k+1}(d) = g_{k+1}^T d + \frac{1}{2} d^T B_{k+1} d, \tag{9.43}$$

where B_{k+1} is an approximation of the Hessian $\nabla^2 f(x_{k+1})$ and d is the direction which follows to be determined. The direction d_{k+1} is computed as

$$d_{k+1} = -g_{k+1} + a_k s_k + b_k y_k \tag{9.44}$$

where the scalars a_k and b_k are determined as solution of the following minimizing problem

$$\min_{a_k \in \mathbb{R}, b_k \in \mathbb{R}} \Phi_{k+1}(d_{k+1}) \tag{9.45}$$

Introducing d_{k+1} from (9.44) in the minimizing problem (9.45), then a_k and b_k are obtained as solution of the following linear algebraic system

$$a_k(s_k^T B_{k+1} s_k) + b_k(s_k^T B_{k+1} y_k) = g_{k+1}^T B_{k+1} s_k - s_k^T g_{k+1}, \tag{9.46a}$$

$$a_k(s_k^T B_{k+1} y_k) + b_k(y_k^T B_{k+1} y_k) = g_{k+1}^T B_{k+1} y_k - y_k^T g_{k+1}. \tag{9.46b}$$

Having in view that B_{k+1} is an approximation of $\nabla^2 f(x_{k+1})$ and $\nabla^2 f(x_{k+1}) s_k \approx y_k$, B_{k+1} can be considered to satisfy the secant equation $B_{k+1} s_k = y_k$. Therefore, the system (9.46) can be written as

$$a_k(s_k^T y_k) + b_k \|y_k\|^2 = g_{k+1}^T y_k - s_k^T g_{k+1}, \tag{9.47a}$$

$$a_k \|y_k\|^2 + b_k(y_k^T B_{k+1} y_k) = g_{k+1}^T B_{k+1} y_k - y_k^T g_{k+1}. \tag{9.47b}$$

In order to solve the system (9.47), the quantities $\eta_k \equiv y_k^T B_{k+1} y_k$ and $\omega_k \equiv g_{k+1}^T B_{k+1} y_k$ must be evaluated. Suppose that B_{k+1} is positive definite. Now, using the secant equation $B_{k+1} s_k = y_k$, it is clear that

$$\eta_k = y_k^T B_{k+1} y_k = \frac{y_k^T B_{k+1} y_k s_k^T B_{k+1} s_k}{(y_k^T B_{k+1} s_k)^2} \frac{(y_k^T B_{k+1} s_k)^2}{s_k^T B_{k+1} s_k}$$

$$= \frac{y_k^T B_{k+1}^{\frac{1}{2}} B_{k+1}^{\frac{1}{2}} y_k s_k^T B_{k+1}^{\frac{1}{2}} B_{k+1}^{\frac{1}{2}} s_k}{(y_k^T B_{k+1}^{\frac{1}{2}} B_{k+1}^{\frac{1}{2}} s_k)^2} \frac{(y_k^T y_k)^2}{y_k^T s_k} = \frac{\left\| B_{k+1}^{\frac{1}{2}} y_k \right\|^2 \left\| B_{k+1}^{\frac{1}{2}} s_k \right\|^2}{\left((B_{k+1}^{\frac{1}{2}} y_k)^T (B_{k+1}^{\frac{1}{2}} s_k) \right)^2} \frac{(y_k^T y_k)^2}{y_k^T s_k}$$

$$= \frac{1}{\cos^2 < B_{k+1}^{\frac{1}{2}} y_k, B_{k+1}^{\frac{1}{2}} s_k >} \frac{(y_k^T y_k)^2}{y_k^T s_k}. \tag{9.48}$$

Since B_{k+1} is unknown, the quantity $\cos^2 < B_{k+1}^{\frac{1}{2}} y_k, B_{k+1}^{\frac{1}{2}} s_k >$ in (9.48) is also unknown. However, since the mean value of $\cos^2 \xi = 1/2$, then in (9.48), it seems reasonable to replace the above quantity $\cos^2 < B_{k+1}^{\frac{1}{2}} y_k, B_{k+1}^{\frac{1}{2}} s_k >$ by 1/2. Therefore, η_k can be computed as

$$\eta_k = 2 \frac{(y_k^T y_k)^2}{y_k^T s_k}. \tag{9.49}$$

Next, to compute ω_k, the BFGS update initialized with the identity matrix can be used, thus obtaining

$$\omega_k = g_{k+1}^T B_{k+1} y_k = g_{k+1}^T \left[I + \frac{y_k y_k^T}{y_k^T s_k} - \frac{s_k s_k^T}{s_k^T s_k} \right] y_k$$

$$= g_{k+1}^T y_k + \frac{(g_{k+1}^T y_k)(y_k^T y_k)}{y_k^T s_k} - \frac{(g_{k+1}^T s_k)(s_k^T y_k)}{s_k^T s_k}. \tag{9.50}$$

Another way to compute ω_k is to use the BFGS update initialized from the scaling matrix $((s_k^T y_k)/\|s_k\|^2)I$. However, using this variant our numerical tests did not prove any improvement of the algorithm.

Using (9.49) and (9.50), the linear algebraic system (9.47) can be written as

$$a_k (s_k^T y_k) + b_k \|y_k\|^2 = g_{k+1}^T y_k - s_k^T g_{k+1}, \tag{9.51a}$$

$$a_k \|y_k\|^2 + b_k \eta_k = \omega_k - y_k^T g_{k+1}. \tag{9.51b}$$

Using (9.49), the determinant of the system (9.51) is

$$\Delta_k = (s_k^T y_k)\eta_k - (y_k^T y_k)^2 = (y_k^T y_k)^2 \geq 0. \tag{9.52}$$

Supposing that $\Delta_k > 0$, then the solution of the linear system (9.51) is obtained as

$$a_k = \frac{1}{\Delta_k}\left[\eta_k(y_k^T g_{k+1} - s_k^T g_{k+1}) - \|y_k\|^2(\omega_k - y_k^T g_{k+1})\right], \tag{9.53}$$

$$b_k = \frac{1}{\Delta_k}\left[(y_k^T s_k)(\omega_k - y_k^T g_{k+1}) - \|y_k\|^2(y_k^T g_{k+1} - s_k^T g_{k+1})\right]. \tag{9.54}$$

Therefore, if $\Delta_k > 0$, then the search direction is computed as in (9.44), where the scalars a_k and b_k are computed as in (9.53) and (9.54), respectively.

If the line search is exact, that is $s_k^T g_{k+1} = 0$, then from (9.53) and (9.54), it results that $a_k = (y_k^T g_{k+1})/(y_k^T s_k)$ and $b_k = 0$, i.e., the search direction is computed as

$$d_{k+1} = -g_{k+1} + \frac{y_k^T g_{k+1}}{y_k^T s_k} s_k, \tag{9.55}$$

which is exactly the HS conjugate gradient algorithm.

Proposition 9.5 *Suppose that $B_{k+1} > 0$. Then, d_{k+1} given by (9.44), where a_k and b_k are computed as in (9.53) and (9.54,) respectively, is a descent direction.*

Proof From (9.43), observe that $\Phi_{k+1}(0) = 0$. Since $B_{k+1} > 0$ and d_{k+1} given by (9.44), (9.53), and (9.54) is the solution of (9.45), it follows that $\Phi_{k+1}(d_{k+1}) \leq 0$. Therefore,

$$g_{k+1}^T d_{k+1} \leq -\frac{1}{2} d_{k+1}^T B_{k+1} d_{k+1} < 0, \tag{9.56}$$

i.e., d_{k+1} is a descent direction. ♦

Proposition 9.6 *Suppose that the search direction d_{k+1} is given by (9.44), where a_k and b_k satisfy the linear algebraic system (9.51). Then, the direction d_{k+1} satisfies the Dai–Liao conjugacy condition $y_k^T d_{k+1} = -s_k^T g_{k+1}$.*

Proof Since d_{k+1} is given by (9.44), it follows that $y_k^T d_{k+1}$ is given by (9.51a), which is exactly the Dai–Liao conjugacy condition. ♦

Taking into consideration the acceleration scheme presented in Remark 5.1, where the acceleration factor ξ_k is computed as in (5.24), according to the value of the parameter "*acceleration*" (true or false) the following algorithms TTS and TTSa can be presented. TTSa is the accelerated version of TTS.

Algorithm 9.2 Three-term subspace minimization: TTS/TTSa

1.	Select a starting point $x_0 \in dom f$ and compute: $f_0 = f(x_0)$ and $g_0 = \nabla f(x_0)$. Select $\varepsilon_A > 0$ sufficiently small and some positive values $0 < \rho < \sigma < 1$ used in Wolfe line search. Set $d_0 = -g_0$ and $k = 0$
2.	Test a criterion for stopping the iterations. If the test is satisfied, then stop; otherwise continue with step 3
3.	Determine the stepsize α_k using the Wolfe line search conditions (9.19) and (9.20). Update the variables $x_{k+1} = x_k + \alpha_k d_k$. Compute f_{k+1}, g_{k+1} and $s_k = x_{k+1} - x_k$, $y_k = g_{k+1} - g_k$
4.	If the parameter *acceleration* is true, then (a) Compute: $z = x_k + \alpha_k d_k$, $g_z = \nabla f(z)$ and $y_k = g_k - g_z$ (b) Compute: $\bar{a}_k = \alpha_k g_k^T d_k$, and $\bar{b}_k = -\alpha_k y_k^T d_k$ (c) If $\lvert \bar{b}_k \rvert \geq \varepsilon_A$, then compute $\xi_k = -\bar{a}_k / \bar{b}_k$ and update the variables as $x_{k+1} = x_k + \xi_k \alpha_k d_k$. Compute f_{k+1} and g_{k+1}. Compute $s_k = x_{k+1} - x_k$ and $y_k = g_{k+1} - g_k$
5.	Compute η_k, ω_k and Δ_k as in (9.49), (9.50) and (9.52), respectively
6.	Compute a_k and b_k as in (9.53) and (9.54) respectively
7.	Compute the search direction as: $d_{k+1} = -g_{k+1} + a_k s_k + b_k y_k$
8.	Powell restart criterion. If $\lvert g_{k+1}^T g_k \rvert > 0.2 \lVert g_{k+1} \rVert^2$ then set $d_{k+1} = -g_{k+1}$
9.	Consider $k = k + 1$ and go to step 2 ♦

Convergence analysis. Suppose that the Assumption CG holds. Under this assumption on f, there exists a constant $\Gamma \geq 0$ so that $\lVert \nabla f(x) \rVert \leq \Gamma$ for all $x \in S$, where S is the level set of function f. As in Proposition 9.3, the above three-term conjugate gradient method satisfies the Zoutendijk condition under the standard Wolfe line search (9.19) and (9.20). For strongly convex functions, it is easy to prove that the norm of the direction d_{k+1} generated by (9.44), (9.53), and (9.54) is bounded above. Therefore, by Proposition 9.4, the following theorem can be proved.

Theorem 9.3 *Suppose that the Assumption CG holds and consider the algorithm (9.17) and (9.44) with (9.53) and (9.54), where d_k is a descent direction and α_k is computed by the Wolfe line search (9.19) and (9.20). Suppose that ∇f satisfies the Lipschitz condition and f is a strongly convex function on S, i.e., there exists a constant $\mu > 0$ so that*

$$(\nabla f(x) - \nabla f(y))^T (x - y) \geq \mu \lVert x - y \rVert^2 \tag{9.57}$$

for all $x, y \in S$, then

$$\liminf_{k \to \infty} \lVert g_k \rVert = 0. \tag{9.58}$$

Proof From the Lipschitz continuity: $\|y_k\| \leq L\|s_k\|$. On the other hand, from the strong convexity, it follows that $y_k^T s_k \geq \mu\|s_k\|^2$. Now, using the Cauchy inequality, from the Lipschitz continuity and the strong convexity, it follows that

$$\left|y_k^T g_{k+1} - s_k^T g_{k+1}\right| \leq \left|y_k^T g_{k+1}\right| + \left|s_k^T g_{k+1}\right| \leq L\Gamma\|s_k\| + \Gamma\|s_k\| = \Gamma(L+1)\|s_k\|. \tag{9.59}$$

On the other hand,

$$\begin{aligned} \left|\omega_k - y_k^T g_{k+1}\right| &\leq \frac{\left|y_k^T g_{k+1}\right|\left|y_k^T y_k\right|}{\left|y_k^T s_k\right|} + \frac{\left|s_k^T g_{k+1}\right|\left|y_k^T s_k\right|}{\left|s_k^T s_k\right|} \\ &\leq \frac{\Gamma\|y_k\|^3}{\mu\|s_k\|^2} + \frac{\Gamma\|s_k\|^2\|y_k\|}{\|s_k\|^2} \leq \frac{\Gamma L^3\|s_k\|}{\mu} + \Gamma L\|s_k\| \\ &= \left(\frac{\Gamma L^3}{\mu} + \Gamma L\right)\|s_k\|. \end{aligned} \tag{9.60}$$

From the strong convexity and the Cauchy inequality observe that $\mu\|s_k\|^2 \leq y_k^T s_k \leq \|y_k\|\|s_k\|$, i.e.,

$$\mu\|s_k\| \leq \|y_k\|. \tag{9.61}$$

From (9.53), using (9.61), the following estimation is obtained

$$\begin{aligned} |a_k| &\leq \frac{1}{\|y_k\|^4}\left[2\frac{\|y_k\|^4}{\left|y_k^T s_k\right|}\left|y_k^T g_{k+1} - s_k^T g_{k+1}\right| + \|y_k\|^2\left|\omega_k - y_k^T g_{k+1}\right|\right] \\ &\leq \frac{2}{\left|y_k^T s_k\right|}\left|y_k^T g_{k+1} - s_k^T g_{k+1}\right| + \frac{1}{\|y_k\|^2}\left|\omega_k - y_k^T g_{k+1}\right| \\ &\leq \frac{2}{\left|y_k^T s_k\right|}\Gamma(L+1)\|s_k\| + \frac{1}{\|y_k\|^2}\left(\frac{\Gamma L^3}{\mu} + \Gamma L\right)\|s_k\| \\ &\leq \frac{2}{\mu\|s_k\|^2}\Gamma(L+1)\|s_k\| + \frac{1}{\mu^2\|s_k\|^2}\left(\frac{\Gamma L^3}{\mu} + \Gamma L\right)\|s_k\| \\ &\leq \left(\frac{2\Gamma(L+1)}{\mu} + \frac{\Gamma L^3}{\mu^3} + \frac{\Gamma L}{\mu^2}\right)\frac{1}{\|s_k\|} \equiv M_1\frac{1}{\|s_k\|}. \end{aligned} \tag{9.62}$$

But, from (9.54), since $\|s_k\| \le \|y_k\|/\mu$, it follows that

$$\begin{aligned}|b_k| &\le \frac{1}{\|y_k\|^4}\left[\left|y_k^T s_k\right|\left|\omega_k - y_k^T g_{k+1}\right| + \|y_k\|^2\left|y_k^T g_{k+1} - s_k^T g_{k+1}\right|\right] \\ &\le \frac{\|y_k\|\|s_k\|}{\|y_k\|^4}\left(\frac{\Gamma L^3}{\mu} + \Gamma L\right)\|s_k\| + \frac{1}{\|y_k\|^2}\Gamma(L+1)\|s_k\| \\ &\le \left(\frac{\Gamma L^3}{\mu^3} + \frac{\Gamma L}{\mu^2} + \frac{\Gamma(L+1)}{\mu}\right)\frac{1}{\|y_k\|} \equiv M_2\frac{1}{\|y_k\|}.\end{aligned} \tag{9.63}$$

Therefore, from (9.44),

$$\|d_{k+1}\| \le \|g_{k+1}\| + |a_k|\|s_k\| + |b_k|\|y_k\| \le \Gamma + M_1 + M_2.$$

From Proposition 9.4, it is easy to see that (9.58) is true. ♦

Numerical study. Figure 9.6 shows the performance profiles of the accelerated TTS method (TTSa) versus the unaccelerated TTS for solving the problems from the UOP collection (Andrei, 2018g), where for each problem 10 numerical experiments have been executed with the number of variables $n = 1000, \ldots, 10000$. All the numerical experiments are given in the context of Remark 1.1. Observe that TTSa is more robust than TTS and the difference is substantial. This shows the importance of the acceleration of the conjugate gradient methods.

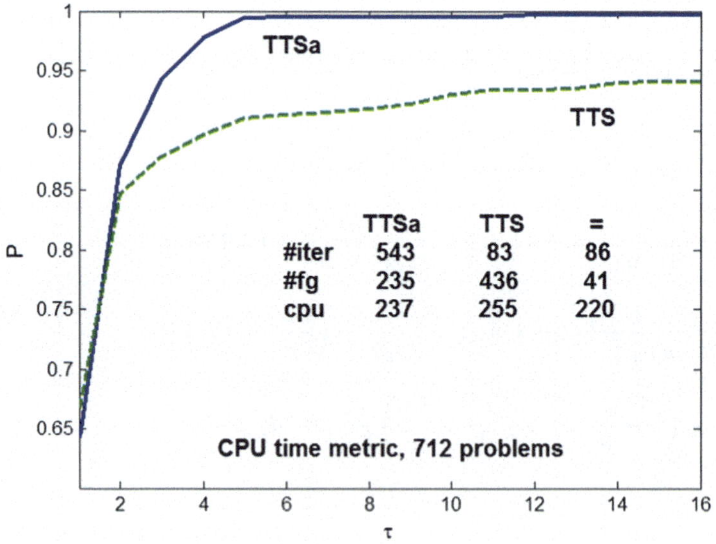

Figure 9.6 Performance profiles of TTS versus TTSa

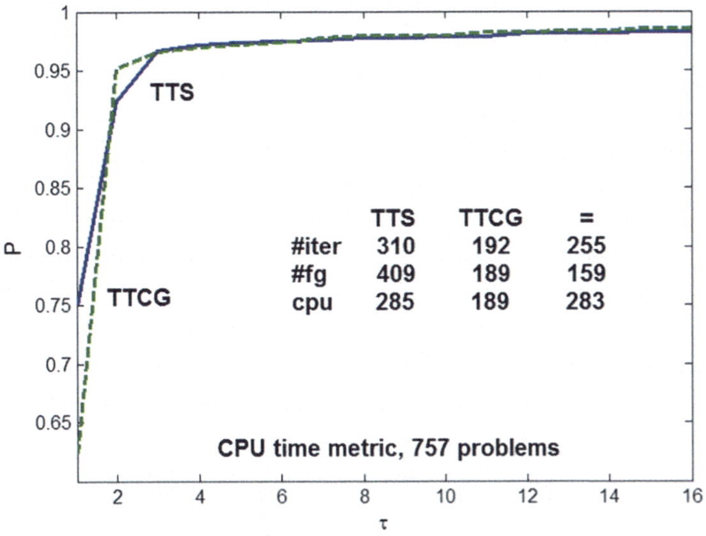

Figure 9.7 Performance profiles of TTS versus TTCG

Figure 9.7 presents the performance profiles of TTS versus TTCG. Observe that the performance profiles of these methods are very close to each other, TTS being slightly more efficient. Both TTS and TTCG are three-term conjugate gradient methods. The search direction in TTCG satisfies the descent condition (see Proposition 9.1) and also the Dai–Liao conjugacy condition with $t_k > 0$ (see Proposition 9.2), being modifications of the HS or of the CG-DESCENT methods. On the other hand, the search direction in TTS is defined by two parameters and is determined to minimize the quadratic approximation of the minimizing function in x_{k+1}. In fact, the search direction in TTS is also descent (see Proposition 9.5) and satisfies the Dai–Liao conjugacy conditions with $t = 1$ (see Proposition 9.6). The convergence of both these methods is established by using the Zoutendijk condition (9.31) and Proposition 9.4.

The performance profiles of TTS versus DL ($t = 1$), DL+ ($t = 1$), CG-DESCENT (version 1.4), and DESCONa are illustrated in Figure 9.8. Notice that TTS is more efficient and more robust than both DL ($t = 1$) and DL+ ($t = 1$). This is not a surprise, as DL is a simple modification of the HS method. Both CG-DESCENT and DESCONa are much more efficient and more robust than TTS.

Figure 9.9 also shows that CONMIN is more robust than TTS, but TTS is top performer in comparison with SCALCG (θ_k spectral).

Figure 9.10 shows that TTS is more efficient than L-BFGS ($m = 5$). However, L-BFGS ($m = 5$) is more robust. Compared with TN, TTS is clearly the best.

Observe that, compared to CG-DESCENT or to DESCONa, the three-term conjugate gradient method based on the subspace minimization TTS in its *basic form* has modest performances. However, some variants of this method using the

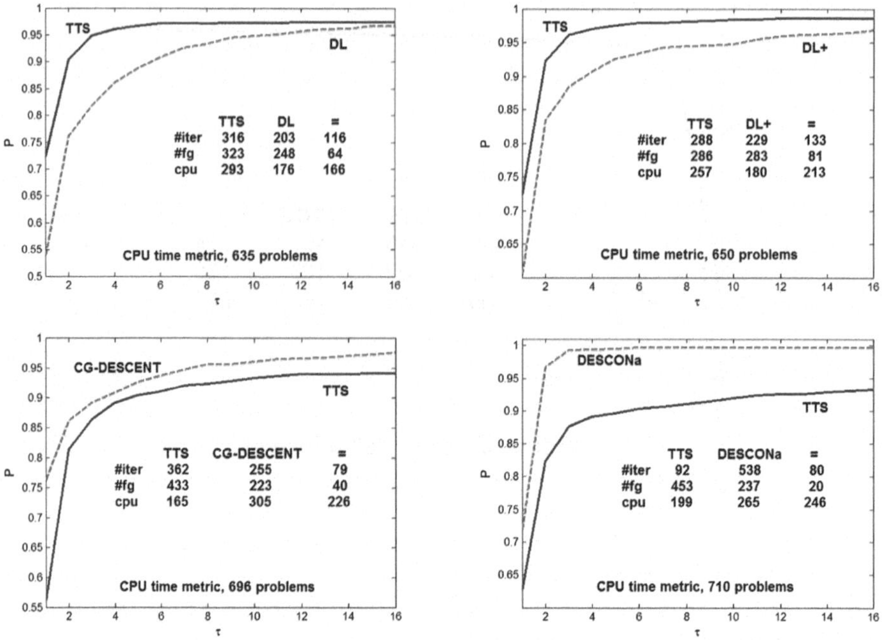

Figure 9.8 Performance profiles of TTS versus DL ($t = 1$), DL+ ($t = 1$), CG-DESCENT, and DESCONa

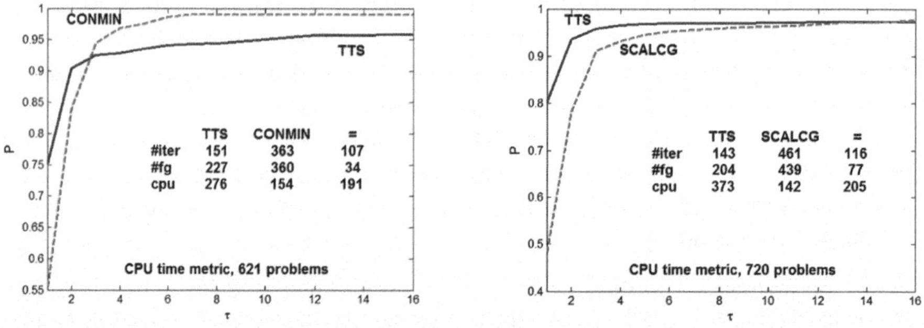

Figure 9.9 Performance profiles of TTS versus CONMIN and versus SCALCG (spectral)

subspace minimization are more efficient and more robust than CG-DESCENT. Indeed, a new subspace minimization conjugate gradient algorithm with non-monotone line search was developed by Li, Liu, and Liu (2018). The search direction is obtained by minimizing the function f on the subspace $span\{-g_{k+1}, s_k, s_{k-1}\}$, or $span\{-g_{k+1}, s_k\}$. In their algorithm, they provided three choices of the search direction, in which two of them are obtained by minimizing an approximation to the objective function on the above subspaces and the third one is

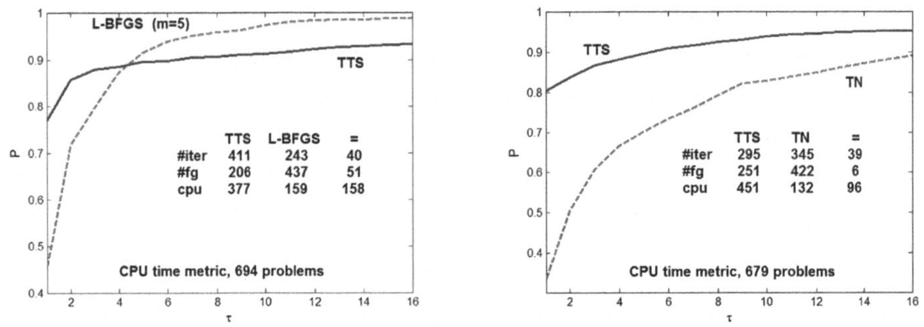

Figure 9.10 Performance profiles of TTS versus L-BFGS ($m = 5$) and versus TN

$-g_{k+1}$. In the first case, the search direction is expressed as $d_{k+1} = \mu g_{k+1} + \nu s_k + \tau s_{k-1}$, where μ, ν, and τ are scalar parameters determined as solution of the model

$$\min_{\mu,\nu,\tau} \begin{pmatrix} \|g_{k+1}\|^2 \\ g_{k+1}^T s_k \\ g_{k+1}^T s_{k-1} \end{pmatrix}^T \begin{pmatrix} \mu \\ \nu \\ \tau \end{pmatrix} + \frac{1}{2} \begin{pmatrix} \mu \\ \nu \\ \tau \end{pmatrix}^T \begin{pmatrix} \rho_k & g_{k+1}^T y_k & g_{k+1}^T y_{k-1} \\ g_{k+1}^T y_k & y_k^T s_k & y_k^T s_{k-1} \\ g_{k+1}^T y_{k-1} & y_k^T s_{k-1} & y_{k-1}^T s_{k-1} \end{pmatrix} \begin{pmatrix} \mu \\ \nu \\ \tau \end{pmatrix}, \tag{9.64}$$

and $\rho_k \approx g_{k+1}^T B_{k+1} g_{k+1}$. If some criteria are not satisfied, then this approximation to the minimizing function is abandoned and the search direction is expressed as $d_{k+1} = \mu g_{k+1} + \nu s_k$, where the scalar parameters μ and ν are determined as solution of the problem

$$\min_{\mu,\nu} \begin{pmatrix} \|g_{k+1}\|^2 \\ g_{k+1}^T s_k \end{pmatrix}^T \begin{pmatrix} \mu \\ \nu \end{pmatrix} + \frac{1}{2} \begin{pmatrix} \mu \\ \nu \end{pmatrix}^T \begin{pmatrix} \rho_k & g_{k+1}^T y_k \\ g_{k+1}^T y_k & y_k^T s_k \end{pmatrix} \begin{pmatrix} \mu \\ \nu \end{pmatrix}. \tag{9.65}$$

Also, they introduced certain ingredients: criteria for choosing the search directions, the initial stepsize computation as well as the approximated line search proposed by Hager and Zhang (2005). Using most of the unconstrained optimization problems from the UOP collection, their algorithm SMCG_NLS is more efficient and more robust than CG-DESCENT (5.3) (Hager & Zhang, 2005) and CGOPT (Dai & Kou, 2013). However, it is unknown whether the performances of SMCG_NLS are better due to its search direction or to the ingredients used.

Another approach has been given by Momeni and Peyghami (2019) who propose an algorithm that try to adjust the positive values for the Dai–Liao parameter by using quadratic and/or cubic regularization models of the objective function (see Chapter 11). The cubic regularization model of the objective function is properly employed when the nonpositive curvature is detected.

9.3 A Three-Term Conjugate Gradient Method with Minimization of One-Parameter Quadratic Model of Minimizing Function (TTDES)

In this section, another approach for getting three-term conjugate gradient algorithms by the minimization of the one-parameter quadratic model of the function f is described (Andrei, 2015a). The idea is to consider the quadratic approximation of the function f in the current point and to determine the search direction by the minimization of this quadratic model. It is assumed that the symmetrical approximation of the Hessian matrix satisfies the general quasi-Newton equation, which depends on a positive parameter. The search direction is obtained by modifying the iteration matrix corresponding to the solution of the quadratic model minimization in order to be symmetric. The parameter in the search direction is determined by the minimization of the condition number of this new iteration matrix.

Let us consider that at the kth iteration an inexact Wolfe line search was executed and hence the stepsize α_k was determined by satisfying (9.19) and (9.20). With this value of the stepsize the following elements $s_k = x_{k+1} - x_k$ and $y_k = g_{k+1} - g_k$ can be determined. Now, consider the quadratic approximate of function f in x_{k+1} as

$$\Phi_{k+1} = f_{k+1} + g_{k+1}^T d + \frac{1}{2} d^T B_{k+1} d, \tag{9.66}$$

where B_{k+1} is a symmetrical approximation of the Hessian $\nabla^2 f(x_{k+1})$ and d is the direction which follows to be determined. The search direction d_{k+1} is computed as

$$d_{k+1} = -g_{k+1} + \beta_k s_k, \tag{9.67}$$

where the scalar β_k is determined as solution of the following minimization problem

$$\min_{\beta_k \in \mathbb{R}} \Phi_{k+1}(d_{k+1}). \tag{9.68}$$

Introducing d_{k+1} from (9.67) in the minimizing problem (9.68), then β_k is obtained as

$$\beta_k = \frac{g_{k+1}^T B_{k+1} s_k - g_{k+1}^T s_k}{s_k^T B_{k+1} s_k}. \tag{9.69}$$

Now, suppose that the symmetric matrix B_{k+1} is an approximation of the Hessian matrix $\nabla^2 f(x_{k+1})$ so that $B_{k+1} s_k = \omega^{-1} y_k$, with $\omega \neq 0$, known as the *generalized quasi-Newton equation*. Therefore, the parameter β_k can be written as

$$\beta_k = \frac{g_{k+1}^T y_k - \omega g_{k+1}^T s_k}{y_k^T s_k}. \tag{9.70}$$

Hence, the search direction d_{k+1} from (9.67) becomes

$$d_{k+1} = -g_{k+1} + \frac{s_k y_k^T - \omega s_k s_k^T}{y_k^T s_k} g_{k+1}. \tag{9.71}$$

Now, using the idea of Perry (1977), (9.71) can be written as

$$d_{k+1} = -\overline{Q}_{k+1} g_{k+1}, \tag{9.72}$$

where

$$\overline{Q}_{k+1} = I - \frac{s_k y_k^T}{y_k^T s_k} + \omega \frac{s_k s_k^T}{y_k^T s_k} = I - \frac{s_k (y_k - \omega s_k)^T}{y_k^T s_k}. \tag{9.73}$$

Remark 9.1 Observe that the solution of the minimization problem (9.68) is the solution of the linear algebraic system of equations

$$B_{k+1} d = -g_{k+1}. \tag{9.74}$$

Using the above approach, observe that the search direction d_{k+1} is as in (9.72) where $\overline{Q}_{k+1}$ is defined by (9.73), which is not a symmetric matrix. ♦

Remark 9.2 From (9.71), the search direction can be expressed as

$$d_{k+1} = -g_{k+1} + \frac{y_k^T g_{k+1}}{y_k^T s_k} s_k - \omega \frac{s_k^T g_{k+1}}{y_k^T s_k} s_k, \tag{9.75}$$

i.e., it is a three-term search direction. ♦

There are a number of choices for the parameter ω in (9.75). For example, if $\omega = y_k^T g_{k+1} / s_k^T g_{k+1}$, then the steepest-descent method can be obtained. When $\omega = 0$, then (9.75) is exactly the Hestenes and Stiefel (1952) search direction. If $\omega = 2\|y_k\|^2 / y_k^T s_k$, then the CG-DESCENT method by Hager and Zhang (2005) is obtained. On the other hand, if ω is a bounded positive constant ($\omega = 0.1$, for example), then the resulting method is that of Dai and Liao (2001). Therefore, (9.75) is a general formula for the search direction calculation, which covers a lot of known conjugate gradient methods.

However, the matrix $\overline{Q}_{k+1}$ in (9.73) determines a crude form of a quasi-Newton method which is not symmetric. Therefore, using $\overline{Q}_{k+1}$, let us slightly modify it and consider the following symmetric matrix

$$Q_{k+1} = I - \frac{s_k y_k^T}{y_k^T s_k} + \frac{y_k s_k^T}{y_k^T s_k} + \omega \frac{s_k s_k^T}{y_k^T s_k}. \tag{9.76}$$

Using (9.76), the following search direction is obtained

$$\begin{aligned} d_{k+1} &= -Q_{k+1} g_{k+1} \\ &= -g_{k+1} + \frac{y_k^T g_{k+1} - \omega s_k^T g_{k+1}}{y_k^T s_k} s_k - \frac{s_k^T g_{k+1}}{y_k^T s_k} y_k, \end{aligned} \tag{9.77}$$

which determines a three-term conjugate gradient method.

Proposition 9.7 *Consider $\omega > 0$ and the stepsize α_k in (9.17) determined by the Wolfe line search (9.19) and (9.20). Then, the search direction (9.77) satisfies the Dai and Liao conjugacy condition $y_k^T d_{k+1} = -t_k(s_k^T g_{k+1})$, where $t_k > 0$.*

Proof By direct computation it follows that

$$y_k^T d_{k+1} = -\left(\omega + \frac{\|y_k\|^2}{y_k^T s_k}\right)(s_k^T g_{k+1}) \equiv t_k(s_k^T g_{k+1}),$$

where $t_k = \omega + \|y_k\|^2/(y_k^T s_k)$. By the Wolfe line search $y_k^T s_k > 0$, therefore $t_k > 0$. ♦

Proposition 9.8 *Consider $\omega > 0$ and the stepsize α_k in (9.17) determined by the Wolfe line search (9.19) and (9.20). Then, the search direction (9.77) satisfies the descent condition $g_{k+1}^T d_{k+1} \leq 0$.*

Proof By direct computation,

$$g_{k+1}^T d_{k+1} = -\|g_{k+1}\|^2 - \omega \frac{(s_k^T g_{k+1})^2}{y_k^T s_k} \leq 0,$$

since $y_k^T s_k > 0$ by the Wolfe line search (9.19) and (9.20). ♦

To define the corresponding algorithm, the only problem is to specify a suitable value for the parameter ω. There are some possibilities. For example, for

$$\omega = 1 + \frac{\|y_k\|^2}{y_k^T s_k}$$

the method reduces to the three-term conjugate gradient method THREECG (Andrei, 2013a). On the other hand, if

$$\omega = 1 + 2\frac{\|y_k\|^2}{y_k^T s_k},$$

then the three-term conjugate gradient method TTCG is obtained (Andrei, 2013b). In the following, the parameter ω is determined by minimizing the condition number of the symmetric matrix Q_{k+1}.

Theorem 9.4 *Let Q_{k+1} be defined by (9.76). If $\omega > 0$, then Q_{k+1} is a nonsingular matrix and its eigenvalues consist of 1 ($n-2$ multiplicity), λ_{k+1}^+ and λ_{k+1}^-, where*

$$\lambda_{k+1}^+ = \frac{1}{2}\left[(a_k + 2) + \sqrt{(a_k + 2)^2 - 4(a_k + b_k)}\right], \tag{9.78}$$

$$\lambda_{k+1}^- = \frac{1}{2}\left[(a_k + 2) - \sqrt{(a_k + 2)^2 - 4(a_k + b_k)}\right] \tag{9.79}$$

and

$$a_k = \omega\frac{\|s_k\|^2}{y_k^T s_k}, \quad b_k = \frac{\|s_k\|^2}{y_k^T s_k}\frac{\|y_k\|^2}{y_k^T s_k} > 1. \tag{9.80}$$

Proof Consider

$$Q_{k+1} = I - \frac{s_k(y_k - \omega s_k)^T}{y_k^T s_k} + \frac{y_k s_k^T}{y_k^T s_k}.$$

Therefore, it follows that (see Appendix A)

$$\det(Q_{k+1}) = \omega\frac{\|s_k\|^2}{y_k^T s_k} + \frac{\|s_k\|^2}{y_k^T s_k}\frac{\|y_k\|^2}{y_k^T s_k} = a_k + b_k.$$

Hence, the matrix Q_{k+1} is nonsingular.

Since for any $\xi \in span\{s_k, y_k\}^{\perp} \subset \mathbb{R}^n$, $Q_{k+1}\xi = \xi$, it follows that Q_{k+1} has the eigenvalue 1 of multiplicity $n-2$, corresponding to the eigenvectors $\xi \in span\{s_k, y_k\}^{\perp}$.

Now, (see Appendix A)

$$tr(Q_{k+1}) = tr\left[I - \frac{s_k y_k^T}{y_k^T s_k} + \frac{y_k s_k^T}{y_k^T s_k} + \omega \frac{s_k s_k^T}{y_k^T s_k}\right]$$
$$= n + \omega \frac{\|s_k\|^2}{y_k^T s_k} = n + a_k.$$

Therefore, by the relationships between the trace and the determinant of a matrix and its eigenvalues, it follows that the other two eigenvalues of Q_{k+1} are the roots of the following quadratic polynomial

$$\lambda^2 - (a_k + 2)\lambda + (a_k + b_k) = 0. \tag{9.81}$$

Thus, the other two eigenvalues of the symmetric matrix Q_{k+1} are determined from (9.81) as (9.78) and (9.79), respectively. $b_k > 1$ follows from the following inequality

$$\frac{y_k^T s_k}{\|s_k\|^2} \leq \frac{\|y_k\|^2}{y_k^T s_k}.$$

♦

Proposition 9.9 *The Q_{k+1} matrix defined by (9.76) is a normal matrix.*

Proof By direct computation, it is easy to see that $Q_{k+1}Q_{k+1}^T = Q_{k+1}^T Q_{k+1}$. ♦

In order to have λ_{k+1}^+ and λ_{k+1}^- as real eigenvalues, from (9.78) and (9.79), the following condition must be fulfilled $(a_k + 2)^2 - 4(a_k + b_k) \geq 0$, out of which the following estimation of the parameter ω can be determined

$$\omega \geq \frac{2}{\|s_k\|^2}\sqrt{\|s_k\|^2\|y_k\|^2 - (y_k^T s_k)^2}. \tag{9.82}$$

Since $b_k > 1$, it follows that the estimation of ω given by (9.82) is well defined (if $\|s_k\| \neq 0$). From (9.81), it results that

$$\lambda_{k+1}^+ + \lambda_{k+1}^- = a_k + 2 \geq 0, \tag{9.83}$$

$$\lambda_{k+1}^+ \lambda_{k+1}^- = a_k + b_k \geq 0. \tag{9.84}$$

Therefore, from (9.83) and (9.84), it follows that both λ_{k+1}^+ and λ_{k+1}^- are positive eigenvalues. Since $(a_k + 2)^2 - 4(a_k + b_k) \geq 0$, from (9.78) and (9.79), it follows that $\lambda_{k+1}^+ \geq \lambda_{k+1}^-$.

By direct computation, from (9.78) and (9.82), it follows that

$$\lambda_{k+1}^{+} \geq 1 + \sqrt{b_k - 1} > 1. \tag{9.85}$$

A simple analysis of equation (9.81) shows that 1 (the eigenvalue of Q_{k+1}) is not into the interval $[\lambda_{k+1}^{-}, \lambda_{k+1}^{+}]$. Since both λ_{k+1}^{+} and λ_{k+1}^{-} are positive, $\lambda_{k+1}^{+} > 1$ and $\lambda_{k+1}^{+} \geq \lambda_{k+1}^{-}$, it follows that $1 \leq \lambda_{k+1}^{-} \leq \lambda_{k+1}^{+}$. Therefore, the maximum eigenvalue of Q_{k+1} is λ_{k+1}^{+} and its minimum eigenvalue is 1.

From Proposition 9.9, Q_{k+1} is a normal matrix. Therefore, the condition number $\kappa(Q_{k+1})$ of Q_{k+1} can be computed as in the following proposition.

Proposition 9.10 *The condition number of the normal, symmetric, matrix Q_{k+1} is*

$$\kappa(Q_{k+1}) = \frac{\lambda_{k+1}^{+}}{1} = \frac{1}{2}\left[(a_k + 2) + \sqrt{(a_k + 2)^2 - 4(a_k + b_k)}\right]. \tag{9.86}$$

$\kappa(Q_{k+1})$ gets its minimum $\sqrt{b_k - 1} + 1$ when $a_k = 2\sqrt{b_k - 1}$.

Proof Observe that $b_k > 1$. By direct computation, the minimum of (9.86) is obtained for $a_k = 2\sqrt{b_k - 1}$, for which $\kappa(Q_{k+1})$ arrives to its minimum $\sqrt{b_k - 1} + 1$. ♦

According to Proposition 9.10, when $a_k = 2\sqrt{b_k - 1}$, the condition number of Q_{k+1} defined by (9.76) arrives at its minimum. Therefore, from (9.80), by using this equality a suitable choice of parameter ω is

$$\omega = \frac{2}{\|s_k\|^2}\sqrt{\|s_k\|^2\|y_k\|^2 - (y_k^T s_k)^2}. \tag{9.87}$$

Since $b_k > 1$, it follows that ω in (9.87) is well defined (if $\|s_k\| \neq 0$). This choice of the parameter ω makes the condition number of Q_{k+1} approach to its minimum.

To conclude, the search direction is given by (9.77) as

$$d_{k+1} = -g_{k+1} + \delta_k s_k - \eta_k y_k, \tag{9.88}$$

where the parameters δ_k and η_k are computed as

$$\delta_k = \frac{y_k^T g_{k+1} - \omega s_k^T g_{k+1}}{y_k^T s_k}, \quad \eta_k = \frac{s_k^T g_{k+1}}{y_k^T s_k}, \tag{9.89}$$

respectively, and ω is computed as in (9.87).

Taking into account the acceleration scheme from Remark 5.1, where the acceleration factor ξ_k is computed as in (5.24), according to the value of the parameter "*acceleration*" (true or false) the following algorithms TTDES and TTDESa can be presented. TTDESa is the accelerated version of TTDES.

Algorithm 9.3 Three-term quadratic model minimization: TTDES/TTDESa

1.	Select a starting point $x_0 \in domf$ and compute: $f_0 = f(x_0)$ and $g_0 = \nabla f(x_0)$. Select $\varepsilon_A > 0$ sufficiently small some positive values $0 < \rho < \sigma < 1$ used in Wolfe line search. Set $d_0 = -g_0$ and $k = 0$
2.	Test a criterion for stopping the iterations. If the test is satisfied, then stop; otherwise continue with step 3
3.	Determine the stepsize α_k by using the Wolfe line search conditions (9.19) and (9.20). Update the variables $x_{k+1} = x_k + \alpha_k d_k$. Compute f_{k+1}, g_{k+1} and $s_k = x_{k+1} - x_k$, $y_k = g_{k+1} - g_k$
4.	If *acceleration* equal true, then (a) Compute: $z = x_k + \alpha_k d_k$, $g_z = \nabla f(z)$ and $y_k = g_k - g_z$ (b) Compute: $a_k = \alpha_k g_k^T d_k$, and $b_k = -\alpha_k y_k^T d_k$ (c) If $\lvert b_k \rvert \geq \varepsilon_A$, then compute $\xi_k = -a_k / b_k$ and update the variables as $x_{k+1} = x_k + \xi_k \alpha_k d_k$. Compute f_{k+1} and g_{k+1}. Compute $s_k = x_{k+1} - x_k$ and $y_k = g_{k+1} - g_k$
5.	Compute ω as in (9.87) and determine δ_k and η_k as in (9.89)
6.	Compute the search direction as: $d_{k+1} = -g_{k+1} + \delta_k s_k - \eta_k y_k$
7.	Powell restart criterion. If $\lvert g_{k+1}^T g_k \rvert > 0.2 \lVert g_{k+1} \rVert^2$ then set $d_{k+1} = -g_{k+1}$
8.	Consider $k = k + 1$ and go to step 2 ♦

Under reasonable assumptions, the Wolfe line search conditions and the Powell restart criterion are sufficient to prove the global convergence of the algorithm TTDES.

Convergence analysis. To prove the global convergence of this nonlinear conjugate gradient algorithm, the Zoutendijk condition is used. The analysis is given under the Assumption CG. Under this Assumption on *f*, there exists a constant $\Gamma \geq 0$ so that $\lVert \nabla f(x) \rVert \leq \Gamma$ for all $x \in S = \{x \in \mathbb{R}^n : f(x) \leq f(x_0)\}$.

Theorem 9.5 *Suppose that the Assumption CG holds and consider the algorithm* (9.17) *with* (9.88) *and* (9.89), *where* ω *is given by* (9.87), d_k *is a descent direction and* α_k *is computed by the strong Wolfe line search* (9.19) *and* (9.21). *Suppose that f is a strongly convex function on S, i.e., there exists a constant* $\mu > 0$ *so that*

$$(\nabla f(x) - \nabla f(y))^T (x - y) \geq \mu \lVert x - y \rVert^2 \tag{9.90}$$

for all $x, y \in S$, *then*

$$\lim_{k \to \infty} \lVert g_k \rVert = 0. \tag{9.91}$$

Proof From the Lipschitz continuity, $\|y_k\| \le L\|s_k\|$. On the other hand, from the strong convexity and the Cauchy inequality, it follows that $\mu\|s_k\|^2 \le y_k^T s_k \le \|y_k\|\|s_k\|$, i.e., $\mu\|s_k\| \le \|y_k\|$. Therefore, for strongly convex functions, under the Wolfe line search it follows that $L \ge \mu$ (if $\|s_k\| \ne 0$). Now, from (9.87), by using the Cauchy inequality, the Assumption CG and the above inequalities, it follows that

$$\begin{aligned}|\omega| &\le \frac{2}{\|s_k\|^2}\sqrt{\|s_k\|^2\|y_k\|^2-\mu^2\|s_k\|^4}\\ &= \frac{2}{\|s_k\|}\sqrt{\|y_k\|^2-\mu^2\|s_k\|^2}\\ &\le \frac{2}{\|s_k\|}\sqrt{L^2\|s_k\|^2-\mu^2\|s_k\|^2}\\ &= 2\sqrt{L^2-\mu^2}.\end{aligned} \tag{9.92}$$

But, from (9.89),

$$\begin{aligned}|\delta_k| &\le \frac{|y_k^T g_{k+1}|}{|y_k^T s_k|} + |\omega|\frac{|s_k^T g_{k+1}|}{|y_k^T s_k|}\\ &\le \frac{\Gamma L}{\mu\|s_k\|} + 2\sqrt{L^2-\mu^2}\frac{\Gamma}{\mu\|s_k\|}\\ &= \frac{\Gamma}{\mu}\left(L + 2\sqrt{L^2-\mu^2}\right)\frac{1}{\|s_k\|}.\end{aligned} \tag{9.93}$$

At the same time,

$$|\eta_k| = \frac{|s_k^T g_{k+1}|}{|y_k^T s_k|} \le \frac{\|s_k\|\|g_{k+1}\|}{\mu\|s_k\|^2} \le \frac{\Gamma}{\mu\|s_k\|}. \tag{9.94}$$

Therefore, using (9.93) and (9.94) in (9.88), it can be seen that

$$\begin{aligned}\|d_{k+1}\| &\le \|g_{k+1}\| + |\delta_k|\|s_k\| + |\eta_k|\|y_k\|\\ &\le \Gamma + \frac{2\Gamma}{\mu}\left(L + \sqrt{L^2-\mu^2}\right).\end{aligned} \tag{9.95}$$

Hence, (9.32) in Proposition 9.4 is true. By Proposition 9.4, it follows that (9.33) is true, which for strongly convex functions is equivalent to (9.91). ♦

Numerical study. In the first set of numerical experiments, let us compare the performances of TTDES versus its accelerated version TTDESa. Figure 9.11 presents the Dolan and Moré performance profiles of these methods, subject to the CPU time metric.

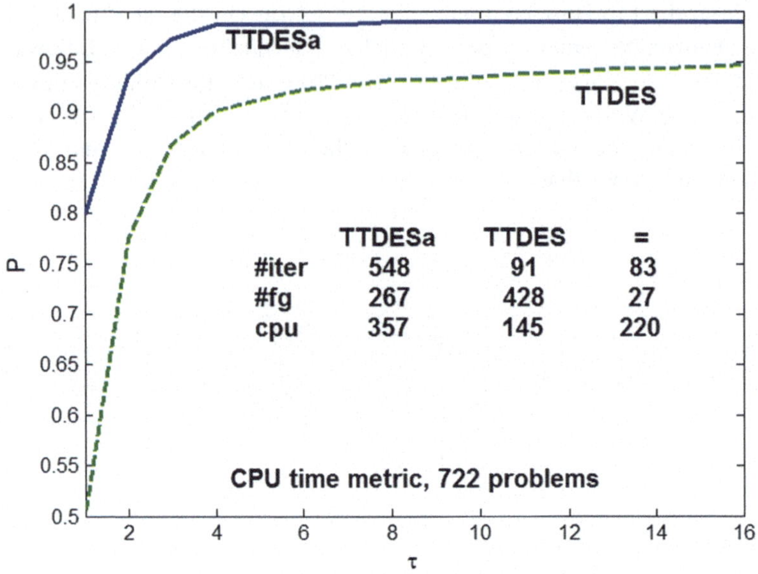

Figure 9.11 Performance profiles of TTDES versus TTDESa

Observe that TTDESa is more efficient and more robust than TTDES and the differences are substantial.

In the second set of numerical experiments, the comparisons of TTDES versus TTCG and TTS are presented in Figure 9.12. Even if these three-term conjugate gradient methods, TTCG, TTS, and TTDES, are based on different strategies, they have similar performances. In all these three-term conjugate gradient methods the search direction simultaneously satisfies both the descent and the conjugacy conditions. However, as we know, the conjugate gradient algorithms satisfying both these conditions are not necessarily the best ones. Besides, in TTS the search directions are computed by minimizing the quadratic approximation of the minimizing function f in a subspace spanned by the vectors: $-g_{k+1}$, s_k and y_k. The weakness of TTS is the formula for updating the quantities $\eta_k = y_k^T B_{k+1} y_k$ and $\omega_k = g_{k+1}^T B_{k+1} y_k$. On the other hand, TTDES is based on the minimization of the one-parameter quadratic model of the function f using the generalized secant equation. However, the weakness of TTDES is that the matrix Q_{k+1} given by (9.76) is not able to capture the curvature of the minimizing function along the iterations. The performances of TTDES can be improved by choosing a modification of the $\overline{Q}_{k+1}$ matrix given by (9.73) in such a way so that the curvature of the minimizing function is better captured by the modified matrix Q_{k+1} and the eigenvalues of Q_{k+1}, which determines the search direction (9.77), could be easily obtained.

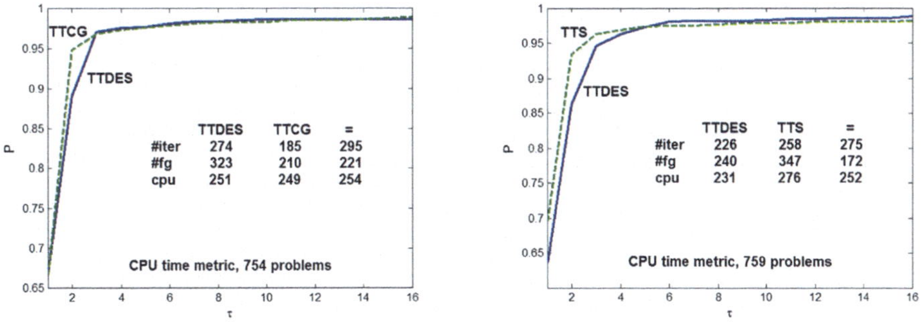

Figure 9.12 Performance profiles of TTDES versus TTCG and versus TTS

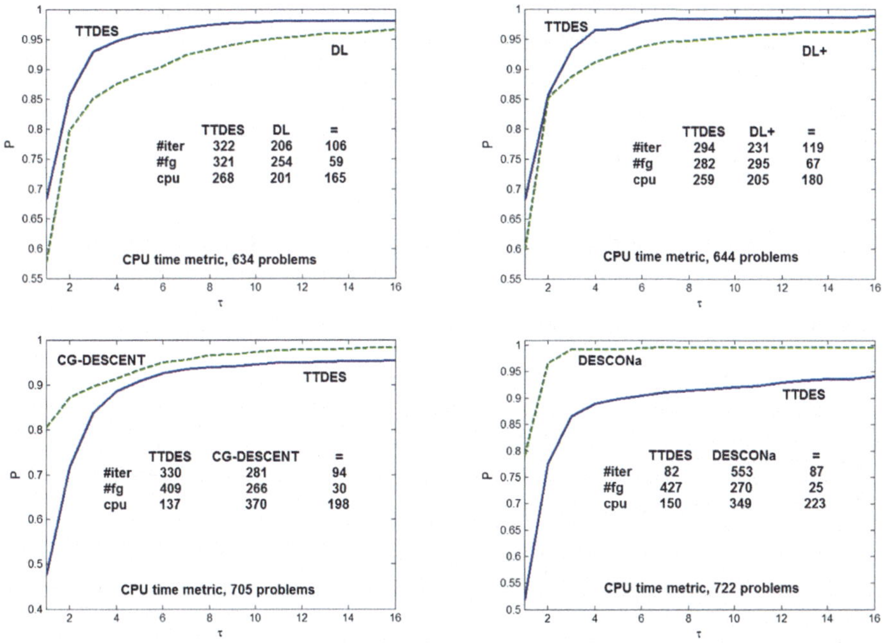

Figure 9.13 Performance profiles of TTDES versus DL ($t = 1$), DL+ ($t = 1$), CG-DESCENT, and DESCONa

In the third set of numerical experiments, the performance profiles of TTDES versus DL ($t = 1$), DL+ ($t = 1$), CG-DESCENT (version 1.4), and DESCONa are presented in Figure 9.13. Observe that TTDES is more efficient and more robust than DL ($t = 1$) and DL+ ($t = 1$). However, both CG-DESCENT and DESCONa are way more efficient and more robust versus TTDES.

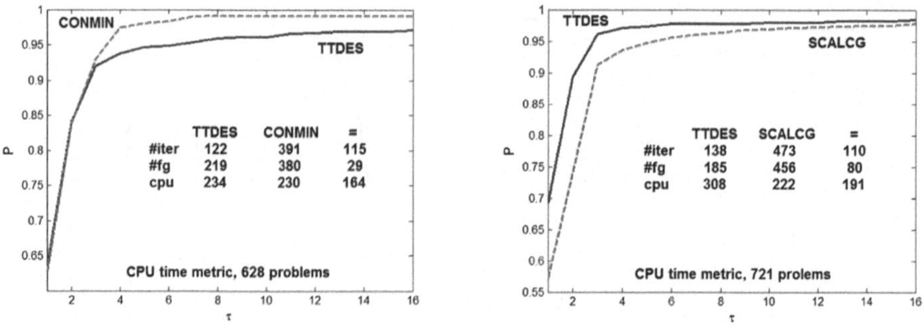

Figure 9.14 Performance profiles of TTDES versus CONMIN and versus SCALCG

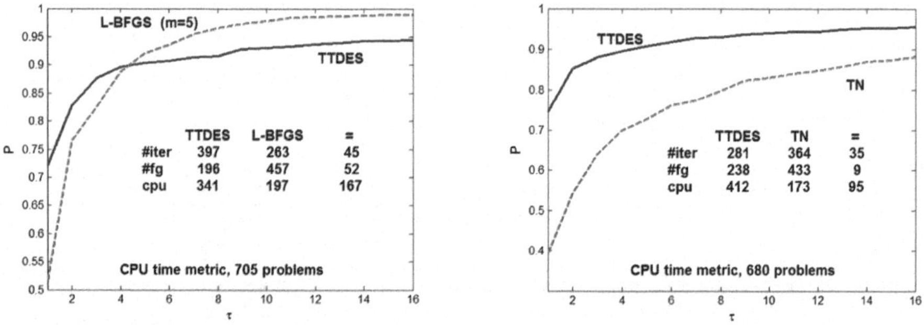

Figure 9.15 Performance profiles of TTDES versus L-BFGS ($m = 5$) and versus TN

In the fourth set of numerical experiments, the performance profiles of TTDES versus CONMIN and versus SCALCG are presented in Figure 9.14.

In the following, the performance profiles of TTDES versus L-BFGS ($m = 5$) and versus TN are presented in Figure 9.15.

Table 9.1 contains the performances of TTCG, TTS, and TTDES for solving the applications from the MINPACK-2 collection, where the number of variables for each one is 40,000.

The entries across the last row of Table 9.1 demonstrate that all these three-term conjugate gradient methods, TTCG, TTS, and TTDES, have similar performances, TTS being slightly more efficient. This is in agreement with the results from Figure 9.12. Observe that these three-term conjugate gradient methods, in one way or another, use a modified BFGS update. The search direction in TTCG is artificially introduced in such a way so that the iteration matrix from (9.28) should be close to the BFGS approximation of the Hessian. Only TTS and TTDES are based on the principle of minimizing the quadratic approximation of the function f in x_{k+1}. The search direction in TTS uses two parameters determined in such a way so that both the descent and the conjugacy conditions in this method should be satisfied. The search direction in TTDES depends on one parameter, determined to

Table 9.1 Performances of TTCG, TTS and TTDES for solving five applications from the MINPACK-2 collection

		TTCG			TTS			TTDES		
	n	*#iter*	*#f*	*cpu*	*#iter*	*#fg*	*cpu*	*#iter*	*#fg*	*cpu*
A1	40,000	428	693	7.24	433	680	6.26	426	671	6.74
A2	40,000	833	1346	14.79	988	1543	15.04	736	1176	11.35
A3	40,000	7371	11116	237.59	4253	6435	138.73	4302	6709	141.83
A4	40,000	551	896	46.08	654	1036	39.82	1085	1716	81.26
A5	40,000	442	720	11.67	368	557	7.94	357	570	9.41
Total	-	9625	14771	317.37	6696	10251	207.79	6906	10842	250.59

Table 9.2 The total performances of L-BFGS ($m = 5$), TN, TTCG, TTS, and TTDES for solving five applications from the MINPACK-2 collection with 40,000 variables

Algorithms	*#iter*	*#fg*	*cpu*
L-BFGS ($m = 5$)	4842	4987	102.92
TN	153	3714	104.57
TTCG	9625	14771	317.37
TTS	6696	10251	207.79
TTDES	6906	10842	250.59

minimize the condition number of the iteration matrix from (9.76). For all these methods, the convergence has been proved under classical assumptions. The fact that these methods can be proved to converge does not necessarily imply that they are good methods. Their limitation is that the iteration matrices do not capture in a proper way the curvature of the minimizing function in the current point.

Table 9.2 contains the total performances of L-BFGS ($m = 5$) (See Table 1.2), of TN (see Table 1.3), of TTCG, TTS, and TTDES (see Table 9.1) for solving all five applications from the MINPACK-2 collection, each of them with 40,000 variables. Subject to the CPU time metric, both L-BFGS and TN are top performers.

Notes and References

Three-term conjugate gradient methods are interesting innovations introduced by Beale (1972) and Nazareth (1977). Plenty of three-term conjugate gradient algorithms are known. In this chapter, only three of them have been presented, the ones based on different concepts including: satisfying the descent and the conjugacy conditions, the subspace minimization and the minimization of one-parameter quadratic model of the minimizing function. For the set of unconstrained optimization problems included in the UOP collection they have similar performances, TTS based on the subspace minimization being slightly more efficient. In this class of algorithms, the subspace minimization approach proved to be one of the best.

The subspace minimization is a very active area of research in nonlinear optimization generating three-term conjugate gradient algorithms. Branch, Coleman, and Li (1999) developed a subspace, interior point, and conjugate gradient method for large-scale bound-constrained minimization problems. A great deal of effort was devoted to relate the trust-region method to the subspace technique. Wang and Yuan (2006) developed a subspace implementation of quasi-Newton trust region methods for unconstrained optimization. In order to study the idea of solving the trust-region problem in a small subspace, while still obtaining globally and locally fast convergence, Bellavia and Morini (2006) introduced a prototype subspace trust-region method for large bound-constrained nonlinear systems. Erway and Gill (2009) developed a subspace minimization method that solves the inequality constrained trust-region subproblem over a sequence of evolving low-dimensional subspaces. Wei and Yang (2016) presented a new limited-memory symmetric-rank-1 (SR1) trust-region algorithm on compact Riemannian manifolds by using the subspace technique. Yang, Chen, and Lu (2017) proposed a subspace three-term conjugate gradient method in which the direction is generated by minimizing a quadratic approximation of the objective function in a subspace. Carlberg, Forstall, and Tuminaro (2016) presented a Krylov-subspace-recycling method for efficiently solving sequences of linear algebraic systems of equations characterized by varying right-hand-sides and symmetric positive definite matrices. Hager and Zhang (2013) presented the limited-memory conjugate gradient method by solving the corresponding subspace problem in which the space is spanned by the recent prior search directions. Various kinds of subspace techniques used to generate methods for nonlinear optimization problems are summarized by Yuan (2014).

Recently, a new subspace minimization conjugate gradient method based on the tensor model for unconstrained optimization has been presented by Wang, Liu, and Liu (2019). In this method, if the objective function is close to a quadratic, then, to generate the search direction, a quadratic approximation model in a two-dimensional subspace is constructed; otherwise, a tensor model is developed. Numerical comparisons proved that this algorithm is competitive with CGOPT (Dai & Kou, 2013) and CG-DESCENT (Hager & Zhang, 2005). Further, Li, Liu, and Liu (2019) developed a subspace minimization conjugate gradient method based on a conic model for the minimizing function. The search direction is computed by minimizing a selected approximate model in a two-dimensional subspace. That is, if the objective function is not close to a quadratic, the search direction is generated by a conic model. Otherwise, a quadratic model is considered. For unconstrained strictly convex problems, a variant of conjugate gradient algorithm with a subspace minimization problem on each iteration, related to earlier work by Nemirovsky and Yudin (1983), was developed by Karimi and Vavasis (2012). (See also the Ph.D. Thesis by Karimi, 2013). Their algorithm attains a theoretical complexity bound of $O(\log(1/\varepsilon)\sqrt{L/l})$, where the ratio L/l characterizes the strong convexity of the objective function and ε is the desired relative accuracy, that is,

$(f(x_n) - f(x^*))/(f(x_0) - f(x^*)) \leq \varepsilon$, where x_0 is the starting point, x^* is the optimizer, and x_n is the final iterate.

Anyway, three-term conjugate gradient algorithms are always a very active area of research, with various possibilities of development.

Chapter 10
Preconditioning of the Nonlinear Conjugate Gradient Algorithms

Preconditioning is a technique to accelerate the conjugate gradient algorithms. In Chapter 2, the preconditioning of the linear conjugate gradient algorithm has been presented. For linear systems $Ax = b$, preconditioning modifies the system of equations in order to improve the eigenvalue distribution of A. Instead of $Ax = b$, another system $(C^{-T}AC^{-1})y = C^{-T}b$, where C is a nonsingular matrix and $y = Cx$ is solved. In practice, however, the C matrix is never directly used. Instead, a constant symmetric positive definite preconditioning matrix $P = CC^T$ is constructed so that $P^{-1} \approx A^{-1}$ and $P^{-1}A \approx I$. The exact sense in which the preconditioned matrix $P^{-1}A$ should approximate the identity matrix is not very well defined. For example, one would like $\rho(I - P^{-1}A) \ll 1$ in order to achieve fast asymptotic convergence, where $\rho(I - P^{-1}A)$ is the spectrum of the matrix $I - P^{-1}A$. Another interpretation is given by $\|I - P^{-1}A\| \ll 1$ to achieve large error reduction at each step.

For linear systems, this process and choices of preconditioning matrices C are well understood. For example, for linear systems, preconditioning matrices can be divided into the following three categories:

- Preconditioners for general classes of matrices like Jacobi, Gauss–Seidel, and SOR preconditioners, the incomplete Cholesky and modified incomplete Cholesky preconditioners;
- Preconditioners for board classes of problems like elliptic partial differential equations (multigrid and domain decomposition preconditioners);
- Preconditioners for a specific matrix or for some problems like diffusion or transport equation.

A thorough discussion on preconditioned algorithms for linear systems including comparisons among preconditioners was given by Greenbaum (1997).

However, extension of the process of preconditioning to nonlinear conjugate gradient methods remains an open question with a lot of interpretations. In the following, we shall present some theoretical developments of preconditioning the

N. Andrei, *Nonlinear Conjugate Gradient Methods for Unconstrained Optimization*, Springer Optimization and Its Applications 158,
https://doi.org/10.1007/978-3-030-42950-8_10

nonlinear conjugate gradient algorithms, as described by Hager and Zhang (2006b). Preconditioning of nonlinear conjugate gradient methods means to make a change of variables $x = Cy$, where $C \in \mathbb{R}^{n \times n}$ is an invertible matrix chosen to accelerate the convergence of the algorithm. After writing the conjugate gradient algorithm in the transformed variable y and converting back to the x variable, the iteration is

$$x_{k+1} = x_k + \alpha_k d_k, \tag{10.1}$$

$$d_{k+1} = -Pg_{k+1} + \bar{\beta}_k d_k, \tag{10.2}$$

$d_0 = -Pg_0$ where $P = CC^T$. In this case, the update parameter $\bar{\beta}_k$ in (10.2) is the same as β_k in the original conjugate gradient, but with g_k and d_k replaced by $C^T g_k$ and $C^{-1}d_k$, respectively. For example, for the FR, PRP, and CD methods, the new formulae of the preconditioned conjugate gradient parameters are

$$\bar{\beta}_k^{FR} = \frac{g_{k+1}^T P g_{k+1}}{g_k^T P g_k}, \quad \bar{\beta}_k^{PRP} = \frac{g_{k+1}^T P y_k}{g_k^T P g_k}, \quad \bar{\beta}_k^{CD} = \frac{g_{k+1}^T P g_{k+1}}{-d_k^T g_k}.$$

Of course, at every iteration, the preconditioning matrix P could be changed as P_k, thus obtaining a *dynamic preconditioning* of conjugate gradient algorithms.

In order to get some insights into preconditioning, let us see how the convergence speed of the conjugate gradient method depends on the eigenvalues of the Hessian of the problem. Suppose that the minimizing function f is quadratic

$$f(x) = \frac{1}{2}x^T Qx + b^T x, \tag{10.3}$$

where Q is a symmetric matrix with eigenvalues $\lambda_1 \geq \lambda_2 \geq \ldots \geq \lambda_n \geq 0$. Under the exact line search, the error in the kth iteration of the conjugate gradient method satisfies the following bound (Stiefel, 1958)

$$(x_k - x^*)^T Q(x_k - x^*) \leq \min_{p \in P_{k-1}} \max_{1 \leq i \leq n} (1 + \lambda_i p(\lambda_i))^2 (x_0 - x^*)^T Q(x_0 - x^*),$$

where P_k denotes the set of polynomials of degree at most k. Therefore, given some integer $q \in [1, k]$, it follows that if $p \in P_{k-1}$ is chosen so that the degree k polynomoal $1 + \lambda p(\lambda)$ vanishes with multiplicity 1 at λ_i, $1 \leq i \leq q-1$, and with multiplicity $k - q$ at $(\lambda_q + \lambda_n)/2$, then it results that

$$(x_k - x^*)^T Q(x_k - x^*) \leq \left(\frac{\lambda_q - \lambda_n}{\lambda_q + \lambda_n}\right)^{2(k-q+1)} (x_0 - x^*)^T Q(x_0 - x^*). \tag{10.4}$$

Now, after the change of variables $x = Cy$ in (10.3), it follows that

$$f(Cy) = \frac{1}{2} y^T C^T Q C y + b^T C y.$$

The matrix $C^T QC$ associated with the quadratic in y is similar to the matrix $QCC^T = QP$. Therefore, the best preconditioner is $P = Q^{-1}$, which leads to convergence in one single step since the eigenvalues of $C^T QC$ are all 1.

Therefore, when f is a general nonlinear function, *a good preconditioner is any matrix that approximates the inverse Hessian* $\nabla^2 f(x^*)^{-1}$. There are a lot of possibilities for choosing the preconditioning matrix C with this property and this makes preconditioning of nonlinear conjugate gradient methods an open question. For example, a possible preconditioning strategy for general nonlinear functions, discussed by Nazareth (1979) and Buckley (1978a), is to take $P_k = B_k$, where B_k is an approximation to the inverse Hessian $\nabla^2 f(x^*)^{-1}$ obtained by a quasi-Newton update formula, like the Broyden family

$$B_{k+1} = \left(I - \frac{s_k y_k^T}{y_k^T s_k} \right) B_k \left(I - \frac{y_k s_k^T}{y_k^T s_k} \right) + \frac{s_k s_k^T}{y_k^T s_k} + \gamma v_k v_k^T,$$

where $\gamma \geq 0$ is a parameter and

$$v_k = \left(y_k^T B_k y_k \right)^{1/2} \left(\frac{B_k y_k}{y_k^T B_k y_k} - \frac{s_k}{y_k^T s_k} \right).$$

Nazareth (1979) showed that when the function f is quadratic and the exact line search is used, then preconditioned conjugate gradient with a fixed preconditioner $P = B_0$ is identical to preconditioned conjugate gradient with $P = B_k$ at iteration k provided B_k is generated by the BFGS formula. On the other hand, Buckley (1978a) showed that if the quasi-Newton preconditioner B_k is randomly updated by the BFGS formula, then the iterates are identical to preconditioned conjugate gradient with fixed preconditioner $P = B_0$. In the same realm of research, the infrequent quasi-Newton updates were considered by Buckley (1978b), where a quasi-Newton step is performed and the preconditioner is updated when

$$\left| \frac{g_k^T P g_{k+1}}{g_k^T P g_k} \right| \leq \rho,$$

where $\rho \in (0, 1)$ is a constant. Buckley reported that these infrequent updates led to improvements over the unpreconditioned conjugate gradient.

Another general preconditioning strategy is to use the matrix generated from the limited-memory update L-BFGS formula of Liu and Nocedal (1989). This was implemented by Hager and Zhang (2013) in their limited-memory L-CG-DESCENT algorithm. A nice survey on the relationship between preconditioned conjugate gradient and quasi-Newton methods was given by Nazareth (1986).

10.1 Preconditioners Based on Diagonal Approximations to the Hessian

Let us present another preconditioning, easier to be implemented, obtained by using a diagonal approximation to the Hessian. In this case, the preconditioning matrix P_k is dynamically updated as the inverse of diagonally updating of the Hessian methods presented in Section 1.4.5. For example,

$$P_{k+1} = \left(diag(b_{k+1}^1, \ldots, b_{k+1}^n)\right)^{-1}, \tag{10.5}$$

where the elements $b_{k+1}^i, \quad i = 1, \ldots, n$, are computed as in (1.88)

$$b_{k+1}^i = b_k^i - \frac{(b_k^i)^2 (s_k^i)^2}{\sum_{i=1}^n b_k^i (s_k^i)^2} + \frac{(y_k^i)^2}{y_k^T s_k}. \tag{10.6}$$

This diagonal approximation to the Hessian was first proposed by Gilbert and Lemaréchal (1989) in the context of a *sparse initialization* of the BFGS update. Observe that if $y_k^T s_k > 0$, that P_k is well defined.

In the following, let us present the numerical results of preconditioning some modern conjugate gradient algorithms with preconditioners computed as in (10.5) and (10.6) by using the standard Wolfe line search.

Note that the conjugate gradient algorithms implemented in these numerical experiments are not the exact original algorithms. We implemented only the parameter β_k defining the algorithms and compared them independently of any accompanying specialized line search like the approximate Wolfe line search or the improved Wolfe line search, previously discussed in the chapters of this book. The interest in our numerical studies is not to exhaustively compare the variants of the algorithms to each other, but instead, to show that the preconditioning can improve the performances of the conjugate gradient algorithms.

Example 10.1 Firstly, let us consider the Hager and Zhang (2005) conjugate gradient algorithm defined by (7.46)–(7.49) with standard Wolfe line search, which we call HZ+. Now, let us present the performances of the preconditioned conjugate gradient method of Hager and Zhang for solving the unconstarined optimization problems from the UOP collection, where at each iteration the preconditioner P_{k+1} is computed as in (10.5) and (10.6), which we call HZ+p. In this case, the preconditioned HZ+p algorithm is defined as:

$$d_{k+1} = -P_{k+1}g_{k+1} + \bar{\beta}_k^{HZ+} d_k, \tag{10.7}$$

where

$$\bar{\beta}_k^{HZ+} = \max\{\bar{\beta}_k^{HZ}, \eta_k\}, \tag{10.8}$$

$$\bar{\beta}_k^{HZ} = \frac{g_{k+1}^T P_{k+1} y_k}{y_k^T d_k} - 2\frac{y_k^T P_{k+1} y_k}{y_k^T d_k}\frac{d_k^T g_{k+1}}{y_k^T d_k}, \tag{10.9}$$

$$\eta_k = \frac{-1}{\|d_k\| \min\{0.01, \|g_k\|\}}. \tag{10.10}$$

For each test function, ten numerical experiments with the number of variables $n = 1000, 2000, \ldots, 10000$ have been considered The maximum number of iterations is limited to 2000. The comparisons of algorithms are given in the context of Remark 1.1.

Figure 10.1 presents the performances of HZ+ versus the accelerated version of HZ+ (HZ+a), where the acceleration is described as in Chapter 5 (see Remark 5.1); the performances of HZ+ versus the preconditioned version of HZ+ (HZ+p) (see (10.7)–(10.10)); the performances of HZ+a versus HZ+p and the performances of HZ+a versus the accelerated version of HZ+p (HZ+pa) for solving this set of 800 problems from the UOP collection. Figure 10.1 shows that the accelerated HZ+a is more efficient and more robust versus the HZ+ and the difference is significant. Also, from Figure 10.1, observe that the performances of the preconditioned HZ+p with the preconditioner given by a diagonal approximation to the Hessian (10.5) and (10.6) are similar to the performances of HZ+. On the other hand, the accelerated HZ+a is top performer versus the preconditioned HZ+p. Finally, notice that the accelerated variant of the preconditioned HZ+ (HZ+pa) is more efficient and more robust than the accelerated variant of HZ+ (HZ+a).

Example 10.2 Let us now consider the Dai and Kou conjugate gradient algorithm defined as in (8.114) and (8.118) with standard Wolfe line search, which we call DK+. The preconditioned version of DK+, which we call DK+p, is defined as

$$d_{k+1} = -P_{k+1}g_{k+1} + \bar{\beta}_k^{DK+} d_k, \tag{10.11}$$

where

$$\bar{\beta}_k^{DK+} = \max\{\bar{\beta}_k^{DK}, \eta_k\}, \tag{10.12}$$

$$\bar{\beta}_k^{DK} = \frac{g_{k+1}^T P_{k+1} y_k}{y_k^T d_k} - \frac{y_k^T P_{k+1} y_k}{y_k^T d_k}\frac{d_k^T g_{k+1}}{y_k^T d_k}, \tag{10.13}$$

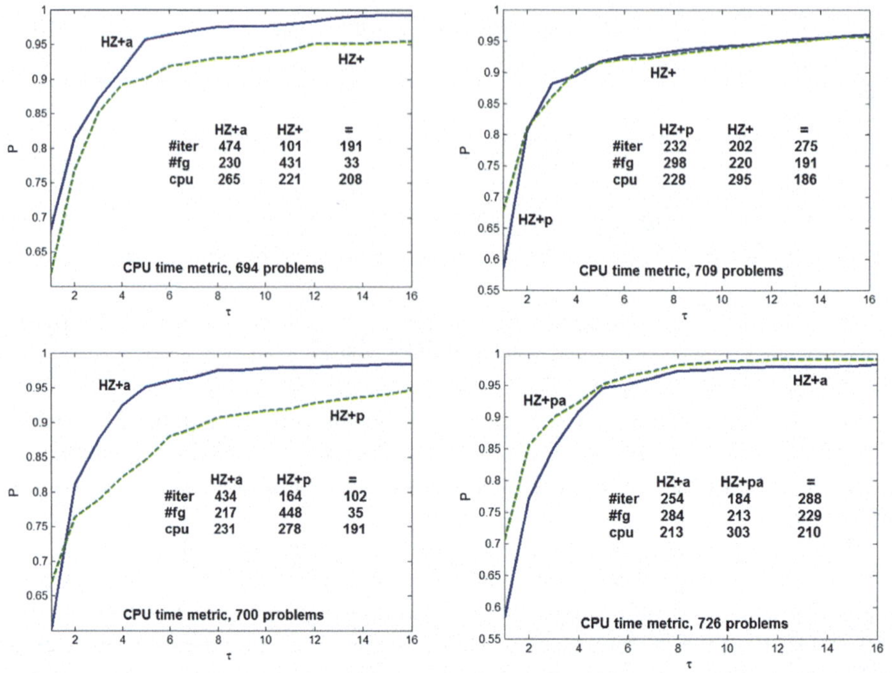

Figure 10.1 Performance profiles of HZ+ versus HZ+a; HZ+ versus HZ+p; HZ+a versus HZ+p and HZ+a versus HZ+pa

$$\eta_k = 0.5\frac{d_k^T g_{k+1}}{\|d_k\|^2}. \tag{10.14}$$

In preconditioned DK+p, the preconditioner P_{k+1} is computed as in (10.5) and (10.6). Figure 10.2 shows the performance of DK+ versus the accelerated version of DK+ (DK+a), where the acceleration is described as in Chapter 5 (see Remark 5.1); the performances of DK+ versus the preconditioned version of DK+ (DK+p) (see(10.11)–(10.14)); the performances of DK+a versus DK+p and the performances of DK+a versus the accelerated version of DK+p (DK+pa) for solving this set of 800 problems from the UOP collection. Observe that DK+a is top performer versus DK+. On the other hand, the preconditioned version DK+p, where the preconditioner is computed as in (10.5) and (10.6) as a diagonal matrix, is less efficient than DK+. Also, Figure 10.2 shows the computational evidence that DK+a is top performer versus DK+p. The accelerated DK+a is more efficient than DK+pa.

Figure 10.3 presents the performance profiles of the preconditioned and accelerated version of HZ+ (HZ+pa) versus HZ+ and the performance profiles of the preconditioned and accelerated version of DK+ (DK+pa) versus DK+.

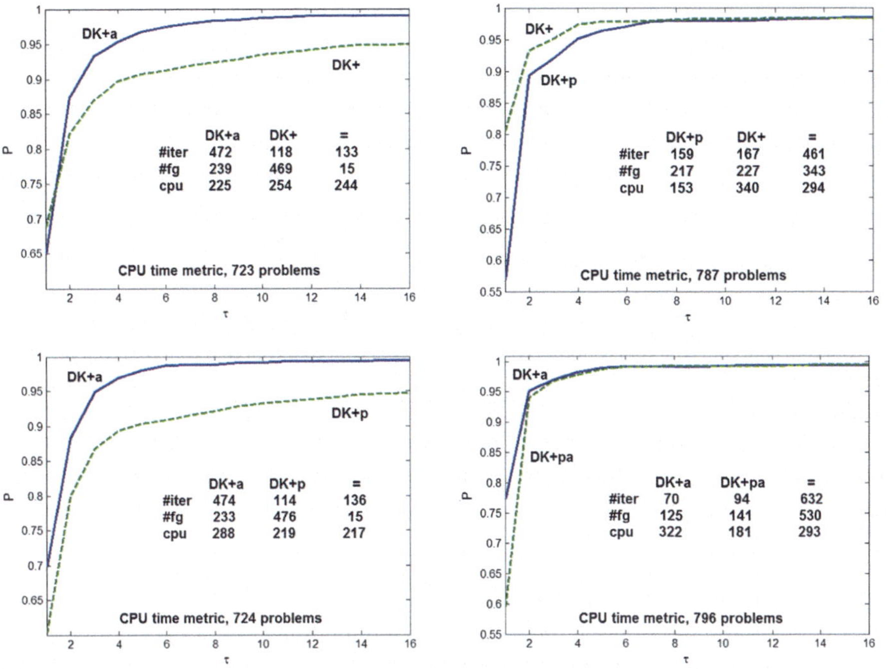

Figure 10.2 Performance profiles of DK+ versus DK+a; DK+ versus DK+p; DK+a versus DK+p and DK+a versus DK+pa

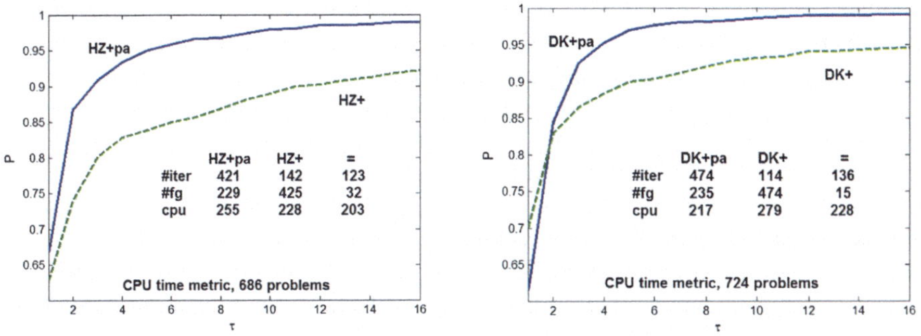

Figure 10.3 Performance profiles of HZ+pa versus HZ+ and of DK+pa versus DK+

Observe that subject to the CPU time metric the preconditioned and accelerated version of HZ+ (HZ+pa) is top performer versus HZ+. On the other hand, the preconditioned and accelerated version of DK+ (DK+pa) is more robust than DK+. Therefore, considered together, the preconditioning and the accelerating in the sense of Remark 5.1 improve the performances of the conjugate gradient algorithms.

The acceleration of conjugate gradient algorithms using a modification of the stepsize α_k as in Remark 5.1 proves to be more benefic than the preconditioning of the problem by using a diagonal approximation to the Hessian. However, this is not a definitive conclusion. Other preconditioners can be obtained by using the diagonal approximation to the Hessian, as those presented in (1.90), (1.91), (1.93) or (1.96). Moreover, using the Sherman–Morrison formula (see Appendix A), other preconditioners may be obtained by using the approximations of the Hessian obtained by scaling the terms on the right-hand side of the BFGS update (1.58), (1.63), (1.65), or (1.70), presented in Section 1.4.4.

Other preconditioners may be obtained by using, for example, the limited-memory BFGS updating of Nocedal (1980) through a limited number m of stored pairs $\{s_i, y_i\}$, $i = 1, \ldots, m$. Observe that in the preconditioned conjugate gradient parameter presented in the examples above (HZ+p and DK+p), only the product $P_k y_k$ is needed. But, these products may be computed during the updating process of the inverse Hessian by performing a sequence of inner products and vector summations involving only g_{k+1} or y_k and the stored pairs $\{s_i, y_i\}$. This is to be investigated.

It may be worthwhile studying and analyzing different preconditioners with different (diagonal or not diagonal) approximations to the inverse Hessian versus the acceleration scheme based on a multiplicative modification of the stepsize given by (5.11). It is interesting to see the preconditioning of the original algorithms CG-DESCENT or CGOPT by using different preconditioners (diagonal or not diagonal) and different line searches.

More sophisticated preconditioning methods are known. For preconditioning the conjugate gradient methods, the quasi-Newton or the limited-memory quasi-Newton updates are used. For example, Caliciotti, Fasano, and Roma (2017, 2018) investigated quasi-Newton updates derived from the modified secant equations as preconditioners for nonlinear conjugate gradient methods. Dener, Denchfield, and Munson (2019) presented the preconditioning of the nonlinear conjugate gradient methods with diagonalized quasi-Newton updates, where the diagonal elements of the Hessian are computed as in (10.6). Livieris, Karlos, Tampakas, and Pintelas (2017) developed preconditioning based on self-scaling memoryless BFGS updates. Also, developing limited-memory quasi-Newton conjugate gradient methods that utilize iteration history like those presented by Buckley and LeNir (1983) or Hager and Zhang (2013), are of great interest. Anyway, a preconditioner generated by a quasi-Newton update, at least in the special case of BFGS and a quadratic objective function, is expected to improve the preconditioning problem for inexact arithmetic or for a general nonlinear function. Actually, the purpose of preconditioning is to improve the structure of the eigenvalues of the inverse Hessian, an old and yet such an actual problem.

10.2 Criticism of Preconditioning the Nonlinear Conjugate Gradient Algorithms

We emphasize that in (10.2) there must be a *balance concerning the quality of the preconditioner* (i.e., the closeness to the inverse Hessian), namely, if the definition of the preconditioner P contains useful information about the inverse Hessian of the objective function, it is better to use the search direction $d_{k+1} = -Pg_{k+1}$, since the addition of the last term $\bar{\beta}_k d_k$ may prevent $d_{k+1} = -Pg_{k+1} + \bar{\beta}_k d_k$ from being an efficient descent direction, unless the line search is sufficiently accurate.

For example, let us consider the HZ+p defined by (10.7)–(10.10), with standard Wolfe line search, where this time the preconditioner P_{k+1} in (10.7) is given by the *self-scaling memoryless BFGS update of Perry and Shanno* (8.104) and the scaling parameter $\tau_k = \tau_k^{OL}$ is computed as in (8.111). Figure 10.4 shows the performance profiles of HZ+pa (accelerated version of HZ+p, where the acceleration is as in Remark 5.1) in which the search direction is computed as

$$d_{k+1} = -P_{k+1}g_{k+1} + \bar{\beta}_k^{HZ+} d_k, \tag{10.15}$$

where

$$P_{k+1} = \frac{1}{\tau_k}\left(I - \frac{s_k y_k^T + y_k s_k^T}{y_k^T s_k}\right) + \left(1 + \frac{1}{\tau_k}\frac{\|y_k\|^2}{y_k^T s_k}\right)\frac{s_k s_k^T}{y_k^T s_k}, \tag{10.16}$$

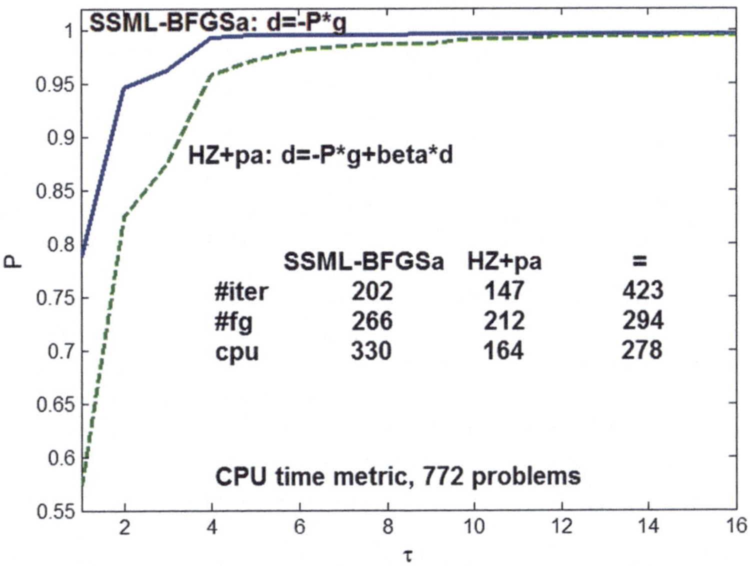

Figure 10.4 Performance profiles of HZ+pa versus SSML-BFGSa

with $\tau_k = \tau_k^{OL}$ and $\overline{\beta}_k^{HZ+}$ computed as in (10.8), where P_{k+1} in (10.9) is given by (10.16), versus the performances of the accelerated self-scaling memoryless BFGS (SSML-BFGSa) update in which the search direction is computed as

$$d_{k+1} = -P_{k+1} g_{k+1}, \tag{10.17}$$

where P_{k+1} is given by (10.16).

Observe that the accelerated self-scaling memoryless BFGS algorithm (10.17) where the scaling parameter is computed in the variant given by Oren and Luenberger (SSML-BFGSa) is more efficient than the preconditioned and the accelerated conjugate gradient algorithm HZ+pa (10.15). Subject to the CPU time metric, out of 800 problems only for 772 problems criterion (1.118) is satisfied. SSML-BFGSa was faster in 330 problems and HZ+pa was faster in 164, etc. In other words, it is not necessary to have a preconditioner very close to the inverse Hessian to improve the performances of the preconditioned conjugate gradient algorithms.

Notes and References

The idea of accelerating nonlinear conjugate gradient algorithms by preconditioning with quasi-Newton information was first considered by Buckley (1978a) and Nazareth (1979) in the context of exploring the connections between conjugate gradient and quasi-Newton methods. Later on, Andrei (2009c) (see also Andrei (2006a)) used this connection in accelerating nonlinear conjugate gradient methods with a scalar-scaling based on quasi-Newton updates (see also Andrei (2007b) and (2010b)).

Preconditioning the linear conjugate gradient methods is a well-understood concept, trying to reduce the condition number of the constant coefficient matrix in the quadratic objective. For nonlinear problems, preconditioning the conjugate gradient methods seeks symmetric positive definite matrices that approximate the inverse of the Hessian at each iteration. A detailed discussion on preconditioning nonlinear conjugate gradient methods was given by Hager and Zhang (2006b) and by Dener, Denchfield, and Munson (2019). The motivation of preconditioning is the requirement to solve large-scale problems, particularly in optical tomography (Abdoulaev, Ren, & Hielscher, 2005), seismic inversion (Epanomeritakis, Akçelik, Ghattas, & Bielak, 2008), and weather forecasting (Fisher, Nocedal, Trémolet, & Wright, 2009), (Navon & Legler, 1987). As already mentioned, the choices of preconditioners are well understood for linear problems. However, although there are plenty of papers on this subject, preconditioning the nonlinear conjugate gradient methods remains an open question with very little consensus. The question is how to construct an inverse approximation to the Hessian that determines a good eigenvalues distribution of the preconditioned problem. Some developments include: preconditioning the nonlinear conjugate gradient algorithms using a diagonalized quasi-Newton update (Dener, Denchfield, & Munson, 2019); preconditioners based on quasi-Newton updates for nonlinear conjugate gradient methods (Caliciotti, Fasano, & Roma, 2017); preconditioning based on a modified secant equation (Caliciotti, Fasano, & Roma, 2018); preconditioning using L-BFGS update used in limited-memory L-CG-DESCENT (Hager & Zhang, 2013), described in the next chapter.

Chapter 11
Other Conjugate Gradient Methods

As already seen, the conjugate gradient algorithms presented so far use some principles based on: hybridization or modifications of the standard schemes, the memoryless or the scaled memoryless BFGS preconditioned or the three-term concept. The corresponding conjugate gradient algorithms are defined by the descent condition, the "pure" conjugacy or the Dai–Liao conjugacy conditions or by the minimization of the quadratic approximation with one or two parameters of the objective function. There are a number of convergence results, mainly based on the Zoutendijk and on the Nocedal conditions under the Wolfe line search (Dai, 2011). These algorithms have good numerical performances, being able to solve large-scale unconstrained optimization problems and applications. However, in the frame of conjugate gradient methods, which is a very active area of research, some other computational schemes were introduced in order to improve their numerical performances. They are too numerous to be presented in this study. However, a short description of some of them is as follows.

Two modified scaled conjugate gradient methods based on the hybridization of the memoryless BFGS preconditioned conjugate gradient method suggested by Shanno and the spectral conjugate gradient method suggested by Birgin and Martínez based on a modified secant equation suggested by Yuan were proposed by Babaie-Kafaki (2014). Zhang (2009a) suggested two new variants of the Dai–Yuan algorithm by using the modified BFGS updating of Li and Fukushima (2001a) on the one hand, or by using a variant of the PRP method developed by Wei, Yao, and Liu (2006) combined with a technique of the modified FR method of Zhang, Zhou, and Li (2006b), on the other hand.

Conjugate gradient algorithms based on the modified secant equation of Zhang, Deng, and Chen (1999), Zhang and Xu (2001) (see 1.77) or Wei, Yu, Yuan, and Lian (2004) (see 1.75) were developed *inter alia* by Yabe and Takano (2004), Yabe and Sakaiwa (2005), Zhou and Zhang (2006), Li, Tang, and Wei (2007), Babaie-Kafaki, Ghanbari, and Mahdavi-Amiri (2010), Andrei (2010a), Babaie-Kafaki (2011), Babaie-Kafaki and Mahdavi-Amiri (2013), Livieris and Pintelas (2013), Kou (2014), and Babaie-Kafaki (2014).

N. Andrei, *Nonlinear Conjugate Gradient Methods for Unconstrained Optimization*,
Springer Optimization and Its Applications 158,
https://doi.org/10.1007/978-3-030-42950-8_11

Conjugate gradient methods with a fixed stepsize α_k defined by a formula were defined by Sun and Zhang (2001). Conjugate gradient algorithms with the search direction modified to fulfill the quadratic termination property were developed by Lukšan, Matonoha, and Vlcek (2008). A conjugate gradient algorithm with finite difference Hessian/vector product approximation for unconstrained optimization was presented by Andrei (2009d).

Other developments of conjugate gradient algorithms concentrate on the stepsize computation. Generally, the stepsize computation is based on the Wolfe line search conditions, but the most efficient conjugate gradient algorithms implement the approximate Wolfe line search by Hager and Zhang (2005) or the improved Wolfe line search by Dai and Kou (2013).

It is obvious that there is a large variety of conjugate gradient methods which combine different ingredients subject to the search direction or to the stepsize computation.

As usual, for solving the nonlinear unconstrained optimization problem

$$\min\{f(x) : x \in \mathbb{R}^n\}, \tag{11.1}$$

where $f : \mathbb{R}^n \to \mathbb{R}$ is a continuously differentiable function bounded from below, a nonlinear conjugate gradient method generates a sequence $\{x_k\}$ as

$$x_{k+1} = x_k + \alpha_k d_k, \tag{11.2}$$

$k = 0, 1, \ldots$, where $\alpha_k > 0$ is obtained by line search and the directions d_k are generated as

$$d_{k+1} = -g_{k+1} + \beta_k s_k, \quad d_0 = -g_0, \tag{11.3}$$

for $k \geq 1$, where $s_k = x_{k+1} - x_k$. The line search in the conjugate gradient algorithms is often based on the standard Wolfe conditions

$$f(x_k + \alpha_k d_k) - f(x_k) \leq \rho \alpha_k g_k^T d_k, \tag{11.4}$$

$$g_{k+1}^T d_k \geq \sigma g_k^T d_k, \tag{11.5}$$

where d_k is a descent direction and the scalar parameters ρ and σ are so that $0 < \rho \leq \sigma < 1$. Here $g_k = \nabla f(x_k)$.

For solving (11.1), this chapter describes some approaches by developing conjugate gradient algorithms based on different principles. The first approach is a more general viewpoint on the eigenvalues and singular values distribution of the iteration matrix, making a comparison between conjugate gradient algorithms with clustering the eigenvalues and conjugate gradient algorithms with minimizing the condition number of the iteration matrix (Andrei, 2017a, 2018b). Both clustering the eigenvalues of the iteration matrix and minimizing its condition number are two important ingredients for improving the performances of conjugate gradient algorithms. The second approach develops an algorithm which guarantees both the descent and the conjugacy conditions (Andrei, 2012). This is an interesting idea, and however, the performances of this algorithm are unexpectedly modest, proving that the algorithms that satisfy the sufficient and the conjugacy conditions are not necessarily the best ones. Some other ingredients have to be considered in these algorithms in order to improve their performances. In this chapter, we develop a simple combination between this conjugate gradient algorithm and the limited-memory BFGS algorithm. The idea is to interlace the iterations of the conjugate gradient algorithm with the iterations of the L-BFGS method according to some criteria. The criteria for triggering from an algorithm to another one are the stepsize or the closeness of the objective function to a quadratic. Finally, the limited-memory conjugate gradient method L-CG-DESCENT (see Hager & Zhang, 2013) and subspace minimization conjugate gradient algorithms based on cubic regularization (see Zhao, Liu, and Liu (2019)) are discussed.

11.1 Eigenvalues Versus Singular Values in Conjugate Gradient Algorithms (CECG and SVCG)

For solving the unconstrained optimization problem (11.1), let us consider the Algorithm (11.2), where the search directions d_k are computed by using the updating formula

$$d_{k+1} = -g_{k+1} + u_{k+1}. \tag{11.6}$$

Here, $u_{k+1} \in \mathbb{R}^n$ is a vector to be determined. Observe that (11.6) is a general updating formula for the search direction computation. The following particularizations of (11.6) can be presented. If $u_{k+1} = 0$, then the steepest descent algorithm is obtained. The Newton method is obtained if $u_{k+1} = (I - \nabla^2 f(x_{k+1})^{-1})g_{k+1}$. Besides, if $u_{k+1} = (I - B_{k+1}^{-1})g_{k+1}$, where B_{k+1} is an approximation of the Hessian $\nabla^2 f(x_{k+1})$, then the quasi-Newton methods are obtained. On the other hand, if $u_{k+1} = \beta_k d_k$, where β_k is a scalar and $d_0 = -g_0$, the family of conjugate gradient algorithms is generated.

In the following, a procedure for u_{k+1} computation by minimizing the quadratic approximation of the function f in x_{k+1} and by using a special representation of the inverse Hessian which depends on a positive parameter is presented (Andrei, 2017a). The parameter in the matrix representing the search direction is determined in two different ways. The first one is based on the eigenvalues analysis of the matrix by trying to minimize the largest eigenvalue. This idea, taken from the linear conjugate gradient, is to cluster the eigenvalues of the matrix representing the search direction. The second way to determine the value of the parameter is based on the fact that if the matrix defining the search direction is ill-conditioned, then, even for small relative errors in the gradient, the relative errors in the search direction may be large. Therefore, the second way is to use the singular value analysis by minimizing the condition number of the matrix representing the search direction of the algorithm.

The basic algorithm
Let us describe the basic algorithm and its properties. For this, consider that at the kth iteration of the algorithm an inexact Wolfe line search is executed, that is, the stepsize α_k is determined. With these, the following elements $s_k = x_{k+1} - x_k$ and $y_k = g_{k+1} - g_k$ are computed. Now, let us take the quadratic approximate of function f in x_{k+1} as

$$\Phi_{k+1}(d) = f_{k+1} + g_{k+1}^T d + \frac{1}{2} d^T B_{k+1} d, \tag{11.7}$$

where B_{k+1} is an approximation to the Hessian $\nabla^2 f(x_{k+1})$ of function f and d is the direction to be determined. The search direction d_{k+1} is computed as in (11.6), where u_{k+1} is determined as solution of the following minimizing problem

$$\min_{u_{k+1} \in \mathbb{R}^n} \Phi_{k+1}(d_{k+1}). \tag{11.8}$$

Introducing d_{k+1} from (11.6) in the minimizing problem (11.8), then u_{k+1} is obtained as

$$u_{k+1} = (I - B_{k+1}^{-1}) g_{k+1}. \tag{11.9}$$

Obviously, using different approximations B_{k+1} of the Hessian $\nabla^2 f(x_{k+1})$, different search directions d_{k+1} can be obtained. In this context, the following expression of B_{k+1}^{-1} is selected

$$B_{k+1}^{-1} = I - \frac{s_k y_k^T - y_k s_k^T}{y_k^T s_k} + \omega_k \frac{s_k s_k^T}{y_k^T s_k}, \tag{11.10}$$

where ω_k is a positive parameter which follows to be determined. Observe that B_{k+1}^{-1} is the sum of a skew symmetric matrix with zero diagonal elements $(s_k y_k^T - y_k s_k^T)/y_k^T s_k$ and a pure symmetric and positive definite matrix $I + \omega_k (s_k s_k^T)/(y_k^T s_k)$. Again, observe that (11.10) is a small modification of the memoryless BFGS updating formula used by Shanno (1978a).

Now, from (11.9),

$$u_{k+1} = \left[\frac{s_k y_k^T - y_k s_k^T}{y_k^T s_k} - \omega_k \frac{s_k s_k^T}{y_k^T s_k} \right] g_{k+1}. \tag{11.11}$$

Denote $H_{k+1} = B_{k+1}^{-1}$. Therefore, using (11.11) in (11.6), the search direction can be expressed as

$$d_{k+1} = -H_{k+1} g_{k+1}, \tag{11.12}$$

where

$$H_{k+1} = I - \frac{s_k y_k^T - y_k s_k^T}{y_k^T s_k} + \omega_k \frac{s_k s_k^T}{y_k^T s_k}. \tag{11.13}$$

Remark 11.1 Observe that H_{k+1} given by (11.13) is identical with Q_{k+1} from (9.76). However, Q_{k+1} is obtained by minimizing the quadratic approximation of the minimizing function in x_{k+1} by using the generalized quasi-Newton equation and slightly modifying it in a canonical way to get a symmetric matrix. On the other hand, H_{k+1} defined by (11.13) is obtained by an *arbitrary* selection of B_{k+1}^{-1} as in (11.10). The motivation of selecting B_{k+1}^{-1} as in (11.10) is that for H_{k+1} defined by (11.13), a very simple analysis of its eigenvalues and of its singular values can be obtained, as it may be seen in the following. ♦

Observe that the search direction (11.12), where H_{k+1} is given by (11.13), is as follows

$$d_{k+1} = -g_{k+1} + \left(\frac{y_k^T g_{k+1}}{y_k^T s_k} - \omega_k \frac{s_k^T g_{k+1}}{y_k^T s_k} \right) s_k - \frac{s_k^T g_{k+1}}{y_k^T s_k} y_k. \tag{11.14}$$

Proposition 11.1 *Consider $\omega_k \geq 0$ and the stepsize α_k in (11.2) determined by the Wolfe line search conditions (11.4) and (11.5). Then, the search direction (11.14) satisfies the descent condition $g_{k+1}^T d_{k+1} \leq 0$.*

Proof By direct computation, since $\omega_k \geq 0$,

$$g_{k+1}^T d_{k+1} = -\|g_{k+1}\|^2 - \omega_k \frac{(g_{k+1}^T s_k)^2}{y_k^T s_k} \leq 0.$$ ♦

Proposition 11.2 *Consider $\omega_k \geq 0$ and the stepsize α_k in (11.2) determined by the Wolfe line search conditions (11.4) and (11.5). Then, the search direction (11.14) satisfies the Dai and Liao conjugacy condition $y_k^T d_{k+1} = -v_k (s_k^T g_{k+1})$, where $v_k \geq 0$.*

Proof By direct computation,

$$y_k^T d_{k+1} = -\left[\omega_k + \frac{\|y_k\|^2}{y_k^T s_k}\right](s_k^T g_{k+1}) \equiv -v_k(s_k^T g_{k+1}),$$

where $v_k \equiv \omega_k + \|y_k\|^2 / y_k^T s_k$. By the Wolfe line search conditions (11.4) and (11.5) it follows that $y_k^T s_k > 0$, therefore $v_k > 0$. ♦

Although we have considered the expression of the inverse Hessian as the one given by (11.10), which is a nonsymmetric matrix, the search direction (11.14) obtained in this way satisfies both the descent condition and the Dai and Liao conjugacy condition. Therefore, the search direction (11.14) is a genuine conjugate gradient algorithm. The expression (11.10) of the inverse Hessian is only a technical argument to get the search direction (11.14). This approach is very general. *Considering other expressions to the inverse Hessian, with parameters, other search directions are obtained.* Observe that the method given by (11.2) and (11.12) can be considered as a quasi-Newton method in which the inverse Hessian is expressed by the nonsymmetric matrix H_{k+1} at each iteration. Moreover, the algorithm based on the search direction given by (11.14) can be considered as a three-term conjugate gradient algorithm. In this point, to define the algorithm, the only problem we face is to specify a suitable value for the positive parameter ω_k. A variant of the algorithm based on the eigenvalues analysis and another variant based on the singular values of H_{k+1} are presented as follows.

The algorithm based on clustering the eigenvalues of H_{k+1}
The idea of this variant of the algorithm is to determine ω_k by clustering the eigenvalues of H_{k+1}, i.e., by minimizing the largest eigenvalue of the matrix H_{k+1} from the spectrum of this matrix. The structure of the eigenvalues of the matrix H_{k+1} is given by the following theorem.

Theorem 11.1 *Let* H_{k+1} *be defined by* (11.13). *Then,* H_{k+1} *is a nonsingular matrix and its eigenvalues consist of 1 (* $n-2$ *multiplicity),* λ_{k+1}^{+} *and* λ_{k+1}^{-}, *where*

$$\lambda_{k+1}^{+} = \frac{1}{2}\left[(2+\omega_k b_k) + \sqrt{\omega_k^2 b_k^2 - 4a_k + 4}\right], \tag{11.15}$$

$$\lambda_{k+1}^{-} = \frac{1}{2}\left[(2+\omega_k b_k) - \sqrt{\omega_k^2 b_k^2 - 4a_k + 4}\right] \tag{11.16}$$

and

$$a_k = \frac{\|y_k\|^2 \|s_k\|^2}{(y_k^T s_k)^2} > 1, \quad b_k = \frac{\|s_k\|^2}{y_k^T s_k} \geq 0. \tag{11.17}$$

Proof By the Wolfe line search conditions (11.4) and (11.5), it follows that $y_k^T s_k > 0$. Therefore, the vectors y_k and s_k are nonzero vectors. Let V be the vector space spanned by $\{s_k, y_k\}$. Clearly, $\dim(V) \leq 2$ and $\dim(V^{\perp}) \geq n-2$. Thus, there exists a set of mutually unit orthogonal vectors $\{u_k^i\}_{i=1}^{n-2} \subset V^{\perp}$ so that

$$s_k^T u_k^i = y_k^T u_k^i = 0, i = 1, \ldots, n-2,$$

which from (11.13) leads to

$$H_{k+1} u_k^i = u_k^i, i = 1, \ldots, n-2.$$

Therefore, the matrix H_{k+1} has $n-2$ eigenvalues equal to 1, which corresponds to $\{u_k^i\}_{i=1}^{n-2}$ as eigenvectors.

Now, we are interested in finding the rest of the two remaining eigenvalues, denoted as λ_{k+1}^+ and λ_{k+1}^-, respectively. Since (see Appendix A)

$$\det(I + pq^T + uv^T) = (1 + q^T p)(1 + v^T u) - (p^T v)(q^T u),$$

where $p = \frac{y_k + \omega_k s_k}{y_k^T s_k}$, $q = s_k$, $u = -\frac{s_k}{y_k^T s_k}$ and $v = y_k$, it follows that

$$\det(H_{k+1}) = \frac{\|s_k\|^2 \|y_k\|^2}{(y_k^T s_k)^2} + \omega_k \frac{\|s_k\|^2}{y_k^T s_k} \equiv a_k + \omega_k b_k. \tag{11.18}$$

But, $a_k > 1$ and $b_k \geq 0$, therefore, H_{k+1} is a nonsingular matrix.

On the other hand, by direct computation (see Appendix A),

$$tr(H_{k+1}) = n + \omega_k \frac{\|s_k\|^2}{y_k^T s_k} \equiv n + \omega_k b_k. \tag{11.19}$$

By the relationships between the determinant and the trace of a matrix and its eigenvalues, it follows that the other eigenvalues of H_{k+1} are the roots of the following quadratic polynomial

$$\lambda^2 - (2 + \omega_k b_k)\lambda + (a_k + \omega_k b_k) = 0. \tag{11.20}$$

Clearly, the other two eigenvalues of the matrix H_{k+1} are determined from (11.20) as (11.15) and (11.16), respectively. Observe that $a_k > 1$ follows from the Wolfe conditions and from the inequality

$$\frac{y_k^T s_k}{\|s_k\|^2} \leq \frac{\|y_k\|^2}{y_k^T s_k}.$$

♦

In order to have both λ_{k+1}^{+} and λ_{k+1}^{-} as real eigenvalues, from (11.15) and (11.16), the following condition must be fulfilled $\omega_k^2 b_k^2 - 4a_k + 4 \geq 0$, out of which the following estimation of the parameter ω_k can be determined

$$\omega_k \geq \frac{2\sqrt{a_k - 1}}{b_k}. \tag{11.21}$$

Since $a_k > 1$, if $\|s_k\| > 0$, it follows that the estimation of ω_k given in (11.21) is well defined. From (11.20), it follows that

$$\lambda_{k+1}^{+} + \lambda_{k+1}^{-} = 2 + \omega_k b_k > 0, \tag{11.22}$$

$$\lambda_{k+1}^{+} \lambda_{k+1}^{-} = a_k + \omega_k b_k > 0. \tag{11.23}$$

Therefore, from (11.22) and (11.23), both λ_{k+1}^{+} and λ_{k+1}^{-} are positive eigenvalues. Since $\omega_k^2 b_k^2 - 4a_k + 4 \geq 0$, from (11.15) and (11.16) observe that $\lambda_{k+1}^{+} \geq \lambda_{k+1}^{-}$. By direct computation, from (11.15) using (11.21), it results that

$$\lambda_{k+1}^{+} \geq 1 + \sqrt{a_k - 1} > 1. \tag{11.24}$$

A simple analysis of Equation (11.20) shows that $1 \leq \lambda_{k+1}^{-} \leq \lambda_{k+1}^{+}$. Therefore, the maximum eigenvalue of H_{k+1} is λ_{k+1}^{+} and its minimum eigenvalue is 1.

Proposition 11.3 *The largest eigenvalue*

$$\lambda_{k+1}^{+} = \frac{1}{2}\left[(2 + \omega_k b_k) + \sqrt{\omega_k^2 b_k^2 - 4a_k + 4}\right] \tag{11.25}$$

gets its minimum $1 + \sqrt{a_k - 1}$ *when* $\omega_k = \frac{2\sqrt{a_k - 1}}{b_k}$. ◆

Proof Observe that $a_k > 1$. By direct computation the minimum of (11.25) is obtained for $\omega_k = (2\sqrt{a_k - 1})/b_k$, for which its minimum value is $1 + \sqrt{a_k - 1}$. ◆

Therefore, according to Proposition 11.3, when $\omega_k = (2\sqrt{a_k - 1})/b_k$, the largest eigenvalue of H_{k+1} arrives at the minimum value, i.e., the spectrum of H_{k+1} is clustered. In fact, for $\omega_k = (2\sqrt{a_k - 1})/b_k$, $\lambda_{k+1}^{+} = \lambda_{k+1}^{-} = 1 + \sqrt{a_k - 1}$. Therefore, from (11.17), the following estimation of ω_k can be obtained

$$\omega_k = 2\frac{y_k^T s_k}{\|s_k\|^2}\sqrt{a_k - 1} \leq 2\frac{\|y_k\|}{\|s_k\|}\sqrt{a_k - 1}. \tag{11.26}$$

From (11.17) $a_k > 1$, hence if $\|s_k\| > 0$, it follows that the estimation of ω_k given by (11.26) is well defined. However, the minimum of λ_{k+1}^{+} obtained for $\omega_k = (2\sqrt{a_k - 1})/b_k$ is given by $1 + \sqrt{a_k - 1}$. Therefore, if a_k is large, then the largest eigenvalue of the matrix H_{k+1} will be large. This motivates the parameter ω_k to be truncated as

$$\omega_k = \begin{cases} 2\sqrt{\tau - 1}\frac{\|y_k\|}{\|s_k\|}, & \text{if } a_k \geq \tau, \\ 2\sqrt{a_k - 1}\frac{\|y_k\|}{\|s_k\|}, & \text{otherwise}, \end{cases} \tag{11.27}$$

where $\tau > 1$ is a positive constant. Hence, the algorithm is an adaptive conjugate gradient algorithm in which the value of the parameter ω_k in the search direction (11.14) is computed as in (11.27), trying to cluster all the eigenvalues of H_{k+1}.

To attain a good computational performance of the algorithm, the idea of Powell (1984a) is applied by considering the following modification of the search direction given by (11.14):

$$d_{k+1} = -g_{k+1} + \max\left(\frac{y_k^T g_{k+1} - \omega_k s_k^T g_{k+1}}{y_k^T s_k}, 0\right) s_k - \frac{s_k^T g_{k+1}}{y_k^T s_k} y_k, \tag{11.28}$$

where ω_k is computed as in (11.27).

Using the procedure of accelerating the conjugate gradient algorithms according to the value of the parameter "*acceleration*" (true or false) and taking into consideration the above developments, the following algorithms with clustering the eigenvalues can be presented. CECGa is the accelerated version of CECG.

Algorithm 11.1 Clustering the eigenvalues: CECG/CECGa

1.	Select a starting point $x_0 \in \mathbb{R}^n$ and compute: $f(x_0)$, $g_0 = \nabla f(x_0)$. Select $\varepsilon_A > 0$ sufficiently small and some positive values $0<\rho<\sigma<1$ used in Wolfe line search conditions. Consider a positive value for the parameter τ. ($\tau > 1$) Set $d_0 = -g_0$ and $k = 0$
2.	Test a criterion for stopping the iterations. If this test is satisfied, then stop; otherwise continue with step 3
3.	Determine the stepsize α_k by the Wolfe line search (11.4) and (11.5). Update the variables $x_{k+1} = x_k + \alpha_k d_k$. Compute f_{k+1}, g_{k+1} and $s_k = x_{k+1} - x_k$, $y_k = g_{k+1} - g_k$
4.	If *acceleration* equal true, then (a) Compute: $z = x_k + \alpha_k d_k$, $g_z = \nabla f(z)$ and $y_k = g_k - g_z$ (b) Compute: $a_k = \alpha_k g_k^T d_k$, and $b_k = -\alpha_k y_k^T d_k$ (c If $\lvert b_k \rvert \geq \varepsilon_A$, then compute $\xi_k = -a_k/b_k$ and update the variables as $x_{k+1} = x_k + \xi_k \alpha_k d_k$. Compute f_{k+1} and g_{k+1}. Compute $y_k = g_{k+1} - g_k$ and $s_k = x_{k+1} - x_k$)
5.	Compute ω_k as in (11.27)
6.	Compute the search direction as in (11.28)
7.	Powell restart criterion. If $\lvert g_{k+1}^T g_k \rvert > 0.2\|g_{k+1}\|^2$, then set $d_{k+1} = -g_{k+1}$
8.	Consider $k = k+1$ and go to step 2 ♦

For *strongly convex functions*, the norm of the direction d_{k+1} computed as in (11.28) with (11.27) is bounded above. Therefore, by Theorem 3.5, the following theorem may be proved. Let $S = \{x \in \mathbb{R}^n : f(x) \le f(x_0)\}$ be the level set.

Theorem 11.2 *Suppose that the Assumption CG holds. Consider the algorithm CECG where the search direction d_k is given by (11.28) and ω_k is computed as in (11.27). Suppose that d_k is a descent direction and α_k is computed by the strong Wolfe line search given by (11.4) and by $|\nabla f(x_k + \alpha_k d_k)^T d_k| \le -\sigma d_k^T g_k$. Suppose that f is a strongly convex function on S, i.e., there exists a constant $\mu > 0$ so that*

$$(\nabla f(x) - \nabla f(y))^T (x - y) \ge \mu \|x - y\|^2 \tag{11.29}$$

for all $x, y \in N$, where $N \subset S$. Then

$$\lim_{k \to \infty} \|g_k\| = 0. \tag{11.30}$$

Proof From the Lipschitz continuity, $\|y_k\| \le L\|s_k\|$. On the other hand, from the strong convexity it follows that $y_k^T s_k \ge \mu \|s_k\|^2$. Now, from (11.27),

$$\omega_k = 2\sqrt{\tau - 1}\frac{\|y_k\|}{\|s_k\|} \le 2\sqrt{\tau - 1}\frac{L\|s_k\|}{\|s_k\|} = 2L\sqrt{\tau - 1}.$$

On the other hand, from (11.28), it follows that

$$\begin{aligned}
\|d_{k+1}\| &\le \|g_{k+1}\| + \frac{|y_k^T g_{k+1}|}{y_k^T s_k}\|s_k\| + \omega_k \frac{|s_k^T g_{k+1}|}{y_k^T s_k}\|s_k\| + \frac{|s_k^T g_{k+1}|}{y_k^T s_k}\|y_k\| \\
&\le \Gamma + \frac{\|y_k\|\Gamma\|s_k\|}{\mu\|s_k\|^2} + 2L\sqrt{\tau - 1}\frac{\|s_k\|\Gamma\|s_k\|}{\mu\|s_k\|^2} + \frac{\|s_k\|\Gamma\|y_k\|}{\mu\|s_k\|^2} \\
&\le \Gamma + 2\frac{L\Gamma}{\mu} + 2L\sqrt{\tau - 1}\frac{\Gamma}{\mu},
\end{aligned}$$

showing that the Nocedal condition is true. By Theorem 3.5, it follows that $\liminf_{k \to \infty} \|g_k\| = 0$ is true, which for strongly convex functions is equivalent to (11.30). ♦

The algorithm based on minimizing the condition number of H_{k+1}

The convergence rate of the nonlinear conjugate gradient algorithms depends on the structure of the eigenvalues of the Hessian. From (11.12), it is clear that the numerical performances and the efficiency of the quasi-Newton methods are based on the condition number of the successive approximations of the inverse Hessian. If the matrix H_{k+1} is ill-conditioned, then even for small values of the relative error of g_{k+1}, the relative error of d_{k+1} may be large. Hence, when the condition number of H_{k+1} is large, the system (11.12) is potentially very sensitive to perturbations in g_{k+1}. In other words, the ill-conditioned matrices H_{k+1} may produce instability in the iterative numerical computation with them. Therefore, the idea of this variant of the algorithm is to minimize the condition number of the matrix H_{k+1} by using its singular values. For this, let us briefly present the singular value analysis. The following theorem is extracted from Watkins (2002).

Theorem 11.3 *Let $A \in \mathbb{R}^{n \times m}$ be a nonzero matrix with rank r. Then $\mathbb{R}^m$ has an orthonormal basis $v_1, \ldots, v_m$, $\mathbb{R}^n$ has an orthonormal basis $u_1, \ldots, u_n$ and there exist the scalars $\sigma_1 \geq \sigma_2 \geq \cdots \geq \sigma_r > 0$ so that*

$$Av_i = \begin{cases} \sigma_i u_i, & i = 1, \ldots, r, \\ 0, & i = r+1, \ldots, m, \end{cases} \quad \textit{and} \quad A^T u_i = \begin{cases} \sigma_i v_i, & i = 1, \ldots, r, \\ 0, & i = r+1, \ldots, n. \end{cases}$$

The scalars $\sigma_1, \ldots, \sigma_r$ from Theorem 11.3 are called the singular values of the matrix A. Based on this theorem, for any nonzero matrix $A \in \mathbb{R}^{n \times m}$ with rank r it follows that

$$\|A\|_F^2 = \sigma_1^2 + \cdots + \sigma_r^2, \tag{11.31}$$

where $\|.\|_F$ represents the Frobenius norm. If $r = m = n$, then

$$|\det(A)| = \sigma_1 \times \sigma_2 \times \cdots \times \sigma_n.$$

For an arbitrary nonsingular matrix A, the scalar $\kappa(A) = \|A\|\|A^{-1}\|$ is called the condition number of A. If $A \in \mathbb{R}^{n \times n}$ is a nonsingular matrix with the singular values $\sigma_1 \geq \sigma_2 \geq \cdots \geq \sigma_n > 0$, then $\kappa(A) = \sigma_1/\sigma_n$. The condition number computed as above is called the spectral condition number. In our analysis, we need to find the singular values of the matrix H_{k+1}.

Theorem 11.4 *Let H_{k+1} be defined by (11.13). Then H_{k+1} has $n-2$ singular values equal to 1 and the remaining singular values σ_{k+1}^+ and σ_{k+1}^- are given by*

$$\sigma_{k+1}^+ = \frac{1}{2}\left[\sqrt{(\omega_k b_k + 2)^2 + 4(a_k - 1)} + \omega_k b_k\right], \tag{11.32}$$

$$\sigma_{k+1}^- = \frac{1}{2}\left[\sqrt{(\omega_k b_k + 2)^2 + 4(a_k - 1)} - \omega_k b_k\right], \tag{11.33}$$

where a_k and b_k are given by (11.17).

Proof By the Wolfe line search conditions (11.4) and (11.5), it follows that $y_k^T s_k > 0$. Therefore, the vectors y_k and s_k are nonzero vectors. Since $y_k^T s_k \neq 0$, there exists a set of mutually orthonormal vectors $\{u_k^i\}_{i=1}^{n-2}$ so that

$$s_k^T u_k^i = y_k^T u_k^i = 0, i = 1, \ldots, n-2,$$

which from (11.13) leads to

$$H_{k+1}u_k^i = H_{k+1}^T u_k^i = u_k^i, i = 1, \ldots, n-2.$$

Therefore, the matrix H_{k+1} has $n-2$ singular values equal to 1. Next, let us find the rest of the two remaining singular values, denoted as σ_{k+1}^+ and σ_{k+1}^-, respectively. But, by direct computation,

$$tr(H_{k+1}^T H_{k+1}) = n - 2 + 2\omega_k b_k + \omega_k^2 b_k^2 + 2a_k.$$

Since $\|H_{k+1}\|_F^2 = tr(H_{k+1}^T H_{k+1})$, from (11.31) it follows that

$$(\sigma_{k+1}^+)^2 + (\sigma_{k+1}^-)^2 = \omega_k^2 b_k^2 + 2\omega_k b_k + 2a_k. \tag{11.34}$$

As in Theorem 11.1 above (see (11.18)), the determinant of the iteration matrix H_{k+1} is the product of the singular values σ_{k+1}^+ and σ_{k+1}^-, i.e.

$$\sigma_{k+1}^+ \sigma_{k+1}^- = a_k + \omega_k b_k. \tag{11.35}$$

Now, from (11.34) and (11.35), the singular values σ_{k+1}^+ and σ_{k+1}^- are the solution of the following quadratic equation

$$\sigma^2 - \left(\sqrt{\omega_k^2 b_k^2 + 4\omega_k b_k + 4a_k}\right)\sigma + (a_k + \omega_k b_k) = 0,$$

expressed as in (11.32) and (11.33), respectively. ♦

Obviously, $\sigma_{k+1}^+ \geq \sigma_{k+1}^-$. But, $\sigma_{k+1}^- \geq 1$. Therefore, $\kappa(H_{k+1}) = \sigma_{k+1}^+$. By direct computation, $\kappa(H_{k+1})$ attains its minimum value $\sqrt{a_k}$ if and only if $\omega_k = 0$. Hence, minimizing the condition number of the matrix H_{k+1} given by (11.13) leads to the following search direction

$$d_{k+1} = -g_{k+1} + \frac{y_k^T g_{k+1}}{y_k^T s_k} s_k - \frac{s_k^T g_{k+1}}{y_k^T s_k} y_k. \tag{11.36}$$

Observe that (11.36) is a simple modification of the Hestenes and Stiefel conjugate gradient algorithm. At the same time, (11.36) is exactly the search direction of the three-term conjugate gradient method proposed by Zhang, Zhou, and Li (2007).

The following algorithms SVCG and SVCGa can be presented. SVCGa is the accelerated version of SVCG.

Algorithm 11.2 Singular values minimizing the condition number: SVCG/SVCGa

1.	Select a starting point $x_0 \in \mathbb{R}^n$ and compute: $f(x_0)$, $g_0 = \nabla f(x_0)$. Select $\varepsilon_A > 0$ sufficiently small and some positive values $0<\rho<\sigma<1$ used in Wolfe line search conditions. Set $d_0 = -g_0$ and $k = 0$
2.	Test a criterion for stopping the iterations. If this test is satisfied, then stop; otherwise continue with step 3
3.	Determine the stepsize α_k by using the Wolfe line search (11.4) and (11.5). Update the variables $x_{k+1} = x_k + \alpha_k d_k$. Compute f_{k+1}, g_{k+1} and $s_k = x_{k+1} - x_k$, $y_k = g_{k+1} - g_k$
4.	If *acceleration* equal true, then (a) Compute: $z = x_k + \alpha_k d_k$, $g_z = \nabla f(z)$ and $y_k = g_k - g_z$ (b) Compute: $a_k = \alpha_k g_k^T d_k$, and $b_k = -\alpha_k y_k^T d_k$ (c) If $\lvert b_k\rvert \geq \varepsilon_A$, then compute $\xi_k = -a_k/b_k$ and update the variables as $x_{k+1} = x_k + \xi_k\alpha_k d_k$. Compute f_{k+1} and g_{k+1}. Compute $y_k = g_{k+1} - g_k$ and $s_k = x_{k+1} - x_k$
5.	Compute the search direction as in (11.36)
6.	Powell restart criterion. If $\lvert g_{k+1}^T g_k\rvert > 0.2\lVert g_{k+1}\rVert^2$, then set $d_{k+1} = -g_{k+1}$
7.	Consider $k = k+1$ and go to step 2 ♦

Form (11.36), $g_{k+1}^T d_{k+1} = -\lVert g_{k+1}\rVert^2$, i.e., the search direction (11.36) satisfies the sufficient descent condition. Besides, $y_k^T d_{k+1} = -(\lVert y_k\rVert^2/y_k^T s_k)(s_k^T g_{k+1})$, i.e., the search direction (11.36) satisfies the Dai and Liao conjugacy condition. For *strongly convex functions* the norm of the direction d_{k+1} computed as in (11.36) is bounded above. Therefore, by Theorem 3.5, the following theorem can be proved.

Theorem 11.5 *Suppose that the Assumption CG holds. Consider the algorithm SVCG where the search direction d_k is given by (11.36). Suppose that d_k is a descent direction and α_k is computed by the strong Wolfe line search given by (11.4) and by $\lvert\nabla f(x_k + \alpha_k d_k)^T d_k\rvert \leq -\sigma d_k^T g_k$. Suppose that f is a strongly convex function on S, i.e., there exists a constant $\mu > 0$ so that $(\nabla f(x) - \nabla f(y))^T(x-y) \geq \mu\lVert x-y\rVert^2$ for all $x, y \in S$. Then*

$$\lim_{k\to\infty}\lVert g_k\rVert = 0. \tag{11.37}$$

Proof From (11.36), the following estimation is obtained

$$\begin{aligned}\lVert d_{k+1}\rVert &\leq \lVert g_{k+1}\rVert + \frac{\lvert y_k^T g_{k+1}\rvert}{y_k^T s_k}\lVert s_k\rVert + \frac{\lvert s_k^T g_{k+1}\rvert}{y_k^T s_k}\lVert y_k\rVert \\ &\leq \Gamma + \frac{\lVert y_k\rVert\Gamma\lVert s_k\rVert}{\mu\lVert s_k\rVert^2} + \frac{\lVert s_k\rVert\Gamma\lVert y_k\rVert}{\mu\lVert s_k\rVert^2} \leq \Gamma + 2\frac{L\Gamma}{\mu},\end{aligned}$$

showing that the norm of the search direction is bounded. Therefore, $\liminf_{k\to\infty}\|g_k\| = 0$ is true, which for strongly convex functions is equivalent to (11.37). ♦

Remark 11.2 Suppose that the Assumption CG holds. Consider the algorithm CECG with parameter ω_k defined as in (11.27). It can be shown that there exists a positive constant Ω so that $0 \le \omega_k \le \Omega$. Hence, if the search directions computed as in (11.28) are descent directions and the stepsizes are determined to satisfy the strong Wolfe conditions, then Theorem 3.6 of Dai and Liao (2001) ensures the global convergence of the method for general objective functions. The convergence of the SVCG algorithm for general objective functions can be proved by following the methodology given by Gilbert and Nocedal (1992) and by Theorem 3.6 of Dai and Liao (2001). ♦

Numerical study. In the following, let us present the performances of the CECG and SVCG conjugate gradient algorithms for solving the problems from the UOP collection (Andrei, 2018g). In this collection, there are 80 unconstrained optimization problems. For each of them, 10 numerical experiments have been done with $n = 1000, \ldots, 10000$ variables. Figure 11.1 shows the performances of CECG with $\tau = 10$ or $\tau = 100$ versus SVCG. By comparison with the minimization of the condition number of the iteration matrix, observe that the clustering of its eigenvalues yields a more efficient algorithm.

From Figure 11.1, observe that the CECG algorithm is very little sensitive to the values of the parameter $\tau > 1$. In fact, for $a_k \ge \tau$, from (11.28) it follows that

$$\frac{\partial d_{k+1}}{\partial \tau} = -\frac{1}{\sqrt{\tau - 1}} \frac{\|y_k\|}{\|s_k\|} \frac{s_k^T g_{k+1}}{y_k^T s_k} s_k, \tag{11.38}$$

where $\tau > 1$. Therefore, since the gradient of the function f is Lipschitz continuous and the quantity $s_k^T g_{k+1}$ is going to zero, it follows that $\partial d_{k+1}/\partial \tau$ tends to zero along the iterations, showing that along the iterations the search direction is less and

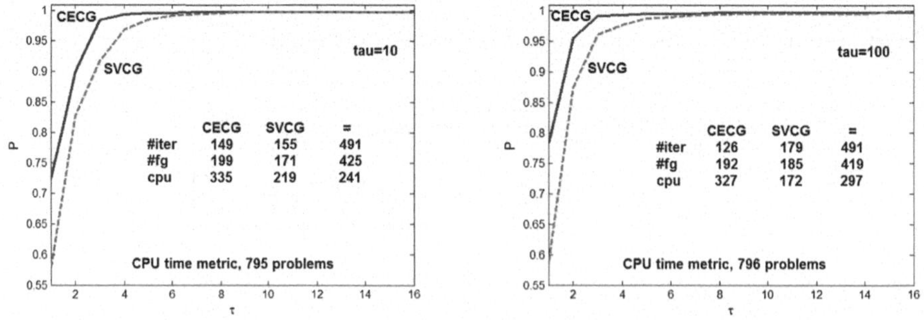

Figure 11.1 Performance profiles of CECG ($\tau = 10$) and CECG ($\tau = 100$) versus SVCG

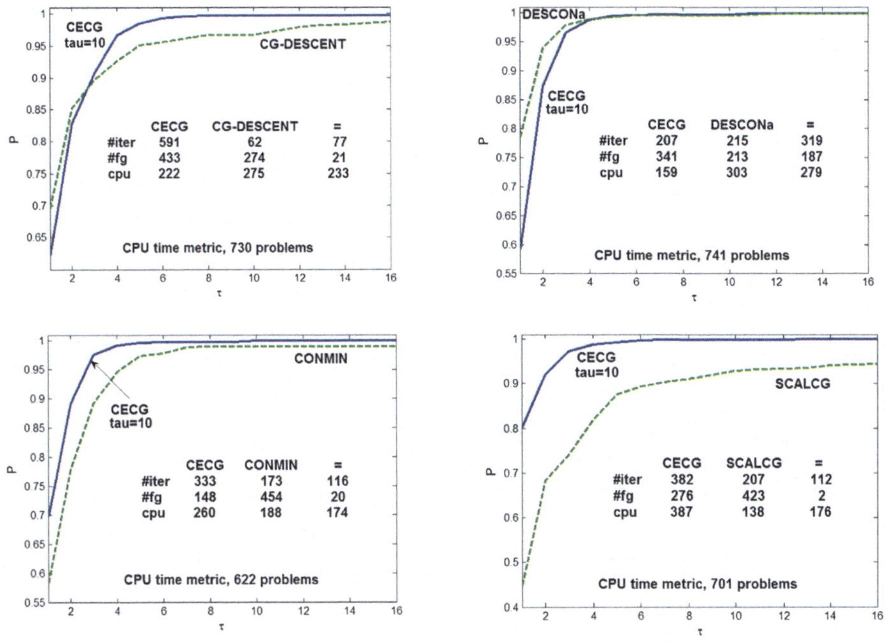

Figure 11.2 Performance profiles of CECG ($\tau = 10$) versus CG-DESCENT, DESCONa, CONMIN and SCALCG

less sensitive subject to the value of the parameter τ. For strongly convex functions, using the Assumption CG it follows that

$$\left\| \frac{\partial d_{k+1}}{\partial \tau} \right\| \leq \frac{1}{\sqrt{\tau - 1}} \frac{L\Gamma}{\mu}. \tag{11.39}$$

For example, for larger values of τ, the variation of d_{k+1} subject to τ decreases, showing that the CECG algorithm is very little sensitive to the values of the parameter τ. This is illustrated in Figure 11.1 where the performance profiles have the same allure for different values of the parameter $\tau > 1$.

Figure 11.2 shows the performances of CECG with $\tau = 10$ versus CG-DESCENT (version 1.4), DESCONa, CONMIN, and SCALCG. CG-DESCENT is slightly more efficient than CECG with $\tau = 10$, but CECG with $\tau = 10$ is more robust. DESCONa is a top performer in this comparison. CECG with $\tau = 10$ is more efficient and more robust than CONMIN and SCALCG.

Figure 11.3 illustrates the performances of CECG with $\tau = 10$ versus DK+w and versus DK+aw (DK+ with approximate Wolfe line search).

Observe that if CECG with $\tau = 10$ is a top performer in comparison with DK+w, DK+aw (DK+ with approximate Wolfe line search) is significantly more efficient.

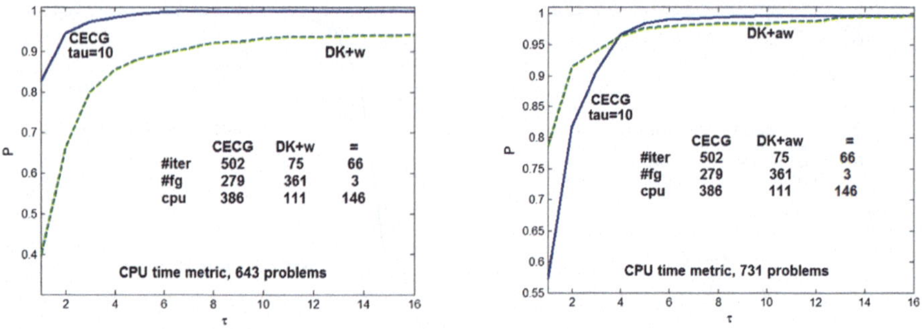

Figure 11.3 Performance profiles of CECG ($\tau = 10$) versus DK+w and versus DK+aw

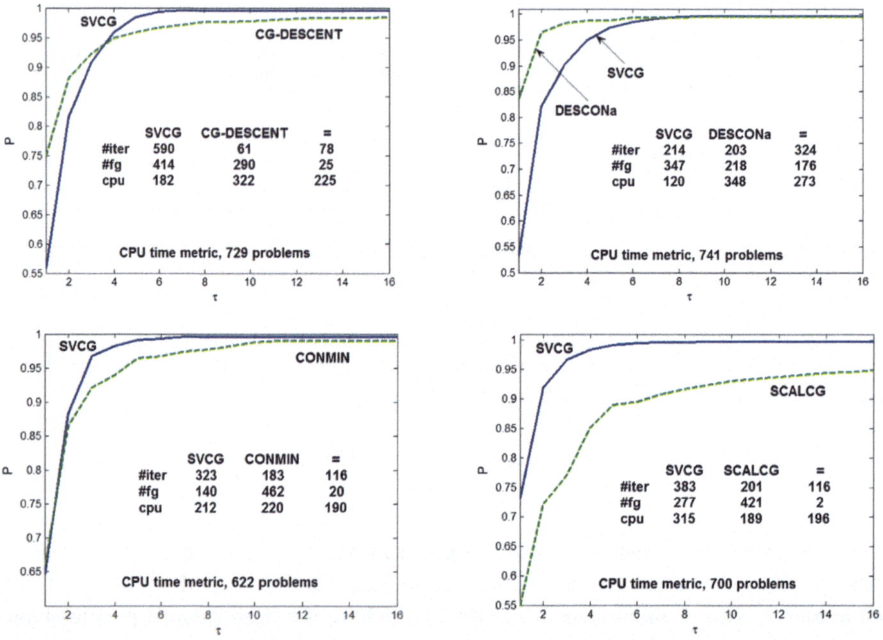

Figure 11.4 Performance profiles of SVCG versus CG-DESCENT, DESCONa, CONMIN, and SCALCG

This emphasizes once again the importance of the line search procedure in conjugate gradient algorithms.

Figures 11.4 and 11.5 show the performances of SVCG versus the performances of the same algorithms considered in the above numerical experiments. Observe that only DESCONa is more efficient than SVCG. SCALCG is much less efficient and less robust than SVCG.

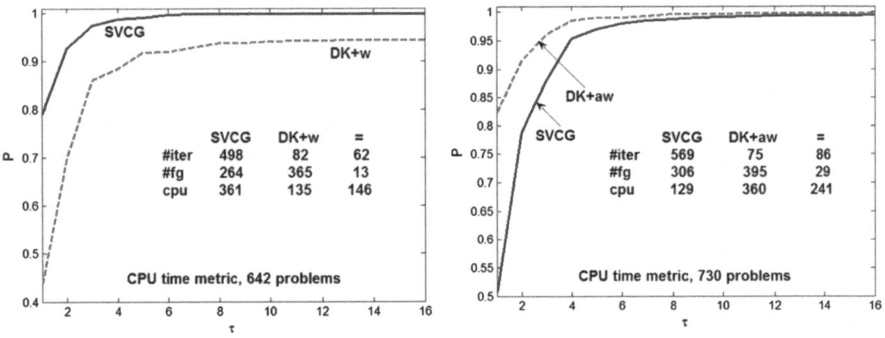

Figure 11.5 Performance profiles of SVCG versus DK+w and versus DK+aw

Figures 11.2 and 11.4 have a lot in common. They illustrate that clustering the eigenvalues or minimizing the condition number of the iteration matrix yields more efficient algorithms. Observe that the DK+ algorithm is obtained by seeking the search direction closest to the Perry–Shanno search direction. In a way, this is an artificial approach without any justification. The Perry–Shanno search direction is obtained from the self-scaling memoryless BFGS update, where at every step the updating is initialized with a scaled identity matrix $(1/\tau_k)I$, $\tau_k \geq 0$ being a scaling parameter. On the other hand, SVCG is obtained by minimizing the condition number of the iteration matrix H_{k+1}, which has a very strong theoretical justification. The weakness of this variant of the SVCG algorithm is the form and structure of the H_{k+1} matrix (11.13). By considering other approximations to the inverse Hessian closer to $\nabla^2 f(x_k)$, more efficient algorithms can be hopefully obtained. For example, consider the approximation to the inverse Hessian as the self-scaling memoryless BFGS update given by (8.104), which includes the parameter τ_k. Using the determinant and the trace of H_{k+1} given by (8.126) and (8.127), respectively, the values of the parameter τ_k can be determined, which cluster the eigenvalues of H_{k+1} or minimize its condition number.

By comparing SVCG versus DK+aw in Figure 11.5, it is obvious that DK+aw is a top performer. However, SVCG is much more efficient and more robust than DK+w. The approximate Wolfe line search proves to be an important ingredient in improving the performances of conjugate gradient algorithms.

11.2 A Conjugate Gradient Algorithm with Guaranteed Descent and Conjugacy Conditions (CGSYS)

In the following, let us present a conjugate gradient algorithm in which for all $k \geq 1$ both the descent and the conjugacy conditions are guaranteed (Andrei, 2012). As it is known, the conjugate gradient algorithm (11.2) and (11.3) with exact line search always satisfies the condition $g_{k+1}^T d_{k+1} = -\|g_{k+1}\|^2$, which is in a direct connection with the sufficient descent condition

$$g_{k+1}^T d_{k+1} \leq -t\|g_{k+1}\|^2 \tag{11.40}$$

for some positive constant $t > 0$. The sufficient descent condition has often been used to analyze the global convergence of the conjugate gradient algorithms with inexact line search based on the strong Wolfe conditions. The sufficient descent condition is not needed in the convergence analysis of the Newton or quasi-Newton algorithms. However, it is necessary for the global convergence of conjugate gradient algorithms.

Dai and Liao (2001) extended the conjugacy condition

$$d_{k+1}^T y_k = 0 \tag{11.41}$$

and proposed the following new conjugacy condition

$$d_{k+1}^T y_k = -u g_{k+1}^T s_k, \tag{11.42}$$

where $u \geq 0$ is a scalar. Minimizing a convex quadratic function in a subspace spanned by a set of mutually conjugate directions is equivalent to minimizing this function along each conjugate direction in turn. This is a very good idea, but the performance of these algorithms is dependent on the accuracy of the line search. However, the inexact line search is always used in conjugate gradient algorithms. Hence, when the line search is not exact, the "pure" conjugacy condition (11.41) may have disadvantages. Therefore, it seems more reasonable to use the conjugacy condition (11.42). When the algorithm is convergent, observe that $g_{k+1}^T s_k$ tends to zero along the iterations and therefore the conjugacy condition (11.42) tends to the pure conjugacy condition (11.41).

For solving the minimization problem (11.1), suppose that the search direction is computed as

$$d_{k+1} = -\theta_k g_{k+1} + \beta_k s_k \tag{11.43}$$

$k = 0, 1, \ldots, d_0 = -g_0$, where θ_k and β_k are scalar parameters which are to be determined. Algorithms of this form or variations of them were studied by many authors. For example, Birgin and Martínez (2001) proposed a spectral conjugate gradient method, where $\theta_k = s_k^T s_k / s_k^T y_k$. Also, Andrei (2007a, 2007b, 2007c) considered a preconditioned conjugate gradient algorithm where the preconditioner is a scaled memoryless BFGS matrix and the parameter scaling the gradient is selected as the spectral gradient. Stoer and Yuan (1995) studied the conjugate gradient algorithm on a subspace where the search direction d_{k+1} at the kth iteration ($k \geq 1$) is taken from the subspace $span\{-g_{k+1}, d_k\}$. Recently, Li, Liu, and Liu

(2018) developed a new subspace minimization conjugate gradient algorithm with nonmonotone Wolfe line search in which the search direction is in the subspace $\Omega_{k+1} = span\{g_{k+1}, s_k, s_{k-1}\}$. Also, Zhao, Liu, and Liu (2019) introduced a new subspace minimization conjugate gradient algorithm based on regularization model where the search direction is computed as in (11.43).

In the algorithm which follows to be presented, for all $k \geq 0$, the scalar parameters θ_k and β_k in (11.43) are determined from the *descent condition*

$$g_{k+1}^T d_{k+1} = -\theta_k g_{k+1}^T g_{k+1} + \beta_k g_{k+1}^T s_k = -t\|g_{k+1}\|^2 \tag{11.44}$$

and the *conjugacy condition* (11.42), which is

$$y_k^T d_{k+1} = -\theta_k y_k^T g_{k+1} + \beta_k y_k^T s_k = -u(s_k^T g_{k+1}), \tag{11.45}$$

where $t > 0$ and $u > 0$ are scalar parameters. Observe that in (11.44) the classical sufficient descent condition (11.40) is modified with equality. It is worth pointing out that the main condition in any conjugate gradient algorithm is the descent condition $g_k^T d_k < 0$ or the sufficient descent condition (11.40). The conjugacy condition (11.41) or its modification (11.42) is not so stringent. In fact, it is satisfied by very few conjugate gradient algorithms.

If $u = 0$, then (11.45) is the "pure" conjugacy condition. However, in order to accelerate the algorithm and incorporate the second-order information, let us consider $u > 0$.

Now, let us define

$$\Delta_k \equiv (y_k^T g_{k+1})(s_k^T g_{k+1}) - \|g_{k+1}\|^2 (y_k^T s_k). \tag{11.46}$$

Supposing that $\Delta_k \neq 0$, then, from the linear algebraic system given by (11.44) and (11.45), the following values for θ_k and β_k are obtained

$$\theta_k = \frac{-(y_k^T s_k)\|g_{k+1}\|^2 t + (s_k^T g_{k+1})^2 u}{\Delta_k}, \tag{11.47}$$

$$\beta_k = \frac{-(y_k^T g_{k+1})\|g_{k+1}\|^2 t + (s_k^T g_{k+1})\|g_{k+1}\|^2 u}{\Delta_k}. \tag{11.48}$$

If the line search is exact, that is $s_k^T g_{k+1} = 0$, then $\Delta_k = -\|g_{k+1}\|^2 (y_k^T s_k) < 0$ if the line search satisfies the Wolfe condition (11.5) and if $g_{k+1} \neq 0$. Therefore, from (11.47) and (11.48), it follows that $\theta_k = t$ and $\beta_k = (y_k^T g_{k+1})t/(y_k^T s_k)$, i.e.,

$$d_{k+1} = t\left(-g_{k+1} + \frac{y_k^T g_{k+1}}{y_k^T s_k} s_k\right) = t d_{k+1}^{HS},$$

where d_{k+1}^{HS} is the Hestenes and Stiefel search direction.

Proposition 11.4 *If*

$$\sigma \le \frac{\|g_{k+1}\|^2}{|y_k^T g_{k+1}| + \|g_{k+1}\|^2}, \tag{11.49}$$

then for all $k \ge 1, \Delta_k < 0$.

Proof Observe that

$$s_k^T g_{k+1} = s_k^T y_k + s_k^T g_k < s_k^T y_k. \tag{11.50}$$

The Wolfe condition (11.5) gives

$$g_{k+1}^T s_k \ge \sigma g_k^T s_k = -\sigma y_k^T s_k + \sigma g_{k+1}^T s_k. \tag{11.51}$$

Since $\sigma < 1$, (11.51) can be rearranged to obtain

$$g_{k+1}^T s_k \ge \frac{-\sigma}{1-\sigma} y_k^T s_k. \tag{11.52}$$

Now, let us combine this lower bound for $g_{k+1}^T s_k$ with the upper bound (11.50) to obtain

$$|g_{k+1}^T s_k| \le |y_k^T s_k| \max\left\{1, \frac{\sigma}{1-\sigma}\right\}. \tag{11.53}$$

Again, observe that the Wolfe condition gives $y_k^T s_k > 0$ (if $g_k \ne 0$). Therefore, if σ is bounded as in (11.49), then

$$|g_{k+1}^T s_k|\,|g_{k+1}^T y_k| \le |y_k^T s_k|\,|g_{k+1}^T y_k| \max\left\{1, \frac{\sigma}{1-\sigma}\right\} \le |y_k^T s_k|\,\|g_{k+1}\|^2.$$

i.e., $\Delta_k < 0$ for all $k \ge 1$. ♦

From (11.49), observe that $\sigma < 1$. Since $g_k^T s_k = -t\|g_k\|^2 < 0$, i.e., d_k is a descent direction, it follows that $|g_{k+1}^T y_k| \to \|g_{k+1}\|^2$. Therefore, $\sigma \to 1/2$, i.e., $0 < \rho < \sigma < 1$, since ρ is usually selected small enough to ensure the reduction of the function values along the iterations.

In the following, let us prove the convergence of the algorithm assuming that $\nabla^2 f(x)$ is bounded, that is for all $x \in S$ there is a positive constant M so that $\nabla^2 f(x) \le MI$, i.e., $MI - \nabla^2 f(x)$ is a positive semidefinite matrix, which implies that $x^T \nabla^2 f(x) x \le M\|x\|^2$. To prove the convergence, a limiting behavior of the algorithm when $k \to \infty$ is considered. This is motivated by the fact that at every iteration k, the search direction d_k is a descent one (see the condition (11.44)) and the stepsize is obtained by the strong Wolfe line search.

Theorem 11.6 *Suppose that the Assumption CG holds. Consider the conjugate gradient algorithm (11.2), where the direction d_{k+1} is given by (11.43), (11.46)–(11.48) and the step length α_k is obtained by the strong Wolfe line search conditions. Assume that $\nabla^2 f(x)$ is bounded, i.e., $\nabla^2 f(x) \le MI$, where M is a positive constant. Then, $\liminf_{k\to\infty} \|g_k\| = 0$.*

Proof Since $\nabla^2 f(x)$ is bounded, there is an index k_0 so that for all $k > k_0$

$$y_k^T s_k = (g_{k+1} - g_k)^T s_k = s_k^T \nabla^2 f(\bar{x}_k) s_k \le M\|s_k\|^2 = O(\|s_k\|^2),$$

where $\bar{x}_k$ is a point on the line segment connecting x_k and x_{k+1}. As above, observe that

$$s_k^T g_{k+1} \le \|s_k\| \, \|g_{k+1}\| \le \Gamma\|s_k\| = O(\|s_k\|),$$
$$y_k^T g_{k+1} \le \|y_k\| \, \|g_{k+1}\| \le L\Gamma\|s_k\| = O(\|s_k\|).$$

Hence, for all $k > k_0$,

$$(s_k^T g_{k+1})(y_k^T g_{k+1}) = O(\|s_k\|^2). \tag{11.54}$$

Therefore, from (11.46), for all sufficiently large k, i.e., for $k > k_0$,

$$\Delta_k = \max\left\{O(\|s_k\|^2), O(\|s_k\|^2)\right\} = O(\|s_k\|^2). \tag{11.55}$$

On the other hand, since t and u are positive constants, for $k > k_0$,

$$-(y_k^T s_k)\|g_{k+1}\|^2 t + (s_k^T g_{k+1})^2 u = \max\left\{O(\|s_k\|^2), O(\|s_k\|^2)\right\} = O(\|s_k\|^2),$$
$$-(y_k^T g_{k+1})\|g_{k+1}\|^2 t + (s_k^T g_{k+1})\|g_{k+1}\|^2 u = \max\{O(\|s_k\|), O(\|s_k\|)\} = O(\|s_k\|).$$

Therefore, for all sufficiently large k, i.e., for $k > k_0$,

$$\theta_k = \frac{O(\|s_k\|^2)}{O(\|s_k\|^2)} = O(1) \quad \text{and} \quad \beta_k = \frac{O(\|s_k\|)}{O(\|s_k\|^2)} = \frac{1}{O(\|s_k\|)}. \tag{11.56}$$

From (11.43), it follows that

$$\|d_{k+1}\| \le |\theta_k|\|g_{k+1}\| + |\beta_k|\|s_k\| \le \Gamma O(1) + \frac{1}{O(\|s_k)\|}\|s_k\| = O(1). \tag{11.57}$$

Therefore, there is an index k_0 and a positive constant D so that for all $k \geq k_0$, $\|d_k\| \leq D$, i.e., $\sum_{k \geq 1} \frac{1}{\|d_k\|^2} = \infty$. By Theorem 3.5, since d_k is a descent direction, it follows that $\liminf_{k \to \infty} \|g_k\| = 0$. ♦

Observe that from (11.47) and (11.48), the search direction may be written as

$$\begin{aligned} d_{k+1} = {} & \left[\frac{(y_k^T s_k)\|g_{k+1}\|^2}{\Delta_k} g_{k+1} - \frac{(y_k^T g_{k+1})\|g_{k+1}\|^2}{\Delta_k} s_k \right] t \\ & + \left[\frac{(s_k^T g_{k+1})\|g_{k+1}\|^2}{\Delta_k} s_k - \frac{(s_k^T g_{k+1})^2}{\Delta_k} g_{k+1} \right] u. \end{aligned} \tag{11.58}$$

Since the algorithm is convergent, i.e., $\{x_k\} \to x^*$, where x^* is the local optimal point of (11.1), it follows that $\lim_{k \to \infty} \|s_k\| = 0$. On the other hand, $s_k^T g_{k+1} \to 0$ for $k \to \infty$. Therefore, the coefficient of u in (11.58) tends to zero, i.e., the algorithm is not very much sensitive to the values of parameter u.

However, since $s_k^T g_{k+1} \to 0$ for $k \to \infty$, it follows that

$$\frac{t(y_k^T s_k)\|g_{k+1}\|^2}{\Delta_k} \to -t,$$

showing that the descent condition (11.44) is more important than the conjugacy condition (11.45). However, the conjugacy condition is important in the economy of the algorithm since it includes the information of the second order.

Now, taking into consideration the acceleration scheme presented in Remark 5.1, the following algorithms CGSYS and CGSYSa can be presented. CGSYSa is the accelerated version of CGSYS.

Algorithm 11.3 Guaranteed descent and conjugacy conditions: CGSYS/CGSYSa

1.	Select a starting point $x_0 \in dom f$ and compute: $f_0 = f(x_0)$ and $g_0 = \nabla f(x_0)$. Select $\varepsilon_A > 0$ sufficiently small and positive values $0 < \rho < \sigma < 1$ used in Wolfe line search conditions. Select some positive values for t and u. Set $d_0 = -g_0$ and $k = 0$
2.	Test a criterion for stopping the iterations. If the test is satisfied, then stop; otherwise continue with step 3
3.	Using the Wolfe line search conditions determine the stepsize α_k. Update the variables $x_{k+1} = x_k + \alpha_k d_k$. Compute f_{k+1}, g_{k+1} and $s_k = x_{k+1} - x_k$, $y_k = g_{k+1} - g_k$
4.	If *acceleration* equal true, then (a) Compute: $z = x_k + \alpha_k d_k$, $g_z = \nabla f(z)$ and $y_k = g_k - g_z$ (b) Compute: $a_k = \alpha_k g_k^T d_k$, and $b_k = -\alpha_k y_k^T d_k$ (c) If $\lvert b_k \rvert \geq \varepsilon_A$, then compute $\xi_k = -a_k / b_k$ and update the variables as $x_{k+1} = x_k + \xi_k \alpha_k d_k$. Compute f_{k+1} and g_{k+1}. Compute $s_k = x_{k+1} - x_k$ and $y_k = g_{k+1} - g_k$
5.	Determine θ_k and β_k as in (11.47) and (11.48) respectively, where Δ_k is computed as in (11.46)

(continued)

Algorithm 11.1 (continued)

6.	Compute the search direction as: $d_{k+1} = -\theta_k g_{k+1} + \beta_k s_k$
7.	Restart criterion. If $\left\|g_{k+1}^T g_k\right\| > 0.2\|g_{k+1}\|^2$ then set $d_{k+1} = -g_{k+1}$
8.	Consider $k = k+1$ and go to step 2 ♦

Numerical study. The performances of the above algorithms for solving 80 unconstrained optimization problems from the UOP collection, where for each problem 10 experiments have been taken with the number of variables $n = 1000, 2000, \ldots, 10000$ are presented as follows. The algorithm implements the standard Wolfe line search conditions (11.4) and (11.5) with $\rho = 0.0001$ and $\sigma = \|g_{k+1}\|^2/(\left|y_k^T g_{k+1}\right| + \|g_{k+1}\|^2)$. If $\sigma < \rho$, then $\sigma = 0.8$ is set. If $\Delta_k \geq \varepsilon_m$, where ε_m is the epsilon machine, then θ_k and β_k are computed as in (11.47) and (11.48), respectively. Otherwise, set $\theta_k = 1$ and $\beta_k = \|g_{k+1}\|^2/y_k^T s_k$, i.e., the Dai–Yuan conjugate gradient algorithm is used. In CGSYS and CGSYSa, $t = 7/8$ and $u = 0.01$. The maximum number of iterations is limited to 2000.

Figure 11.6 presents the performance profiles of CGSYS and its accelerated version CGSYSa for solving the unconstrained optimization problems from the UOP collection.

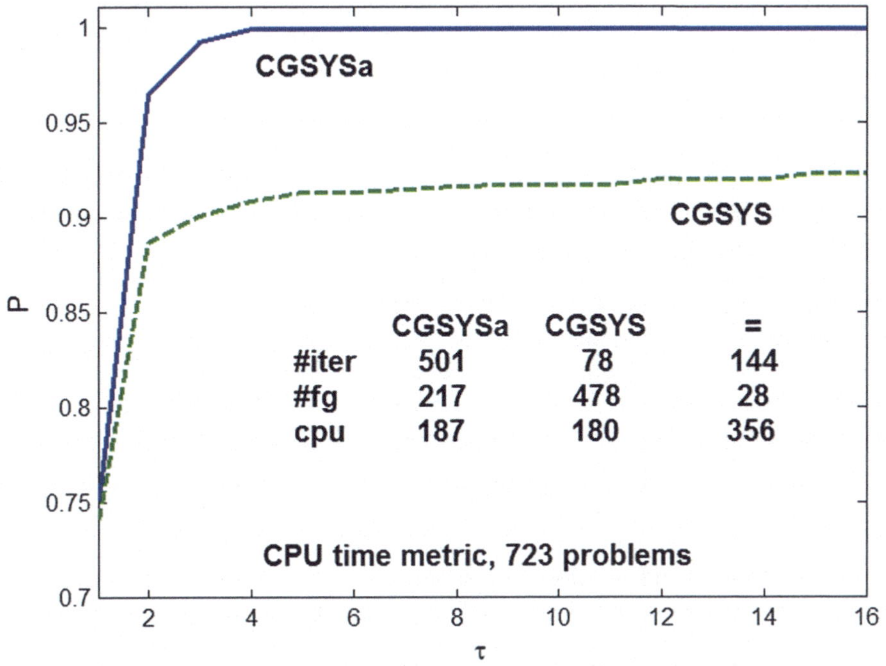

Figure 11.6 Performance profiles of CGSYS versus CGSYSa

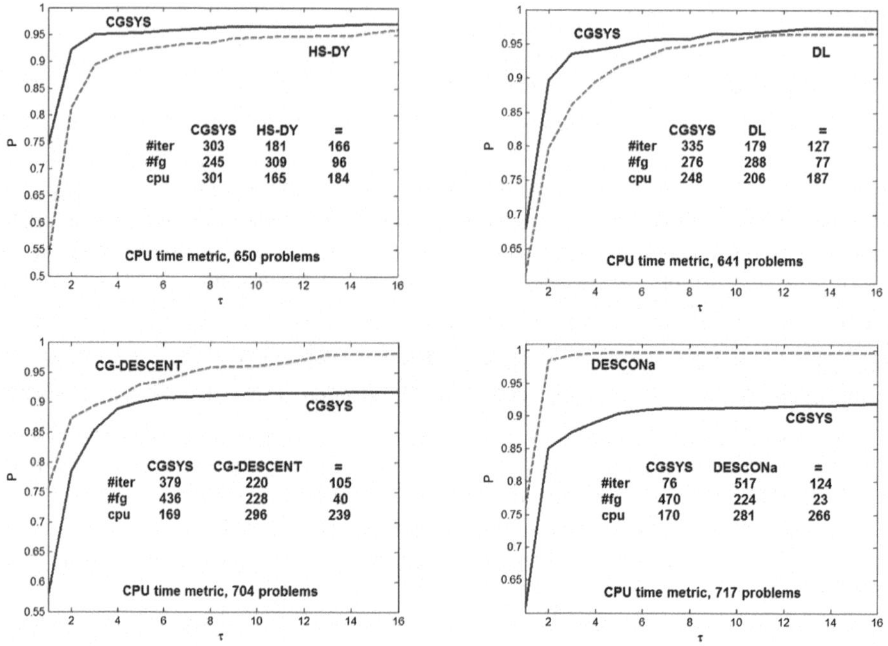

Figure 11.7 Performance profiles of CGSYS versus HS-DY, DL ($t = 1$), CG-DESCENT, and DESCONa

Compared to CGSYS, Figure 11.6 shows that CGSYSa is a top performer. They have the same efficiency, but CGSYSa is much more robust than CGSYS.

Figure 11.7 illustrates the performance profiles of CGSYS versus HS-DY, DL ($t = 1$), CG-DESCENT (version 1.4), and DESCONa. By using both the sufficient descent and the conjugacy conditions, CGSYS is more efficient and more robust than the hybrid conjugate gradient HS-DY and than the Dai–Liao DL ($t = 1$) algorithms. Observe that both CG-DESCENT and DESCONa are much more efficient and more robust than CGSYS. We know that DESCONa outperforms CG-DESCENT (see Figure 7.8). In Figure 11.7, observe that the difference between the performance profiles of CGSYS and DESCONa is bigger than the difference between the performance profiles of CGSYS and CG-DESCENT.

The next set of numerical experiments presents comparisons of CGSYS versus the memoryless BFGS preconditioned algorithms CONMIN and SCALCG. Figure 11.8 shows the performance profiles of these algorithms.

Both CONMIN and SCALCG are more robust than CGSYS. The machinery behind the memoryless BFGS preconditioned algorithms CONMIN and SCALCG is quite complex. By using the memoryless BFGS preconditioning, these algorithms are able to better capture the curvature of the objective function and this is the reason why they are more robust. *Observe that the sufficient and the conjugacy conditions used in CGSYS are not sufficient to get a good algorithm.* It is worth

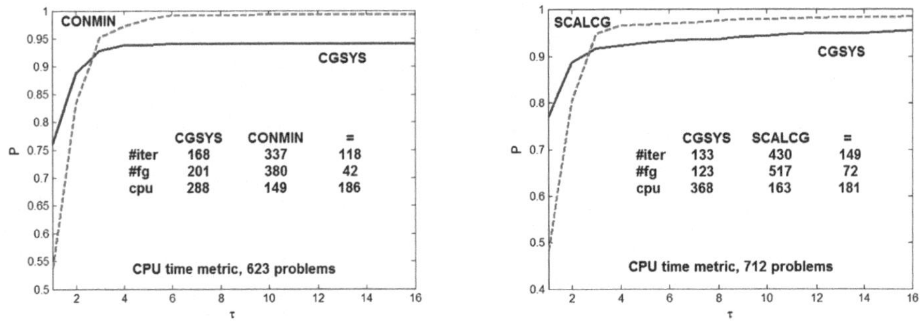

Figure 11.8 Performance profiles of CGSYS versus CONMIN and versus SCALCG

seeing the performance profiles of CGSYS versus the three-term conjugate gradient algorithms TTCG and TTDES. Figure 11.9 illustrates these performance profiles.

Both three-term conjugate gradient algorithms TTCG and TTDES are more robust than CGSYS.

11.3 Combination of Conjugate Gradient with Limited-Memory BFGS Methods

In CGSYS, both the sufficient descent and the conjugacy conditions are satisfied. However, the performances of CGSYS are modest. The conclusion of the above numerical experiments with CGSYS is that satisfying both the sufficient descent and the conjugacy conditions does not mean that the algorithm is efficient. Some additional ingredients are necessary for it to be performing.

Observe that the search direction in CGSYS and DESCONa satisfies both the sufficient descent and the conjugacy conditions. However, as illustrated in Figure 11.7, DESCONa is far more efficient and more robust than CGSYS. The difference between CGSYS and DESCONa is that in DESCONa the modified second Wolfe line search condition (7.84) is used. This is a crucial ingredient for DESCONa to perform best.

In the following, *some simple combinations of the conjugate gradient methods with the limited-memory BFGS method are presented*, as an ingredient to improve the performances of the conjugate gradient algorithms for which the search direction satisfies both the sufficient descent and the conjugacy conditions, like CGSYS. The motivation for selecting the L-BFGS in this combination is that for highly nonlinear problems, L-BFGS is the best performer. Firstly, three combinations of CGSYS with L-BFGS are discussed, after which the combination of CG-DESCENT with L-BFGS is detailed (see Hager & Zhang, 2013)

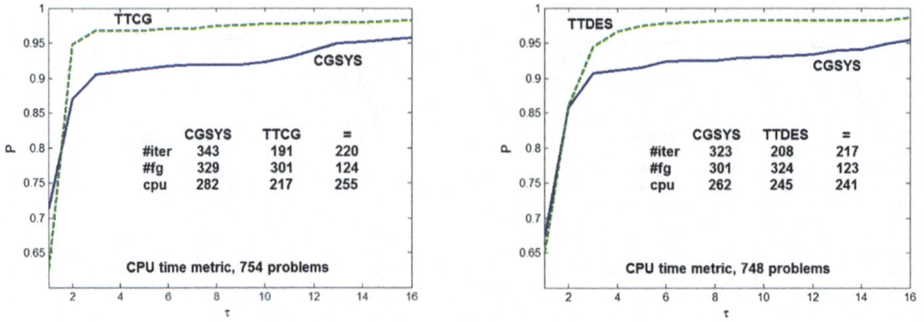

Figure 11.9 Performance profiles of CGSYS versus TTCG and versus TTDES

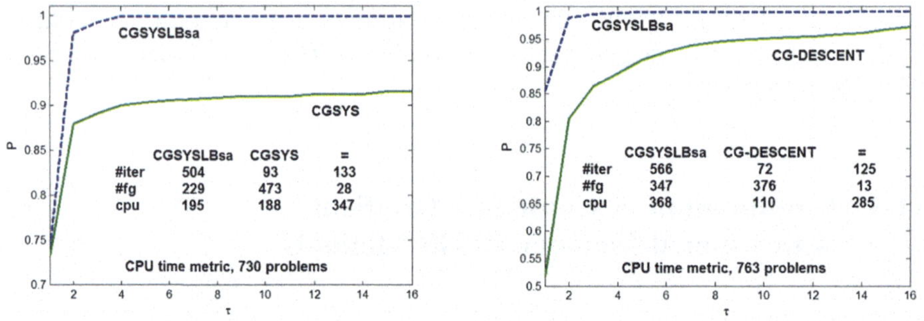

Figure 11.10 Performance profiles of CGSYSLBsa versus CGSYS and versus CG-DESCENT

Combination of the conjugate gradient CGSYS with L-BFGS based on the stepsize The idea is to combine the CGSYS algorithm with the limited-memory L-BFGS algorithm by interlacing iterations of the CGSYS with iterations of the L-BFGS algorithms. In this algorithm, which we called CGSYSLBs, the iterations of CGSYS are performed only if the stepsize is less or equal to a prespecified threshold. Otherwise, the iterations of L-BFGS ($m = 5$) are performed. This simple procedure for triggering between CGSYS and L-BFGS proved to be very profitable. Figure 11.10 presents the performances of CGSYSLBsa (the accelerated version of CGSYSLBs) versus CGSYS and CG-DESCENT (version 1.4). Observe that CGSYSLBsa is more efficient and more robust than these algorithms. In Figure 11.11, we present the performance profiles of CGSYSLBsa versus DESCONa and DK+w. Again, we can see that CGSYSLBsa is a top performer versus the accelerated conjugate gradient with guaranteed descent and conjugacy conditions and a modified Wolfe line search (DESCONa) and versus the Dai–Kou conjugate gradient algorithm with standard Wolfe line search (DK+w).

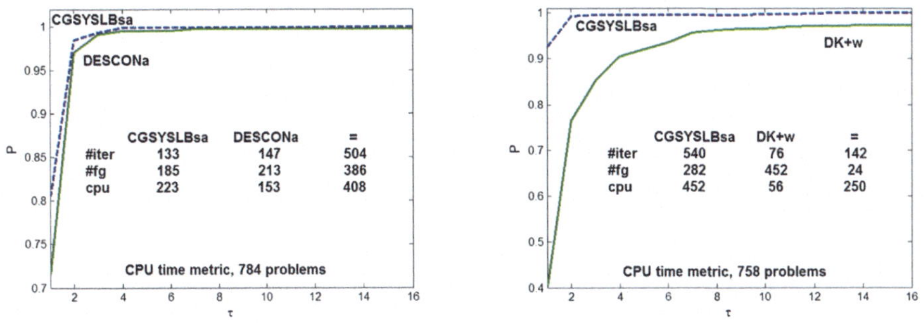

Figure 11.11 Performance profiles of CGSYSLBsa versus DESCONa and versus DK+w

Combination of the conjugate gradient CGSYS with L-BFGS based on the closeness of the minimizing function to a quadratic

Consider the one-dimensional line search function $\varphi_k(\alpha) = f(x_k + \alpha d_k)$, $\alpha \geq 0$, where f is the minimizing function. Using the values of φ_k at α_k and 0, a new quantity may be introduced showing how φ_k is close to a quadratic function. Specifically, $p(\varphi_k(0), \varphi_k'(0), \varphi_k'(\alpha_k))$ denotes the quadratic interpolating function by $\varphi_k(0)$, $\varphi_k'(0)$ and $\varphi_k'(\alpha_k)$. If the value of this polynomial p at α_k is very close to the real function value $\varphi_k(\alpha_k)$, it follows that φ_k is inclined to be a quadratic function on the line connecting x_k and $x_{k+1} = x_k + \alpha_k d_k$. With this, Yuan (1991) introduced the parameter

$$t_k = \left| \frac{2(f_k - f_{k+1} + g_{k+1}^T s_k)}{y_k^T s_k} - 1 \right|, \tag{11.59}$$

which describes the difference between $p(\alpha_k)$ and $\varphi_k(\alpha_k)$. If t_k is close to zero, then φ_k is regarded as a quadratic function, otherwise it is not. In other words, if $t_k \leq c$, where c is a small positive constant ($c = 10^{-8}$), it can be concluded that φ_k is close to a quadratic function. Motivated by this idea and having in view that for most of the highly nonlinear problems L-BFGS is one of the best algorithms, CGSYS and L-BFGS are combined in the following way. In this algorithm, which we call CGSYSLBq, if $t_k \leq c$, then the CGSYS iterations are performed, otherwise the L-BFGS ($m = 5$) iterations are considered. Figure 11.12 presents the performances of CGSYSLBqa (the accelerated version of CGSYSLBq) versus CGSYS and versus CG-DESCENT (version 1.4). We can see that CGSYSLBqa is more robust than CGSYS and more efficient and more robust than CG-DESCENT (version 1.4). Figure 11.13 shows the performances of CGSYSLBqa versus DESCONa and versus DK+w. Only DESCONa is slightly more efficient than CGSYSLBqa.

Combination of the conjugate gradient CGSYS with L-BFGS based on the orthogonality of the current gradient to the previous search direction

As it is known, in theory, for the quadratic problems, the gradient at each iteration of either the conjugate gradient method or L-BFGS should be orthogonal to the

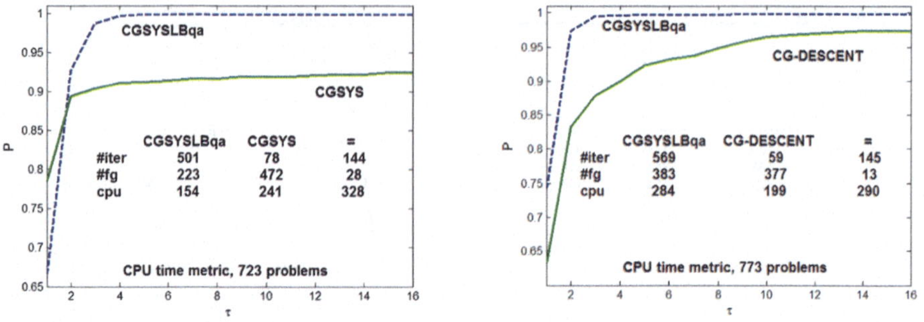

Figure 11.12 Performance profiles of CGSYSLBqa versus CGSYS and versus CG-DESCENT

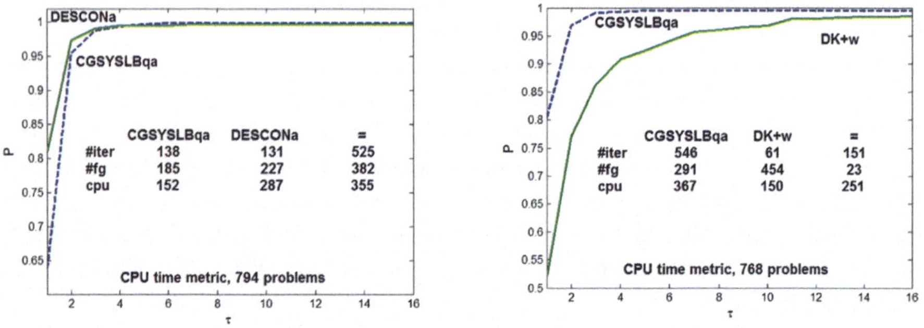

Figure 11.13 Performance profiles of CGSYSLBqa versus DESCONa and versus DK+w

space spanned by the previous search direction. For general nonlinear functions, the gradients in the conjugate gradient method may lose orthogonality and after a number of iterations, the gradient essentially lies in the space spanned by the previous search direction. On the other hand, the L-BFGS method preserves this orthogonality. This is the motivation to combine the conjugate gradient CGSYS with L-BFGS by monitoring the loss of orthogonality of the current gradient to the previous search direction. In other words, in our algorithm, we call CGSYSLBo the CGSYS and L-BFGS methods are combined as follows: if $g_{k+1}^T d_k \leq c$, where c is a small positive constant ($c = 10^{-5}$), then the CGSYS iterations are performed, otherwise the L-BFGS ($m = 5$) iterations are considered. Figure 11.14 presents the performances of CGSYSLBoa (the accelerated version of CGSYSLBo) versus CGSYS and versus CG-DESCENT (version 1.4). We can see that CGSYSLBoa is more efficient and more robust than CGSYS and than CG-DESCENT (version 1.4). Figure 11.15 shows the performances of CGSYSLBoa versus DESCONa and versus DK+w. Only DESCONa is more efficient than CGSYSLBoa.

The interlacing of the iterations of CGSYS and L-BFGS is very profitable. Figures 11.16 and 11.17 present the performances of CGSYSLBsa, CGSYSLBqa, and CGSYSLBoa versus L-BFGS ($m = 5$).

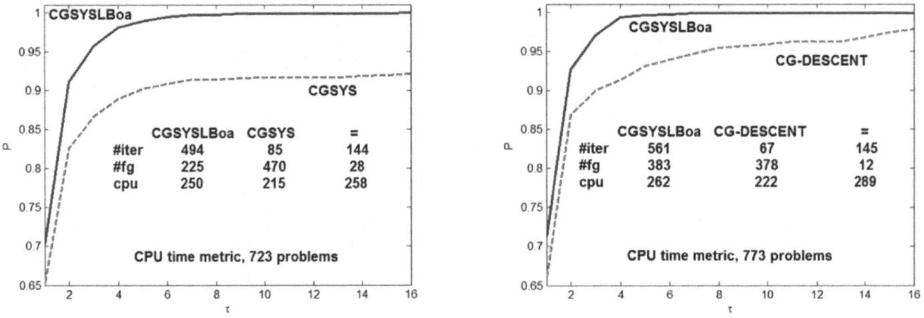

Figure 11.14 Performance profiles of CGSYSLBoa versus CGSYS and versus CG-DESCENT

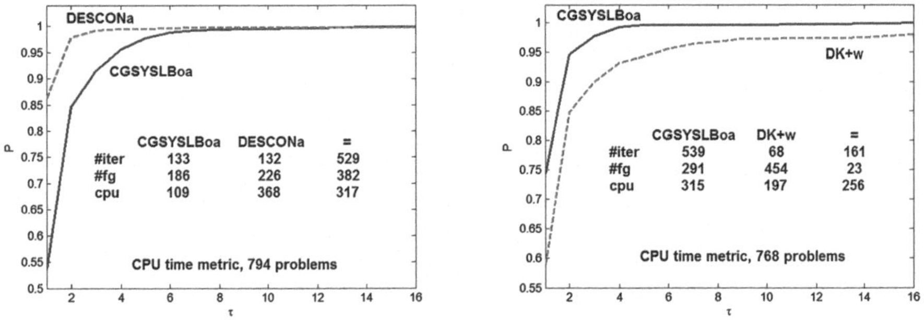

Figure 11.15 Performance profiles of CGSYSLBoa versus DESCONa and versus DK+w

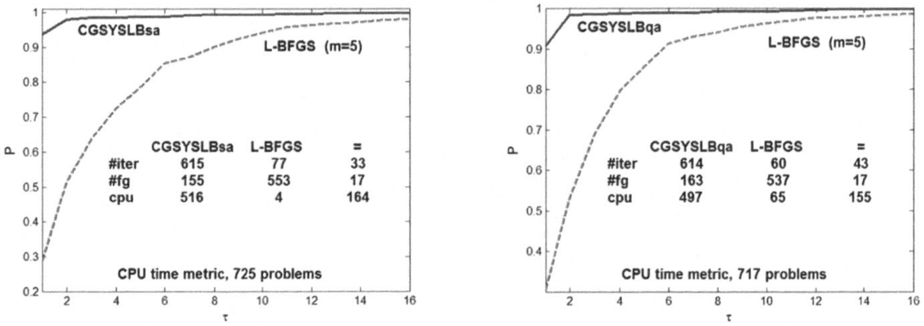

Figure 11.16 Performance profiles of CGSYSLBsa and CGSYSLBqa versus L-BFGS ($m = 5$)

The combination of solvers, when carefully applied, may greatly improve the convergence properties of nonlinear optimization algorithms (Brune, Knepley, Smith, & Tu, 2015). In our approach, we combined CGSYS and L-BFGS in a simple way by using the stepsize or the deviation of the minimizing function from a

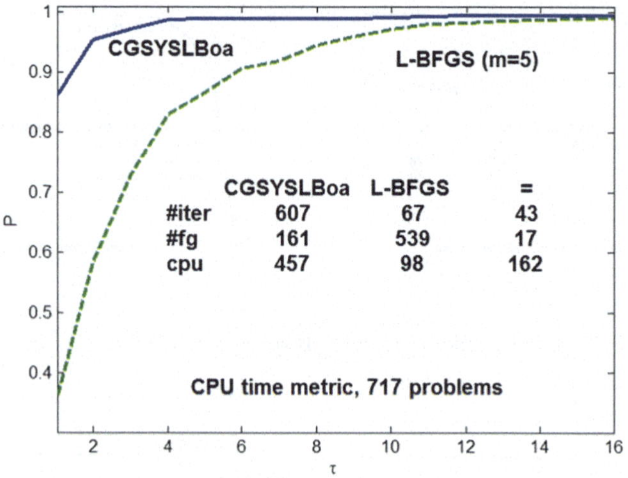

Figure 11.17 Performance profiles of CGSYSLBoa versus L-BFGS ($m = 5$)

quadratic or by monitoring the orthogonality of the current gradient to the previous search direction as criteria of triggering between methods. Obviously, some other conjugate gradient algorithms may be combined with L-BFGS. Other more refined machinery combining CG-DESCENT with L-BFGS is presented in the following, as the limited-memory L-CG-DESCENT method.

Limited-Memory L-CG-DESCENT
A more sophisticated combination of the conjugate gradient and the L-BFGS algorithms can be obtained by monitoring the loss of orthogonality of successive gradients in the conjugate gradient method, like in the limited-memory L-CG-DESCENT algorithm developed by Hager and Zhang (2013). As it is known, the linear conjugate gradient method has the property that after k iterations the gradient is orthogonal to the previous search directions $d_0, \ldots, d_{k-1}$. (see Propositions 2.2 and 2.3) This is an important property of the linear conjugate gradient, known as the finite termination property, and in recent years, it has been extended to get more efficient nonlinear conjugate gradient algorithms (See the papers by Hager and Zhang (2013), Fatemi (2016a, 2016b, 2017), or by Livieris and Pintelas (2016).).

Using the CUTE collection, Hager and Zhang (2013) intensively studied the performance of the CG-DESCENT algorithm. They observed that for an ill-conditioned positive definite quadratic optimization problem the convergence of CG-DESCENT was much slower than expected, even if the dimension of the problem was small. An ill-conditioned problem is characterized by the fact that the condition number of its Hessian is very large. As it is known, for quadratic problems, the conjugate gradient method and the limited-memory BFGS method (L-BFGS) should generate the same iterates and at each iterate, the gradient of either method should be orthogonal to the space spanned by the previous search directions. However, for some quadratic problems like the one considered by Hager and Zhang (PALMER1C), it was observed that the L-BFGS method preserves the orthogonality property, while the conjugate gradient method loses orthogonality at about the same time when the iterate error grows substantially. Therefore, the performances of the conjugate gradient method heavily depend not only on the problem conditioning, but also on the preservation of the orthogonality property.

To correct the loss of orthogonality that can occur in ill-conditioned optimization problems, Hager and Zhang (2013) developed the limited-memory conjugate gradient methods. The idea is to test the distance between the current gradient and the space S_k spanned by the recent prior search directions. When this distance becomes small enough, the orthogonality property has been lost, and in this case, the objective function f in (11.1) is minimized over S_k until a gradient that is approximately orthogonal to S_k has been achieved. This approximate orthogonality condition is eventually fulfilled by the first-order optimality conditions for a local minimizer in the subspace.

The development of the limited-memory conjugate gradient algorithm is given in the context of CG-DESCENT. In this algorithm, the search directions are updated as

$$d_{k+1} = -g_{k+1} + \beta_k d_k, \tag{11.60}$$

$$\beta_k = \frac{1}{y_k^T d_k}\left(y_k - \theta_k \frac{\|y_k\|^2}{y_k^T d_k} d_k \right)^T g_{k+1}. \tag{11.61}$$

Here, $\theta_k > 1/4$ is a parameter associated to the CG-DESCENT family. In CG-DESCENT, $\theta_k = 2$. The limited-memory conjugate gradient algorithm uses a preconditioned version of (11.60)–(11.61). The idea of preconditioning is to make a change of variables $x = Cy$, where $C \in \mathbb{R}^{n\times n}$ is a nonsingular matrix, in order to improve the condition number of the objective function. The goal of preconditioning is to choose C in such a way that the eigenvalues of the Hessian of $f(Cy)$, i.e., the eigenvalues of $\nabla_y^2 f(Cy) = C^T \nabla^2 f(x) C$ are roughly the same, i.e., clustered. Since $C^T \nabla^2 f(x) C$ is similar to $\nabla^2 f(x) CC^T$, it follows that the product CC^T is usually chosen to approximate the inverse Hessian $\nabla^2 f(x)^{-1}$. The product $P = CC^T$ is usually called the preconditioner.

The preconditioner in the preconditioned CG-DESCENT is changed at each iteration. If P_k denotes a symmetric, positive semidefinite preconditioner, then the search directions for the preconditioned CG-DESCENT are updated as

$$d_{k+1} = -P_k g_{k+1} + \beta_k d_k, \tag{11.62}$$

where

$$\beta_k = \frac{y_k^T P_k g_{k+1}}{y_k^T d_k} - \theta_k \frac{y_k^T P_k y_k}{y_k^T d_k} \frac{d_k^T g_{k+1}}{y_k^T d_k}. \tag{11.63}$$

Observe that $P_k = I$ corresponds to the update formula (11.61) used in CG-DESCENT. To ensure the global convergence when β_k becomes too small, it must be truncated as

$$\beta_k^+ = \max\{\beta_k, \eta_k\}, \ \eta_k = \eta\left(\frac{d_k^T g_k}{d_k^T P_k^{-1} d_k}\right), \tag{11.64}$$

where η is a positive parameter ($\eta = 0.4$ in numerical experiments) and P_k^{-1} is the inverse of P_k. Therefore, with (11.64), the preconditioned search direction is

$$d_{k+1} = -P_k g_{k+1} + \beta_k^+ d_k. \tag{11.65}$$

Dai and Yuan (1999, 2001a) proved that the standard Wolfe line search is sufficient to prove the global convergence of conjugate gradient methods. Therefore, if $\theta_k = \theta > 1/4$ and the smallest and the largest eigenvalues of the preconditioner P_k are uniformly bounded away from 0 and ∞, then the CG-DESCENT family is globally convergent under the standard Wolfe line search.

The limited-memory conjugate gradient algorithm is in close connection with both L-BFGS of Nocedal (1980) and Liu and Nocedal (1989) and with the reduced Hessian algorithm of Gill and Leonard (2001, 2003). In the limited-memory conjugate gradient algorithm of Hager and Zhang (2013) the memory is used to monitor the orthogonality of the search directions. When orthogonality is lost, the memory is used to generate a new orthogonal search direction.

Let $m > 0$ denote the number of vectors in the memory and let S_k denote the subspace spanned by the previous m search directions

$$S_k = span\{d_{k-1}, d_{k-2}, \ldots .d_{k-m}\}.$$

If g_k is nearly contained in S_k, then it means that the algorithm has lost its orthogonality property, the conjugate gradient iterations are interrupted and the following minimization problem is considered

$$\min_{z \in S_k} f(x_k + z). \tag{11.66}$$

Proposition 11.5 (*Subspace optimality*) *Consider the problem*

$$\min\{f(x) : x \in x_0 + S\}, \tag{11.67}$$

where the minimizing function f is continuously differentiable and S is the subspace $S = span\{v_1, \ldots, v_m\}$. *If* $\hat{x}$ *is a solution of problem (11.67), then* $\nabla f(\hat{x}) \perp S$.

Proof If $V \in \mathbb{R}^{n \times m}$ has columns $v_1, \ldots, v_m$, then (11.67) is equivalent to

$$\min\{f(x_0 + Vz) : z \in \mathbb{R}^m\}. \tag{11.68}$$

Now, let us set $\hat{f}(z) = f(x_0 + Vz)$. If $\hat{z}$ is a solution to (11.68), then

$$V^T \nabla f(x_0 + V\hat{z}) = \nabla \hat{f}(\hat{z}) = 0.$$

Observe that $\hat{x} = x_0 + V\hat{z}$ is a solution to (11.67) if and only if $\hat{z}$ is a solution to (11.68). Therefore, $V^T \nabla f(\hat{x}) = 0$, or equivalently, $v_i^T \nabla f(\hat{x}) = 0$ for all $i = 1, \ldots, m$. In other words, $\nabla f(\hat{x}) \in span\{v_1, \ldots, v_m\}^{\perp}$ (See Appendix A). ♦

From Proposition 11.5, if z_k is a solution of (11.66) and $x_{k+1} = x_k + z_k$, then by the first-order optimality conditions for (11.66) it follows that $d^T g_{k+1} = 0$ for all $d \in S_k$. To implement the subspace minimization process, Hager and Zhang introduced two parameters η_0 and η_1, where $0 < \eta_0 < \eta_1 < 1$. ($\eta_0 = 0.001$, $\eta_1 = 0.900$). If the condition

$$dist\{g_k, S_k\} \leq \eta_0 \|g_k\| \tag{11.69}$$

is satisfied, then the algorithm switches to the subspace problem (11.66). The iterations inside the subspace are continued until the gradient becomes sufficiently orthogonal to the subspace to satisfy the condition

$$dist\{g_{k+1}, S_k\} \geq \eta_1 \|g_{k+1}\|, \tag{11.70}$$

where $dist\{x, S\} = \inf\{\|y - x\| : y \in S\}$. If Z is a matrix whose columns are an orthogonal basis for S_k, then the conditions (11.69) and (11.70) can be expressed as

$$(1 - \eta_0^2)\|g_k\|^2 \leq \left\|g_k^T Z\right\|^2 \quad \text{and} \quad (1 - \eta_1^2)\|g_{k+1}\|^2 \geq \left\|g_{k+1}^T Z\right\|^2. \tag{11.71}$$

The subspace problem is solved by means of a quasi-Newton method. The quasi-Newton iteration applied to the subspace problem (11.66) can be a special case of CG-DESCENT with a preconditioner of the form $P_k = Z\hat{H}_{k+1}Z^T$, where $\hat{H}_{k+1}$ is the quasi-Newton matrix in the subspace. The search direction $\hat{d}_{k+1}$ in the subspace is computed as $\hat{d}_{k+1} = -\hat{H}_{k+1}\hat{g}_{k+1}$, where $\hat{g}_{k+1} = Z^T g_{k+1}$ is the gradient in the subspace.

Let $\hat{P}_k$ be the preconditioner in the subspace, which can be considered as an approximation to the inverse Hessian in the subspace. If Z is the matrix whose columns are an orthogonal basis for the subspace S_k, then the following preconditioner for the conjugate gradient iteration (11.65)

$$P_k = Z\hat{P}_k Z^T + \sigma_k \bar{Z}\bar{Z}^T,$$

can be considered, where $\bar{Z}$ is a matrix whose columns are an orthogonal basis for the complement of S_k and $\sigma_k I$ is the safe-guarded Barzilai–Borwein approximation to the inverse Hessian given by

$$\sigma_k = \max\left\{\sigma_{\min}, \min\left\{\sigma_{\max}, \frac{y_k^T s_k}{y_k^T y_k}\right\}\right\}, \quad 0 < \sigma_{\min} \leq \sigma_{\max} < \infty. \tag{11.72}$$

Since $\hat{P}_k$ is an approximation to the inverse Hessian in the subspace, then $Z\hat{P}_k Z^T$ can be viewed as an approximation to the full Hessian restricted to the subspace. Since outside the subspace there is no information about the Hessian, then the Barzilai–Borwein approximation $\sigma_k \bar{Z}\bar{Z}^T$ in the complement of S_k may be used. But, $\bar{Z}\bar{Z}^T = I - ZZ^T$. Therefore, the preconditioned search direction (11.65) can be expressed as

$$\begin{aligned} d_{k+1} &= -P_k g_{k+1} + \beta_k^+ d_k \\ &= -Z\hat{P}_k Z^T g_{k+1} - \sigma_k (I - ZZ^T) g_{k+1} + \beta_k^+ d_k \\ &= -Z(\hat{P}_k - \sigma_k I)\hat{g}_{k+1} - \sigma_k g_{k+1} + \beta_k^+ d_k, \end{aligned} \tag{11.73}$$

where $\hat{g}_{k+1} = Z^T g_{k+1}$ is the gradient in the subspace. Observe that the first term in (11.73) is the subspace contribution to the search direction, while the remaining terms are a scaled conjugate gradient direction.

The conjugate gradient parameter is computed as

$$\beta_k^+ = \max\{\beta_k, \eta_k\}, \quad \eta_k = \eta\left(\frac{d_k^T g_k}{d_k^T P_k^{-1} d_k}\right) = \eta\left(\frac{s_k^T g_k}{d_k^T y_k}\right). \tag{11.74}$$

The limited-memory conjugate gradient algorithm L-CG-DESCENT developed by Hager and Zhang (2013) has three stages:

(1) *Standard conjugate gradient iteration.* This is defined by (11.65) with $P_k = I$ as long as $dist\{g_k, S_k\} > \eta_0 \|g_k\|$. When the subspace condition $dist\{g_k, S_k\} \leq \eta_0 \|g_k\|$ is satisfied, then go to the subspace iteration.

(2) *Subspace iteration.* Solve the subspace problem (11.66) by CG-DESCENT with the preconditioner $P_k = Z\hat{P}_k Z^T$, where Z is a matrix whose columns are an orthonormal basis for the subspace S_k and $\hat{P}_k$ is a preconditioner in the subspace. Stop at the first iteration where $dist\{g_k, S_k\} \geq \eta_1 \|g_k\|$ and go to the preconditioning step.

(3) *Preconditioning step.* When the subspace iteration terminates and returns to the full space standard conjugate gradient iteration, the convergence can be accelerated by performing a single preconditioned iteration. In the special case $\hat{P}_k = \hat{H}_{k+1}$, where $\hat{H}_{k+1}$ is a quasi-Newton matrix, an appropriate

preconditioned step corresponds to the search direction (11.73), where σ_k is given by (11.72), Z is a matrix whose columns are an orthonormal basis for the subspace S_k and β_k^+ is given by (11.74). After the preconditioning stage, the algorithm continues with the standard conjugate gradient iteration.

Observe that along the iterations of the limited-memory conjugate gradient algorithm, three different preconditioners could be used, which correspond to the three parts (stages) of the algorithm: $P_k = I$; $P_k = Z\hat{P}_k Z^T$, where $\hat{P}_k$ is the subspace preconditioner and Z is a matrix whose columns are an orthonormal basis for the subspace S_k; $P_k = Z\hat{P}_k Z^T + \sigma_k \bar{Z}\bar{Z}^T$, where $\bar{Z}$ is a matrix whose columns are an orthogonal basis for the complement of S_k and $\sigma_k I$ is the safe-guarded Barzilai–Borwein approximation to the inverse Hessian given by (11.72).

The convergence of the preconditioned conjugate gradient algorithm given by (11.63)–(11.65) is shown by Hager and Zhang (2013). *Suppose that the Assumption CG holds. If* $\theta_k > 1/4$, *the line search satisfies the standard Wolfe conditions* (11.4) *and* (11.5), *and for all k the preconditioner* P_k *satisfies the conditions*

$$\|P_k\| \leq \gamma_0, \; g_{k+1}^T P_k g_{k+1} \geq \gamma_1 \|g_{k+1}\|^2, \; d_k^T P_k^{-1} d_k \geq \gamma_2 \|d_k\|^2,$$

where γ_0, γ_1, *and* γ_2 *are positive constants, then either* $g_k = 0$ *for some k, or*

$$\liminf_{k \to \infty} \|g_k\| = 0.$$

Moreover, if P_k is expressed in terms of a subspace matrix $\hat{P}_k$ and of a matrix Z with orthonormal columns that form a basis for the subspace $S_k : P_k = Z\hat{P}_k Z^T$, then the algorithm in the subspace is also convergent (stage 2 of L-CG-DESCENT algorithm). *Suppose that the Assumption CG holds. If* $\theta_k > 1/4$, *the line search satisfies the standard Wolfe conditions* (11.4) *and* (11.5) *and for all k the preconditioner* $\hat{P}_k$ *satisfies the conditions*

$$\|\hat{P}_k\| \leq \hat{\gamma}_0, \; \hat{g}_{k+1}^T \hat{P}_k \hat{g}_{k+1} \geq \hat{\gamma}_1 \|\hat{g}_{k+1}\|^2, \; \hat{d}_k^T \hat{P}_k^{-1} \hat{d}_k \geq \hat{\gamma}_2 \|\hat{d}_k\|^2,$$

where $\hat{\gamma}_0, \hat{\gamma}_1$ *and* $\hat{\gamma}_2$ *are positive constants, then either* $g_k = 0$ *for some k, or*

$$\liminf_{k \to \infty} \|g_k\| = 0.$$

The L-CG-DESCENT algorithm is implemented in the context of the CG-DESCENT algorithm and it is known as CG-DESCENT 6.0. Three algorithms, L-CG-DESCENT, L-BFGS, and CG-DESCENT version 5.3 correspond to different parameter settings in the CG-DESCENT version 6.0. The number of search directions in the subspace S_k is controlled by the parameter *memory*. When the *memory* parameter is zero, CG-DESCENT 6.0 reduces to CG-DESCENT 5.3. If the parameter *LBFGS* in CG-DESCENT 6.0 is TRUE, then CG-DESCENT 6.0 reduces to L-BFGS. Therefore, all three algorithms employ the same CG-DESCENT line searches: standard Wolfe or approximate Wolfe, developed by Hager and Zhang (2013). The line search in the L-BFGS algorithm implemented in CG-DESCENT 6.0 is different from the MCSRCH line search of Moré and Thuente (1994) implemented in the L-BFGS algorithm by Liu and Nocedal (1989).

L-CG-DESCENT includes a number of 55 parameters concerning: the search direction and the line search computations, the control of the orthogonality of the gradient to the subspace S_k, the stopping conditions, the printing facilities, etc.

Example 11.1 (PALMER1C problem) In the following, let us see the performances of L-CG-DESCENT versus DY (Dai & Yuan, 1999), versus DESCONa (Andrei, 2013c), versus L-BFGS (Liu & Nocedal, 1989) and versus CG-DESCENT 5.3 for solving the problem PALMER1C (see Andrei, 2019e). This is a positive definite quadratic optimization problem with 8 variables. The eigenvalues of its Hessian are all positive and range from 2×10^{-4} up to 2×10^{8}. Therefore, the condition number of this problem is 10^{12}. In theory, the conjugate gradient algorithm should solve this problem in 8 iterations. However, with the standard Wolfe line search (see Figure 5.1), the Dai–Yuan conjugate gradient algorithm where $\beta_k^{DY} = g_{k+1}^T g_{k+1}/d_k^T y_k$, (see Table 4.1) needs over 300,000 iterations to reduce the max norm of the gradient to 10^{-5}. The DESCONa algorithm with modified standard Wolfe line search needs 937 iterations, 5005 evaluations of the function and its gradient and 0.02 s to reduce the max norm of the gradient to 10^{-3}. The L-BFGS code with $m = 5$ of Liu and Nocedal (1989) with the MCSRCH line search of Moré and Thuente (1994) needs 5350 iterations, 6511 evaluations of the function and its gradient and 0.05 s to reduce the max norm of the gradient to 10^{-3}. All these algorithms obtained the same optimal value of the function: 0.097594. The performances of L-CG-DESCENT for solving this problem with different values of the parameter *memory* are presented in Table 11.1, where *#iter* is the number of iterations, *#f* and *#g* represent the number of function and its gradient evaluations, respectively, and *cpu(s)* is the CPU time in seconds for obtaining a solution.

In Table 11.1, the entries across the first line show the performances of L-CG-DESCENT with the line search implemented in CG-DESCENT, when the parameter LBFGS is TRUE to get a solution of the problem for which the max norm of the gradient is reduced to 10^{-7}. When the parameter memory is assigned to 0, then the problem is solved with CG-DESCENT 5.3, with Wolfe line search. The number of iterations used by version 5.3 for solving this problem was 51302, while the number of iterations used by L-CG-DESCENT with *memory* = 9 was 12. Also, there is a big difference between L-CG-DESCENT with *memory* = 5 and with *memory* = 9. Observe that L-CG-DESCENT with *memory* = 9 has the best performances. If the number of stored search directions in the subspace S_k increases, the performances of L-CG-DESCENT are the same. For *memory* = 5, L-CG-DESCENT needs a number of 5791 subspace iterations. On the other hand, for *memory* = 7, L-CG-DESCENT needs a number of 190 subspace iterations. ♦

Table 11.1 Performances of L-CG-DESCENT for solving PALMER1C problem

	#iter	*#f*	*#g*	*cpu(s)*
TRUE	51302	83296	143343	147.82
Memory = 0	51302	83296	143343	138.10
Memory = 5	14242	17103	32229	36.41
Memory = 7	480	579	1106	1.32
Memory = 9	12	23	24	0.04

Table 11.2 Performances of L-CG-DESCENT for solving 10 problems from the UOP collection. $n = 10,000$; Wolfe line search; *memory* = 5

Problem name	*#iter*	*#f*	*#g*	*#si*	*cpu* (s)
Freudenstein & Roth	10	27	19	9	0.11
Extended Rosenbrock	63	153	94	62	0.57
BDQRTIC	124	574	551	77	7.23
CUBE	3721	7770	4136	112	7.52
NONDQUAR	1978	3958	1980	419	17.51
EDENSCH	23	71	63	0	0.45
ARWHEAD	8	19	11	7	0.14
DQDRTIC	23	135	133	19	0.66
DENSCHNB	6	13	7	5	0.07
DENSCHNF	8	20	12	7	0.12
TOTAL	5964	12740	7006	717	34.4

Numerical study. In the following, let us present the performances of L-CG-DESCENT for solving 10 problems from the UOP collection (Andrei, 2018g). Tables 11.2–11.5 show the performances of L-CG-DESCENT for different values of the parameter *memory*, as well as the comparison versus the L-BFGS ($m = 5$) of Liu and Nocedal. The number of variables for each problem considered in this numerical study was assigned to 10,000. In all numerical experiments, the standard Wolfe line search was used. In these tables, *#si* represents the number of subspace iterations. Comparing L-CG-DESCENT with *memory* = 5 (Table 11.2) versus the same algorithm with *memory* = 9 (Table 11.3) observe that they have similar performances, L-CG-DESCENT with *memory* = 9 being slightly more efficient. Comparing L-CG-DESCENT with the parameter LBFGS = TRUE versus the L-BFGS of Liu and Nocedal (1989) in Table 11.4, at least for this set of 10 unconstrained optimization problems L-BFGS of Liu and Nocedal is faster. Now, a comparison between the performances of L-CG-DESCENT with *memory* = 5 (Table 11.2) or with *memory* = 9 (Table 11.3) versus the performances of L-CG-DESCENT with *memory* = 0 (Table 11.5), i.e., versus CG-DESCENT 5.3, shows that L-CG-DESCENT with *memory* = 5 or *memory* = 9 is faster. For example, for the problem NONDQUAR with $n = 10,000$ variables, L-CG-DESCENT with *memory* = 5 needs only 17.51 s to get the solution,

Table 11.3 Performances of L-CG-DESCENT for solving 10 problems from the UOP collection. n = 10,000; Wolfe Line search; *memory* = 9

Problem name	*#iter*	*#f*	*#g*	*#si*	*cpu* (s)
Freudenstein & Roth	9	25	18	8	0.10
Extended Rosenbrock	62	148	91	61	0.54
BDQRTIC	85	351	336	42	4.48
CUBE	2303	4808	2548	1207	4.73
NONDQUAR	2453	4908	2455	1205	22.60
EDENSCH	23	71	63	0	0.47
ARWHEAD	8	19	11	7	0.14
DQDRTIC	22	143	139	18	0.71
DENSCHNB	6	13	7	5	0.05
DENSCHNF	8	20	12	7	0.13
Total	4979	10506	5680	2560	33.9

Table 11.4 Performances of L-CG-DESCENT versus L-BFGS ($m = 5$) of Liu and Nocedal for solving 10 problems from the UOP collection. n = 10,000; Wolfe Line search; Wolfe = TRUE in L-CG-DESCENT

	L-CG-DESCENT				L-BFGS ($m = 5$)		
Problem name	*#iter*	*#f*	*#g*	*cpu* (s)	*#iter*	*#fg*	*cpu* (s)
Freudenstein & Roth	11	27	17	0.12	17	20	0.20
Extended Rosenbrock	61	134	75	0.61	64	86	0.10
BDQRTIC	205	630	650	8.53	183	255	0.36
CUBE	1088	2282	1213	2.20	4010	5001	8.84
NONDQUAR	2760	5524	2765	28.35	3267	3638	5.91
EDENSCH	18	64	55	0.36	22	47	0.80
ARWHEAD	9	90	86	0.75	12	15	0.20
DQDRTIC	21	87	82	0.48	13	23	0.20
DENSCHNB	7	15	8	0.07	18	22	0.03
DENSCHNF	9	19	10	0.11	28	40	0.09
Total	4189	8872	4961	41.62	7634	9147	16.73

L-CG-DESCENT with LBFGS = TRUE needs 28.35 s. But, L-BFGS of Liu and Nocedal needs 5.91 s. However, for ill-conditioned problems, L-CG-DESCENT is much more efficient (faster). L-CG-DESCENT is one of the most respectable conjugate gradient algorithms with a very sophisticated implementation in computer code, designed to solve difficult (ill-conditioned) problems and having much better practical performances.

It is worth seeing the performances of DESCONa for solving the above 10 problems with $\varepsilon = 10^{-7}$ in the criterion for stopping the iterations. Table 11.6 shows the performances of DESCONa.

Table 11.5 Performances of L-CG-DESCENT for solving 10 problems from the UOP collection. n = 10,000; Wolfe Line search; *memory* = 0 (CG-DESCENT 5.3)

Problem name	*#iter*	*#f*	*#g*	*cpu* (s)
Freudenstein & Roth	13	67	58	0.29
Extended Rosenbrock	52	137	94	0.56
BDQRTIC	133	537	495	6.65
CUBE	3420	7151	3793	6.88
NONDQUAR	2563	5128	2565	22.65
EDENSCH	23	72	64	0.45
ARWHEAD	11	81	76	0.69
DQDRTIC	64	243	220	1.32
DENSCHNB	8	17	9	0.08
DENSCHNF	11	24	13	0.15
Total	6298	13457	7387	39.7

Table 11.6 Performances of DESCONa for solving 10 problems from the UOP collection. n = 10,000; modified Wolfe Line search

Problem name	*#iter*	*#fg*	*cpu* (s)
Freudenstein & Roth	9	35	0.02
Extended Rosenbrock	60	215	0.07
BDQRTIC	105	745	1.43
CUBE	1657	5001	2.34
NONDQUAR	1754	5001	10.15
EDENSCH	24	124	0.22
ARWHEAD	4	20	0.03
DQDRTIC	5	16	0.03
DENSCHNB	10	33	0.05
DENSCHNF	11	42	0.09
Total	3629	11232	14.43

Enriched methods

Another idea is to combine CGSYS (or any other conjugate gradient algorithm) and L-BFGS in a more sophisticated way by performing a prespecified number, say p, of L-BFGS iterations and a prespecified number q of CGSYS iterations. The algorithm should start with the L-BFGS iterations and the matrix obtained at the end of the p L-BFGS iterations is used to precondition the first of the q CGSYS iterations (Morales & Nocedal, 2002). This follows to be investigated.

11.4 Conjugate Gradient with Subspace Minimization Based on Regularization Model of the Minimizing Function

Conjugate gradient methods are based on the conjugacy condition $d_{k+1}^T y_k = 0$, or $d_{k+1}^T y_k = t(g_{k+1}^T s_k)$, where $t > 0$ is a parameter. The main reason for generating conjugate directions is that the minimization of a convex quadratic function in a subspace spanned by a set of mutually conjugate directions is equivalent to the minimization of the objective function along each conjugate direction in turn. This is a very good idea, but it works only when the line searches are exact. When the line searches are not exact, the conjugacy property may have disadvantages in the sense that the error in the current iteration can not be eliminated in the following iterations as long as the following search directions are conjugate to the current search direction. Therefore, the conjugacy condition is not so strict.

As known, for quadratic functions, at each iteration of the conjugate gradient method the gradient should be orthogonal to the space spanned by the previous search directions. For some ill-conditioned problems, the orthogonality property is quickly lost and the convergence is much slower than expected. A solution to deal with these ill-conditioned problems was given by Hager and Zhang (2013), who introduced the limited-memory conjugate gradient method L-CG-DESCENT. Another solution is to solve a *p-regularized subproblem,* where $p > 2$ is an integer. For a minimizing function f, its p-regularization model is constructed by adding a p th regularization term to the quadratic estimation of f. The idea is to construct and minimize a local quadratic approximation of the minimizing function with a weighted regularization term $(\sigma_k/p)\|x\|^p$, $p > 2$. The most common choice to regularize the quadratic approximation is the p-regularization with $p = 3$, which is known as the *cubic regularization.* The idea of using the cubic regularization into the context of the Newton method first appeared in Griewank (1981) and was later developed by many authors, proving its convergence and complexity (e.g., see: (Nesterov & Polyak, 2006), (Cartis, Gould, & Toint, 2011a, 2011b), (Gould, Porcelli, & Toint, 2012), (Bianconcini, Liuzzi, Morini, & Sciandrone, 2013), (Bianconcini & Sciandrone, 2016), (Hsia, Sheu, & Yuan, 2017)). Griewank proved that any accumulation point of the sequence generated by minimizing the p-regularized subproblem is a second-order critical point of f, i.e., a point $\bar{x} \in \mathbb{R}^n$ satisfying $\nabla f(\bar{x}) = 0$ and $\nabla^2 f(\bar{x})$ semipositive definite. Later, Nesterov and Polyak (2006) proved that the cubic regularization method has a better global iteration complexity bound than the one for the steepest descent method. Based on these results, Cartis, Gould, and Toint (2011a, 2011b) proposed an adaptive cubic regularization method for minimizing the function f, where the sequence of the regularization parameter $\{\sigma_k\}$ is dynamically determined and the p-regularized subproblems are inexactly solved. In their adaptive cubic regularization method, the minimizing function f is approximated by the model

$$m_k(d) = f(x_k) + g_k^T d + \frac{1}{2} d^T B_k d + \frac{1}{3} \sigma_k \|d\|^3, \tag{11.75}$$

where σ_k is a positive parameter (regularization parameter) dynamically updated in a specific way and B_k is an approximation to the Hessian of the objective function. The adaptive cubic regularization method for the unconstrained optimization was further developed by Bianconcini, Liuzzi, Morini, and Sciandrone, (2013). The idea was to compute the trial step as a suitable approximate minimizer of the above cubic model of the minimizing function by using the nonmonotone globalization techniques of Grippo and Sciandrone (2002). Another approach was presented by Gould, Porcelli, and Toint (2012), who presented new updating strategies for the regularization parameter σ_k based on interpolation techniques, which improved the overall numerical performance of the algorithm. New subspace minimization conjugate gradient methods based on p-regularization models, with $p = 3$ and $p = 4$, were developed by Zhao, Liu, and Liu (2019). A complete theory of the p-regularized subproblems for $p > 2$, including the solution of these problems was presented by Hsia, Sheu, and Yuan (2017).

In the following, let us develop a variant of the conjugate gradient algorithm with subspace minimization ((Stoer & Yuan, 1995), (Andrei, 2014), (Li, Liu, & Liu, 2019)) based on the regularization model (Zhao, Liu, & Liu, 2019). The algorithm combines the minimization of a p-regularized model (11.75) of the minimizing function with the subspace minimization. The main objective is to elaborate numerical algorithms based on the p-regularized model (11.75) with inexact line searches in which the search direction is a linear combination of the steepest descent direction and the previous search direction. If the minimizing function is close to a quadratic, then a quadratic approximation model in a two-dimensional subspace is minimized to generate the search direction, otherwise a p-regularization model is minimized.

The p-regularized subproblem
In the following, by using a special scaled norm, the p-regularized subproblem is introduced and then its solution techniques are presented.

The general form of the p-regularized subproblem is

$$\min_{x \in \mathbb{R}^n} h(x) = c^T x + \frac{1}{2} x^T B x + \frac{\sigma}{p} \|x\|^p, \tag{11.76}$$

where $p > 2$, $\sigma > 0$, $c \in \mathbb{R}^n$, and $B \in \mathbb{R}^{n \times n}$ is a symmetric matrix. Because of the regularization term $\sigma \|x\|^p / p$, it follows that $h(x)$ is a coercive function, that is, $\lim_{\|x\| \to \infty} h(x) = +\infty$, i.e., the p-regularized subproblem can always attain the global minimum, even for nonpositive definite B (see Appendix A). The solution of this subproblem is given by the following theorem, proved by Hsia, Sheu, and Yuan (2017).

Theorem 11.7 *For $p > 2$ the point x^* is a global minimizer of (11.76) if and only if*

$$\left(B+\sigma\|x^*\|^{p-2}I\right)x^* = -c, \quad B+\sigma\|x^*\|^{p-2}I\geq 0. \tag{11.77}$$

Moreover, the l_2 norms of all global minimizers are equal. ♦

Another form of the p-regularized subproblem with a scaled norm can be

$$\min_{x\in\mathbb{R}^n} h(x) = c^T x+\frac{1}{2}x^T Bx+\frac{\sigma}{p}\|x\|_A^p, \tag{11.78}$$

where $A \in \mathbb{R}^{n\times n}$ is a symmetric and positive definite matrix and $\|x\|_A= \sqrt{x^T Ax}$, known as l_A norm. Considering $y = A^{1/2}x$, (11.78) can be rewritten as

$$\min_{y\in\mathbb{R}^n} h(y) = (A^{-1/2}c)^T y+\frac{1}{2}y^T(A^{-1/2}BA^{-1/2})y+\frac{\sigma}{p}\|y\|^p. \tag{11.79}$$

From Theorem 11.7, the point y^* is a global minimizer of (11.79) if and only if

$$\left(A^{-1/2}BA^{-1/2}+\sigma\|y^*\|^{p-2}I\right)y^* = -A^{-1/2}c, \tag{11.80a}$$

$$A^{-1/2}BA^{-1/2}+\sigma\|y^*\|^{p-2}I\geq 0. \tag{11.80b}$$

Let $V \in \mathbb{R}^{n\times n}$ be an orthogonal matrix such that $V^T(A^{-1/2}BA^{-1/2})V = Q$, where Q is a diagonal matrix with the elements on the main diagonal as the eigenvalues $0\leq\mu_1\leq\cdots\leq\mu_n$ of $A^{-1/2}BA^{-1/2}$. Let us introduce a vector $a \in \mathbb{R}^n$ such that $y = Va$. Defining $z = \|y^*\|$ and premultiplying (11.80a) by V^T it follows that

$$(Q+\sigma z^{p-2}I)a = -\beta, \tag{11.81}$$

where $\beta = V^T(A^{-1/2}c)$. After some simple algebraic manipulations, (11.81) is equivalent to

$$a_i = \frac{-\beta_i}{\mu_i+\sigma z^{p-2}}, \quad i = 1,\ldots,n,$$

where a_i, $i = 1,\ldots,n$, and β_i, $i = 1,\ldots,n$, are the components of the vectors a and β, respectively.

Observe that

$$z^2 = y^T y = a^T a = \sum_{i=1}^{n}\frac{\beta_i^2}{(\mu_i+\sigma z^{p-2})^2}. \tag{11.82}$$

Let us define:

$$\Phi(z) = \sum_{i=1}^{n} \frac{\beta_i^2}{(\mu_i + \sigma z^{p-2})^2} - z^2.$$

For $p > 2$, $z > 0$ and $\sigma > 0$ observe that $\Phi'(z) < 0$. Therefore, on $[0, +\infty)$ it follows that $\Phi(z)$ is monotonically decreasing. Besides, when $\beta \neq 0$, $\Phi(0) > 0$ and $\lim_{z\to\infty} \Phi(z) = -\infty$. Hence, when $\beta \neq 0$, the Equation (11.82) has a unique positive solution. On the other hand, if $\beta = 0$, it follows that $z = 0$ is the only solution of (11.82), i.e., $x^* = 0$ is the only global solution of (11.78). Therefore, by using the above developments, the following theorem presents the global solution of the p-regularized subproblem (11.78).

Theorem 11.8 *The point x^* is a global minimizer of the p-regularized subproblem with a scaled norm* (11.78) *for $p > 2$ if and only if*

$$\left(B + \sigma (z^*)^{p-2} A\right) x^* = -c, \quad B + \sigma (z^*)^{p-2} A \geq 0, \tag{11.83}$$

where z^* is the unique nonnegative root of the equation

$$z^2 - \sum_{i=1}^{n} \frac{\beta_i^2}{(\mu_i + \sigma z^{p-2})^2} = 0. \tag{11.84}$$

Moreover, the l_A norms of all global minimizers are equal. ♦

In the following, let us consider the case in which B is symmetric and positive definite and $A = B$. In this case, since $\sigma > 0$ and $z \geq 0$, it follows that $B + \sigma(z^{p-2})B$ is always a positive definite matrix. Therefore, the global minimizer of the p-regularized subproblem with a scaled norm (11.78) is unique. In conclusion, the following remark is true:

Remark 11.3 Let $B > 0$ and $A = B$, then the point

$$x^* = \frac{-1}{1 + \sigma (z^*)^{p-2}} B^{-1} c \tag{11.85}$$

is the only global minimizer of (11.78) for $p > 2$ where z^* is the unique nonnegative solution of the equation

$$\sigma z^{p-1} + z - \sqrt{c^T B^{-1} c} = 0. \tag{11.86}$$

Concerning the Equation (11.86), observe that for $c = 0$ the equation is $z(\sigma z^{p-2} + 1) = 0$. Since $\sigma > 0$ it follows that $z^* = 0$ is the unique nonnegative solution of (11.86).

On the other hand, for $c \neq 0$ defining the function $\varphi(z) = \sigma z^{p-1} + z - \sqrt{c^T B^{-1} c}$, it is easy to see that $\varphi'(z) = \sigma(p-1)z^{p-2} + 1 > 0$, which proves that $\varphi(z)$ is monotonically increasing. Since $\varphi(0) < 0$ and $\varphi(\sqrt{c^T B^{-1} c}) > 0$, it follows that z^* is the unique positive solution of (11.86). ♦

The p-regularized subproblem in two-dimensional subspace
Consider the quadratic approximation of f in x_{k+1} as

$$\bar{h}_{k+1}(d) = g_{k+1}^T d + \frac{1}{2} d^T B_{k+1} d,$$

where B_{k+1} is a symmetric and positive definite approximation to the Hessian of f in x_{k+1} which satisfies the secant equation $B_{k+1} s_k = y_k$, with $s_k = x_{k+1} - x_k$ and $y_k = g_{k+1} - g_k$. Consider that g_{k+1} and s_k are two linearly independent vectors and define $\Omega_k = \{d_{k+1} : d_{k+1} = \mu_k g_{k+1} + \eta_k s_k\}$, where μ_k and η_k are real scalars. The corresponding p-regularized subproblem is defined as

$$\min_{d_{k+1} \in \Omega_k} h_{k+1}(d_{k+1}) = g_{k+1}^T d_{k+1} + \frac{1}{2} d_{k+1}^T B_{k+1} d_{k+1} + \frac{\sigma_k}{p} \|d_{k+1}\|_{B_{k+1}}^p, \tag{11.87}$$

where $\sigma_k > 0$ is the regularized parameter. Having in view that $d_{k+1} \in \Omega_k$ the p-regularized subproblem in the two-dimensional subspace can be expressed as

$$\min_{\mu_k, \eta_k \in \mathbb{R}} \begin{pmatrix} \|g_{k+1}\|^2 \\ g_{k+1}^T s_k \end{pmatrix}^T \begin{pmatrix} \mu_k \\ \eta_k \end{pmatrix} + \frac{1}{2} \begin{pmatrix} \mu_k \\ \eta_k \end{pmatrix}^T M_k \begin{pmatrix} \mu_k \\ \eta_k \end{pmatrix} + \frac{\sigma_k}{p} \left\| \begin{pmatrix} \mu_k \\ \eta_k \end{pmatrix} \right\|_{M_k}^p, \tag{11.88}$$

where

$$M_k = \begin{bmatrix} \rho_k & g_{k+1}^T y_k \\ y_k^T g_{k+1} & s_k^T y_k \end{bmatrix}, \quad \rho_k = g_{k+1}^T B_{k+1} g_{k+1}. \tag{11.89}$$

Observe that M_k is a symmetric and positive definite matrix since B_{k+1} is symmetric and positive definite and the vectors g_{k+1} and s_k are linear independent.

By Remark 11.3, the unique solution of (11.88) is

$$\begin{pmatrix} \mu_k^* \\ \eta_k^* \end{pmatrix} = \frac{-1}{1 + \sigma_k (z^*)^{p-2}} M_k^{-1} \begin{pmatrix} \|g_{k+1}\|^2 \\ g_{k+1}^T s_k \end{pmatrix}, \tag{11.90}$$

where z^* is the unique nonnegative solution of the equation

$$\sigma_k z^{p-1} + z - \sqrt{\begin{pmatrix} \|g_{k+1}\|^2 \\ g_{k+1}^T s_k \end{pmatrix}^T M_k^{-1} \begin{pmatrix} \|g_{k+1}\|^2 \\ g_{k+1}^T s_k \end{pmatrix}} = 0. \tag{11.91}$$

Denote

$$\delta_k = \frac{1}{1 + \sigma_k (z^*)^{p-2}}.$$

Therefore, from (11.90) the solution of the p-regularized subproblem in the two-dimensional subspace (11.88) is

$$\mu_k^* = \frac{\delta_k}{\Delta_k} \left[(g_{k+1}^T y_k)(g_{k+1}^T s_k) - (s_k^T y_k) \| g_{k+1} \|^2 \right], \tag{11.92a}$$

$$\eta_k^* = \frac{\delta_k}{\Delta_k} \left[(g_{k+1}^T y_k) \| g_{k+1} \|^2 - \rho_k (g_{k+1}^T s_k) \right], \tag{11.92b}$$

where $\Delta_k = \rho_k (s_k^T y_k) - (g_{k+1}^T y_k)^2$ is the determinant of M_k.

For the ρ_k computation, some procedures are known. One of them, given by Stoer and Yuan (1995), is

$$\rho_k = 2 \frac{(g_{k+1}^T y_k)^2}{s_k^T y_k}. \tag{11.93}$$

Using the Barzilai–Borwein method, another procedure for the ρ_k computation was given by Dai and Kou (2016)

$$\rho_k = \frac{3}{2} \frac{\| y_k \|^2}{s_k^T y_k} \| g_{k+1} \|^2. \tag{11.94}$$

Another simple way is to let B_{k+1} be a self-scaling memoryless BFGS with parameter τ_k given as

$$\rho_k = g_{k+1}^T \left[\tau_k I - \tau_k \frac{s_k s_k^T}{\| s_k \|^2} + \frac{y_k y_k^T}{s_k^T y_k} \right] g_{k+1}, \tag{11.95}$$

(see (1.53) with $B_k = I\tau_k$), where τ_k can be chosen to be any of τ_k^{OS} (8.107), τ_k^{OL} (8.111), $\bar{\tau}_k^H$ or $\bar{\tau}_k^B$ (8.112).

For the σ_k computation, there are a number of procedures. For example, Cartis, Gould, and Toint (2011a) suggested a procedure based on the trust-region ratio. Another procedure using an interpolation condition was given by Zhao, Liu, and Liu (2019). In our algorithm, let us define

$$r_k = \frac{f(x_k) - f(x_{k+1})}{f(x_k) - h_{k+1}(s_k)},$$

which measures the actual decrease in the objective function $f(x_k) - f(x_{k+1})$ versus the predicted model decrease $f(x_k) - h(s_k)$. The regularized parameter σ_k is updated as follows:

$$\sigma_{k+1} = \begin{cases} \max\{\min\{\sigma_k, \|g_{k+1}\|\}, \varepsilon_M\}, & \text{if } r_k > \lambda_2, \\ \sigma_k + \|g_{k+1}\|^2, & \text{if } \lambda_1 \le r_k \le \lambda_2, \\ 3\left|f_k - f_{k+1} + s_k^T g_{k+1} - 0.5 y_k^T s_k\right| / (y_k^T s_k)^{p/2}, & \text{otherwise}, \end{cases} \quad (11.96)$$

where $\sigma_0 = 1$, ε_M is the relative machine precision, $\lambda_1 = 10^{-5}$ and $\lambda_2 = 0.5$. Of course, this is a suggestion which proved to be successful in our numerical experiments, but some other proposals may be considered as well.

Strategies for search direction computation
In this algorithm, if the objective function is close to a quadratic, then a quadratic approximation model in a two-dimensional subspace is used to generate the search direction, otherwise a p-regularization model in a two-dimensional subspace is to be considered. Indeed, to see how the function $f(x)$ is close to a quadratic function on the line segment connecting x_{k-1} and x_k, Yuan (1991) introduced the parameter

$$t_k = \left| \frac{2(f_{k-1} - f_k + g_k^T s_{k-1})}{s_{k-1}^T y_{k-1}} - 1 \right|. \quad (11.97)$$

On the other hand, the ratio

$$\theta_k = \frac{f_{k-1} - f_k}{0.5 s_{k-1}^T y_{k-1} - g_k^T s_{k-1}} \quad (11.98)$$

shows the difference between the actual reduction of the function values and the predicted reduction given by the quadratic model.

The strategy for using the quadratic approximation or the p-regularization model of the minimizing function is as follows. If the conditions

$$t_k \le c_1 \text{ or } |\theta_k - 1| \le c_2 \quad (11.99)$$

hold, where c_1 and c_2 are positives constants ($c_1 = 10^{-4}$ and $c_2 = 10^{-5}$), then the function $f(x)$ might be very close to a quadratic on the line segment connecting x_{k-1} and x_k. In this case, for the search direction, the quadratic approximation model in a two-dimensional subspace is selected, which corresponds to (11.88) with $\sigma_k = 0$. Therefore, in our algorithm, the parameters μ_k and η_k which define the search direction d_{k+1} are computed as

$$\mu_k^* = \frac{1}{\Delta_k}\left[(g_{k+1}^T y_k)(g_{k+1}^T s_k) - (s_k^T y_k)\|g_{k+1}\|^2\right], \quad (11.100a)$$

$$\eta_k^* = \frac{1}{\Delta_k}\left[(g_{k+1}^T y_k)\|g_{k+1}\|^2 - \rho_k(g_{k+1}^T s_k)\right], \tag{11.100b}$$

where ρ_k is computed as in (11.94). On the other hand, if $t_k > c_1$ and $|\theta_k - 1| > c_2$, then the parameters μ_k and η_k which define the search direction d_{k+1} are computed as in (11.92), where ρ_k and σ_k are computed as in (11.94) and (11.96), respectively.

Of course, some other variants of the algorithm may have regard for it, where for the computation of $\rho_k = g_{k+1}^T B_{k+1} g_{k+1}$, (11.93) or (11.95) may be used. In our numerical experiments, the variant proposed by Dai and Kou (2016) given by (11.94) proved to be the most efficient. The other crucial ingredient of our algorithm is the computation of the regularization parameter σ_k. Here, σ_k is computed as in (11.96), but some other strategies may be implemented. For example, Cartis, Gould, and Toint (2011a) proposed a procedure for this parameter computation by analogy with the trust-region method. In such a framework, σ_k could be regarded as the reciprocal of the trust-region radius. Thus, σ_k is increased if insufficient decrease is obtained, but it is decreased or unchanged otherwise. Other procedures for updating the regularization parameter σ_k for minimizing the p-regularization model is discussed by Gould, Porcelli, and Toint (2012).

However, finding a global minimizer of the $h_{k+1}(.)$ defined by (11.87) may not be essential in practice. Therefore, the global minimization problem (11.87) of the p-regularized subproblem may be relaxed by letting d_{k+1} be an approximation to such a minimizer.

With these developments, taking into consideration the acceleration scheme presented in Remark 5.1, according to the value of the parameter "acceleration" (true or false), the following algorithms CUBIC and CUBICa may be presented. Clearly, CUBICa is the accelerated version of CUBIC.

Algorithm 11.4 Subspace minimization based on cubic regularization CUBIC/CUBICa

1.	Select a starting point $x_0 \in dom f$ and compute: $f_0 = f(x_0)$ and $g_0 = \nabla f(x_0)$. Select $\varepsilon_A > 0$ sufficiently small and positive values $0 < \rho < \sigma < 1$ used in Wolfe line search conditions. Select some positive values for: $c_1, c_2, \lambda_1, \lambda_2$. Set $d_0 = -g_0$ and $k = 0$
2.	Test a criterion for stopping the iterations. If the test is satisfied, then stop; otherwise continue with step 3
3.	Using the Wolfe line search conditions determine the stepsize α_k. Update the variables $x_{k+1} = x_k + \alpha_k d_k$. Compute f_{k+1}, g_{k+1} and $s_k = x_{k+1} - x_k$, $y_k = g_{k+1} - g_k$
4.	If *acceleration* equal true, then (a) Compute: $z = x_k + \alpha_k d_k$, $g_z = \nabla f(z)$ and $y_k = g_k - g_z$ (b) Compute: $a_k = \alpha_k g_k^T d_k$, and $b_k = -\alpha_k y_k^T d_k$ (c) If $\|b_k\| \geq \varepsilon_A$, then compute $\xi_k = -a_k/b_k$ and update the variables as $x_{k+1} = x_k + \xi_k \alpha_k d_k$. Compute f_{k+1} and g_{k+1}. Compute $s_k = x_{k+1} - x_k$ and $y_k = g_{k+1} - g_k$
5.	If $t_k > c_1$ and $\|\theta_k - 1\| > c_2$, then the search direction is computed as $d_{k+1} = \mu_k g_{k+1} + \eta_k s_k$, where the parameters μ_k and η_k are computed as in (11.92), with ρ_k and σ_k are computed as in (11.94) and (11.96), respectively

(continued)

Algorithm 11.1 (continued)

6.	If $t_k \le c_1$ or $\lvert\theta_k - 1\rvert \le c_2$, then the search direction is computed as $d_{k+1} = \mu_k g_{k+1} + \eta_k s_k$, where the parameters μ_k and η_k are computed as in (11.100), with ρ_k computed as in (11.94)
7.	Restart criterion. If $\lvert g_{k+1}^T g_k\rvert > 0.2\lVert g_{k+1}\rVert^2$ then set $d_{k+1} = -g_{k+1}$
8.	Consider $k = k+1$ and go to step 2 ♦

This is a variant of the subspace minimization conjugate gradient algorithm based on the cubic regularization model of the unconstrained optimization problem. Some other variants may be generated by selecting different procedures for ρ_k and σ_k computation, as well as for the restarting criterion.

Convergence analysis

Suppose that the search direction d_{k+1} in the CUBIC algorithm is calculated under the following conditions

$$\xi_1 \le \frac{s_k^T y_k}{\lVert s_k\rVert^2} \le \frac{\lVert y_k\rVert^2}{s_k^T y_k} \le \xi_2, \tag{11.101}$$

where ξ_1 and ξ_2 are positive constants ($\xi_1 = 10^{-7}$, $\xi_2 = 10^5$). For general nonlinear functions, if (11.101) holds, then the condition number of the Hessian of the minimizing function might not be very large. In this case, both the quadratic and the p-regularization models may be used.

Proposition 11.6 *Under the conditions (11.101), the search direction $d_{k+1} = \mu_k g_{k+1} + \eta_k s_k$, where the parameters μ_k and η_k are computed as in (11.92), with ρ_k and σ_k computed as in (11.94) and (11.96), respectively, satisfies the sufficient descent condition $g_{k+1}^T d_{k+1} \le -\bar{c}\lVert g_{k+1}\rVert^2$, where $\bar{c}$ is a positive constant.*

Proof Since $\sigma_k \ge 0$ and $z^* \ge 0$, it follows that $\delta_k < 1$. Therefore,

$$g_{k+1}^T d_{k+1} \le -\frac{\lVert g_{k+1}\rVert^4}{\Delta_k}\left[(s_k^T y_k) - 2(g_{k+1}^T y_k)\frac{g_{k+1}^T s_k}{\lVert g_{k+1}\rVert^2} + \rho_k\left(\frac{g_{k+1}^T s_k}{\lVert g_{k+1}\rVert^2}\right)^2\right].$$

Denote the term in the square brackets of the above inequality by χ_k and consider it as a function of the variable $g_{k+1}^T s_k/\lVert g_{k+1}\rVert^2$. Now, taking minimization of χ_k, it follows that $\chi_k \ge \Delta_k/\rho_k$. Therefore, from (11.94) and since $s_k^T y_k/\lVert y_k\rVert^2 \ge \xi_2^{-1}$, it follows that

$$g_{k+1}^T d_{k+1} \le -\frac{\|g_{k+1}\|^4}{\rho_k} = -\frac{2}{3}\frac{s_k^T y_k}{\|y_k\|^2}\|g_{k+1}\|^2 \le -\frac{2}{3\xi_2}\|g_{k+1}\|^2 = -\bar{c}\|g_{k+1}\|^2,$$

where $\bar{c} = 2/(3\xi_2)$. ♦

Proposition 11.7 *Under the conditions (11.101), the search direction* $d_{k+1} = \mu_k g_{k+1} + \eta_k s_k$, *where the parameters* μ_k *and* η_k *are computed as in (11.92), with* ρ_k *and* σ_k *computed as in (11.94) and (11.96), respectively, satisfies* $\|d_{k+1}\| \le \tilde{c}\|g_{k+1}\|$, where $\tilde{c}$ is a positive constant.

Proof Firstly, from (11.101) and (11.94), the following lower bound of Δ_k is obtained

$$\Delta_k = \rho_k(s_k^T y_k) - (g_{k+1}^T y_k)^2 = (s_k^T y_k)\left(\rho_k - \frac{(g_{k+1}^T y_k)^2}{(s_k^T y_k)}\right)$$

$$\ge \xi_1\|s_k\|^2\left(\rho_k - \frac{(g_{k+1}^T y_k)^2}{(s_k^T y_k)}\right) \ge \frac{1}{2}\xi_1\|s_k\|^2\frac{\|y_k\|^2}{s_k^T y_k}\|g_{k+1}\|^2. \tag{11.102}$$

From the triangle inequality, the Cauchy–Schwarz inequality, (11.94), (11.102) and since $\delta_k < 1$ it follows that

$$\begin{aligned}
\|d_{k+1}\| &= \|\mu_k g_{k+1} + \eta_k s_k\| \\
&\le \frac{1}{\Delta_k}\|((g_{k+1}^T y_k)(g_{k+1}^T s_k) - (s_k^T y_k)\|g_{k+1}\|^2)g_{k+1} + ((g_{k+1}^T y_k)\|g_{k+1}\|^2 - \rho_k(g_{k+1}^T s_k))s_k\| \\
&\le \frac{1}{\Delta_k}\left[|(g_{k+1}^T y_k)(g_{k+1}^T s_k) - (s_k^T y_k)\|g_{k+1}\|^2|\|g_{k+1}\| + |(g_{k+1}^T y_k)\|g_{k+1}\|^2 - \rho_k(g_{k+1}^T s_k)|\|s_k\|\right] \\
&\le \frac{\|g_{k+1}\|^3}{\Delta_k}\left[3\|y_k\|\|s_k\| + \rho_k\frac{\|s_k\|^2}{\|g_{k+1}\|^2}\right] \le \frac{2\|g_{k+1}\|(s_k^T y_k)}{\xi_1\|s_k\|^2\|y_k\|^2}\left[3\|y_k\|\|s_k\| + \rho_k\frac{\|s_k\|^2}{\|g_{k+1}\|^2}\right] \\
&\le \|g_{k+1}\|\left[\frac{6(s_k^T y_k)}{\xi_1\|s_k\|\|y_k\|} + \frac{3}{\xi_1}\right].
\end{aligned}$$

Now, from the Cauchy–Schwarz inequality it follows that

$$\|d_{k+1}\| \le \frac{9}{\xi_1}\|g_{k+1}\| = \tilde{c}\|g_{k+1}\|,$$

where $\tilde{c} = 9/\xi_1$ is a positive constant. ♦

Theorem 11.9 *Suppose that the Assumption CG holds. If the sequence* $\{x_k\}$ *is generated by the algorithm CUBIC, then*

$$\liminf_{k\to\infty}\|g_k\| = 0.$$

Proof Firstly, observe that under the Assumption CG from (11.5) it follows that

$$\alpha_k \geq \frac{1-\sigma}{L}\frac{|g_k^T d_k|}{\|d_k\|^2}.$$

Now, from (11.4),

$$f_{k+1} \leq f_k - \rho\frac{(1-\sigma)}{L}\frac{(g_k^T d_k)^2}{\|d_k\|^2}.$$

From Propositions 11.6 and 11.7,

$$f_{k+1} \leq f_k - \rho\frac{(1-\sigma)\bar{c}^2}{L\tilde{c}^2}\|g_k\|^2.$$

Denote $\omega = \rho(1-\sigma)\bar{c}^2/(L\tilde{c}^2)$. Therefore,

$$f_{k+1} \leq f_k - \omega\|g_k\|^2.$$

By summing this expression over all indices less than or equal to k, it follows that

$$f_{k+1} \leq f_0 - \omega\sum_{i=0}^{k}\|g_i\|^2. \tag{11.103}$$

Since f is bounded from below, it results that $f_0 - f_{k+1}$ is less than some positive constant for all k. Hence, by taking limits in (11.103), we get

$$\sum_{k=0}^{\infty}\|g_k\|^2 < \infty,$$

which concludes the proof. ♦

Example 11.2 (PALMER1C problem) It is interesting to see that for solving the ill-conditioned problem PALMER1C, this variant of the CUBICa algorithm needs 510 iterations, 5002 evaluations of the function and of its gradient and 0.05 s to reduce the max norm of the gradient to 10^{-2}. ♦

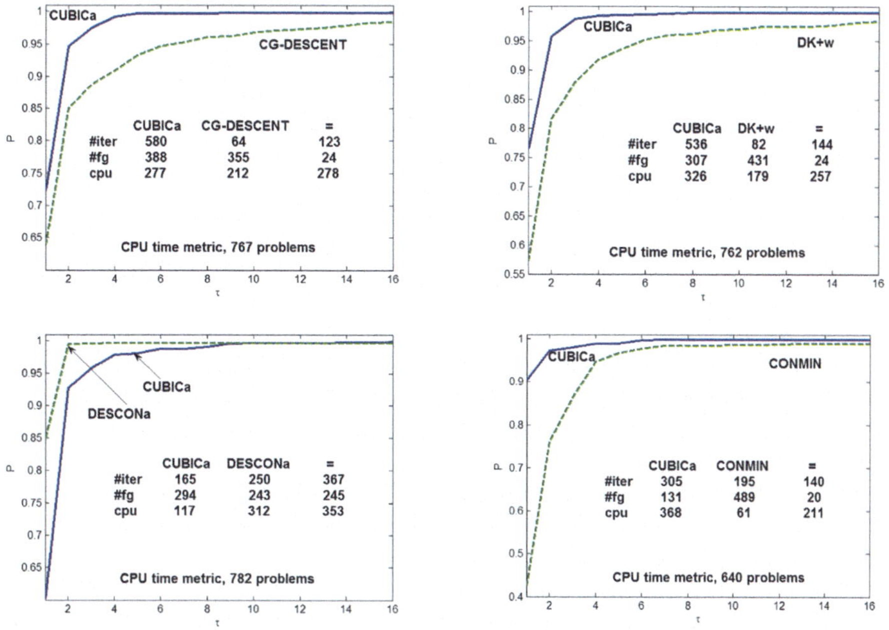

Figure 11.18 Performance profiles of CUBICa versus CG-DESCENT, DK+w, DESCONa and CONMIN

Numerical study. The performances of CUBICa for solving 80 unconstrained optimization problems from the UOP collection, where 10 numerical experiments have been taken for each problem, with the number of variables $n = 1000, 2000, \ldots, 10000$, are as follows. The standard Wolfe line search conditions (11.4) and (11.5) are implemented with $\rho = 0.0001$ and $\sigma = 0.8$. Figure 11.18 shows the Dolan and Moré performance profiles of CUBICa versus CG-DESCENT, DK+w, DESCONa, and CONMIN.

From Figure 11.18, CUBICa proves to be more efficient and more robust than CG-DESCENT, DK+w, and CONMIN. Note that all these algorithms implement the standard Wolfe line search (11.4) and (11.5) with the same values of the parameters ρ and σ.

A more sophisticated variant of the subspace minimization conjugate gradient based on the regularization model was presented by Zhao, Liu, and Liu (2019). It implemented the dynamic restart strategy, different procedures for choosing the initial stepsize, the nonmonotone line search proposed by Hager and Zhang, the conditions (11.101) as well as a more complex strategy for the search direction computation for $p = 3$ or $p = 4$. Intensive numerical experiments using the CUTE collection showed that their algorithm SMCG_PR is more efficient and more robust than CG-DESCENT (Hager & Zhang, 2005), CGOPT (Dai & Kou, 2013), SMCG_BB (Liu & Liu, 2019), and SMCG_Conic (Li, Liu, & Liu, 2019).

Table 11.7 Performances of CGSYS for solving five applications from the MINPACK-2 collection

		CGSYS		
	n	*#iter*	*#fg*	*cpu*
A1	40,000	423	667	10.69
A2	40,000	1155	1830	36.95
A3	40,000	4309	6609	145.39
A4	40,000	897	1412	87.22
A5	40,000	318	489	5.98
Total	–	7102	11007	286.23

Now, let us see the performances of these algorithms for solving the applications from the MINPACK2 collection. Table 11.7 contains the performances of CGSYS for solving the applications from the MINPACK-2 collection, each of them with 40,000 variables.

Table 11.8 presents the performances of CGSYSLBsa, GCSYSLBqa, and CGSYSLBoa for solving the applications from the MINPACK-2, where each application has 40,000 variables.

From Table 11.8, observe that CGSYSLBsa is 4.73 times faster than CGSYS. Also, CGSYSLBqa is 5.33 faster than CGSYS, etc. It is worth seeing the distribution of iterations between CGSYS and L-BFGS in the frame of CGSYSLBqa. For example, for solving the application A4 in CGSYSLBqa the iterations are distributed as: 244 iterations for CGSYS and 75 for L-BFGS ($m = 5$).

Table 11.9 contains the performances of CECG ($\tau = 10$) and of SVCG for solving the applications from the MINPACK-2, where each application has 40,000 variables.

Table 11.10 contains the performances of CUBICa for solving the applications from MINPACK-2, where each application has 40,000 variables

Table 9.2 contains the total performances of L-BFGS ($m = 5$) (102.92 s), TN (104.57 s), TTCG (317.37 s), TTS (207.79 s), and TTDES (250.59 s) for solving all five applications from the MINPACK-2 collection, each of them with 40,000 variables. In Tables 11.7–11.10 the performances of CGSYS (286.23 s),

Table 11.8 Performances of CGSYSLBsa, CGSYSLBqa, and CGSYSLBoa for solving five applications from the MINPACK-2 collection

		CGSYSLBsa			CGSYSLBqa			CGSYSLBoa		
	n	*#iter*	*#f*	*cpu*	*#iter*	*#fg*	*cpu*	*#iter*	*#fg*	*cpu*
A1	40,000	241	510	3.21	350	420	5.35	298	642	9.81
A2	40,000	631	1297	8.96	886	891	10.27	722	1489	24.60
A3	40,000	1020	2070	24.80	922	925	15.05	830	1685	37.76
A4	40,000	299	632	18.28	319	588	17.91	332	702	41.74
A5	40,000	283	586	5.17	302	528	5.11	173	378	9.50
Total	–	2474	5095	60.42	2779	3352	53.69	2355	4896	123.41

Table 11.9 Performances of CECG ($\tau = 10$) and SVCG for solving five applications from the MINPACK-2 collection

		CECG			SVCG		
	n	*#iter*	*#fg*	*cpu*	*#iter*	*#fg*	*cpu*
A1	40,000	241	510	5.62	241	510	5.56
A2	40,000	638	1311	12.30	634	1303	16.52
A3	40,000	1034	2096	35.97	985	1995	41.18
A4	40,000	299	632	32.57	299	632	44.90
A5	40,000	285	591	9.43	274	569	11.50
Total	–	2497	5140	95.89	2433	5009	119.66

Table 11.10 Performances of CUBICa for solving five applications from the MINPACK-2 collection

		CUBICa		
	n	*#iter*	*#fg*	*cpu*
A1	40,000	241	510	3.16
A2	40,000	555	1145	8.11
A3	40,000	1021	2070	24.39
A4	40,000	299	632	17.92
A5	40,000	284	588	5.23
Total	–	2400	4945	58.81

CGSYSLBsa (60.42 s), CGSYSLBqa (53.69), CGSYSLBoa (123.41), CECG (95.89 s), SVCG (119.66 s), and CUBICa (58.81) for solving the same applications can be seen. Observe that this simple combination of CGSYS and L-BFGS based on the closeness of the minimizing function to a quadratic implemented in CGSYSLBqa is clearly top performer.

Notes and References

The eigenvalue clustering of the iteration matrix and the minimization of the condition number of the same iteration matrix in the context of conjugate gradient methods were presented by Andrei (2017a). An intensive numerical study proved that *both these approaches are two different ways to basically pursue similar ideas based on eigenvalues or on singular values of the iteration matrix.* CECG has modest performances in comparison with CG-DESCENT, CONMIN, DK+w and DESCONa. This is because the iteration matrix (11.13) in CECG is a very crude approximation to the inverse Hessian of the minimizing function. Clustering the eigenvalues and minimizing the condition number of the iteration matrix are two important ingredients in getting efficient conjugate gradient algorithms. However, the approximation to the Hessian plays a crucial role in capturing the curvature of the function along the search direction. In this avenue of researches, it would be interesting to see the performances of an algorithm based on clustering the eigenvalues or

Table 11.11 Performances of CONOPT, KNITRO, IPOPT and MINOS for solving the problem PALMER1C

CONOPT			KNITRO			IPOPT			MINOS		
#iter	*#fg*	*cpu*	*#iter*	*#fg*	*cpu*	*#iter*	*#fg*	*cpu*	*#iter*	*#fg*	*cpu*
10	15	0.051	1	2	0.008	1	2	0.018	1/25	58	0.02

on minimizing the condition number of the self-scaling memoryless BFGS method by Perry and Shanno (8.104), subject to the value of the parameter τ_k.

The combination of solvers, a new research area in unconstrained optimization, may improve the convergence properties of nonlinear optimization algorithms. In our presentation, only the combination of CGSYS and CG-DESCENT with limited-memory L-BFGS was discussed. The resulting algorithms proved to be more efficient and more robust, being able to solve complex unconstrained optimization problems. L-CG-DESCENT is one of the most respectable algorithms for solving ill-conditioned unconstrained optimization problems, which combine the CG-DESCENT iterations with iterations of L-BFGS. For solving the PALMER1C problem, Table 11.1 shows that the best performances are obtained when *memory* = 9. Larger values of the parameter *memory* do not improve the performances of L-CG-DESCENT. It is worth seeing the performances of the optimizers CONOPT, KNITRO, IPOPT, and MINOS for solving this ill-conditioned unconstrained optimization problem. Table 11.11 presents the performances of these optimizers.

These optimizers included in the GAMS technology are described in (Andrei, 2017c). They are designed for solving large-scale continuous constrained optimization problems. For solving the problem PALMER1C, MINOS needs 1 major iteration and 25 minor iterations. From Tables 11.1 and 11.11, observe that L-CG-DESCENT is competitive with the optimizers included in the GAMS technology. More details are given in (Andrei, 2019e).

Finally, the recent new research area based on subspace minimization conjugate gradient methods using the regularization model of the minimizing function is indeed most promising.

Chapter 12
Discussions, Conclusions, and Large-Scale Optimization

Having a very simple computational scheme with a very well elaborated convergence theory and requiring modest computational resources for their implementation in computer codes, the conjugate gradient methods are of prime importance for solving large-scale unconstrained optimization problems and real applications. Developments of the limited-memory quasi-Newton method by Nocedal (1980) and by Liu and Nocedal (1989) and of the truncated Newton methods by Dembo and Steihaug (1983), Nash (1985), and by Schlick and Fogelson (1992a, 1992b) have narrowed in a way the class of problems for which the conjugate gradient methods are recommended. However, the modern, recently developed conjugate gradient methods remain the best choice for solving very large problems with relatively inexpensive objective functions.

The search direction in conjugate gradient algorithms is computed as

$$d_{k+1} = -g_{k+1} + \beta_k d_k, \quad k = 0, 1, \ldots, \tag{12.1}$$

where $d_0 = -g_0$ and β_k is a scalar which determines the algorithm. As a result of the major intellectual efforts of Hestenes and Stiefel (1952), Zoutendijk (1970), Wolfe (1969, 1971), Powell (1976b, 1984a), Al-Baali (1985), Nocedal (1992), Gilbert and Nocedal (1992), Dai (2010, 2011), Dai and Yuan (1996c, 1999, 2001a, 2003), the theory behind these methods is separately developed for conjugate gradient methods with strong Wolfe line search and for conjugate gradient methods with standard Wolfe line search. The most important is the Zoutendijk condition, referring to the angle between the search direction of the algorithm and the steepest direction. The Nocedal condition referring to the growth in the length of the search direction along the iterations is equally important. Both these conditions work under the assumptions that the level set of the minimizing function is bounded and its gradient is Lipschitz continuous. Both of them give some conditions on the conjugate gradient parameter β_k to ensure the convergence. A sufficient condition for d_{k+1} to be a descent direction is that β_k satisfies the inequality

N. Andrei, *Nonlinear Conjugate Gradient Methods for Unconstrained Optimization*, Springer Optimization and Its Applications 158,
https://doi.org/10.1007/978-3-030-42950-8_12

$$\beta_k \leq \|g_{k+1}\|^2/(d_k^T y_k). \tag{12.2}$$

The conjugate gradient algorithms for which the parameter β_k satisfies (12.2) and the stepsize is determined by the standard Wolfe line search are convergent. Dai (2010) established another necessary and sufficient condition for ensuring the convergence of the conjugate gradients under the standard Wolfe line search, namely the scalars β_k are so that

$$\sum_{k=1}^{\infty} \prod_{j=0}^{k} \beta_j^{-2} = \infty. \tag{12.3}$$

Two important properties which give some conditions on the parameter β_k to ensure the convergence of the corresponding conjugate gradient method are Property (*) established by Gilbert and Nocedal (1992) and Property (#) established by Dai (2010). The idea of these properties is that in addition to $\beta_k \geq 0$, β_k should be small when the step $s_k = x_k - x_{k-1}$ is small. If the gradients are bounded away from zero and if the method has Property (*) or Property (#), then the algorithm is convergent.

These methods are "matrix free," i.e., they do not require operations with matrices. Therefore, they are suitable for solving large-scale problems. Even if the second-order information contained in the quasi-Newton matrices or in the memoryless BFGS is included in the definition of the conjugate gradient methods, the search direction of these methods is free of matrices. The search directions in conjugate gradient algorithms have simple algebraic expressions. The parameter β_k involves only some elements like: $y_k^T s_k$, $s_k^T g_{k+1}$, $y_k^T g_{k+1}$, $\|s_k\|$, $\|y_k\|$, $\|g_k\|$ or $\|g_{k+1}\|$. Therefore, their implementation in computer codes is relatively simple.

There is a large diversity of conjugate gradient methods. All of them center on the idea that the search direction is determined in such a way so as to satisfy the descent or the sufficient descent condition. This is in sharp contrast with the Newton or quasi-Newton methods, where if the Hessian or its approximation is positive definite, then the search direction is always descent. The standard conjugate gradient methods have been modified in many respects to get algorithms with better convergence properties. Thus, the hybrid methods can be mentioned, which dynamically adjust the formula for β_k calculation as the iterations progress. Some modifications of the standard conjugate gradient methods (mainly of the HS method) have led to new conjugate gradient algorithms like CG-DESCENT (Hager & Zhang, 2005) and DESCON (Andrei, 2013c). The methods based on updating the parameter β_k with close connections to the memoryless quasi-Newton methods, like CONMIN (Shanno, 1983), SCALCG (Andrei, 2007a, 2007b, 2007c) and DK (CGOPT (Dai & Kou, 2013)), are among the best algorithms. The three-term conjugate gradient methods are algorithms in which the search direction includes three terms. A large variety of three-term conjugate gradient is known. All of them have more or less similar performances. Other conjugate gradient methods are based on clustering the eigenvalues of the iteration matrix, on satisfying both the conjugacy and the sufficient descent conditions or on the modified secant equation. It seems that the conjugate gradient algorithms are less dependent on the accuracy of the secant equation. New conjugate gradient algorithms are based on subspace minimization, regularization, or combination of these two ingredients. However, the best conjugate gradient methods are those able to better capture the curvature of the minimizing function along the search direction.

A very important ingredient in conjugate gradient methods is the acceleration scheme introduced by Andrei (2009c). This computational scheme is motivated by the fact that in conjugate gradient algorithms, the stepsizes computed by the Wolfe line search may differ from 1 in a very unpredictable way. They can be larger or smaller than 1 depending on how the problem is scaled along the search direction. This is in very sharp contrast to the Newton and the quasi-Newton methods, including the limited-memory quasi-Newton methods and the truncated Newton method, which accept the unit stepsize most of the time along the iterations and therefore usually require only few function evaluations per search direction. The majority of the conjugate gradient algorithms presented have two variants: a basic variant and an accelerated one. The accelerated version of the conjugate gradient algorithms outperforms the unaccelerated algorithms. However, the acceleration is advantageous only if the evaluation of the gradient of the minimizing function is not too time consuming, otherwise it is better not to use acceleration.

Another form of accelerating the conjugate gradient algorithms is preconditioning. By preconditioning, the original problem is modified by a linear transformation of variables in order to improve the structure of the eigenvalues of the Hessian of the modified problem. Instead of minimizing $f(x)$, by preconditioning a transformed function $f(y)$ is minimized with respect to the modified variables $y = Cx$, where C is a nonsingular linear transformation. At every step, different linear transformation matrices C_k are computed. By this transformation, the search direction is obtained as

$$\bar{d}_{k+1} = -\bar{g}_{k+1} + \beta_k \bar{d}_k, \tag{12.4}$$

where $\bar{g}_k = C_k^T g_k$ and $\bar{d}_k = C_k^{-1} d_k$. In the original variables, the search direction becomes

$$d_{k+1} = -P_k g_{k+1} + \bar{\beta}_k d_k, \tag{12.5}$$

where $P_k = C_k C_k^T$ and $\bar{\beta}_k$ is the conjugate gradient parameter for the preconditioned problem. Therefore, by preconditioning, symmetric positive definite matrices P_k that approximate the inverse Hessian at each iteration are sought so that $P_k \nabla^2 f(x_k) \approx I$. The process of choosing the preconditioners is well understood for linear algebraic systems. For nonlinear optimization problems, the preconditioning remains an open question (Dener, Denchfield, & Munson, 2019). An effective preconditioning combined with subspace minimization was developed by Hager and Zhang (2013) in their limited-memory preconditioned L-CG-DESCENT algorithm.

The conjugate gradient methods are very sensitive to the stepsize computation. Usually, the stepsize is computed by means of the Wolfe line search. A major result is that Dai (2003a) proved the convergence of conjugate gradient methods with standard Wolfe line search conditions. Numerical experiments showed that the conjugate gradient methods with standard Wolfe line search are top performers versus those methods implementing the strong Wolfe line search (see Figure 3.1 and Table 3.1). A great deal of effort was made to get efficient line search algorithms based on Wolfe line search conditions. Let us mention a few papers on this subject matter: Lemaréchal (1981), Shanno (1983), Al-Baali and Fletcher (1984), Moré and Sorensen (1984), Hager (1989), Lukšan (1992), Moré and Thuente (1990, 1994), Hager and Zhang (2005), Dai and Kou (2013), Yuan, Wei, and Lu (2017), Yuan, Sheng, Wang, Hu, and Li, (2018), Dehmiry (2019). One of the best line search algorithms implements the so-called *approximate* Wolfe line search conditions, developed by Hager and Zhang. Conjugate gradient algorithms with approximate Wolfe conditions are the best in this class of algorithms. In Chapter 7, it is showed that CG-DESCENT with approximate Wolfe line search is more efficient and more robust than its version with standard Wolfe line search. In Chapter 8, there is the computational evidence that DK+ with approximate Wolfe line search is much more efficient and more robust than the same algorithm with standard Wolfe conditions. Dai and Kou (2013) introduced the so-called *improved* Wolfe line search. They claimed that their CGOPT (DK+) with improved Wolfe line search is the fastest conjugate gradient algorithm. Andrei (2013c) introduced the *modified* Wolfe line search used in DESCON, one of the best conjugate gradient algorithms. Anyway, the procedure for the stepsize computation is the crucial (weak) point in conjugate gradient algorithms. The line search algorithm that incorporates some ingredients like: sufficient decrease and curvature conditions, finite-precision arithmetic, bracketing of the stepsizes, interval update rules, interpolating conditions, bisection step, stopping tests, etc., is difficult to code. This is in contrast to the Newton or the quasi-Newton methods, for which the crucial (weak) point is storing and handling the Hessian or the approximations to the Hessian. Observe that both the approximate Wolfe line search implemented in CG-DESCENT (Hager & Zhang, 2005) and the improved Wolfe line search implemented in CGOPT (Dai & Kou, 2013) modify the first Wolfe condition with parameter $\rho > 0$ which is responsible with a sufficient reduction (often very small) of the objective function. On the other hand, the modified Wolfe line search implemented in DESCON (Andrei, 2013c) modifies the second Wolfe condition with parameter $\sigma > \rho$. It is worth emphasizing that in the convergence analysis of the conjugate gradient methods with standard or strong Wolfe line search, it is the second Wolfe condition, i.e., parameter σ, which has a crucial role.

Also, the initial stepsize computation has a great importance on the performances of conjugate gradient algorithms. For Newton or quasi-Newton methods, the initial stepsize may always be considered as $\alpha_k^0 = 1$. For optimization methods which do not generate well-scaled search directions, like the steepest descent or the conjugate gradient methods, it is important to consider the current information about the problem and the elements of the algorithm to compute an initial guess of the stepsize. Some algorithms for the computation of the initial stepsize are described by Shanno and Phua (1980), Hager and Zhang (2005), Nocedal and Wright (2006), Li, Liu, and Liu (2019).

Assuming that the first-order change in the function at iterate x_k will be the same as the one obtained at the previous step, then an acceptable initial stepsize should be computed as

$$\alpha_k^0 = \alpha_{k-1} \frac{g_{k-1}^T d_{k-1}}{g_k^T d_k}, \tag{12.6}$$

where α_{k-1} is the previous stepsize at the iterate x_{k-1}.

Another strategy is to interpolate a quadratic to the data f_{k-1}, f_k and $g_{k-1}^T d_{k-1}$ and to define α_k^0 as the minimizer of this quadratic. This strategy yields

$$\alpha_k^0 = \frac{2(f_k - f_{k-1})}{g_k^T d_k}. \tag{12.7}$$

The advantage of using this value for the initial stepsize is that if $x_k \to x^*$ superlinearly, then the ratio in the above expression converges to 1. Therefore, taking $\alpha_k^0 = \min\{1, \ 1.01\alpha_k^0\}$ ensures that $\alpha_k^0 = 1$ will be tried and accepted for large k.

A popular strategy for the initial stepsize computation proposed by Shanno (1983) is

$$\alpha_k^0 = \alpha_{k-1} \frac{\|d_{k-1}\|}{\|d_k\|}. \tag{12.8}$$

At the first iteration, Shanno suggests $\alpha_k^0 = 1/\|g_k\|$. In our numerical experiments, we have considered this strategy.

Some more sophisticated procedures for the initial stepsize computation have been proposed. For example, Hager and Zhang (2005) proposed the following initial stepsize

$$\alpha_k^0 = \begin{cases} \arg\min p(\varphi_k(0), \ \varphi_k'(0), \ \varphi_k'(\psi_1\alpha_{k-1})), & \text{if } \varphi_k(\psi_1\alpha_{k-1}) \le \varphi_k(0), \\ \psi_2\alpha_{k-1}, & \text{otherwise}, \end{cases} \tag{12.9}$$

where ψ_1 and ψ_2 are positive constants $(\psi_1 = 0.1, \ \psi_2 = 2)$ and $p(\varphi_k(0), \ \varphi_k'(0), \ \varphi_k'(\psi_1\alpha_{k-1}))$ denotes the interpolation function by $\varphi_k(0)$, $\varphi_k'(0)$ and $\varphi_k'(\psi_1\alpha_{k-1})$.

Dai and Kou (2013) proposed another procedure for the initial stepsize as

$$\alpha_k^0 = \begin{cases} \alpha, & \text{if } |\varphi_k(\alpha) - \varphi_k(0)|/(\varepsilon_1 + |\varphi_k(0)|) > \varepsilon_2, \\ \arg\min p(\varphi_k(0), \varphi_k'(0), \varphi_k'(\alpha), & \text{otherwise}, \end{cases} \tag{12.10}$$

where $\alpha = \max\{\varepsilon_3\alpha_{k-1}, -2|f_k - f_{k-1}|/g_k^T d_k\}$ and ε_1, ε_2, and ε_3 are positive parameters ($\varepsilon_1 = 10^{-3}$, $\varepsilon_2 = 100$, $\varepsilon_3 = 5$).

If the search direction switches to the negative gradient $-g_k$, then Li, Liu, and Liu (2019) adopted a new adaptive choice of the initial stepsize as

$$\alpha_k^0 = \begin{cases} \max\left\{\alpha_{\min}, \min\{\alpha_{\max}, s_k^T y_k/\|y_k\|^2\}\right\}, & \text{if } g_{k+1}^T s_k \geq 0, \\ \max\left\{\alpha_{\min}, \min\{\alpha_{\max}, \|s_k\|^2/s_k^T y_k\}\right\}, & \text{otherwise}, \end{cases} \tag{12.11}$$

where $\alpha_{\min}$ and $\alpha_{\max}$ are positive constants ($\alpha_{\min} = 10^{-30}$, $\alpha_{\max} = 10^{30}$). At the first iteration, for the initial stepsize α_0^0 Li, Liu, and Liu (2019) proposed an adaptive strategy as

$$\alpha_0^0 = \begin{cases} 2|f_0|/\|g_0\|, & \textit{if } \|x_0\|_\infty < 10^{-30} \textit{ and } |f_0| \geq 10^{-30}, \\ 1.0, & \textit{if } \|x_0\|_\infty < 10^{-30} \textit{ and } |f_0| < 10^{-30}, \\ \min\{1.0, \max\{\|x_0\|_\infty/\|g_0\|_\infty, 1/\|g_0\|_\infty\}\}, & \textit{if } \|x_0\|_\infty \geq 10^{-30} \textit{ and } \|g_0\|_\infty \geq 10^7, \\ \min\{1.0, \|x_0\|_\infty/\|g_0\|_\infty\}, & \textit{if } \|x_0\|_\infty \geq 10^{-30} \textit{ and } \|g_0\|_\infty < 10^7, \end{cases}$$

where $\|.\|_\infty$ denotes the maximum absolute component of a vector.

Observe that there is great diversity of procedures for the initial stepsize computation for which we do not have a clear and distinct conclusion on their importance and of their impact on the performances of conjugate gradient algorithms.

Another important aspect in conjugate gradient methods is the restart of the algorithms, i.e., to restart the iteration at every n steps by setting $\beta_k = 0$ in (12.1). The convergence rate of the conjugate gradient algorithms may be improved from linear to n-step quadratic if the algorithm is restarted with negative gradient at every n steps. n-step quadratic convergence means that

$$\|x_{k+n} - x^*\| = O\left(\|x_k - x^*\|^2\right). \tag{12.12}$$

In conjugate gradient algorithms, the Powell restart criterion: "if $|g_{k+1}^T g_k| > 0.2\|g_{k+1}\|^2$ then set $d_{k+1} = -g_{k+1}$" is often used. However, Dai and Kou (2013) introduced another criterion for restarting the algorithm with negative gradient. The idea behind this criterion is to see how the minimizing function is close to a quadratic function on the segment connecting x_{k-1} and x_k. Their restarting strategy is as follows. Compute the quantity

$$r_{k-1} = \frac{2(f_k - f_{k-1})}{\alpha_{k-1}(g_{k-1}^T d_{k-1} + g_k^T d_{k-1})}. \tag{12.13}$$

If r_{k-1} is close to 1, then the minimizing function is close to a quadratic, otherwise it is not. More exactly, "if there is continuously a maximum prespecified number of iterations so that the corresponding quantities r_k are close to 1, then the algorithm is restarted with steepest descent direction." This strategy, discussed by Dai and Zhang (2001), known as *dynamic restart strategy*, is implemented in CGOPT.

Although the result (12.12) is interesting from the theoretical viewpoint, it may not be relevant in the practical implementations of conjugate gradient algorithms. This is because nonlinear conjugate gradient algorithms are recommended for solving large-scale problems. Therefore, restarts may never occur in such problems since an approximate local solution of such large-scale problems may often be determined in fewer than n iterations. Hence, conjugate gradient methods are often implemented without restarts, or they include strategies for restarting based on considerations other than iteration counts. For example, a restart strategy makes use of the observation that the gradient is orthogonal to the previous search directions (see Propositions 2.2 and 2.3). Often, the truncation of the conjugate gradient parameter $\beta_k^+ = \max\{\beta_k, 0\}$ is viewed as a restarting strategy, because the search direction is replaced by the steepest descent direction.

The conjugate gradient methods are designed for solving large-scale unconstrained optimization problems. Most of the numerical experiments considered so far have involved only problems of different complexities up to 10,000 variables and applications from the MINAPCK-2 collection with 40,000 variables. A close inspection of the performances of the algorithms described in this book shows that CUBICa with 58.81 s (see Table 11.10) is one of the fastest algorithms in this class of conjugate gradient algorithms for solving the applications from the MINPACK-2 collection with 40,000 variables. DESCONa with 78.99 s (see Table 7.3) is also one of the fastest for solving these problems. But, although there is no solid theory behind it, the best is CGSYSLBqa, with 53.69 s (see Table 11.8). Observe that DESCON is four times faster than TTCG.

Numerical study. In the following, let us see the performances of the above described conjugate gradient algorithms for solving large-scale applications from the MINPACK-2 collection, each with 250,000 variables ($nx = 500$ and $ny = 500$). Table 12.1 presents the characteristics of the applications.

Table 12.2 shows the performances of L-BFGS ($m = 5$) (Liu & Nocedal, 1989) and TN (Nash, 1985) for solving these applications.

From Table 12.2 observe that subject to the CPU time metric, both L-BFGS ($m = 5$) and TN are comparable, TN being faster. L-BFGS and TN use different principles to compute the search direction, but both of them use the cubic interpolation to obtain a stepsize satisfying the strong Wolfe line search. The arithmetic

Table 12.1 Characteristics of the MINPACK-2 applications

	Applications	Parameters
A1	Elastic plastic torsion	$c = 5$
A2	Pressure distribution in a journal bearing	$b = 10,\ \varepsilon = 0.1$
A3	Optimal design with composite materials	$\lambda = 0.008$
A4	Steady-state combustion	$\lambda = 5$
A5	Minimal surface with Enneper conditions	–

Table 12.2 Performances of L-BFGS ($m = 5$) and of TN for solving five large-scale applications from the MINPACK-2 collection

		L-BFGS ($m = 5$)			TN		
	n	*#iter*	*#fg*	*cpu*	*#iter*	*#fg*	*cpu*
A1	250,000	1398	1448	171.07	12	649	72.42
A2	250,000	2805	2902	437.19	56	1933	247.67
A3	250,000	3504	3535	681.78	139	4205	1285.00
A4	250,000	2157	2235	864.33	29	943	363.98
A5	250,000	1431	1461	300.70	16	703	99.55
Total	–	11295	11581	2455.07	252	8433	2068.62

costs used by these algorithms are drastically different. L-BFGS uses a fixed, low-cost formula requiring no extra derivative information, whereas TN uses an elaborated and quite sophisticated variable-cost iteration with partial second-derivative information.

Tables 12.3, 12.4, 12.5, 12.6, 12.7, 12.8, 12.9, 12.10, 12.11, 12.12, 12.13 and 12.14 present the performances of the conjugate gradient algorithms described in this book for solving five applications from the MINPACK-2 collection, each of them having 250,000 variables. By comparing the performances of HS versus PRP, both with standard Wolfe line search, from Table 12.3 notice that HS is top performer and the difference is significant. Observe that both HS and PRP belong to the same class of standard conjugate gradient algorithms with $y_k^T g_{k+1}$ in the numerator of β_k. These algorithms automatically adjust β_k to avoid jamming, and this explains their performances. However, both L-BFG ($m = 5$) and TN are clearly faster.

Table 12.4 shows the performances of the hybrid conjugate gradient algorithms CCPRPDY versus NDPRPDY for solving the applications from the MINPACK-2 collection, each of them with 250,000 variables. By comparing the performances of the standard conjugate gradient HS and PRP versus the hybrid conjugate gradient algorithms CCPRPDY and NDPRPDY, it follows that the hybrid algorithms are clearly more efficient. It is interesting to see the performances of the accelerated CCPRPDY (CCPRPDYa) and of the accelerated NDPRPDY (NDPRPDYa). For solving all five applications, CCPRPDYa needs 5817 iterations, 11,790 evaluations

Table 12.3 Performances of HS and of PRP for solving five large-scale applications from the MINPACK-2 collection

		HS			PRP		
	n	*#iter*	*#fg*	*cpu*	*#iter*	*#fg*	*cpu*
A1	250,000	919	1178	86.63	851	1211	88.52
A2	250,000	4268	5535	469.26	2912	4123	358.32
A3	250,000	7070	8056	1207.48	12759	17023	2426.86
A4	250,000	2267	2917	964.40	1838	2627	864.87
A5	250,000	1423	1773	242.18	1977	2810	299.34
Total	–	15947	19459	2969.95	20337	27794	4037.91

Table 12.4 Performances of CCPRPDY and of NDPRPDY for solving five large-scale applications from the MINPACK-2 collection

		CCPRPDY			NDPRPDY		
	n	*#iter*	*#fg*	*cpu*	*#iter*	*#fg*	*cpu*
A1	250,000	864	1075	85.38	831	915	75.42
A2	250,000	4095	5066	494.76	4097	4766	471.45
A3	250,000	2359	2425	350.70	2469	2521	366.36
A4	250,000	2652	3351	1140.37	1925	2105	721.73
A5	250,000	750	794	92.79	712	737	86.68
Total	–	10720	12711	2164.00	10034	11044	1721.64

Table 12.5 Performances of DL $(t = 1)$ and of DL+ $(t = 1)$ for solving five large-scale applications from the MINPACK-2 collection

		DL $(t = 1)$			DL+ $(t = 1)$		
	n	*#iter*	*#fg*	*cpu*	*#iter*	*#fg*	*cpu*
A1	250,000	904	1022	61.30	952	1088	66.21
A2	250,000	4129	5061	535.89	4138	5125	532.96
A3	250,000	5661	6255	878.79	5218	5714	832.36
A4	250,000	3000	3487	1134.63	2752	3179	1042.37
A5	250,000	1135	1253	140.30	1317	1467	162.17
Total	–	14829	13939	2750.91	14377	16573	2636.07

of function and its gradient, and a total of 1822.72 s. On the other hand, NDPRPDYa needs 5815 iterations, 11,786 evaluations of function and its gradient, and a total of 1773.04 s. The hybrid conjugate gradient based on the convex combination of PRP and DY using the Newton direction is faster than the corresponding hybrid conjugate gradient algorithm using the conjugacy condition. Compared with the standard conjugate gradient algorithms (HS and PRP), the

Table 12.6 Performances of CG-DESCENT and of CG-DESCENTaw for solving five large-scale applications from the MINPACK-2 collection

		CG-DESCENT			CG-DESCENTaw		
	n	*#iter*	*#f*	*cpu*	*#iter*	*#f*	*cpu*
A1	250,000	610	1221	92.42	610	1221	136.47
A2	250,000	1752	3505	382.78	1752	3505	448.94
A3	250,000	2370	4742	878.06	2370	4742	943.39
A4	250,000	925	1851	902.03	925	1851	961.31
A5	250,000	635	1271	145.72	635	1271	196.56
Total	–	6292	12590	2401.01	6292	12590	2686.67

Table 12.7 Performances of DESCON and of DESCONa for solving five large-scale applications from the MINPACK-2 collection

		DESCON			DESCONa		
	n	*#iter*	*#fg*	*cpu*	*#iter*	*#fg*	*cpu*
A1	250,000	602	950	53.01	591	1209	69.58
A2	250,000	2578	4056	309.55	1495	3021	261.97
A3	250,000	5001	7626	1134.73	2342	4727	607.68
A4	250,000	1644	2577	868.62	727	1489	468.01
A5	250,000	1070	1674	216.98	655	1334	130.31
Total	–	10895	16883	2582.89	5810	11780	1537.55

Table 12.8 Performances of CONMIN for solving five large-scale applications from the MINPACK-2 collection

		CONMIN		
	n	*#iter*	*#fg*	*cpu*
A1	250,000	657	1332	126.12
A2	250,000	1863	3771	417.97
A3	250,000	2539	5174	869.54
A4	250,000	1372	2775	977.43
A5	250,000	796	1614	209.30
Total	–	7227	14666	2600.36

hybrid conjugate gradient algorithms CCPRPDY and NDPRPDY (unaccelerated or accelerated) are top performers.

Table 12.5 presents the performances of DL ($t = 1$) and DL+ ($t = 1$), both implementing the standard Wolfe line search for the stepsize computation. Recall that DL and DL+ are modifications of the numerator of the HS update parameter. Both algorithms have similar performances. For solving these five applications, both DL and DL+ are faster than HS and than PRP.

CG-DESCENT (Hager & Zhang, 2005) and DESCON (Andrei, 2013c) are conjugate gradient algorithms devised to ensure sufficient descent, independent of

Table 12.9 Performances of SCALCG (spectral) and of SCALCGa (spectral) for solving five large-scale applications from the MINPACK-2 collection

		SCALCG			SCALCGa		
	n	*#iter*	*#fg*	*cpu*	*#iter*	*#fg*	*cpu*
A1	250,000	821	1061	178.62	590	1208	152.04
A2	250,000	1970	2572	345.53	1495	3023	412.04
A3	250,000	3873	4745	879.95	2321	4696	836.41
A4	250,000	1668	2141	810.20	726	1489	589.23
A5	250,000	1359	1768	259.69	959	1942	268.66
Total	–	9691	12287	2473.99	6091	12358	2258.38

Table 12.10 Performances of DK+w and of DK+aw for solving five large-scale applications from the MINPACK-2 collection

		DK+w			DK+aw		
	n	*#iter*	*#fg*	*cpu*	*#iter*	*#fg*	*cpu*
A1	250,000	693	1093	107.58	613	1227	145.21
A2	250,000	2299	3650	413.70	1762	3525	457.22
A3	250,000	4001	6257	1048.29	2354	4710	1007.74
A4	250,000	1396	2211	846.27	923	1847	937.49
A5	250,000	931	1455	227.40	622	1245	202.24
Total	–	9320	14666	2643.24	6274	12554	2749.90

Table 12.11 (**a**) Performances of TTCG and of TTS for solving five large-scale applications from the MINPACK-2 collection. (**b**) Performances of TTDES for solving five large-scale applications from the MINPACK-2 collection

(a)

		TTCG			TTS		
	n	*#iter*	*#fg*	*cpu*	*#iter*	*#fg*	*cpu*
A1	250,000	718	1152	78.88	659	1019	145.21
A2	250,000	2070	3048	311.86	1591	2514	454.65
A3	250,000	1001	14625	2092.13	9926	15002	2133.74
A4	250,000	1107	1779	596.37	1792	2814	915.85
A5	250,000	1266	2060	213.43	909	1396	157.21
Total	–	15162	22664	3292.67	14877	22745	3806.66

(b)

		TTDES		
	n	*#iter*	*#fg*	*cpu*
A1	250,000	629	999	86.46
A2	250,000	2014	3209	323.98
A3	250,000	9620	15001	2023.25
A4	250,000	1317	2074	692.19
A5	250,000	1181	1874	200.80
Total	–	14761	23157	3326.68

Table 12.12 Performances of CGSYS and of CGSYSLBsa for solving five large-scale applications from the MINPACK-2 collection

		CGSYS			CGSYSLBsa		
	n	*#iter*	*#fg*	*cpu*	*#iter*	*#fg*	*cpu*
A1	250,000	588	917	81.88	591	1209	46.75
A2	250,000	2725	4299	405.36	1495	3021	141.90
A3	250,000	10001	15117	2102.99	2001	4041	293.63
A4	250,000	1486	2324	820.11	727	1489	272.58
A5	250,000	1285	2014	226.11	646	1313	72.69
Total	–	16085	24671	3636.45	5460	11073	827.55

Table 12.13 Performances of CECG ($\tau = 10$) and of SVCG for solving five large-scale applications from the MINPACK-2 collection

		CECG			SVCG		
	n	*#iter*	*#fg*	*cpu*	*#iter*	*#fg*	*cpu*
A1	250,000	591	1209	93.61	591	1209	148.58
A2	250,000	1495	3021	281.75	1495	3021	533.61
A3	250,000	2358	4764	653.95	2263	4549	616.67
A4	250,000	727	1489	501.86	727	1489	488.90
A5	250,000	639	1302	153.06	644	1307	146.60
Total	–	5810	11785	1684.23	5720	11575	1934.36

Table 12.14 Performances of CUBICa for solving five large-scale applications from the MINPACK-2 collection

		CUBICa		
	n	*#iter*	*#fg*	*cpu*
A1	250,000	591	1209	94.67
A2	250,000	1290	2613	222.24
A3	250,000	2351	4748	669.73
A4	250,000	727	1489	450.53
A5	250,000	704	1427	142.56
Total	–	5663	11486	1579.73

the accuracy of the line search procedure. Both these algorithms are modifications of the HS method, even if CG-DESCENT may be interpreted as a modification of the self-scaling memoryless BFGS method. The stepsize in CG-DESCENT is computed by using the standard Wolfe line search or the approximated Wolfe line search introduced by Hager and Zhang (2005). In DESCON, the stepsize is computed by means of the standard Wolfe line search, where the parameter σ in the second Wolfe line search condition is adaptively updated. From Table 12.6 observe that for solving these large-scale applications with 250,000 variables, CG-DESCENT with Wolfe line search needs 2401.01 s, while CG-DESCENT with approximate Wolfe line search (CG-DESCENTaw) needs 2686.67 s. Table 12.7 shows that DESCONa with 1537.55 s outperforms CG-DESCENT with 2401.01 s. Moreover, Table 7.4 shows that DESCONa also outperforms CG-DESCENT for solving the applications from the MINPACK-2 collection, each of them with 40,000 variables. Besides, the plots in Figure 7.8 show that DESCONa also outperforms CG-DESCENT for solving 800 unconstrained optimization problems of different structure (of their Hessian) and complexities, with the number of variables in the range [1000, 10,000].

Observe that the acceleration in DESCON plays a crucial role. Even if the acceleration at each iteration involves one additional evaluation of the gradient of the minimizing function, the efficiency of the algorithm endowed with the acceleration scheme is significantly improved.

Both CONMIN (Shanno, 1983) and SCALCG (Andrei, 2007a) are conjugate gradient algorithms, memoryless BFGS preconditioned. The subroutine CONMIN incorporates two nonlinear optimization methods: a conjugate gradient algorithm BFGS preconditioned and the quasi-Newton method BFGS with an initial scaling. These algorithms may be selected according to a value of a parameter at the choice of the user. Both in CONMIN and in SCALCG, the line search implements the standard Wolfe line search with Davidon's cubic interpolation (see Figure 5.1). The performances in Table 12.8 refer to the performances of the conjugate gradient algorithm implemented in CONMIN. This is the Beale restarted memoryless BFGS quasi-Newton method. Table 12.9 presents the performances of SCALCG and its accelerated version SCALCGa. In SCALCG, the preconditioner is a scaled memoryless BFGS matrix which is reset when the Powell restart criterion holds. In fact, SCALCG includes a double quasi-Newton update scheme. The scaling factor in the preconditioner is selected as spectral gradient or as a scalar computed by using the information in two successive points of the iterative process.

In Table 12.9 observe that there is not a spectacular difference between the performances of SCALCG and those of SCALCGa. The reason is as follows. A close inspection of Algorithm 8.2 shows that the acceleration scheme is implemented in Steps 7 and 13. Therefore, at every iteration two additional evaluations of the minimizing function and its gradient are needed for acceleration. Even so, for these applications, SCALCGa is faster than SCALCG.

By comparing Tables 12.6 and 12.9, notice that subject to the CPU time metric, SCALCG is comparable with CG-DESCENT. Also, Table 12.8 shows that SCALCG is faster than CONMIN. However, with 1537.55 s DESCONa (see Table 12.7) is top performer among these algorithms.

Table 12.10 presents the performances of two algorithms: DK+w (DK+ with Wolfe line search) and DK+aw (DK+ with approximate Wolfe line search) (Andrei, 2019a). DK+aw implements the approximate Wolfe line search. Subject to the CPU time metric, these algorithms, DK+w and DK+aw, are comparable, DK+w being slightly faster. By comparing Tables 12.8 and 12.9 versus Table 12.10, observe that both CONMIN and SCALCG are faster than DK+aw. The algorithm implementing the approximate Wolfe line search conditions is more expensive. Anyway, subject to the search direction computation, DK differs from CG-DESCENT only in a constant coefficient in the second term of the Hager–Zhang family of the conjugate gradient methods. Similarly, Tables 12.6 and 12.10 show that CG-DESCENT is faster than both DK+w and DK+aw.

Tables 12.11a, b show the performances of the three-term conjugate gradient algorithms TTCG, TTS, and TTDES for solving the large-scale applications from the MINPACK-2 collection. Strictly speaking, from these numerical experiments, it results that all these three-term conjugate gradient algorithms are less efficient. Among them, TTCG is slightly faster. Table 12.3 shows that HS is faster than all of them. Rather unexpectedly, the performance of PRP with 4037.91 s is less efficient than all these three algorithms. It seems that the modification of the search direction to include three terms does not lead to more efficient conjugate gradient algorithms. Observe that the negative gradient in the search direction of these algorithms (see (9.18)) is modified by the last term that includes the vector y_k. The drawback is that if $-g_{k+1} - a_k s_k$ is a good descent direction, then it is better to use it as a search direction, since the addition of the last term $-b_k y_k$ may prevent d_{k+1} from being a descent direction unless the line search is sufficiently accurate. Observe that the convergence of these three-term conjugate gradient algorithms is proved for uniformly convex functions under the strong Wolfe line search.

Table 12.12 shows the performances of CGSYS and of CGSYSLBsa. CGYSS is a conjugate gradient algorithm with guaranteed sufficient descent and conjugacy conditions. On the other hand, CGSYSLBsa is a combination of CGSYS with the limited-memory L-BFGS ($m = 5$) algorithm by interlacing iterations of the CGSYS with iterations of the L-BFGS algorithms subject to the stepsize. In both algorithms, the stepsize is computed by the standard Wolfe line search.

Observe that this simple interlacing of the iterations between CGSYS and L-BFGS yields a more efficient algorithm. Subject to CPU computing time, CGSYSLBsa is 4.39 times faster than CGSYS.

Table 12.13 includes the performances of CECG with $\tau = 10$ and SVCG. Among all these algorithms considered in this numerical study, CECG with 1684.23 s is closed to DESCONa with 1537.55 s. Observe that the algorithm with clustering the eigenvalues is more efficient than the algorithm with minimizing the condition number of the iteration matrix. Moreover, from the numerical experiments with CECG for solving the problems from the UOP collection, it follows that CECG is more efficient than SVCG, CONMIN, SCALCG, and DK+w and is more robust than CG-DESCENT, CONMIN, SCALCG, and DK+w. Theoretically, clustering the eigenvalues and minimizing the condition number of the iteration matrix are similar. However, in practical implementation, clustering the eigenvalues proves to be more efficient.

Table 12.14 presents the performances of CUBICa, which is a simple variant of the subspace minimization conjugate gradient algorithm based on cubic regularization, for solving the applications from the MINPACK-2 collection, each of them with 250,000 variables. The subspace minimization conjugate gradient algorithm based on cubic regularization implemented in the CUBIC algorithm depends on the procedures for $\rho_k = g_{k+1}^T B_{k+1} g_{k+1}$ and σ_k computation. In CUBIC, for the ρ_k

computation we adopted the formula proposed by Dai and Kou (2016), which is a good estimation of $g_{k+1}^T B_{k+1} g_{k+1}$. For the regularized parameter σ_k, an ad hoc formula was proposed (see (11.96)), which is a combination of the formulae suggested by Cartis, Gould, and Toint (2011a) and by Zhao, Liu, and Liu (2019). Observe that in the CUBIC algorithm, the regularized parameter σ_k is scaling the search direction (see (11.92)), an idea dating back to Fletcher (1987). Besides, CUBIC depends on a number of parameters, their tuning leading to different performances of it. For the set of parameters implemented in our algorithm (c_1, c_2, λ_1, λ_2), subject to CPU time metric, CUBICa with 1579.73 s is immediately after DESCONa with 1537.55 s.

The results obtained so far may be assembled as in Table 12.15. A close inspection at the entries across the columns of Table 12.15 demonstrates that CGSYSLBsa is the most efficient for solving large-scale unconstrained

Table 12.15 Total performances of L-BFGS ($m = 5$), TN, HS, PRP, CCPRPDY, NDPRPDY, CCPRPDYa, NDPRPDYa, DL ($t = 1$), DL+ ($t = 1$), CG-DESCENT, CG-DESCENTaw, DESCON, DESCONa, CONMIN, SCALCG, SCALCGa, DK+w, DK+aw, TTCG, TTS, TTDES, CGSYS, CGSYSLBsa, CECG, SVCG, and CUBICa for solving all five large-scale applications from the MINPACK-2 collection with 250,000 variables each

Algorithms	*#iter*	*#fg*	*cpu*
LBFGS ($m = 5$)	11295	11581	2455.07
TN	252	8433	2068.62
HS	15947	19459	2969.95
PRP	20337	27794	4037.91
CCPRPDY	10720	12711	2164.00
NDPRPDY	1034	11044	1721.64
CCPRPDYa	5817	11790	1822.72
NDPRPDYa	5815	11786	1773.04
DL ($t = 1$)	14829	13939	2750.91
DL+ ($t = 1$)	14377	16573	2636.07
CG-DESCENT	6292	12590	2401.01
CG-DESCENTaw	6292	12590	2686.67
DESCON	10895	16883	2582.89
DESCONa	5810	11780	1537.55
CONMIN	7227	14666	2600.36
SCALCG	9691	12287	2473.99
SCALCGa	6091	12358	2258.38
DK+w	9320	14666	2643.24
DK+aw	6274	12554	2749.90
TTCG	15162	22664	3292.67
TTS	14877	22745	3806.66
TTDES	14761	23157	3326.68
CGSYS	16085	24671	3636.45
CGSYSLBsa	5460	11073	827.55
CECG	5810	11785	1684.23
SVCG	5720	11575	1934.36
CUBICa	5663	11486	1579.73

optimization problems. However, this is not a genuine conjugate gradient algorithm. Although there is not solid theoretical development of the combination of the CGSYS iterations with the L-BFGS iterations based on the stepsize, the computational experiments show the superiority of the CGSYSLBsa algorithm. The CGSYSLBqa algorithm has similar performances. As a genuine conjugate gradient algorithm, DESCONa is on the first place. This is in agreement with the results obtained for solving these applications with 40,000 variables. CUBICa which is a subspace minimization conjugate gradient algorithm based on the cubic regularization is immediately close to DESCONa. The least efficient, as already mentioned above, are the three-term conjugate gradient algorithms TTCG, TTS, TTDES, and CGSYS. It is worth mentioning that L-BFGS ($m = 5$) and TN are less efficient than CGSYSLBsa, DESCONa, and CUBICa. Even if both L-BFGS and TN take into account the curvature of the minimizing function along the search direction, they are not able to get better results under the Wolfe line search with cubic interpolation.

Notes and References
We have presented plenty of numerical results using the UOP collection of 80 artificially unconstrained optimization test problems and five applications from the MINPACK-2 collection. From the above numerical experiments and comparisons, we have the computational evidence that conjugate gradient algorithms considered in these numerical studies are able to solve a large variety of large-scale unconstrained optimization problems of different nonlinear complexity and with different structures of their Hessian matrix. *Apparently, some algorithms are more efficient or faster than others. But this is not a definitive conclusion.* This behavior is obtained by means of a relatively large collection of artificial unconstrained optimization problems used in our numerical studies. It is quite clear that there are an infinite number of artificial unconstrained optimization test problems in front of us, from which it is always possible to assemble a set of problems for which completely different conclusions are obtained, regarding the efficiency and robustness of the algorithms considered in these numerical studies. This is the weakness of the conclusions obtained from the numerical studies that use artificial optimization test problems, even if they are of different nonlinear complexity and with different structures of their Hessian matrix.

Therefore, in order to get a fairly true conclusion, real unconstrained optimization applications must be used in numerical experiments and comparisons. The main characteristic of real optimization applications is that their mathematical model is written on the basis of the conservation laws. In this respect, Noether's theorem (1918) shows that the conservation laws are direct consequences of symmetries. But, at any time and in any place we are surrounded by concepts that appear in dual-symmetric pairs. Therefore, the conservation laws have very solid fundamentals, directly transmitted to the mathematical models of the real applications. This is the main reason why real optimization applications give true insights into the behavior and performances of optimization algorithms.

Finally, we may conclude that conjugate gradient methods represent a major contribution to solving large-scale unconstrained optimization problems. In the last decade, they have diversified in an unexpected way, with lots of variants and developments. The efforts have been toward two points: to get search directions which better capture the curvature of the objective function and to develop accurate line search algorithms for stepsize computation. Both of these points are important and still remain active subjects for further research studies.

Appendix A
Mathematical Review

A.1 Elements of Linear Algebra

Vectors
Define a column n-vector to be an array of n numbers, denoted as

$$x = \begin{bmatrix} x_1 \\ x_2 \\ \vdots \\ x_n \end{bmatrix}.$$

The number $x_i, i = 1, \ldots, n$, is called the i-th *component* of the vector x. Define by $\mathbb{R}$ the set of real numbers. The space of the real vectors of length n is denoted by $\mathbb{R}^n$. Vectors are always column vectors. The *transpose* of x is denoted by x^T. Therefore, x^T is a row vector. Given the vectors $x, y \in \mathbb{R}^n$, the *scalar product* is defined by

$$x^T y = \sum_{i=1}^{n} x_i y_i.$$

The vectors $x, y \in \mathbb{R}^n$ are *orthogonal* (*perpendicular*) if $x^T y = 0$. This is denoted by writing $x \perp y$. If x and y are orthogonal and $x^T x = 1$ and $y^T y = 1$, then we say that x and y are *orthonormal*.

A set of vectors $v_1, \ldots, v_k$ is said to be *linearly dependent* if there are the scalars $\lambda_1, \ldots, \lambda_k$, not all zero, so that $\sum_{i=1}^{k} \lambda_i v_i = 0$. If no such set of scalars exists, then the vectors are said to be *linearly independent*. A *linear combination* of the vectors $v_1, \ldots, v_k$ is a vector of the form $\sum_{i=1}^{k} \lambda_i v_i$, where all λ_i are scalars.

N. Andrei, *Nonlinear Conjugate Gradient Methods for Unconstrained Optimization*,
Springer Optimization and Its Applications 158,
https://doi.org/10.1007/978-3-030-42950-8

Let $\{x_1, \ldots, x_n\}$ be a set of vectors. The *span* of this set of vectors, denoted $span\{x_1, \ldots, x_n\}$, is the set of all vectors that can be expressed as a linear combination of $\{x_1, \ldots, x_n\}$. That is,

$$span\{x_1, \ldots, x_n\} = \left\{ v : v = \sum_{i=1}^{n} \alpha_i x_i,\ \alpha_i \in \mathbb{R} \right\}.$$

If $\{x_1, \ldots, x_n\}$ is a set of n linearly independent vectors where each $x_i \in \mathbb{R}^n$, then $span\{x_1, \ldots, x_n\} = \mathbb{R}^n$. In other words, any vector $v \in \mathbb{R}^n$ can be written as a linear combination of $x_1, \ldots, x_n$. A linearly independent set of vectors that span $\mathbb{R}^n$ is said to be a *basis* for $\mathbb{R}^n$.

Norms of vectors
For a vector $x \in \mathbb{R}^n$ the following *norms* can be defined:

$$\|x\|_1 = \sum_{i=1}^{n} |x_i|, \quad \|x\|_2 = (x^T x)^{1/2}, \quad \|x\|_\infty = \max_{i=1,\ldots,n} |x_i|.$$

The norm $\|.\|_2$ is often called the *Euclidean norm* or l_2 norm. On the other hand, $\|.\|_1$ is referred as the l_1 norm and $\|.\|_\infty$ as the l_∞ norm. All these norms measure the *length* of the vector in some sense, and they are equivalent, i.e., each one is bounded above and below by a multiple of the other. More exactly, for all $x \in \mathbb{R}^n$ it follows that

$$\|x\|_\infty \le \|x\|_2 \le \sqrt{n}\|x\|_\infty \text{ and } \|x\|_\infty \le \|x\|_1 \le n\|x\|_\infty.$$

In general, a norm is any mapping $\|.\|$ from $\mathbb{R}^n$ to the nonnegative real numbers that satisfies the following properties:

1. For all $x, y \in \mathbb{R}^n$, $\|x + y\| \le \|x\| + \|y\|$, with equality if and only if one of the vectors x and y is a nonnegative scalar multiple of the other.
2. $\|x\| = 0 \Rightarrow x = 0$,
3. $\|\alpha x\| = |\alpha| \|x\|$, for all $\alpha \in \mathbb{R}$ and $x \in \mathbb{R}^n$.

The *magnitude* of a vector x is $\|x\|_2 = (x^T x)^{1/2}$. The *angle* between nonzero vectors $x, y \in \mathbb{R}^n$ is defined to be the number $\theta \in [0, \pi]$ so that $\cos\theta = x^T y / \|x\| \|y\|$.

For the Euclidian norm, the *Cauchy–Schwarz inequality* holds $|x^T y| \le \|x\| \|y\|$, with equality if and only if one of these vectors is a nonnegative multiple of the other one. In particular,

$$|x^T y| = \left| \sum_i x_i y_i \right| \le \sum_i |x_i||y_i| \le \left(\max_i |x_i| \right) \left(\sum_i |y_i| \right) = \|x\|_\infty \|y\|_1.$$

The *Hölder inequality*, a generalization of the Cauchy–Schwarz inequality, states that for all $a_i > 0$, $b_i > 0$, $i = 1,\ldots,n$, $p, q > 0$ so that $1/p + 1/q = 1$

$$\sum_{i=1}^{n} a_i b_i \leq \left(\sum_{i=1}^{n} a_i^p\right)^{1/p} \left(\sum_{i=1}^{n} b_i^q\right)^{1/q}.$$

Matrices
A *matrix* is a rectangular array of numbers with m rows and n columns specified by its elements a_{ij}, $i = 1,\ldots,m$, $j = 1,\ldots,n$. The space of the real $m \times n$ matrices is denoted by $\mathbb{R}^{m\times n}$. A *submatrix* of a given matrix A is an array obtained by deleting any combination of rows and columns from A. The *leading* $j \times j$ *principal submatrix* of A is denoted as $A(1:j, 1:j)$. The *transpose* of $A \in \mathbb{R}^{m\times n}$, denoted by A^T, is the $n \times m$ matrix with elements a_{ji}. In other words, (i,j)-th entry of A^T is the (j,i)-th entry of A. Therefore, if $A \in \mathbb{R}^{m\times n}$, then $A^T \in \mathbb{R}^{n\times m}$. The matrix A is *squared* if $m = n$. For a square matrix $A = (a_{ij}) \in \mathbb{R}^{n\times n}$, the elements $a_{11}, a_{22},\ldots,a_{nn}$ define the *main diagonal* of the matrix. A squared matrix is *symmetric* if $A = A^T$. A matrix $A \in \mathbb{R}^{n\times n}$ is *diagonal* if $a_{ij} = 0$ for all $i \neq j$. The *identity matrix,* denoted by I, is the square diagonal matrix whose diagonal elements are all 1. A square matrix $A = (a_{ij})$ is said to be *lower triangular* if $a_{ij} = 0$ for $i<j$. A *unit lower triangular* matrix is a lower triangular matrix with all diagonal elements equal to 1. The matrix A is said to be *upper triangular* if $a_{ij} = 0$ for $i > j$. A matrix $A \in \mathbb{R}^{n\times n}$ is *tridiagonal* if $a_{ij} = 0$ for $|i-j| > 1$. A matrix $A \in \mathbb{R}^{n\times n}$ is *pentadiagonal* if $a_{ij} = 0$ for $|i-j| > 2$. A matrix A is *normal* if $A^TA = AA^T$.

Subspaces
For a function $f : \mathbb{R}^n \to \mathbb{R}^m$, let $R(f)$ denote the *range* of f. That is, $R(f) = \{f(x) : x \in \mathbb{R}^n\} \subseteq \mathbb{R}^m$ is the set of all "*images*" when x varies over $\mathbb{R}^n$. The *range* of a matrix $A \in \mathbb{R}^{m\times n}$, denoted $R(A)$, is the span of the columns of A. That is

$$R(A) = \{v \in \mathbb{R}^m : v = Ax,\ x \in \mathbb{R}^n\}.$$

Therefore, $R(A)$ is the space spanned by the columns of A (*column space*). The range of A^T is the span of the columns of A^T. But, the columns of A^T are just the rows of A. Therefore,

$$R(A^T) = \left\{w \in \mathbb{R}^n : w = A^Ty, y \in \mathbb{R}^m\right\}$$

is the space spanned by the rows of A (*row space*).

The dimension of $R(A)$ is the *rank* of A, denoted *rank*(A). The rank of a matrix A is equal to the maximum number of linearly independent columns in A. This number is also equal to the maximum number of linearly independent rows in A. The rank of $A \in \mathbb{R}^{m\times n}$ can never be greater than the minimum of m and n. The $m \times n$ matrix A is said to be of *full rank* if the rank of A is equal to the minimum of m and n.

The *nullspace of a matrix* $A \in \mathbb{R}^{m \times n}$ is the set

$$N(A) = \{x : Ax = 0\} \subseteq \mathbb{R}^n.$$

In other words, $N(A)$ is the set of all solutions to the homogeneous system $Ax = 0$. For $A \in \mathbb{R}^{m \times n}$, the set $N(A^T) = \{y \in \mathbb{R}^m : A^T y = 0\} \subseteq \mathbb{R}^m$ is called the *left-hand nullspace* of A, because $N(A^T)$ is the set of all solutions to the left-hand homogeneous system $y^T A = 0^T$. Observe that vectors in $R(A)$ are of size m, while vectors in $N(A)$ are of size n. Therefore, vectors in $R(A^T)$ and $N(A)$ are both in $\mathbb{R}^n$. The following equations are true:

1. $\{w : w = u + v, u \in R(A^T), v \in N(A)\} = \mathbb{R}^n$.
2. $R(A^T) \cap N(A) = \{0\}$.

In other words, $R(A^T)$ and $N(A)$ are disjoint subsets that together span the entire space of $\mathbb{R}^n$. The *fundamental theorem of linear algebra* states that

$$N(A) \oplus R(A^T) = \mathbb{R}^n,$$

where n is the number of columns of A and $\oplus$ denotes the *direct sum* of two sets (If S_1 and S_2 are two sets, then $S_1 \oplus S_2 = \{u + v : u \in S_1, v \in S_2\}$.). Often sets of this type are called *orthogonal complements*, and we write that as $R(A^T) = N(A)^{\perp}$.

If $A \in \mathbb{R}^{m \times n}$, then:

1. $N(A) = \{0\}$ if and only if $rank(A) = n$.
2. $N(A^T) = \{0\}$ if and only if $rank(A) = m$.

For $A \in \mathbb{R}^{m \times n}$, the following statements are true:

1. $R(A^T A) = R(A^T)$ and $R(AA^T) = R(A)$.
2. $N(A^T A) = N(A)$ and $N(AA^T) = N(A^T)$.

For all matrices $A \in \mathbb{R}^{m \times n}$, $\dim R(A) + \dim N(A) = n$. Traditionally, $\dim N(A)$ is known as *nullity* of A.

Inverse of a matrix

A squared $n \times n$ matrix A is *nonsingular* if for any vector $b \in \mathbb{R}^n$ there exists $x \in \mathbb{R}^n$ so that $Ax = b$. For nonsingular matrices A, there exists a unique $n \times n$ matrix B so that $AB = BA = I$. The matrix B is denoted by A^{-1} and is called the *inverse* of A. For nonsingular matrices A and B, the following properties hold:

1. $(A^{-1})^{-1} = A$,
2. If the product AB exists and it is nonsingular, then $(AB)^{-1} = B^{-1}A^{-1}$,
3. $(A^T)^{-1} = (A^{-1})^T$.
4. $(cA)^{-1} = c^{-1}A^{-1}$, for any nonzero scalar c.
5. If A is nonsingular and symmetric, then A^{-1} is symmetric.

6. If $A \in \mathbb{R}^{n \times n}$ is nonsingular, then $rank(A) = n$.
7. $\det(A) \neq 0$, where $\det(A)$ is the determinant of A.

Sherman–Morrison formula. Let $a, b \in \mathbb{R}^n$ be two vectors so that $1 + b^T a \neq 0$. It is straightforward to verify by direct multiplication that

$$(I + ab^T)^{-1} = I - \frac{ab^T}{1 + b^T a}.$$

Let $A \in \mathbb{R}^{n \times n}$ be a nonsingular matrix and $a, b \in \mathbb{R}^n$ two vectors so that $1 + b^T A^{-1} a \neq 0$. Then, the inverse of the matrix $B = A + ab^T$ is

$$\begin{aligned} B^{-1} &= (A + ab^T)^{-1} = (A(I + A^{-1}ab^T))^{-1} = (I + A^{-1}ab^T)^{-1}A^{-1} \\ &= \left(I - \frac{A^{-1}ab^T}{1 + b^T A^{-1} a}\right)A^{-1} = A^{-1} - \frac{A^{-1}ab^T A^{-1}}{1 + b^T A^{-1} a}. \end{aligned}$$

If $1 + b^T A^{-1} a = 0$, then B is a singular matrix. This is often called the Sherman–Morrison rank-one update formula, because when $a \neq b \neq 0$, then $rank(ab^T) = 1$.

A generalization of the Sherman–Morrison formula is as follows. If $C, D \in \mathbb{R}^{n \times p}$ so that $(I + D^T A^{-1} C)^{-1}$ exists, then

$$(A + CD^T)^{-1} = A^{-1} - A^{-1}C(I + D^T A^{-1} C)^{-1} D^T A^{-1}.$$

Some results for the quasi-Newton BFGS methods in unconstrained optimization.

(1) Let

$$B_{k+1} = B_k - \frac{B_k s_k s_k^T B_k}{s_k^T B_k s_k} + \frac{y_k y_k^T}{y_k^T s_k},$$

be the BFGS updating formula, where $B_k \in \mathbb{R}^{n \times n}$ is invertible and $s_k, y_k \in \mathbb{R}^n$ so that $y_k^T s_k > 0$. If $H_k = B_k^{-1}$, then the inverse of B_{k+1}, denoted by H_{k+1}, is computed by twice applying the Sherman–Morrison update formula as

$$H_{k+1} = H_k - \frac{H_k y_k s_k^T + s_k y_k^T H_k}{y_k^T s_k} + \left(1 + \frac{y_k^T H_k y_k}{y_k^T s_k}\right)\frac{s_k s_k^T}{y_k^T s_k}.$$

(2) Let

$$B_{k+1} = \delta_k \left[B_k - \frac{B_k s_k s_k^T B_k}{s_k^T B_k s_k}\right] + \gamma_k \frac{y_k y_k^T}{y_k^T s_k}$$

be the *scaled* BFGS updating formula, where $B_k \in \mathbb{R}^{n \times n}$ is invertible, $s_k, y_k \in \mathbb{R}^n$ so that $y_k^T s_k > 0$ and $\delta_k, \gamma_k \in \mathbb{R}$ are two known nonzero scalar parameters.

If $H_k = B_k^{-1}$, then the inverse of B_{k+1}, denoted by H_{k+1}, is computed by twice applying the Sherman–Morrison update formula as

$$H_{k+1} = \frac{1}{\delta_k}\left[H_k - \frac{H_k y_k s_k^T + s_k y_k^T H_k}{y_k^T s_k} + \left(\frac{\delta_k}{\gamma_k} + \frac{y_k^T H_k y_k}{y_k^T s_k}\right)\frac{s_k s_k^T}{y_k^T s_k}\right].$$

(3) Let

$$B_{k+1} = \delta_k\left[I - \frac{s_k s_k^T}{\|s_k\|^2}\right] + \frac{y_k y_k^T}{y_k^T s_k}$$

where $s_k, y_k \in \mathbb{R}^n$ so that $y_k^T s_k > 0$, $s_k \neq 0$ and $\delta_k \in \mathbb{R}$ is a known nonzero scalar parameter. Then, the inverse of B_{k+1}, denoted by H_{k+1}, is computed by twice applying the Sherman–Morrison update formula as

$$H_{k+1} = \frac{1}{\delta_k} I - \frac{1}{\delta_k}\frac{s_k y_k^T + y_k s_k^T}{y_k^T s_k} + \left[1 + \frac{1}{\delta_k}\frac{\|y_k\|^2}{y_k^T s_k}\right]\frac{s_k s_k^T}{y_k^T s_k}.$$

(4) Let

$$B_{k+1} = \delta_k\left[I - \frac{s_k s_k^T}{\|s_k\|^2}\right] + \gamma_k \frac{y_k y_k^T}{y_k^T s_k},$$

where $s_k, y_k \in \mathbb{R}^n$ so that $y_k^T s_k > 0$, $s_k \neq 0$ and $\delta_k, \gamma_k \in \mathbb{R}$ are two known nonzero scalar parameters. Then, the inverse of B_{k+1}, denoted by H_{k+1}, is computed by twice applying the Sherman–Morrison update formula as

$$H_{k+1} = \frac{1}{\delta_k} I - \frac{1}{\delta_k}\frac{s_k y_k^T + y_k s_k^T}{y_k^T s_k} + \left[\frac{1}{\gamma_k} + \frac{1}{\delta_k}\frac{\|y_k\|^2}{y_k^T s_k}\right]\frac{s_k s_k^T}{y_k^T s_k}.$$

Orthogonality
A square matrix $Q \in \mathbb{R}^{n \times n}$ is *orthogonal* if it has the property $QQ^T = Q^T Q = I$, where I is the $n \times n$ identity matrix. Therefore, the inverse of an orthogonal matrix is its transpose.

Suppose that $\|u\| = 1$ and let $u^\perp$ denote the space consisting of all vectors that are perpendicular to u. $u^\perp$ is called the *orthogonal complement* of u. The matrix $P = I - uu^T$ is the *orthogonal projector* onto $u^\perp$ in the sense that P maps each x to its orthogonal projection in $u^\perp$. For a subspace $S \subset \mathbb{R}^n$, the orthogonal complement $S^\perp$ of S is defined as the set of all vectors in $\mathbb{R}^n$ that are orthogonal to every vector in S. In this case, $\dim S^\perp = n - \dim S$.

Eigenvalues

A scalar value λ is an *eigenvalue* of the $n \times n$ matrix A if there exists a nonzero vector $u \in \mathbb{R}^n$ so that $Au = \lambda u$. The vector u is called an *eigenvector* of A. The *spectrum* of a matrix is the set of all its eigenvalues. Let $\lambda_1, \ldots, \lambda_n$ be the eigenvalues of the matrix A, real or complex. Then, its *spectral radius* $\rho(A)$ is defined as $\rho(A) = \max\{|\lambda_1|, \ldots, |\lambda_n|\}$. Observe that $\rho(A) \leq \|A\|$ for every matrix norm. The condition number of A can be expressed as $\kappa(A) = \rho(A)\rho(A^{-1})$. A matrix A is nonsingular if all its eigenvalues are different from zero. The eigenvalues of symmetric matrices are all real numbers. The nonsymmetric matrices may have imaginary eigenvalues.

Two matrices $A, B \in \mathbb{R}^{n \times n}$ are *similar* if there exists a nonsingular matrix $P \in \mathbb{R}^{n \times n}$ so that $B = P^{-1}AP$. Similar matrices represent the same linear operator in different bases, with P being the change of the basis matrix. Two similar matrices have the same eigenvalues, even though they will usually have different eigenvectors.

Positive definite matrices

A square matrix A is *positive definite* if and only if $x^T Ax > 0$ for every nonzero $x \in \mathbb{R}^n$. For real symmetric matrices A, the following statements are equivalent:

1. All eigenvalues of A are positive.
2. $A = B^T B$ for some nonsingular B. While B is not unique, there is one and only one upper-triangular matrix R with positive diagonals so that $A = R^T R$. This is the *Cholesky factorization* of A.
3. A has an *LU* (or *LDU*) factorization with all pivots being positive. The *LDU* factorization is of the form $A = LDL^T = R^T R$, where $R = D^{1/2} L^T$ is the Cholesky factor of A.

Any of the statements above can serve as the definition of a positive definite matrix.

A matrix A is *positive semidefinite* if for all $x \in \mathbb{R}^n$, $x^T Ax \geq 0$. The following statements are equivalent and can serve as the definition of a positive semidefinite matrix:

1. All eigenvalues of A are nonnegative.
2. $A = B^T B$ for some B with $rank(B) = r$.

If a matrix is symmetric and positive definite, then its eigenvalues are all positive real numbers. A symmetric matrix can be tested if it is positive definite by computing its eigenvalues and by verifying if they are all positive or by performing a Cholesky factorization.

Gaussian elimination (LU factorization)
For solving the system $Ax = b$, where A is nonsingular, the Gaussian elimination consists of the following four steps:

1. Factorize the matrix A as $A = PLU$, where
   P is a *permutation* matrix,
   L is an *unit lower triangular* matrix,
   U is a *nonsingular upper triangular* matrix.
2. Solve the system $PLUx = b$ subject to LUx by permuting the entries of b, i.e., $LUx = P^{-1}b = P^T b$.
3. Solve the system $LUx = P^{-1}b$ subject to Ux by *forward substitution*, i.e., $Ux = L^{-1}(P^{-1}b)$.
4. Solve the system $Ux = L^{-1}(P^{-1}b)$ subject to x by *backward substitution*, i.e., $x = U^{-1}(L^{-1}(P^{-1}b))$.

The following result is central in Gaussian elimination.
The following two statements are equivalent:

1. *There exists a unique unit lower triangular matrix L and a nonsingular upper triangular matrix U such that A* $= LU$. *This is called LU factorization of A.*
2. *All leading principal submatrices of A are nonsingular.*

LU factorization without pivoting can fail on nonsingular matrices, and therefore, we need to introduce permutations into Gaussian elimination.

If A is a nonsingular matrix, then there exist permutation matrices P_1 *and* P_2, *a unit lower triangular matrix L and a nonsingular upper triangular matrix U such that* $P_1AP_2 = LU$. *Observe that* P_1A *reorders the rows of A.* AP_2 *reorders the columns of A.* P_1AP_2 *reorders both the rows and columns of A.*

The next two results state simple ways to choose the permutation matrices P_1 and P_2 to guarantee that Gaussian elimination will run on nonsingular matrices.

Gaussian elimination with partial pivoting
The permutation matrices $P_2' = I$ *and* P_1' *can be chosen in such a way that* a_{11} *is the largest entry in absolute value in its column. More generally, at step i of the Gaussian elimination, where the i-th column of L is computed, the rows i through n are permuted so that the largest entry in the column is on the diagonal. This is called "Gaussian elimination with partial pivoting," or GEPP for short. GEPP guarantees that all entries of L are bounded by one in absolute value.*

Gaussian elimination with complete pivoting
The permutation matrices P_2' *and* P_1' *are chosen in such a way that* a_{11} *is the largest entry in absolute value in the whole matrix. More generally, at step i of Gaussian elimination, where the i-th column of L is computed the rows and the columns i through n are permuted so that the largest entry in this submatrix is on the diagonal. This is called "Gaussian elimination with complete pivoting," or GECP for short.*

Cholesky factorization
The Cholesky factorization method for solving a symmetric positive definite system $Ax = b$ by using the factorization $A = LL^T$ computes the elements of the lower triangular matrix L as follows. Consider the k-th row of A, then the elements of the k-column of L are computed as:

$$l_{kk} = \sqrt{a_{kk} - \sum_{i=1}^{k-1} l_{ki}^2}, \quad l_{jk} = \frac{a_{kj} - \sum_{i=1}^{k-1} l_{ki} l_{ji}}{l_{kk}} \text{ for } j = k+1, \ldots, n.$$

A complete Cholesky factorization consists of applying the above formulae for $k = 1, \ldots, n$. When A is symmetric and positive definite, then Cholesky factorization requires about $n^3/6$ multiplications per iteration. This process breaks down at a stage i if the computation of l_{ii} involves the square root of a negative number. This is the case if A is not positive definite. If A is indefinite, then the Cholesky factorization may not exist. Even if it does exist, numerically it is unstable when it is applied to such matrices, in the sense that the elements of L can become arbitrarily large. In this case, the modified Cholesky factorization may be used as described in Gill, Murray, and Wright (1981) or in Moré and Sorensen (1984) (see also Nocedal and Wright (2006), pp. 53–54).

Singular value decomposition
Suppose $A \in \mathbb{R}^{m \times n}$ with $rank(A) = r$. Then, A can be factored as $A = U \Sigma V^T$, where $U \in \mathbb{R}^{m \times r}$ satisfies $U^T U = I$, $V \in \mathbb{R}^{n \times r}$ satisfies $V^T V = I$ and $\Sigma = diag(\sigma_1, \ldots, \sigma_r)$ with $\sigma_1 \geq \sigma_2 \geq \cdots \geq \sigma_r > 0$. The columns of U are called *left singular vectors* of A, the columns of V are called *right singular vectors* of A and the numbers σ_i are the *singular value*.

Spectral decomposition (*Symmetric eigenvalue decomposition*)
Suppose $A \in \mathbb{R}^{n \times n}$ is a real symmetric matrix. Then, A can be factored as $A = Q \Lambda Q^T$, where $Q = [q_1, \ldots, q_n] \in \mathbb{R}^{n \times n}$ is an orthogonal matrix with columns q_i, $i = 1, \ldots, n$, as eigenvectors of A and $\Lambda = diag(\lambda_1, \ldots, \lambda_n)$, where λ_i are the eigenvalues of A. When A is positive definite as well as symmetric, this spectral decomposition is identical to the singular value decomposition. In this case, the singular values σ_i and the eigenvalues λ_i coincide.

Matrix norms
The *matrix norms* induced by the vector l_1 norm and the vector l_∞ norm are as follows:

$$\|A\|_1 = \max_{\|x\|_1 = 1} \|Ax\|_1 = \max_j \sum_i |a_{ij}| = \text{the largest absolute column sum.}$$

$$\|A\|_\infty = \max_{\|x\|_\infty = 1} \|Ax\|_\infty = \max_i \sum_j |a_{ij}| = \text{the largest absolute row sum.}$$

The matrix norm induced by the Euclidian vector norm is $\|A\|_2 = \max_{\|x\|_2=1}\|Ax\|_2 = \sqrt{\lambda_{\max}}$, where $\lambda_{\max}$ is the largest eigenvalue of A^TA.

The *Frobenius norm* of $A \in \mathbb{R}^{m\times n}$ is defined as

$$\|A\|_F^2 = \sum_{i,j}|a_{ij}|^2 = tr(A^TA),$$

where for the matrix $A^{n\times n} = (a_{ij})$, $tr(A) = a_{11} + \cdots + a_{nn}$ is the trace of A. The *ellipsoid norm* is defined as $\|x\|_A = (x^TAx)^{1/2}$, where A is a symmetric and positive definite matrix.

Conditioning and stability
These are two terms used in numerical computations when a problem is solved with an algorithm. Conditioning is a property of the problem, irrespective it is a linear algebra problem, an optimization or a differential equation one. A problem is *well-conditioned* if its solution is not affected greatly by small perturbations to the data that define the problem. Otherwise, it is *ill-conditioned.* On the other hand, stability of an algorithm is a property of the algorithm. An algorithm is *stable* if it is guaranteed to generate accurate answers to well-conditioned problems.

The *condition number* of a nonsingular matrix $A \in \mathbb{R}^{n\times n}$ denoted as $cond(A)$ or $\kappa(A)$ is defined as $cond(A) = \|A\|\|A^{-1}\|$. If the 2-norm is used, then $\kappa(A) = \sigma_{\max}(A)/\sigma_{\min}(A)$, where $\sigma_{\max}(A)$ and $\sigma_{\min}(A)$ are the largest and the smallest singular values of A, respectively. For normal matrices, $\kappa(A) = |\lambda_{\max}(A)|/|\lambda_{\min}(A)|$, where $\lambda_{\max}(A)$ and $\lambda_{\min}(A)$ are the largest and the smallest eigenvalues of A, respectively. The matrix A is *well-conditioned* if $\kappa(A)$ is small (close to 1). The matrix A is *ill-conditioned* if $\kappa(A)$ is large.

For general linear systems $Ax = b$, where $A \in \mathbb{R}^{n\times n}$, the condition number of the matrix can be used to see the conditioning of the system. If the matrix A is perturbed to $\bar{A}$ and b to $\bar{b}$ and consider $\bar{x}$ as the solution of the perturbed system $\bar{A}\bar{x} = \bar{b}$, it can be shown that (Golub & Van Loan, 1996)

$$\frac{\|x-\bar{x}\|}{\|x\|} \approx \kappa(A)\left[\frac{\|A-\bar{A}\|}{\|A\|} + \frac{\|b-\bar{b}\|}{\|b\|}\right].$$

Therefore, a large condition number $\kappa(A)$ indicates that the problem $Ax = b$ is ill-conditioned, while a small value shows well-conditioning of the problem.

To see the significance of the stability of an algorithm, let us consider the linear system $Ax = b$ solved by means of the Gaussian elimination with partial pivoting and triangular substitution. It is shown that this algorithm gives a solution $\bar{x}$ whose relative error is approximately

$$\frac{\|x-\bar{x}\|}{\|x\|} \approx \kappa(A)\frac{gr(A)}{\|A\|}u,$$

where $gr(A)$ is the size of the largest element that arises in A during the execution of the Gaussian elimination with partial pivoting and u is the unit roundoff (In double-precision IEEE arithmetic u is about 1.1×10^{-16}.). In the worst case, it can be shown that $gr(A)/\|A\|$ may be around 2^{n-1}, which indicates that the Gaussian elimination with partial pivoting is an unstable algorithm (Demmel, 1997). However, in practice, after decades of numerical experience with Gaussian elimination with partial pivoting algorithm it was noticed that $gr(A)$ is growing slowly as a function of n. In practice, $gr(A)$ is almost always n or less. The average behavior seems to be $n^{2/3}$ or perhaps even just $n^{1/2}$ (Trefethen & Schreiber, 1990). Therefore, the Gaussian elimination with partial pivoting is stable for all practical purposes. However, Gaussian elimination without pivoting is definitely unstable. For system $Ax = b$ where A is a symmetric and positive definite matrix, the Cholesky factorization method with triangular substitution is a stable algorithm.

Determinant of a matrix

The *determinant* is a scalar defined only for square matrices. A *permutation* $p = (p_1, p_2, \ldots, p_n)$ of the numbers $(1, 2, \ldots, n)$ is simply any rearrangement of these numbers. The *sign of a permutation* p is defined to be the number

$$\sigma(p) = \begin{cases} +1, & \text{if } p \text{ can be restored to natural order by an even number of interchanges,} \\ -1, & \text{if } p \text{ can be restrored to natural order by an odd number of interchanges.} \end{cases}$$

Let $A = (a_{ij}) \in \mathbb{R}^{n \times n}$ be an arbitrary matrix, where all its elements a_{ij} are real numbers. The *determinant* of A is defined to be the scalar

$$\det(A) = \sum_{p} \sigma(p) a_{1p_1} a_{2p_2} \ldots a_{np_n},$$

where the sum is taken over the $n!$ permutations $p = (p_1, p_2, \ldots, p_n)$ of $(1, 2, \ldots, n)$. ($n! = 1 \times 2 \times \cdots \times n$. For example, $3! = 1 \times 2 \times 3 = 6$.) Each term $a_{1p_1} a_{2p_2} \ldots a_{np_n}$ contains exactly one entry from each row and from each column of A.

Some properties of determinants:

1. The determinant of a diagonal matrix: $\det[diag(x_1, x_2, \ldots, x_n)] = x_1 x_2 \ldots x_n$.
2. Let I_n be the identity matrix of order n. Then, $\det(I_n) = 1$.
3. The determinant of a triangular matrix is the product of its diagonal entries.
4. For any matrix $A \in \mathbb{R}^{n \times n}$, and constant c, $\det(cA) = c^n \det(A)$.
5. Suppose that B is obtained from A by swapping two of the rows (columns) of A. Then, $\det(B) = -\det(A)$.
6. If there is a row (column) of A all zero, then $\det(A) = 0$.
7. If two rows (columns) of A are equal, then $\det(A) = 0$.
8. $\det(A^T) = \det(A)$.
9. $\det(A^{-1}) = 1/\det(A)$.

10. $\det(AB) = \det(A)\det(B)$.
11. If $\lambda_1, \lambda_2, \ldots, \lambda_n$ are the eigenvalues of $A \in \mathbb{R}^{n\times n}$, then $\det(A) = \lambda_1\lambda_2\cdots\lambda_n$.

For a matrix $A \in \mathbb{R}^{n\times n}$, the polynomial $p(\lambda) = \det(A - \lambda I)$ is called the *characteristic polynomial* of A. The set of all eigenvalues of A is the set of all roots of its characteristic polynomial. The *Cayley–Hamilton theorem* says that $p(A) = 0$.

Let I_n be the identity matrix of order n and $u_1, u_2 \in \mathbb{R}^n$ arbitrary vectors, then

$$\det(I_n + u_1u_2^T) = 1 + u_1^Tu_2.$$

Let I_n be the identity matrix of order n and $u_1, u_2, u_3, u_4 \in \mathbb{R}^n$ arbitrary vectors, then

$$\det(I_n + u_1u_2^T + u_3u_4^T) = (1 + u_1^Tu_2)(1 + u_3^Tu_4) - (u_1^Tu_4)(u_2^Tu_3).$$

Indeed:

$$I_n + u_1u_2^T + u_3u_4^T = (I_n + u_1u_2^T)\left[I_n + (I_n + u_1u_2^T)^{-1}u_3u_4^T\right].$$

Therefore,

$$\begin{aligned}
\det(I_n + u_1u_2^T + u_3u_4^T) &= \det(I_n + u_1u_2^T)\det\left[I_n + (I_n + u_1u_2^T)^{-1}u_3u_4^T\right] \\
&= (1 + u_1^Tu_2)\left[1 + u_4^T(I_n + u_1u_2^T)^{-1}u_3\right] \\
&= (1 + u_1^Tu_2)\left[1 + u_4^T\left(I_n - \frac{u_1u_2^T}{1 + u_1^Tu_2}\right)u_3\right] \\
&= (1 + u_1^Tu_2)(1 + u_3^Tu_4) - (u_1^Tu_4)(u_2^Tu_3).
\end{aligned}$$

Determinant of the quasi-Newton BFGS update

(1) Let

$$B_{k+1} = B_k - \frac{B_ks_ks_k^TB_k}{s_k^TB_ks_k} + \frac{y_ky_k^T}{y_k^Ts_k},$$

be the BFGS updating of the matrix B_k, where $B_k \in \mathbb{R}^{n\times n}$ and $s_k, y_k \in \mathbb{R}^n$ so that $y_k^Ts_k > 0$, then

$$\begin{aligned}
\det(B_{k+1}) &= \det\left[B_k\left(I - \frac{s_ks_k^TB_k}{s_k^TB_ks_k} + \frac{B_k^{-1}y_ky_k^T}{y_k^Ts_k}\right)\right] \\
&= \det(B_k)\det\left(I - s_k\frac{(B_ks_k)^T}{s_k^TB_ks_k} + B_k^{-1}y_k\frac{y_k^T}{y_k^Ts_k}\right) = \det(B_k)\frac{y_k^Ts_k}{s_k^TB_ks_k}.
\end{aligned}$$

(2) Let

$$B_{k+1} = \delta_k \left[B_k - \frac{B_k s_k s_k^T B_k}{s_k^T B_k s_k} \right] + \gamma_k \frac{y_k y_k^T}{y_k^T s_k},$$

where $s_k, y_k \in \mathbb{R}^n$ so that $y_k^T s_k > 0$ and $\delta_k, \gamma_k \in \mathbb{R}$ are two known nonzero scalar parameters. Then

$$\det(B_{k+1}) = \det(B_k) \frac{y_k^T s_k}{s_k^T B_k s_k} \delta_k^{n-1} \gamma_k.$$

Trace of a matrix
The trace of a square matrix $A = (a_{ij}) \in \mathbb{R}^{n \times n}$ is

$$trace(A) = tr(A) = \sum_{i=1}^{n} a_{ii}.$$

The trace satisfies:

1. $tr(A^T) = tr(A)$.
2. $tr(AB) = tr(BA)$.
3. $tr(\alpha A + \beta B) = \alpha tr(A) + \beta tr(B), \alpha, \beta \in \mathbb{R}$.

If $\lambda_1, \lambda_2, \ldots, \lambda_n$ are the eigenvalues of $A \in \mathbb{R}^{n \times n}$, then $trace(A) = \lambda_1 + \lambda_2 + \cdots + \lambda_n$. If $A = (a_{ij}) \in \mathbb{R}^{m \times n}$, then $tr(A^T A) = \sum_{i=1}^{m} \sum_{j=1}^{n} a_{ij}^2$.

Let $B_{k+1} = B_k - \frac{B_k s_k s_k^T B_k}{s_k^T B_k s_k} + \frac{y_k y_k^T}{y_k^T s_k}$, be the BFGS updating of the matrix B_k, where $B_k \in \mathbb{R}^{n \times n}$ and $s_k, y_k \in \mathbb{R}^n$ so that $y_k^T s_k > 0$, then

$$tr(B_{k+1}) = tr(B_k) - \frac{\|B_k s_k\|^2}{s_k^T B_k s_k} + \frac{\|y_k\|^2}{y_k^T s_k}.$$

A.2 Elements of Analysis

Let $\{x_k\}$ be a sequence of points from $\mathbb{R}^n$. A sequence $\{x_k\}$ *converges* to a point x^*, written as $\lim_{k \to \infty} x_k = x^*$, if for any $\varepsilon > 0$ there exists an index K so that $\|x_k - x^*\| \leq \varepsilon$ for all $k \geq K$. Given an index set $\bar{K} \subset \{1, 2, \ldots\}$, a *subsequence* of $\{x_k\}$ corresponding to $\bar{K}$ can be defined and denoted by $\{x_k\}_{k \in \bar{K}}$. Consider a convergent sequence $\{x_k\}$ with limit x^*. Then, any subsequence of $\{x_k\}$ also converges to x^*. A convergent sequence has only one limit. A sequence $\{x_k\}$ in $\mathbb{R}^n$ is *bounded* if there exists a number $B \geq 0$ such that $\|x_k\| \leq B$ for all $k = 1, 2, \ldots$.

Every convergent sequence is bounded. A sequence $\{x_k\}$ in $\mathbb{R}^n$ is *uniformly bounded away from zero* if there exists $\varepsilon > 0$ such that $|x_k| \geq \varepsilon$ for any $k \geq 1$.

Theorem A.2.1 (Bolzano–Weierstrass Theorem) *Each bounded sequence in* $\mathbb{R}^n$ *has a convergent subsequence.* ♦

The point $x^* \in \mathbb{R}^n$ is an *accumulation point* or a *limit point* or a *cluster point* for the sequence $\{x_k\}$ if there is an infinite set of indices $k_1, k_2, k_3, \ldots$ so that the subsequence $\{x_{k_i}\}_{i=1,2,3,\ldots}$ converges to x^*, i.e., $\lim_{i\to\infty} x_{k_i} = x^*$. A sequence is a *Cauchy sequence* if for any $\varepsilon > 0$, there exists an integer $K > 0$ so that $\|x_k - x_m\| \leq \varepsilon$ for all indices $k \geq K$ and $m \geq K$. A sequence converges if and only if it is a Cauchy sequence.

A function $f : \mathbb{R}^n \to \mathbb{R}^m$ is *continuous* at $x \in \mathbb{R}^n$ if for all $\varepsilon > 0$, there exists a $\delta(\varepsilon, x) > 0$ so that for any $y \in \mathbb{R}^n$, $\|y - x\|_2 \leq \delta(\varepsilon, x) \Rightarrow \|f(y) - f(x)\|_2 \leq \varepsilon$. The continuity can be described in terms of limits: whenever the sequence $\{x_k\}$ in $\mathbb{R}^n$ converges to a point $x \in \mathbb{R}^n$, the sequence $\{f(x_k)\}$ in $\mathbb{R}^m$ converges to $f(x)$, i.e., $\lim_{k\to\infty} f(x_k) = f(\lim_{k\to\infty} x_k)$. A function f is continuous if it is continuous at every point in $\mathbb{R}^n$.

A function $f : \mathbb{R}^n \to \mathbb{R}^m$ is *uniformly continuous* at $x \in \mathbb{R}^n$ if for all $\varepsilon > 0$, there exists a $\delta(\varepsilon) > 0$ so that for any $y \in \mathbb{R}^n$, $\|y - x\|_2 \leq \delta(\varepsilon) \Rightarrow \|f(y) - f(x)\|_2 \leq \varepsilon$. It is obvious that a uniformly continuous function is continuous.

If $\{x_k\}$ is a Cauchy sequence and f is uniformly continuous on a convex domain, then $\{f(x_k)\}$ is also a Cauchy sequence.

A function $f : \mathbb{R}^n \to \mathbb{R}^m$ is *bounded* if there exists a constant $C \geq 0$ so that $\|f(x)\| \leq C$ for all $x \in \mathbb{R}^n$.

A continuous function $f : \mathbb{R}^n \to R$ is *coercive* if $\lim_{\|x\|\to\infty} = +\infty$. This means that for any constant M there must be a positive number R_M such that $f(x) \geq M$ whenever $\|x\| \geq R_M$. In particular, the values of $f(x)$ cannot remain bounded on a set in $\mathbb{R}^n$ that is not bounded. For $f(x)$ to be coercive, it is not sufficient that $f(x) \to \infty$ as each coordinate tends to ∞. Rather $f(x)$ must become infinite along any path for which $\|x\|$ becomes infinite. If $f(x)$ is coercive, then $f(x)$ has at least one global minimizer and these minimizers can be found among the critical points of $f(x)$.

Let $f : \mathbb{R} \to \mathbb{R}$ be a real-valued function of a real variable. The *first derivative* is defined by

$$f'(x) = \lim_{\varepsilon\to 0} \frac{f(x+\varepsilon) - f(x)}{\varepsilon}.$$

The *second derivative* is defined by

$$f''(x) = \lim_{\varepsilon\to 0} \frac{f'(x+\varepsilon) - f'(x)}{\varepsilon}.$$

The *directional derivative* of a function $f : \mathbb{R}^n \to \mathbb{R}$ in the direction $p \in \mathbb{R}^n$ is given by

$$D(f(x);p) = \lim_{\varepsilon \to 0} \frac{f(x+\varepsilon p) - f(x)}{\varepsilon}.$$

Let $f : \mathbb{R}^n \to \mathbb{R}$ be a continuously differentiable function. The conditions which characterize a minimum can be expressed in terms of the *gradient* $\nabla f(x)$ with the first partial derivatives defined as

$$\nabla f(x) = \left[\frac{\partial f}{\partial x_1}, \ldots, \frac{\partial f}{\partial x_n}\right]^T$$

and of $n \times n$ *Hessian* matrix $\nabla^2 f(x)$ with the second partial derivatives whose (i,j)-th element is

$$(\nabla^2 f(x))_{ij} = \partial^2 f(x)/\partial x_i \partial x_j, i,j = 1, \ldots, n.$$

When f is twice continuously differentiable, the Hessian matrix is always symmetric.

As a simple example, let us consider the quadratic function $f : \mathbb{R}^n \to \mathbb{R}$, $f(x) = (1/2)x^T Ax + b^T x + a$, where $A \in \mathbb{R}^{n \times n}$ is a symmetric matrix. Then, $\nabla f(x) = Ax + b$. The Hessian of f is given by $\nabla^2 f(x) = A$, i.e., the second-order approximation of a quadratic function is itself.

If f is continuously differentiable in a neighborhood of x, then

$$D(f(x);p) = \nabla f(x)^T p.$$

Theorem A.2.2 (Mean Value Theorem) *Given a continuously differentiable function $f : \mathbb{R} \to \mathbb{R}$ and two real numbers x_1 and x_2 that satisfy $x_2 > x_1$, then*

$$f(x_2) = f(x_1) + f'(\xi)(x_2 - x_1)$$

for some $\xi \in (x_1, x_2)$.

For a multivariate function $f : \mathbb{R}^n \to \mathbb{R}$ the mean value theorem says that for any vector $d \in \mathbb{R}^n$,

$$f(x+d) = f(x) + \nabla f(x + \alpha d)^T d$$

for some $\alpha \in (0, 1)$. ♦

Theorem A.2.3 (Taylor's Theorem) *If f is continuously differentiable in a domain containing the line segment $[x_1, x_2]$, then there is a θ, $0 \leq \theta \leq 1$, so that*

$$f(x_2) = f(x_1) + \nabla f(\theta x_1 + (1-\theta)x_2)^T (x_2 - x_1).$$

Moreover, if f is twice continuously differentiable in a domain containing the line segment $[x_1, x_2]$, *then there is a* θ, $0 \leq \theta \leq 1$, *so that*

$$f(x_2) = f(x_1) + \nabla f(x_1)^T (x_2 - x_1) + \frac{1}{2}(x_2 - x_1)^T \nabla^2 f(\theta x_1 + (1-\theta)x_2)(x_2 - x_1).$$

♦

For twice differentiable functions $f : \mathbb{R}^n \to \mathbb{R}^m$ for any vector $d \in \mathbb{R}^n$, one form of the Taylor theorem is

$$f(x+d) = f(x) + \nabla f(x)^T d + \frac{1}{2} d^T \nabla^2 f(x + \alpha d) d,$$

for some $\alpha \in (0, 1)$.

The *level set* of a function $f : \mathbb{R}^n \to \mathbb{R}$ at level c is the set of points

$$S = \{x : f(x) = c\}.$$

Theorem A.2.4 *Suppose that f is continuously differentiable. Then, the vector* $\nabla f(x_0)$ *is orthogonal to the tangent vector to an arbitrary smooth curve passing through* x_0 *on the level set determined by* $f(x) = f(x_0)$. ♦

In point x_0, the gradient $\nabla f(x_0)$ is the direction of *maximum rate of increase* of f at x_0. Since $\nabla f(x_0)$ is orthogonal to the level set through x_0 determined by $f(x) = f(x_0)$, it follows that the direction of maximum rate of increase of a real-valued differentiable function at a point is orthogonal to the level set of the function through that point.

Rates of convergence

Let $\{x_k\}$ be a sequence from $\mathbb{R}^n$ that converges to $x^* \in \mathbb{R}^n$. This sequence converges *Q-linear* if there is a constant $r \in (0, 1)$ so that

$$\frac{\|x_{k+1} - x^*\|}{\|x_k - x^*\|} \leq r,$$

for all k sufficiently large. The convergence is *Q-superlinear* if

$$\lim_{k \to \infty} \frac{\|x_{k+1} - x^*\|}{\|x_k - x^*\|} = 0.$$

The convergence is *Q-quadratic* if

$$\frac{\|x_{k+1} - x^*\|}{\|x_k - x^*\|^2} \le M$$

for all k sufficiently large, where M is a positive constant, not necessarily smaller than 1.

Typically, under appropriate assumptions the quasi-Newton methods for unconstrained optimization converge Q-superlinearly, whereas the Newton's method converges Q-quadratically. The steepest descent algorithms converge only at a Q-linear rate and when the problem is ill-conditioned, the convergence constant r is close to 1.

Order notation

The *order notation* is a concept used to see how the members of a sequence behave when we get far enough along in the sequence. Let us consider two nonnegative sequences of scalars $\{\eta_k\}$ and $\{\theta_k\}$. $\eta_k = o(\theta_k)$ if the sequence of ratios $\{\eta_k/\theta_k\}$ approaches zero, i.e., $\lim_{k\to\infty} \eta_k/\theta_k = 0$. $\eta_k = O(\theta_k)$ if there is a positive constant c so that $|\eta_k| \le c|\theta_k|$ for all k sufficiently large. If $\eta : \mathbb{R} \to \mathbb{R}$ is a function, then the following is written $\eta(t) = o(t)$ to specify that the ratio $\eta(t)/t$ approaches zero either as $t \to 0$ or $t \to \infty$. Similarly, $\eta(t) = O(t)$ if there is a constant c so that $|\eta(t)| \le c|t|$ for all $t \in \mathbb{R}$. A slight variant of the above definitions is as follows. $\eta_k = o(1)$ to specify that $\lim_{k\to\infty} \eta_k = 0$. Similarly, $\eta_k = O(1)$ to indicate that there is a constant c so that $|\eta_k| \le c$ for all k. Sometimes, in the above definitions there are vectors or matrices quantities as arguments. In these cases, the definitions apply to the norms of these quantities. For instance, if $f : \mathbb{R}^n \to \mathbb{R}^n$, then $f(x) = O(\|x\|)$ if there is a positive constant c so that $\|f(x)\| \le c\|x\|$ for all x in the domain of f.

A.3 Elements of Topology in the Euclidian Space $\mathbb{R}^n$

The *open ball* of radius ε centered at x^* is defined as the set $B(x^*, \varepsilon) = \{x \in \mathbb{R}^n : \|x^* - x\| < \varepsilon\}$ in any norm.

A subset $D \subset \mathbb{R}^n$ is *open* if for every $x \in D$ there exists a positive number $\varepsilon > 0$ so that the ball of radius ε centered at x is contained in D, i.e., $\{y \in \mathbb{R}^n : \|y - x\| \le \varepsilon\} \subset D$. The intersection of a finite number of open sets is open. Any union of open sets is open.

A point $x \in \mathbb{R}^n$ is an *interior point* of the set D if there is an open ball $B(x, \varepsilon)$ so that $B(x, \varepsilon) \subset D$. The *interior* of a set D, denoted by $\operatorname{int} D$, is the set of the interior points of D. The interior of a set is the largest open set contained in D.

A point $x \in \mathbb{R}^n$ is an *exterior point* of D if it is an interior point of $\mathbb{R}^n \backslash D$. Notice that the set D is open if every point of D is an interior point of D. Obviously, if D is open, then $\operatorname{int} D = D$.

A point $\tilde{x}$ is said to be a *limit point* of the set D if every open ball $B(\tilde{x}, \varepsilon)$ contains a point $x \neq \tilde{x}$ so that $x \in D$. Note that $\tilde{x}$ does not necessarily have to be an element of D for being a limit point of D.

The set D is *closed* if for all possible sequences of points $\{x_k\}$ in D all limit points of $\{x_k\}$ are elements of D. The union of a finite number of closed sets is closed. Any intersection of closed sets is closed.

The set D is *bounded* if there is some real number $M > 0$ so that $\|x\| \leq M$ for all $x \in D$.

The set D is *compact* if every sequence $\{x_k\}$ of points in D has at least one limit point and all such limit points are in D. A central result in topology is that in $\mathbb{R}^n$ *the set D is compact if it is both closed and bounded.*

Theorem A.3.1 (Weierstrass Extreme Value Theorem) *Every continuous function on a compact set attains its extreme values on that set.* ♦

The *closure* of the set D is the set $cl(D) = D \cup L$, where L denotes the set of all limit points of D.

For a given point $x \in \mathbb{R}^n$, a *neighborhood* of x is an open set containing x. A useful neighborhood is the open ball of radius ε centered at x.

A point $x \in \mathbb{R}^n$ is a *boundary point* of the set D if every neighborhood of x contains points both inside and outside of D. The set of boundary points of D is denoted by ∂D.

Let $f : D \subset \mathbb{R}^n \to \mathbb{R}^m$. Then, f is *Lipschitz continuous* on an open set $N \subset D$ if there is a constant $0 < L < \infty$ so that

$$\|f(x) - f(y)\| \leq L\|x - y\|$$

for all $x, y \in N$. L is called the *Lipschitz constant*. If $g, h : D \subset \mathbb{R}^n \to \mathbb{R}^m$ are two Lipschitz continuous functions on a set $N \subset D$, then their sum $g + h$ is also Lipschitz continuous, with Lipschitz constant equal to the sum of the Lipschitz constants for f and g, respectively. If $g, h : D \subset \mathbb{R}^n \to \mathbb{R}^m$ are two Lipschitz continuous functions and bounded on a set $N \subset D$, i.e., there is a constant $M > 0$ such that $|g(x)| \leq M$ and $|h(x)| \leq M$ for all $x \in N$, then the product gh is Lipschitz continuous on N.

If f is Lipschitz continuous on a set $D \subset \mathbb{R}^n$, then f is uniformly continuous on D. The reverse is not true.

A.4 Elements of Convexity—Convex Sets and Convex Functions

Convex sets

A set $C \subset \mathbb{R}^n$ is a *convex set* if for every point $x, y \in C$, the point $z = \lambda x + (1 - \lambda)y$ is also in the set C for any $\lambda \in [0, 1]$. The intersection of any family of convex sets is a convex set. An *affine* set in $\mathbb{R}^n$ is the set of all vectors $\{x\} \oplus S$, where $x \in \mathbb{R}^n$

and S is a subspace of $\mathbb{R}^n$. A *cone* is a set V with the property that for all $x \in V$ it follows that $\alpha x \in V$ for all $\alpha > 0$. A *cone generated by* $\{x_1, x_2, \ldots, x_m\}$ is the set of all vectors of the form

$$x = \sum_{i=1}^{m} \alpha_i x_i, \quad \text{where } \alpha_i \geq 0 \text{ for all } i = 1, \ldots, m.$$

Observe that all cones of this form are convex sets. A *convex combination* of a finite set of vectors $\{x_1, x_2, \ldots, x_m\}$ in $\mathbb{R}^n$ is any vector x of the form

$$x = \sum_{i=1}^{m} \alpha_i x_i, \quad \text{where } \sum_{i=1}^{m} \alpha_i = 1, \text{ and } \alpha_i \geq 0 \text{ for all } i = 1, \ldots, m.$$

Convex functions

A function $f : C \to \mathbb{R}$ defined on a convex set $C \subset \mathbb{R}^n$ is a *convex function* if $f(\lambda x + (1 - \lambda)y) \leq \lambda f(x) + (1 - \lambda)f(y)$ for every $x, y \in C$ and every $\lambda \in (0, 1)$. Moreover, f is said to be *strictly convex* if for every $x, y \in C$ and every $\lambda \in (0, 1)$, $f(\lambda x + (1 - \lambda)y) < \lambda f(x) + (1 - \lambda)f(y)$. In other words, this means that if we take any two points x and y, then f evaluated at any convex combination of these two points should be no larger than the same convex combination of $f(x)$ and $f(y)$. A function that is not convex is said to be *nonconvex*. A function f is *concave* if $-f$ is convex. Any linear function of n variables is both convex and concave on $\mathbb{R}^n$. The following result shows why the convex functions are of interest in optimization problems.

Theorem A.4.1 *Any local minimum of a convex function $f : C \to \mathbb{R}$ defined on a convex set $C \subset \mathbb{R}^n$ is also a global minimum on C. Any local minimum of a strictly convex function $f : C \to \mathbb{R}$ defined on a convex set $C \subset \mathbb{R}^n$ is the unique strict global minimum of f on C.* ♦

Strong convexity

A differentiable function f is called *strongly convex* on S with the parameter $\mu > 0$ if for all the points $x, y \in S$,

$$f(y) \geq f(x) + \nabla f(x)^T (y - x) + \frac{\mu}{2} \|y - x\|^2.$$

Intuitively, strong convexity means that there exists a quadratic lower bound on the growth of the function. Observe that a strongly convex function is strictly convex since the quadratic lower bound growth is strictly greater than the linear growth. An equivalent condition for the *strong convexity* of function f on S is

$$(\nabla f(x) - \nabla f(y))^T (x - y) \geq \mu \|x - y\|^2$$

for some $\mu > 0$ and for all $x, y \in S$.

For differentiable strongly convex functions, it is easy to prove that:

1. $\|\nabla f(x)\|^2 \geq 2\mu(f(x) - f(x^*))$ for all $x \in S$, where x^* is a local minimum of function f.
2. $\|\nabla f(x) - \nabla f(y)\| \geq \mu\|x - y\|$ for all $x \in S$.
3. $f(y) \leq f(x) + \nabla f(x)^T (y - x) + \frac{1}{2\mu}\|\nabla f(y) - \nabla f(x)\|^2$, for all $x \in S$.

If the function f is twice continuously differentiable, then it is strongly convex with the parameter $\mu > 0$ on S if and only if $\nabla^2 f(x) \geq \mu I$ for all $x \in S$, where I is the identity matrix and the inequality $\geq$ means that $\nabla^2 f(x) - \mu I$ is positive semi-definite.

Proposition A.4.1 (Convexity of Level Set) *Let C be a convex set in $\mathbb{R}^n$ and let $f : C \to \mathbb{R}$ be a convex function. Then, the level set $C_\alpha = \{x \in C : f(x) \leq \alpha\}$, where α is a real number, is a convex set.*

Proof Let $x_1, x_2 \in C$. Of course, $x_1, x_2 \in C_\alpha$. $f(x_1) \leq \alpha$ and $f(x_2) \leq \alpha$. Now, let $\lambda \in (0, 1)$ and consider $x = \lambda x_1 + (1 - \lambda)x_2$. By convexity of C, it follows that $x \in C$. On the other hand, by convexity of f on C,

$$f(x) \leq \lambda f(x_1) + (1 - \lambda)f(x_2) \leq \lambda\alpha + (1 - \lambda)\alpha = \alpha,$$

i.e., $x \in C_\alpha$. ♦

Proposition A.4.2 (Convexity of a domain defined by a set of convex functions) *Let C be a convex set in $\mathbb{R}^n$ and let $c_i : C \to \mathbb{R}$, $i = 1, \ldots, m$, be convex functions on C. Then, the set defined by $X = \{x \in C : c_i(x) \leq 0, i = 1, \ldots, m\}$ is convex.*

Proof The result follows from Proposition A.4.1 and from the property of the intersection of the convex sets. ♦

The following two propositions give *differential criteria* of checking the convexity of a function.

Proposition A.4.3 (First-order condition for convexity) *Let C be a convex set in $\mathbb{R}^n$ with a nonempty interior. Consider the function $f : C \to \mathbb{R}$ which is continuous on C and differentiable on* $\text{int}(C)$*. Then, f is convex on* $\text{int}(C)$ *if and only if $f(y) \geq f(x) + \nabla f(x)^T (y - x)$ for any points $x, y \in C$.* ♦

Proposition A.4.4 (Second-order condition for convexity) *Let C be a convex set in $\mathbb{R}^n$ with a nonempty interior. Consider the function $f : C \to \mathbb{R}$ which is continuous on C and twice differentiable on* $\text{int}(C)$*. Then, f is convex on* $\text{int}(C)$ *if and only if the Hessian $\nabla^2 f(x)$ is positive semidefinite at each $x \in$* $\text{int}(C)$. ♦

The convexity of the objective function and of the constraints is crucial in nonlinear optimization. The convex programs have very nice theoretical properties which can be used to design efficient optimization algorithms. Therefore, it is important to know how to detect convexity and the operations that preserve the convexity of functions.

Proposition A.4.5 (Linear combination with nonnegative coefficients) *Let* C *be a convex set in* $\mathbb{R}^n$. *If* $f : C \to \mathbb{R}$ *and* $g : C \to \mathbb{R}$ *are convex functions on* C, *then their linear combination* $\lambda f + \eta g$, *where the coefficients* λ *and* η *are nonnegative, is also convex on* C. ♦

Proposition A.4.6 (Composition with affine mapping) *Let* C *and* D *be convex sets in* $\mathbb{R}^m$ *and* $\mathbb{R}^n$, *respectively. If* $g : C \to \mathbb{R}$ *is a convex function on* C *and* $h : D \to \mathbb{R}^m$ *is an affine mapping, i.e.,* $h(x) = Ax + b$ *with* $range(h) \subset C$, *then the composite function* $f : D \to \mathbb{R}$ *defined as* $f(x) = g(h(x))$ *is convex on* D. ♦

Notes and References

The material in this appendix is covered in: (Dennis & Schnabel, 1983), (Peressini, Sullivan, & Uhl, 1988), (Trefethen & Schreiber, 1990), (Bazaraa, Sherali, & Shetty, 1993), (Golub & Van Loan, 1996), (Demmel, 1997), (Trefethen & Bau, 1997), (Meyer, 2000), (Laub, 2005), (Nocedal & Wright, 2006).

Appendix B
UOP: A Collection of 80 Unconstrained Optimization Test Problems

The unconstrained optimization test problems selected in this set, we call *UOP collection*, have different structures and complexities. These problems are used to see the performances of the algorithms described in this book. The name of these problems and the initial point are given in Table 1.1. In this collection, some problems are quadratic and some of them are highly nonlinear. The problems are presented in extended (separable) or generalized (chained) form. The Hessian for the problems in extended form has a block-diagonal structure. On the other hand, the Hessian for the problems in generalized form has a banded structure with small bandwidth, often being tri- or pentadiagonal. For some other optimization problems, from this set, the corresponding Hessian has a sparse structure, or it is a dense (full) matrix. The vast majority of the optimization problems included in this collection is taken from CUTEr (Bongartz, Conn, Gould, & Toint, 1995) collection, others are from (Andrei, 1999), as well as from other publications. The algebraic description of the problems is as follows:

1. **Freudenstein and Roth (CUTE)**

$$f(x) = \sum_{i=1}^{n/2} (-13 + x_{2i-1} + ((5 - x_{2i})x_{2i} - 2)x_{2i})^2 + (-29 + x_{2i-1} + ((x_{2i} + 1)x_{2i} - 14)x_{2i})^2,$$

$$x_0 = [0.5, -2., 0.5, -2., \ldots, 0.5, -2.].$$

2. **Extended White and Holst**

$$f(x) = \sum_{i=1}^{n/2} c\left(x_{2i} - x_{2i-1}^3\right)^2 + (1 - x_{2i-1})^2, \quad x_0 = [-1.2, 1., \ldots, -1.2, 1.],$$

$$c = 1.$$

N. Andrei, *Nonlinear Conjugate Gradient Methods for Unconstrained Optimization*, Springer Optimization and Its Applications 158,
https://doi.org/10.1007/978-3-030-42950-8

3. **Tridiagonal White and Holst**

$$f(x) = \sum_{i=1}^{n-1} c(x_{i+1} - x_i^3)^2 + (1 - x_i)^2, \quad x_0 = [-1.2, 1., \ldots, -1.2, 1.]. \quad c = 4.$$

4. **Extended Beale (CUTE)**

$$f(x) = \sum_{i=1}^{n/2} (1.5 - x_{2i-1}(1 - x_{2i}))^2 + \left(2.25 - x_{2i-1}(1 - x_{2i}^2)\right)^2 + \left(2.625 - x_{2i-1}(1 - x_{2i}^3)\right)^2,$$
$$x_0 = [1., 0.8, \ldots, 1., 0.8].$$

5. **Extended Powell**

$$\begin{aligned} f(x) = \sum_{i=1}^{n/4} & (x_{4i-3} + 10x_{4i-2})^2 + 5(x_{4i-1} - x_{4i})^2 \\ & + (x_{4i-2} - 2x_{4i-1})^4 + 10(x_{4i-3} - x_{4i})^4, \\ x_0 = & [3., -1., 0., 1., \ldots, 3., -1., 0., 1.]. \end{aligned}$$

6. **Extended Maratos**

$$f(x) = \sum_{i=1}^{n/2} x_{2i-1} + c\left(x_{2i-1}^2 + x_{2i}^2 - 1\right)^2, \quad x_0 = [0.1, 0.1, \ldots, 0.1, 0.1], \quad c = 1.$$

7. **Extended Cliff**

$$\begin{aligned} f(x) = & \sum_{i=1}^{n/2} \left(\frac{x_{2i-1} - 3}{100}\right)^2 - (x_{2i-1} - x_{2i}) + \exp(2(x_{2i-1} - x_{2i})), \\ x_0 = & [0.001, 0.001, \ldots, 0.001, 0.001]. \end{aligned}$$

8. **Extended Woods (CUTE)**

$$\begin{aligned} f(x) = & \sum_{i=1}^{n/4} 100\left(x_{4i-3}^2 - x_{4i-2}\right)^2 + (x_{4i-3} - 1)^2 + 90\left(x_{4i-1}^2 - x_{4i}\right)^2 \\ & + (1 - x_{4i-1})^2 + 10.1\left\{(x_{4i-2} - 1)^2 + (x_{4i} - 1)^2\right\} + 19.8(x_{4i-2} - 1)(x_{4i} - 1), \\ x_0 = & [-3., -1., -3., -1., \ldots, -3., -1., -3., -1.]. \end{aligned}$$

9. **Extended Hiebert**

$$f(x) = \sum_{i=1}^{n/2} (x_{2i-1} - 10)^2 + (x_{2i-1}x_{2i} - 500)^2, \quad x_0 = [5.001, 5.001, \ldots, 5.001].$$

10. **Extended Rosenbrock (CUTE)**

$$f(x) = \sum_{i=1}^{n/2} c\left(x_{2i} - x_{2i-1}^2\right)^2 + (1 - x_{2i-1})^2, \quad x_0 = [-1.2, 1., \ldots, -1.2, 1.].$$
$$c = 1000.$$

11. **Generalized Rosenbrock (CUTE)**

$$f(x) = (x_1 - 1)^2 + \sum_{i=2}^{n} 100(x_i - x_{i-1}^2)^2, \quad x_0 = [-1.2, 1., \ldots, -1.2, 1.].$$

12. **Extended Himmelblau—HIMMELBC (CUTE)**

$$f(x) = \sum_{i=1}^{n/2} \left(x_{2i-1}^2 + x_{2i} - 11\right)^2 + \left(x_{2i-1} + x_{2i}^2 - 7\right)^2, \quad x_0 = [1., 1., \ldots, 1.].$$

13. **HIMMELBG (CUTE)**

$$f(x) = \sum_{i=1}^{n/2} \left(2x_{2i-1}^2 + 3x_{2i}^2\right) \exp(-x_{2i-1} - x_{2i}), \quad x_0 = [1.5, 1.5, \ldots, 1.5].$$

14. **HIMMELBH (CUTE)**

$$f(x) = \sum_{i=1}^{n/2} (-3x_{2i-1} - 2x_{2i} + 2 + x_{2i-1}^3 + x_{2i}^2), \quad x_0 = [0.8, 0.8, \ldots, 0.8].$$

15. **Extended Trigonometric ET1**

$$f(x) = \sum_{i=1}^{n} \left(\left(n - \sum_{j=1}^{n} \cos x_j \right) + i(1 - \cos x_i) - \sin x_i \right)^2,$$
$$x_0 = [0.2, 0.2, \ldots, 0.2].$$

16. **Extended Trigonometric ET2**

$$f(x) = \sum_{i=1}^{n} \left(\left(n - \sum_{i=1}^{n} \sin(x_i) \right) + i(1 - \sin(x_i)) - \sin(x_i) \right)^2,$$
$$x_0 = [0.2, 0.2, \ldots, 0.2].$$

17. **Extended Block-Diagonal BD1**

$$f(x) = \sum_{i=1}^{n/2} \left(x_{2i-1}^2 + x_{2i}^2 - 2\right)^2 + (\exp(x_{2i-1}) - x_{2i})^2, \quad x_0 = [1., 1., \ldots, 1.].$$

18. **Extended Tridiagonal 1**

$$f(x) = \sum_{i=1}^{n/2} (x_{2i-1} + x_{2i} - 3)^2 + (x_{2i-1} - x_{2i} + 1)^4, \quad x_0 = [2., 2., \ldots, 2.].$$

19. **Extended Three Exponential Terms**

$$f(x) = \sum_{i=1}^{n/2} (\exp(x_{2i-1} + 3x_{2i} - 0.1) + \exp(x_{2i-1} - 3x_{2i} - 0.1) + \exp(-x_{2i-1} - 0.1)),$$
$$x_0 = [0.1, 0.1, \ldots, 0.1].$$

20. **Generalized Tridiagonal 1**

$$f(x) = \sum_{i=1}^{n-1} (x_i + x_{i+1} - 3)^2 + (x_i - x_{i+1} + 1)^4, \quad x_0 = [2., 2., \ldots, 2.].$$

21. **Generalized Tridiagonal 2**

$$f(x) = ((5 - 3x_1 - x_1^2)x_1 - 3x_2 + 1)^2$$
$$+ \sum_{i=1}^{n-1} ((5 - 3x_i - x_i^2)x_i - x_{i-1} - 3x_{i+1} + 1)^2 + ((5 - 3x_n - x_n^2)x_n - x_{n-1} + 1)^2,$$
$$x_0 = [-1., -1., \ldots, -1.].$$

22. **Tridiagonal Double Border (CUTE)**

$$f(x) = (x_1 - 1)^2 + \sum_{i=1}^{n-1} (x_1 - 0.5x_i^2 - 0.5x_{i+1}^2)^2, \quad x_0 = [1., -1., \ldots, 1., -1.].$$

23. **Broyden Pentadiagonal (CUTE)**

$$f(x) = (3x_1 - 2x_1^2)^2 + \sum_{i=2}^{n-1} (3x_i - 2x_i^2 - x_{i-1} - 2x_{i+1} + 1)^2 + (3x_n + 2x_n^2 - x_{n-1} + 1)^2,$$
$$x_0 = [-1., -1., \ldots, -1.].$$

24. **Extended PSC1**

$$f(x) = \sum_{i=1}^{n/2} (x_{2i-1}^2 + x_{2i}^2 + x_{2i-1}x_{2i})^2 + \sin^2(x_{2i-1}) + \cos^2(x_{2i}),$$
$$x_0 = [3., 0.1, \ldots, 3., 0.1].$$

25. **Perturbed Quadratic PQ1**

$$f(x) = \sum_{i=1}^{n} ix_i^2 + \frac{1}{100}\left(\sum_{i=1}^{n} x_i\right)^2, \quad x_0 = [1., 1., \ldots, 1.]$$

26. **Perturbed Quadratic PQ2**

$$f(x) = \left(\sum_{i=1}^{n} ix_i\right)^2 + \sum_{i=1}^{n} ix_i^2, \quad x_0 = [0.5, 0.5, \ldots, 0.5],$$

27. **Almost Perturbed Quadratic**

$$f(x) = \frac{1}{100}(x_1 + x_n)^2 + \sum_{i=1}^{n} ix_i^2, \quad x_0 = [0.5, 0.5, \ldots, 0.5].$$

28. **Almost Perturbed Quartic**

$$f(x) = \frac{1}{100}(x_1 + x_n)^2 + \sum_{i=1}^{n} ix_i^4, \quad x_0 = [0.5, 0.5, \ldots, 0.5].$$

29. **Extended Penalty Function U52**

$$f(x) = \left(\sum_{i=1}^{n} x_i^2 - 0.25\right)^2 + \sum_{i=1}^{n-1} (x_i - 1)^2, \quad x_0 = [1/100, 2/100, \ldots, n/100].$$

30. **TR-Sum of quadratics**

$$f(x) = \sum_{i=1}^{n-1} x_i^2 + c(x_{i+1} + x_i^2)^2 \quad x_0 = [1., 1., \ldots, 1.]. \quad c = 100000.$$

31. **Quadratic Diagonal Perturbed**

$$f(x) = \left(\sum_{i=1}^{n} x_i\right)^2 + \sum_{i=1}^{n} \frac{i}{100} x_i^2, \quad x_0 = [0.5, 0.5, \ldots, 0.5].$$

32. **Full Hessian FH1**

$$f(x) = \sum_{i=1}^{m} \left(\sum_{j=1}^{n} ijx_j^2 - 1\right)^2, \quad m = 50, \quad x_0 = [1/n, 2/n, \ldots, n/n].$$

33. **Full Hessian FH2**

$$f(x) = \left(\sum_{i=1}^{n} x_i\right)^2 + \sum_{i=1}^{n} \frac{i(\sin(x_i) + \cos(x_i))}{1000}, \quad x_0 = [1., 1., \ldots, 1.].$$

34. **Full Hessian FH3**

$$f(x) = \left(\sum_{i=1}^{n} x_i^2\right)^2 + \sum_{i=1}^{n} \frac{i(\sin(x_i) + \cos(x_i))}{1000}, \quad x_0 = [1., 1., \ldots, 1.].$$

35. **Diagonal Full Border**

$$f(x) = (x_1 - 1)^4 + (x_n^2 - x_1^2)^2 + \sum_{i=1}^{n-2} (\sin(x_{i+1} - x_n) - x_1^2 - x_{i+1}^2)^2,$$

$$x_0 = [0.001, 0.001, \ldots, 0.001].$$

36. **Diagonal Double Border Arrow Up**

$$f(x) = \sum_{i=1}^{n} 4(x_i^2 - x_1)^2 + (x_i - 1)^2, \quad x_0 = [0.4, 1., \ldots, 0.4, 1.].$$

37. **QP1 Extended Quadratic Penalty**

$$f(x) = \left(\sum_{i=1}^{n} x_i^2 - 0.5\right)^2 + \sum_{i=1}^{n-1} (x_i^2 - 2)^2, \quad x_0 = [1., 1., \ldots, 1.].$$

38. **QP2 Extended Quadratic Penalty**

$$f(x) = \left(\sum_{i=1}^{n} x_i^2 - 100\right)^2 + \sum_{i=1}^{n-1} (x_i^2 - \sin(x_i))^2, \quad x_0 = [2., 2., \ldots, 2.].$$

39. **QP3 Extended Quadratic Penalty**

$$f(x) = \left(\sum_{i=1}^{n} x_i^2 - 0.25\right)^2 - \sum_{i=1}^{n-1} (x_i^2 - 1)^2, \quad x_0 = [1., 1., \ldots, 1.].$$

40. **Staircase S1**

$$f(x) = \sum_{i=1}^{n-1} (x_i + x_{i+1} - i)^2, \quad x_0 = [1., 1., \ldots, 1.].$$

41. **Staircase S2**

$$f(x) = \sum_{i=2}^{n} (x_{i-1} + x_i - i)^2, \quad x_0 = [1., 1., \ldots, 1.].$$

42. **Staircase S3**

$$f(x) = \sum_{i=2}^{n} (x_{i-1} + x_i + i)^2, \quad x_0 = [2., 2., \ldots, 2.].$$

43. **NONDQUAR (CUTE)**

$$f(x) = (x_1 - x_2)^2 + \sum_{i=1}^{n-2} (x_i + x_{i+1} + x_n)^4 + (x_{n-1} + x_n)^2,$$
$$x_0 = [1., -1., \ldots, 1., -1.,].$$

44. **TRIDIA (CUTE)**

$$f(x) = \gamma(\delta x_1 - 1)^2 + \sum_{i=2}^{n} i(\alpha x_i - \beta x_{i-1})^2,$$
$$\alpha = 2, \quad \beta = 1, \quad \gamma = 1, \quad \delta = 1, \quad x_0 = [1., 1., \ldots, 1.].$$

45. **ARWHEAD (CUTE)**

$$f(x) = \sum_{i=1}^{n-1} (-4x_i + 3) + \sum_{i=1}^{n-1} (x_i^2 + x_n^2)^2, \quad x_0 = [1., 1., \ldots, 1.].$$

46. **NONDIA (CUTE)**

$$f(x) = (x_1 - 1)^2 + c(x_1 - x_1^2)^2 + \sum_{i=2}^{n} c(x_1 - x_i^2)^2,$$
$$x_0 = [-0.01, -0.01, \ldots, -0.01], \quad c = 100.$$

47. **BDQRTIC (CUTE)**

$$f(x) = \sum_{i=1}^{n-4} (-4x_i + 3)^2 + (x_i^2 + 2x_{i+1}^2 + 3x_{i+2}^2 + 4x_{i+3}^2 + 5x_n^2)^2,$$
$$x_0 = [1., 1., \ldots, 1.].$$

48. **DQDRTIC (CUTE)**

$$f(x) = \sum_{i=1}^{n-2} (x_i^2 + cx_{i+1}^2 + dx_{i+2}^2), \quad c = 1000, \quad d = 1000,$$
$$x_0 = [3., 3., \ldots, 3.].$$

49. **EG2 (CUTE)**

$$f(x) = \sum_{i=1}^{n-1} \sin(x_1 + x_i^2 - 1) + \frac{1}{2}\sin(x_n^2), \quad x_0 = [0.001, 0.001, \ldots, 0.001].$$

50. **EG3**

$$f(x) = \frac{1}{2}\cos(x_n^2) + \sum_{i=1}^{n-1} \cos(x_1 + x_i^2 - 1), \quad x_0 = [0.02, 0.02, \ldots, 0.02].$$

51. **EDENSCH (CUTE)**

$$f(x) = 16 + \sum_{i=1}^{n-1} \left((x_i - 2)^4 + (x_i x_{i+1} - 2x_{i+1})^2 + (x_{i+1} + 1)^2\right),$$
$$x_0 = [0., 0., \ldots, 0.].$$

52. **FLETCHCR (CUTE)**

$$f(x) = \sum_{i=1}^{n-1} c\left(x_{i+1} - x_i + 1 - x_i^2\right)^2, \quad x_0 = [0.5, 0.5, \ldots, 0.5], \quad c = 100.$$

53. **ENGVAL1 (CUTE)**

$$f(x) = \sum_{i=1}^{n-1} (x_i^2 + x_{i+1}^2)^2 + \sum_{i=1}^{n-1} (-4x_i + 3), \quad x_0 = [2., 2., \ldots, 2.].$$

54. **DENSCHNA (CUTE)**

$$f(x) = \sum_{i=1}^{n/2} x_{2i-1}^4 + (x_{2i-1} + x_{2i})^2 + (-1 + \exp(x_{2i}))^2, \quad x_0 = [1., 1., \ldots, 1.].$$

55. **DENSCHNB (CUTE)**

$$f(x) = \sum_{i=1}^{n/2} (x_{2i-1} - 2)^2 + (x_{2i-1} - 2)^2 x_{2i}^2 + (x_{2i} + 1)^2,$$
$$x_0 = [10., 10., \ldots, 10.].$$

56. **DENSCHNC (CUTE)**

$$f(x) = \sum_{i=1}^{n/2} (-2 + x_{2i-1}^2 + x_{2i}^2)^2 + (-2 + \exp(x_{2i-1} - 1) + x_{2i}^3)^2,$$
$$x_0 = [1., 1., \ldots, 1.].$$

57. **DENSCHNF (CUTE)**

$$f(x) = \sum_{i=1}^{n/2} \left(2(x_{2i-1} + x_{2i})^2 + (x_{2i-1} - x_{2i})^2 - 8\right)^2 + \left(5x_{2i-1}^2 + (x_{2i} - 3)^2 - 9\right)^2,$$

$$x_0 = [100., -100.,, \ldots, 100., -100.].$$

58. **SINQUAD (CUTE)**

$$f(x) = (x_1 - 1)^4 + (x_n^2 - x_1^2)^2 + \sum_{i=1}^{n-2} (\sin(x_{i+1} - x_n) - x_1^2 + x_{i+1}^2)^2,$$

$$x_0 = [0., 0., \ldots, 0.].$$

59. **DIXON3DQ (CUTE)**

$$f(x) = (x_1 - 2)^2 + \sum_{i=1}^{n-1} (x_i - x_{i+1})^2 + (x_n - 1)^2,$$

$$x_0 = [-0.1., -0.1., \ldots, -0.1].$$

60. **BIGGSB1 (CUTE)**

$$f(x) = (x_1 - 1)^2 + (1 - x_n)^2 + \sum_{i=2}^{n} (x_i - x_{i-1})^2, \quad x_0 = [0.1, 0.1, \ldots, 0.1].$$

61. **PRODsin**

$$f(x) = \left(\sum_{i=1}^{m} x_i^2\right)\left(\sum_{i=1}^{n} \sin(x_i)\right), \quad x_0 = [0.00001, \ldots, 0.00001]. \quad m = n - 1.$$

62. **PROD1**

$$f(x) = \left(\sum_{i=1}^{m} x_i\right)\left(\sum_{i=1}^{n} x_i\right), \quad x_0 = [1., 1., \ldots, 1.]. \quad m = n.$$

63. **PRODcos**

$$f(x) = \left(\sum_{i=1}^{m} x_i^2\right)\left(\sum_{i=1}^{n} \cos(x_i)\right), \quad x_0 = [1., 0., \ldots, 0.]. \quad m = n - 1.$$

64. **PROD2**

$$f(x) = \left(\sum_{i=1}^{m} x_i^4\right)\left(\sum_{i=1}^{n} i x_i\right), \quad x_0 = [0.00001, \ldots, 0.00001, 1.]. \quad m = 1.$$

65. DIXMAANA (CUTE)

$$f(x) = 1 + \sum_{i=1}^{n} \alpha x_i^2 \left(\frac{i}{n}\right)^{k1} + \sum_{i=1}^{n-1} \beta x_i^2 (x_{i+1} + x_{i+1}^2)^2 \left(\frac{i}{n}\right)^{k2}$$
$$+ \sum_{i=1}^{2m} \gamma x_i^2 x_{i+m}^4 \left(\frac{i}{n}\right)^{k3} + \sum_{i=1}^{m} \delta x_i x_{i+2m} \left(\frac{i}{n}\right)^{k4},$$

$m = n/4, \quad x_0 = [2., 2., , \ldots, 2.],$

$\alpha = 1, \quad \beta = 0, \quad \gamma = 0.125, \quad \delta = 0.125, \quad k1 = 0, \quad k2 = 0, \quad k3 = 0, \quad k4 = 0.$

66. DIXMAANB (CUTE)

$$f(x) = 1 + \sum_{i=1}^{n} \alpha x_i^2 \left(\frac{i}{n}\right)^{k1} + \sum_{i=1}^{n-1} \beta x_i^2 (x_{i+1} + x_{i+1}^2)^2 \left(\frac{i}{n}\right)^{k2}$$
$$+ \sum_{i=1}^{2m} \gamma x_i^2 x_{i+m}^4 \left(\frac{i}{n}\right)^{k3} + \sum_{i=1}^{m} \delta x_i x_{i+2m} \left(\frac{i}{n}\right)^{k4},$$

$m = n/4, \quad x_0 = [2., 2., \ldots, 2.],$

$\alpha = 1, \quad \beta = 0.0625, \quad \gamma = 0.0625, \quad \delta = 0.0625, \quad k1 = 0, \quad k2 = 0, \quad k3 = 0, \quad k4 = 1.$

67. DIXMAANC (CUTE)

$$f(x) = 1 + \sum_{i=1}^{n} \alpha x_i^2 \left(\frac{i}{n}\right)^{k1} + \sum_{i=1}^{n-1} \beta x_i^2 (x_{i+1} + x_{i+1}^2)^2 \left(\frac{i}{n}\right)^{k2}$$
$$+ \sum_{i=1}^{2m} \gamma x_i^2 x_{i+m}^4 \left(\frac{i}{n}\right)^{k3} + \sum_{i=1}^{m} \delta x_i x_{i+2m} \left(\frac{i}{n}\right)^{k4},$$

$m = n/4, \quad x_0 = [2., 2., \ldots, 2.],$

$\alpha = 1, \quad \beta = 0.125, \quad \gamma = 0.125, \quad \delta = 0.125, \quad k1 = 0, \quad k2 = 0, \quad k3 = 0, \quad k4 = 0.$

68. DIXMAAND (CUTE)

$$f(x) = 1 + \sum_{i=1}^{n} \alpha x_i^2 \left(\frac{i}{n}\right)^{k1} + \sum_{i=1}^{n-1} \beta x_i^2 (x_{i+1} + x_{i+1}^2)^2 \left(\frac{i}{n}\right)^{k2}$$
$$+ \sum_{i=1}^{2m} \gamma x_i^2 x_{i+m}^4 \left(\frac{i}{n}\right)^{k3} + \sum_{i=1}^{m} \delta x_i x_{i+2m} \left(\frac{i}{n}\right)^{k4},$$

$m = n/4, \quad x_0 = [2., 2., \ldots, 2.],$

$\alpha = 1, \quad \beta = 0.26, \quad \gamma = 0.26, \quad \delta = 0.26, \quad k1 = 0, \quad k2 = 0, \quad k3 = 0, \quad k4 = 0.$

69. **DIXMAANL (CUTE)**

$$f(x) = 1 + \sum_{i=1}^{n} \alpha x_i^2 \left(\frac{i}{n}\right)^{k1} + \sum_{i=1}^{n-1} \beta x_i^2 (x_{i+1} + x_{i+1}^2)^2 \left(\frac{i}{n}\right)^{k2}$$
$$+ \sum_{i=1}^{2m} \gamma x_i^2 x_{i+m}^4 \left(\frac{i}{n}\right)^{k3} + \sum_{i=1}^{m} \delta x_i x_{i+2m} \left(\frac{i}{n}\right)^{k4},$$
$$m = n/4, \quad x_0 = [1., 1., \ldots, 1.],$$
$$\alpha = 1, \quad \beta = 0.26, \quad \gamma = 0.26, \quad \delta = 0.26, \quad k1 = 2, \quad k2 = 0, \quad k3 = 0, \quad k4 = 2.$$

70. **ARGLINB**

$$f(x) = \sum_{i=1}^{m} \left(\sum_{j=1}^{n} ijx_j - 1 \right)^2, \quad x_0 = [0.01, 0.001, \ldots, 0.01, 0.001]. \quad m = 5.$$

71. **VARDIM (CUTE)**

$$f(x) = \sum_{i=1}^{n} (x_i - 1)^2 + \left(\sum_{i=1}^{n} ix_i - \frac{n(n+1)}{2} \right)^2 + \left(\sum_{i=1}^{n} ix_i - \frac{n(n+1)}{2} \right)^4,$$
$$x_0 = \left[1 - \frac{1}{n}, 1 - \frac{2}{n}, \ldots 1 - \frac{n}{n} \right].$$

72. **DIAG-AUP1**

$$f(x) = \sum_{i=1}^{n} 4(x_i^2 - x_1)^2 + (x_i^2 - 1)^2, \quad x_0 = [4., 4., \ldots, 4.].$$

73. **ENGVAL8 (CUTE)**

$$f(x) = \sum_{i=1}^{n-1} (x_i^2 + x_{i+1}^2)^2 - (7 - 8x_i), \quad x_0 = [2., 2., \ldots, 2.].$$

74. **QUARTIC**

$$f(x) = \sum_{i=1}^{n} (x_i - 1)^4, \quad x_0 = [2., 2., \ldots, 2.].$$

75. **LIARWHD (CUTE)**

$$f(x) = \sum_{i=1}^{n} 4(x_i^2 - x_1)^2 + (x_i - 1)^2, \quad x_0 = [4., 4., \ldots, 4.].$$

76. **NONSCOMP (CUTE)**

$$f(x) = (x_1 - 1)^2 + \sum_{i=2}^{n} 4(x_i - x_{i-1}^2)^2, \quad x_0 = [3., 3., \ldots, 3.].$$

77. **Linear Perturbed**

$$f(x) = \sum_{i=1}^{n} \left(ix_i^2 + \frac{x_i}{100} \right), \quad x_0 = [2., 2., \ldots, 2.].$$

78. **CUBE**

$$f(x) = (x_1 - 1)^2 + \sum_{i=2}^{n} 100(x_i - x_{i-1}^3)^2, \quad x_0 = [-1.2, 1.1, \ldots, -1.2, 1.1].$$

79. **HARKERP**

$$f(x) = \left(\sum_{i=1}^{n} x_i \right)^2 - \sum_{i=1}^{n} \left(x_i + \frac{1}{2} x_i^2 \right), \quad x_0 = [1., 2., \ldots, n].$$

80. **QUARTICM**

$$f(x) = \sum_{i=1}^{n} (x_i - i)^4, \quad x_0 = [2., 2., \ldots, 2.].$$

References

Abdoulaev, G. S., Ren, K., & Hielscher, A. H. (2005). Optical tomography as a PDE-constrained optimization problem. *Inverse Problems, 21*(5), 1507–1530.

Adams, L., & Nazareth, J. L. (1996). Linear and nonlinear conjugate gradient—Related methods. In *AMS-IMS-SIAM Joint Summer Research Conference*. Philadelphia, PA, USA: SIAM.

Akaike, H. (1959). On a successive transformation of probability distribution and its application to the analysis of the optimum gradient method. *Annals of the Institute of Statistical Mathematics Tokyo, 11*(1), 1–16.

Al-Baali, M. (1985). Descent property and global convergence of the Fletcher-Reeves method with inexact line search. *IMA Journal of Numerical Analysis, 5,* 121–124.

Al-Baali, M. (1998). Numerical experience with a class of self-scaling quasi-Newton algorithms. *Journal of Optimization Theory and Applications, 96,* 533–553.

Al-Baali, M., & Fletcher, R. (1984). An efficient line search for nonlinear least squares. *Journal of Optimization Theory and Applications, 48,* 359–377.

Al-Baali, M., & Grandinetti, L. (2009). On practical modifications of the quasi-Newton BFGS method. *AMO-Advanced Modeling and Optimization, 11*(1), 63–76.

Al-Baali, M., Narushima, Y., & Yabe, H. (2015). A family of three-term conjugate gradient methods with sufficient descent property for unconstrained optimization. *Computational Optimization and Applications, 60,* 89–110.

Al-Bayati, A. Y., & Sharif, W. H. (2010). A new three-term conjugate gradient method for unconstrained optimization. *Canadian Journal on Science and Engineering Mathematics, 1*(5), 108–124.

Andrei, N. (1995). *Computational experience with conjugate gradient algorithms for large-scale unconstrained optimization* (Technical Report). Research Institute for Informatics, Bucharest, July 21, 1–14.

Andrei, N. (1999). *Programarea Matematică Avansată. Teorie, Metode Computaţionale, Aplicaţii* [*Advanced mathematical programming. Theory, computational methods, applications*]. Bucureşti: Editura Tehnică.

Andrei, N. (2000). *Optimizare fără Restricţii—Metode de direcţii conjugate* [*Unconstrained optimization—Conjugate direction methods*]. Bucharest: MATRIXROM Publishing House.

Andrei, N. (2004). *A new gradient descent method for unconstrained optimization* (Technical Report). Research Institute for Informatics, Bucharest, March 2004.

Andrei, N. (2006a). An acceleration of gradient descent algorithm with backtracking for unconstrained optimization. *Numerical Algorithms, 42*(1), 63–73.

Andrei, N. (2006b). Performance of conjugate gradient algorithms on some MINPACK-2 unconstrained optimization applications. *Studies in Informatics and Control, 15*(2), 145–168.

N. Andrei, *Nonlinear Conjugate Gradient Methods for Unconstrained Optimization*,
Springer Optimization and Its Applications 158,
https://doi.org/10.1007/978-3-030-42950-8

Andrei, N. (2007a). Scaled conjugate gradient algorithms for unconstrained optimization. *Computational Optimization and Applications, 38*(3), 401–416.

Andrei, N. (2007b). A Scaled BFGS preconditioned conjugate gradient algorithm for unconstrained optimization. *Applied Mathematics Letters, 20,* 645–650.

Andrei, N. (2007c). Scaled memoryless BFGS preconditioned conjugate gradient algorithm for unconstrained optimization. *Optimization Methods and Software, 22*(4), 561–571.

Andrei, N. (2007d). Numerical comparison of conjugate gradient algorithms for unconstrained optimization. *Studies in Informatics and Control, 16*(4), 333–352.

Andrei, N. (2007e). *CGALL—Conjugate gradient algorithms for unconstrained optimization* (Technical Report No. 16). Research Institute for Informatics, Bucharest, March 5, 2007.

Andrei, N. (2007f). *SCALCG—Scaled conjugate gradient algorithms for unconstrained optimization* (Technical Report No. 17). Research Institute for Informatics, Bucharest, March 30, 2007.

Andrei, N. (2008a). A scaled nonlinear conjugate gradient algorithm for unconstrained optimization. *Optimization, 57*(4), 549–570.

Andrei, N. (2008b). Another hybrid conjugate gradient algorithm for unconstrained optimization. *Numerical Algorithms, 47,* 143–156.

Andrei, N. (2008c). A Dai-Yuan conjugate gradient algorithm with sufficient descent and conjugacy condition for unconstrained optimization. *Applied Mathematics Letters, 21*(2), 165–171.

Andrei, N. (2008d). New hybrid conjugate gradient algorithms for unconstrained optimization. In C.A. Floudas & P. Pardalos (Eds.), *Encyclopedia of optimization* (2nd ed., pp. 2560–2571). New York: Springer Science + Business Media.

Andrei, N. (2008e). Performance profiles of conjugate gradient algorithms for unconstrained optimization. In C. A. Floudas & P. Pardalos (Eds.), *Encyclopedia of optimization* (2nd ed., pp. 2938–2953). New York: Springer Science + Business Media.

Andrei, N. (2008f). *40 conjugate gradient algorithms for unconstrained optimization—A survey on their definition* (Technical Report). Research Institute for Informatics-ICI, Bucharest, August 13, 2008.

Andrei, N. (2008g). A hybrid conjugate gradient algorithm for unconstrained optimization as a convex combination of Hestenes-Stiefel and Dai-Yuan. *Studies in Informatics and Control, 17* (1), 55–70.

Andrei, N. (2008h). *Computational experience with L-BFGS—A limited memory BFGS quasi-Newton method for unconstrained optimization* (Technical Report No. 32). Research Institute for Informatics-ICI, Bucharest, October 3–14, 2008.

Andrei, N. (2008i). *HYBRID, HYBRIDM, AHYBRIDM—Conjugate gradient algorithms for unconstrained optimization* (Technical Report No. 35). Research Institute for Informatics-ICI, Bucharest, October 20, 2008.

Andrei, N. (2009a). Hybrid conjugate gradient algorithm for unconstrained optimization. *Journal of Optimization Theory and Applications, 141*(2), 249–264.

Andrei, N. (2009b). Another nonlinear conjugate gradient algorithm for unconstrained optimization. *Optimization Methods and Software, 24*(1), 89–104.

Andrei, N. (2009c). Acceleration of conjugate gradient algorithms for unconstrained optimization. *Applied Mathematics and Computation, 213*(2), 361–369.

Andrei, N. (2009d). Accelerated conjugate gradient algorithm with finite difference Hessian/vector product approximation for unconstrained optimization. *Journal of Computational and Applied Mathematics, 230,* 570–582.

Andrei, N. (2009e). *Critica Raţiunii Algoritmilor de Optimizare fără Restricţii* [*Criticism of the unconstrained optimization algorithms reasoning*]. Bucureşti: Editura Academiei Române.

Andrei, N. (2009f). *Metode Avansate de Gradient Conjugat pentru Optimizare fără Restricţii* [*Advanced conjugate gradient methods for unconstrained optimization*]. Bucureşti: Editura Academiei Oamenilor de Ştiinţă din România.

Andrei, N. (2009g). *ASCALCG—Accelerated scaled memoryless BFGS preconditioned conjugate gradient algorithm for unconstrained optimization* (Technical Report No. 1). Research Institute for Informatics, Bucharest, January 5, 2009.

Andrei, N. (2009h). *CGSYS—Accelerated conjugate gradient algorithm with guaranteed descent and conjugacy conditions for unconstrained optimization* (Technical Report No. 34). Research Institute for Informatics, Bucharest, June 4, 2009.

Andrei, N. (2009i). Accelerated conjugate gradient algorithm with modified secant condition for unconstrained optimization. *Studies in Informatics and Control, 18*(3), 211–232.

Andrei, N. (2010a). Accelerated hybrid conjugate gradient algorithm with modified secant condition for unconstrained optimization. *Numerical Algorithms, 54,* 23–46.

Andrei, N. (2010b). Accelerated scaled memoryless BFGS preconditioned conjugate gradient algorithm for unconstrained optimization. *European Journal of Operational Research, 204,* 410–420.

Andrei, N. (2010c). New accelerated conjugate gradient algorithms as a modification of Dai-Yuan's computational scheme for unconstrained optimization. *Journal of Computational and Applied Mathematics, 234,* 3397–3410.

Andrei, N. (2011a). A modified Polak-Ribiere-Polyak conjugate gradient algorithm for unconstrained optimization. *Optimization, 60*(12), 1457–1471.

Andrei, N. (2011b). Open problems in conjugate gradient algorithms for unconstrained optimization. *Bulletin of the Malaysian Mathematical Sciences Society, 34*(2), 319–330.

Andrei, N. (2012). An accelerated conjugate gradient algorithm with guaranteed descent and conjugacy conditions for unconstrained optimization. *Optimization Methods and Software, 27* (4–5), 583–604.

Andrei, N. (2013a). A simple three-term conjugate gradient algorithm for unconstrained optimization. *Journal of Computational and Applied Mathematics, 241,* 19–29.

Andrei, N. (2013b). On three-term conjugate gradient algorithms for unconstrained optimization. *Applied Mathematics and Computation, 219,* 6316–6327.

Andrei, N. (2013c). Another conjugate gradient algorithm with guaranteed descent and conjugacy conditions for large-scale unconstrained optimization. *Journal of Optimization Theory and Applications, 159,* 159–182.

Andrei, N. (2013d). A numerical study on efficiency and robustness of some conjugate gradient algorithms for large-scale unconstrained optimization. *Studies in Informatics and Control, 22* (4), 259–284.

Andrei, N. (2013e). *Nonlinear optimization applications using the GAMS technology. Springer Optimization and Its Applications Series* (Vol. 81). New York, NY, USA: Springer Science + Business Media.

Andrei, N. (2014). An accelerated subspace minimization three-term conjugate gradient algorithm for unconstrained optimization. *Numerical Algorithms, 65*(4), 859–874.

Andrei, N. (2015a). A new three-term conjugate gradient algorithm for unconstrained optimization. *Numerical Algorithms, 68*(2), 305–321.

Andrei, N. (2015b). *Critica Raţiunii Algoritmilor de Optimizare cu Restricţii* [*Criticism of the constrained optimization algorithms reasoning*]. Bucureşti: Editura Academiei Române.

Andrei, N. (2016). An adaptive conjugate gradient algorithm for large-scale unconstrained optimization. *Journal of Computational and Applied Mathematics, 292,* 83–91.

Andrei, N. (2017a). Eigenvalues versus singular values study in conjugate gradient algorithms for large-scale unconstrained optimization. *Optimization Methods and Software, 32*(3), 534–551.

Andrei, N. (2017b). Accelerated adaptive Perry conjugate gradient algorithms based on the self-scaling memoryless BFGS update. *Journal of Computational and Applied Mathematics, 325,* 149–164.

Andrei, N. (2017c). *Continuous nonlinear optimization for engineering applications in GAMS technology. Springer Optimization and Its Applications Series* (Vol. 121). New York, NY, USA: Springer Science + Business Media.

Andrei, N. (2018a). An adaptive scaled BFGS method for unconstrained optimization. *Numerical Algorithms, 77*(2), 413–432.

Andrei, N. (2018b). A Dai-Liao conjugate gradient algorithm with clustering the eigenvalues. *Numerical Algorithms, 77*(4), 1273–1282.

Andrei, N. (2018c). A double parameter scaled BFGS method for unconstrained optimization. *Journal of Computational and Applied Mathematics, 332,* 26–44.

Andrei, N. (2018d). A double parameter scaling Broyden-Fletcher-Goldfarb-Shanno based on minimizing the measure function of Byrd and Nocedal for unconstrained optimization. *Journal of Optimization Theory and Applications, 178,* 191–218.

Andrei, N. (2018e). A diagonal quasi-Newton method based on minimizing the measure function of Byrd and Nocedal for unconstrained optimization. *Optimization, 67*(9), 1553–1568.

Andrei, N. (2018f). A double parameter scaled modified Broyden-Fletcher-Goldfarb-Shanno method for unconstrained optimization. *Studies in Informatics and Control, 27*(2), 135–146.

Andrei, N. (2018g). *UOP—A collection of 80 unconstrained optimization test problems* (Technical Report No. 7/2018). Research Institute for Informatics, Bucharest, Romania, November 17.

Andrei, N. (2019a). *The conjugate gradient method closest to the scaled memoryless BFGS preconditioned with standard, approximate and improved Wolfe line search* (Technical Report No. 1/2019). Academy of Romanian Scientists, Bucharest, Romania.

Andrei, N. (2019b). *Conjugate gradient algorithms closest to self-scaling memoryless BFGS method based on clustering the eigenvalues of the self-scaling memoryless BFGS iteration matrix or on minimizing the Byrd-Nocedal measure function with different Wolfe line searches for unconstrained optimization* (Technical Report No. 2/2019). Academy of Romanian Scientists, Bucharest, Romania.

Andrei, N. (2019c). A diagonal quasi-Newton updating method for unconstrained optimization. *Numerical Algorithms, 81*(2), 575–590.

Andrei, N. (2019d). A new diagonal quasi-Newton updating method with scaled forward finite differences directional derivative for unconstrained optimization. *Numerical Functional Analysis and Optimization, 40*(13), 1467–1488.

Andrei, N. (2019e). *Performances of DESCON, L-BFGS, L-CG-DESCENT and of CONOPT, KNITRO, MINOS, SNOPT, IPOPT for solving the problem PALMER1C* (Technical Report No. 3/2019). Academy of Romanian Scientists, Bucharest, Romania.

Andrei, N. (2020). New conjugate gradient algorithms based on self-scaling memoryless Broyden–Fletcher–Goldfarb–Shanno method. *Calcolo, 57,* 17. https://doi.org/10.1007/s10092-020-00365-7.

Aris, R. (1975). *The mathematical theory of diffusion and reaction in permeable catalysts*. Oxford, UK: Oxford University Press.

Armijo, L. (1966). Minimization of functions having Lipschitz continuous first partial derivatives. *Pacific Journal of Mathematics, 16*(1), 1–3.

Arnold, D. N. (2001). *A concise introduction to numerical analysis. Lecture Notes*. Pennsylvania State University, MATH 5971-Numerical Analysis, Fall 2001.

Arzam, M. R., Babaie-Kafaki, S., & Ghanbari, R. (2017). An extended Dai-Liao conjugate gradient method with global convergence for nonconvex functions. *Glasnik Matematicki, 52* (72), 361–375.

Averick, B. M., Carter, R. G., & Moré, J. J. (1991). *The MINPACK-2 test problem collection (preliminary version)* (Technical Memorandum No. 150). Mathematics and Computer Science Division, Argonne National Laboratory, 9700 South Cass Avenue, Argonne, Illinois, May 1991.

Averick, B. M., Carter, R. G., Moré, J. J., & Xue, G. L. (1992). *The MINPACK-2 test problem collection*. Mathematics and Computer Science Division, Argonne National Laboratory, 9700 South Cass Avenue, Argonne, Illinois, Preprint MCS-P153-6092, June 1992.

Axelsson, O. (1980). Conjugate gradient type methods for unsymmetric and inconsistent systems of linear equations. *Linear Algebra and Its Applications, 29,* 1–16.

Axelsson, O. (1994). *Iterative solution methods*. Cambridge: Cambridge University Press.

Axelsson, O., & Barker, V. A. (2001). *Finite element solution of boundary value problems. Classics in Applied Mathematics* (Vol. 35). Philadelphia, PA, USA: SIAM.

Axelsson, O., & Lindskog, G. (1986). On the rate of convergence of the preconditioned conjugate gradient method. *Numerische Mathematik, 48,* 499–523.

Babaie-Kafaki, S. (2011). A modified BFGS algorithm based on a hybrid secant equation. *Science China Mathematics, 54*(9), 2019–2036.

Babaie-Kafaki, S. (2012). A note on the global convergence theorem of the scaled conjugate gradient algorithms proposed by Andrei. *Computational Optimization and Applications, 52*(2), 409–414.

Babaie-Kafaki, S. (2013). A modified scaled memoryless BFGS preconditioned conjugate gradient method for unconstrained optimization. *4OR, 11*(4), 361–374.

Babaie-Kafaki, S. (2014). Two modified scaled nonlinear conjugate gradient methods. *Journal of Computational and Applied Mathematics, 261*(5), 172–182.

Babaie-Kafaki, S. (2015). On optimality of the parameters of self-scaling memoryless quasi-Newton updating formulae. *Journal of Optimization Theory and Applications, 167*(1), 91–101.

Babaie-Kafaki, S. (2016). Computational approaches in large-scale unconstrained optimization. In A. Emrouznejad (Ed.), *Big data optimization: Recent developments and challenges. Studies in Big Data* (Vol. 18, pp. 391–417).

Babaie-Kafaki, S., Fatemi, M., & Mahdavi-Amiri, N. (2011). Two effective hybrid conjugate gradient algorithms on modified BFGS updates. *Numerical Algorithms, 58,* 315–331.

Babaie-Kafaki, S., & Ghanbari, R. (2014a). A modified scaled conjugate gradient method with global convergence for nonconvex functions. *Bulletin of the Belgian Mathematical Society Simon Stevin, 21*(3), 465–477.

Babaie-Kafaki, S., & Ghanbari, R. (2014b). The Dai-Liao nonlinear conjugate gradient method with optimal parameter choices. *European Journal of Operational Research, 234,* 625–630.

Babaie-Kafaki, S., & Ghanbari, R. (2015a). A hybridization of the Hestenes-Stiefel and Dai-Yuan conjugate gradient methods based on a least-squares approach. *Optimization Methods and Software, 30*(4), 673–681.

Babaie-Kafaki, S., & Ghanbari, R. (2015b). A hybridization of the Polak-Ribière-Polyak and Fletcher-Reeves conjugate gradient methods. *Numerical Algorithms, 68*(3), 481–495.

Babaie-Kafaki, S., Ghanbari, R., & Mahdavi-Amiri, N. (2010). Two new conjugate gradient methods based on modified secant equations. *Journal of Computational and Applied Mathematics, 234*(5), 1374–1386.

Babaie-Kafaki, S., & Mahdavi-Amiri, N. (2013). Two modified hybrid conjugate gradient methods based on a hybrid secant equation. *Mathematical Modelling and Analysis, 18*(1), 32–52.

Babaie-Kafaki, S., & Rezaee, S. (2018). Two accelerated nonmonotone adaptive trust region line search methods. *Numerical Algorithms, 78*(3), 911–928.

Baluch, B., Salleh, Z., & Alhawarat, A. (2018). A new modified three-term Hestenes–Stiefel conjugate gradient method with sufficient descent property and its global convergence. *Journal of Optimization, 2018,* 13, Article ID 5057096. https://doi.org/10.1155/2018/5057096.

Baptist, P., & Stoer, J. (1977). On the relation between quadratic termination and convergence properties of minimization algorithms. Part II, Applications. *Numerische Mathematik, 28,* 367–392.

Bartholomew-Biggs, M. (2008). *Nonlinear optimization with engineering applications*. New York, NY, USA: Springer Science + Business Media.

Barzilai, J., & Borwein, J. M. (1988). Two-points step size gradient methods. *IMA Journal of Numerical Analysis, 8,* 141–148.

Bazaraa, M. S., Sherali, H. D., & Shetty, C. M. (1993). *Nonlinear programming theory and algorithms* (2nd ed.). New York: Wiley.

Beale, E. M. L. (1972). A derivation of conjugate gradients. In F. A. Lotsma (Ed.), *Numerical methods for nonlinear optimization* (pp. 39–43). New-York: Academic Press.

Bebernes, J., & Eberly, D. (1989). *Mathematical problems from combustion theory. Applied Mathematical Sciences* (Vol. 83). Berlin: Springer.

Bellavia, S., & Morini, B. (2006). Subspace trust-region methods for large bound-constrained nonlinear equations. *SIAM Journal on Numerical Analysis, 44*(4), 1535–1555.

Bellavia, S., & Morini, B. (2015). Strong local convergence properties of adaptive regularized methods for nonlinear least squares. *IMA Journal of Numerical Analysis, 35*(2), 947–968.

Benson, H. Y., & Shanno, D. F. (2014). Interior-point methods for nonconvex nonlinear programming: Cubic regularization. *Computational Optimization and Applications, 58*(2), 323–346.

Benson, H. Y., & Shanno, D. F. (2018). Cubic regularization in symmetric rank-1 quasi-Newton methods. *Mathematical Programming Computation, 10,* 457–486.

Bertsekas, D. P. (1999). *Nonlinear programming* (2nd ed.). Belmont, MA: Athena Scientific.

Bianconcini, T., Liuzzi, G., Morini, B., & Sciandrone, M. (2013). On the use of iterative methods in cubic regularization for unconstrained optimization. *Computational Optimization and Applications, 60*(1), 35–57.

Bianconcini, T., & Sciandrone, M. (2016). A cubic regularization algorithm for unconstrained optimization using line search and nonmonotone techniques. *Optimization Methods and Software, 31,* 1008–1035.

Biggs, M. C. (1971). Minimization algorithms making use of non-quadratic properties of the objective function. *Journal of the Institute of Mathematics and Its Applications, 8,* 315–327.

Biggs, M. C. (1973). A note on minimization algorithms making use of non-quadratic properties of the objective function. *Journal of the Institute of Mathematics and Its Applications, 12,* 337–338.

Birgin, E., & Martínez, J. M. (2001). A spectral conjugate gradient method for unconstrained optimization. *Applied Mathematics & Optimization, 43*(2), 117–128.

Boggs, P. T., & Tolle, J. W. (1994). Convergence properties of a class of rank-two updates. *SIAM Journal on Optimization, 4,* 262–287.

Bongartz, I., Conn, A. R., Gould, N. I. M., & Toint, Ph. L. (1995). CUTE: Constrained and unconstrained testing environments. *ACM Transactions on Mathematical Software, 21,* 123–160.

Branch, M. A., Coleman, T. F., & Li, Y. (1999). A subspace, interior, and conjugate gradient method for large-scale bound-constrained minimization problems. *SIAM Journal on Scientific Computing, 21,* 1–23.

Broyden, C. G. (1970). The convergence of a class of double-rank minimization algorithms. I. General considerations. *Journal of the Institute of Mathematics and Its Applications, 6,* 76–90.

Brune, P. R., Knepley, M. G., Smith, B. F., & Tu, X. (2015). Composing scalable nonlinear algebraic solvers. *SIAM Review, 57*(4), 535–565.

Buckley, A. G. (1978a). Extending the relationship between the conjugate gradient and BFGS algorithms. *Mathematical Programming, 15*(1), 343–348.

Buckley, A. G. (1978b). A combined conjugate-gradient quasi-Newton minimization algorithm. *Mathematical Programming, 15,* 200–210.

Buckley, A. G., & LeNir, A. (1983). QN-like variable storage conjugate gradients. *Mathematical Programming, 27*(2), 155–175.

Bulirsch, R., & Stoer, J. (1980). *Introduction to numerical analysis*. New York: Springer.

Burmeister, W. (1973). Die Konvergenzordnung des Fletcher-Powell Algorithmus. *Zeitschrift für Angewandte Mathematik und Mechanik, 53,* 693–699.

Byrd, R. H., Liu, D. C., & Nocedal, J. (1992). On the behavior of Broyden's class of quasi-Newton methods. *SIAM Journal on Optimization, 2,* 533–557.

Byrd, R. H., & Nocedal, J. (1989). A tool for the analysis of quasi-Newton methods with application to unconstrained minimization. *SIAM Journal on Numerical Analysis, 26,* 727–739.

Byrd, R. H., Nocedal, J., & Yuan, Y. (1987). Global convergence of a class of quasi-Newton methods on convex problems. *SIAM Journal on Numerical Analysis, 24,* 1171–1190.

Byrd, R. H., Schnabel, R. B., & Schultz, G. A. (1985). A family of trust-region-based algorithms for unconstrained minimization with strong global convergence properties. *SIAM Journal on Numerical Analysis, 22,* 47–67.

Byrd, R. H., Schnabel, R. B., & Schultz, G. A. (1988). Approximate solution of the trust-region problem by minimization over two-dimensional subspace. *Mathematical Programming, 40,* 247–263.

Caliciotti, A., Fasano, G., & Roma, M. (2017). Novel preconditioners based on quasi-Newton updates for nonlinear conjugate gradient methods. *Optimization Letters, 11*(4), 835–853.

Caliciotti, A., Fasano, G., & Roma, M. (2018). Preconditioned nonlinear conjugate gradient methods based on a modified secant equation. *Applied Mathematics and Computation, 318*(1), 196–214.

Carlberg, K., Forstall, V., & Tuminaro, R. (2016). Krylov-subspace recycling via the POD-augmented conjugate gradient method. *SIAM Journal on Matrix Analysis and Applications, 37,* 1304–1336.

Cartis, C., Gould, N. I. M., & Toint, Ph. L. (2011a). Adaptive cubic overestimation methods for unconstrained optimization. Part I: Motivation, convergence and numerical results. *Mathematical Programming Series A, 127,* 245–295.

Cartis, C., Gould, N. I. M., & Toint, Ph. L. (2011b). Adaptive cubic overestimation methods for unconstrained optimization. Part II: Worst-case function-evaluation complexity. *Mathematical Programming Series A, 130,* 295–319.

Cătinaş, E. (2019). A survey on the high convergence orders and computational convergence orders of sequences. *Applied Mathematics and Computation, 343,* 1–20.

Cauchy, A. (1847). Méthodes générales pour la resolution des systèmes déquations simultanées. *Comptes Rendus de l'Académie des Sciences Paris, 25*(1), 536–538.

Chachuat, B. C. (2007). *Nonlinear and dynamic optimization—From theory to practice.* IC-31: Winter Semester 2006/2007. École Politechnique Fédérale de Lausanne.

Cheng, W. Y. (2007). A two-term PRP-based descent method. *Numerical Functional Analysis and Optimization, 28,* 1217–1230.

Cheng, W. Y., & Li, D. H. (2010). Spectral scaling BFGS method. *Journal of Optimization Theory and Applications, 146,* 305–319.

Cimatti, G., & Menchi, O. (1978). On the numerical solution of a variational inequality connected with the hydrodynamic lubrication of a complete journal bearing. *Calcolo, 15,* 249–258.

Cohen, A. (1972). Rate of convergence of several conjugate gradient algorithms. *SIAM Journal on Numerical Analysis, 9,* 248–259.

Conn, A. R., Gould, N. I. M., & Toint, P. L. (1988). Testing a class of algorithms for solving minimization problems with simple bounds on the variables. *Mathematics of Computation, 50,* 399–430.

Concus, P., & Golub, G. H. (1976). *A generalized conjugate gradient method for nonsymmetric systems of linear equation.* Preprint for *Lecture Notes in Economic and Mathematical Systems* (Vol. 134, pp. 56–65). Berlin: Springer.

Conn, A. R., Gould, N. I. M., Sartenaer, A., & Toint, Ph. L. (1996). On iterated-subspace minimization methods for nonlinear optimization. In L. Adams & J. L. Nazareth (Eds.), *Linear and nonlinear conjugate gradient related methods* (pp. 50–78). Philadelphia, PA, USA: SIAM.

Conn, A. R., Gould, N. I. M., & Toint, Ph. L. (2000). *Trust-region methods. MPS-SIAM Series on Optimization.* Philadelphia, PA, USA: SIAM.

Contreras, M., & Tapia, R. A. (1993). Sizing the BFGS and DFP updates: A numerical study. *Journal of Optimization Theory and Applications, 78,* 93–108.

Crowder, H. P., & Wolfe, P. (1969). Linear convergence of the conjugate gradient method. *IBM Journal of Research & Development*, 431–433.

Dai, Y. H. (1997). *Analyses of conjugate gradient methods* (Ph.D. thesis). Institute of Computational Mathematics and Scientific/Engineering Computing, Chinese Academy of Sciences.

Dai, Y. H. (2001). New properties of a nonlinear conjugate gradient method. *Numerische Mathematik, 89,* 83–98.

Dai, Y. H. (2002a). A nonmonotone conjugate gradient algorithm for unconstrained optimization. *Journal of Systems Science and Complexity, 15*(2), 139–145.

Dai, Y. H. (2002b). On the nonmonotone line search. *Journal of Optimization Theory and Applications, 112,* 315–330.

Dai, Y. H. (2003a). Convergence properties of the BFGS algorithm. *SIAM Journal on Optimization, 13,* 693–701.

Dai, Y. H. (2003b). A family of hybrid conjugate gradient methods for unconstrained optimization. *Mathematics of Computation, 72*(243), 1317–1328.

Dai, Y. H. (2010). Convergence analysis of nonlinear conjugate gradient methods. In Y. Wang, A. G. Yagola, & C. Yang (Eds.), *Optimization and regularization for computational inverse problems and applications* (Chapter 8, pp. 157–181). Beijing: Higher Education Press; Berlin, Heidelberg: Springer.

Dai, Y. H. (2011). *Nonlinear conjugate gradient methods*. Wiley Encyclopedia of Operations Research and Management Science. https://doi.org/10.1002/9780470400531.eorms0183. Published Online, February 15, 2011.

Dai, Y. H., Hager, W. W., Schittkowski, K., & Zhang, H. (2006). The cyclic Barzilai-Borwein method for unconstrained optimization. *IMA Journal of Numerical Analysis, 26,* 604–627.

Dai, Y. H., Han, J. Y., Liu, G. H., Sun, D. F., Yin, H. X., & Yuan, Y. X. (1999). Convergence properties of nonlinear conjugate gradient methods. *SIAM Journal on Optimization, 10*(2), 345–358.

Dai, Y. H., & Kou, C. X. (2013). A nonlinear conjugate gradient algorithm with an optimal property and an improved Wolfe line search. *SIAM Journal on Optimization, 23*(1), 296–320.

Dai, Y. H., & Kou, C. X. (2016). A Barzilai-Borwein conjugate gradient method. *Science China Mathematics, 59*(8), 1511–1524.

Dai, Y. H., & Liao, L. Z. (2001). New conjugate conditions and related nonlinear conjugate gradient methods. *Applied Mathematics & Optimization, 43,* 87–101.

Dai, Y. H., & Liao, L. Z. (2002). R-linear convergence of the Barzilai and Borwein gradient method. *IMA Journal of Numerical Analysis, 22*(1), 1–10.

Dai, Y. H., Liao, L. Z., & Li, D. (2004). On restart procedures for the conjugate gradient method. *Numerical Algorithms, 35,* 249–260.

Dai, Y. H., & Ni, Q. (2003). Testing different conjugate gradient methods for large-scale unconstrained optimization. *Journal of Computational Mathematics, 22*(3), 311–320.

Dai, Y. H., & Yuan, Y. X. (1996a). Convergence properties of the Fletcher-Reeves method. *IMA Journal of Numerical Analysis, 16,* 155–164.

Dai, Y. H., & Yuan, Y. X. (1996b). Convergence of the Fletcher-Reeves method under a generalized Wolfe search. *Journal of Computational Mathematics, 2,* 142–148.

Dai, Y. H., & Yuan, Y. X. (1996c). Convergence properties of the conjugate descent method. *Advances in Mathematics (China), 26,* 552–562.

Dai, Y. H., & Yuan, Y. X. (1998). Convergence properties of the Beale-Powell restart algorithm. *Sciences in China (Series A), 41*(11), 1142–1150.

Dai, Y. H., & Yuan, Y. (1999). A nonlinear conjugate gradient method with strong global convergence property. *SIAM Journal on Optimization, 10,* 177–182.

Dai, Y. H., & Yuan, Y. (2000). *Nonlinear conjugate gradient methods*. Shanghai, China: Shanghai Science and Technology Publisher.

Dai, Y. H., & Yuan, Y. (2001a). An efficient hybrid conjugate gradient method for unconstrained optimization. *Annals of Operations Research, 103,* 33–47.

Dai, Y. H., & Yuan, Y. (2001b). A three-parameter family of hybrid conjugate gradient method. *Mathematics of Computation, 70,* 1155–1167.

Dai, Y. H., & Yuan, Y. (2002). Modified two-point stepsize gradient methods for unconstrained optimization. *Computational Optimization and Applications, 22,* 103–109.

Dai, Y. H., & Yuan, Y. (2003). A class of globally convergent conjugate gradient methods. *Science China Mathematics Series A, 46*(2), 251–261.

Dai, Y. H., & Zhang, H. (2001). An adaptive two-point stepsize gradient algorithm. *Numerical Algorithms, 27,* 377–385.

Dai, Z., & Wen, F. (2012). Another improved Wei-Yao-Liu nonlinear conjugate gradient method with sufficient descent property. *Applied Mathematics and Computation, 218,* 7421–7430.

Daniel, J. W. (1967). The conjugate gradient method for linear and nonlinear operator equations. *SIAM Journal on Numerical Analysis, 4,* 10–26.

Davidon, W. C. (1959). *Variable metric method for minimization* (Research and Development Report ANL-5990). Argonne National Laboratories.

Davidon, W. C. (1980). Conic approximation and collinear scalings for optimizers. *SIAM Journal on Numerical Analysis, 17*(2), 268–281.

Dehmiry, A. H. (2019). The global convergence of the BFGS method under a modified Yuan-Wei-Lu line search technique. *Numerical Algorithms.* https://doi.org/10.1007/s11075-019-00779-7.

Dembo, R. S., Eisenstat, S. C., & Steihaug, T. (1982). Inexact Newton methods. *SIAM Journal on Numerical Analysis, 19,* 400–408.

Dembo, R. S., & Steihaug, T. (1983). Truncated Newton algorithms for large-scale unconstrained optimization. *Mathematical Programming, 26,* 190–212.

Demmel, J. W. (1997). *Applied numerical linear algebra.* Philadelphia, PA, USA: SIAM.

Dener, A., Denchfield, A., & Munson, T. (2019). *Preconditioning nonlinear conjugate gradient with diagonalized quasi-Newton.* Mathematics and Computer Science Division, Preprint ANL/MCS-P9152-0119, January 2019. Argonne National Laboratory, 9700 South Cass Avenue, Argonne, Illinois 60439.

Deng, N. Y., & Li, Z. (1995). Global convergence of three terms conjugate gradient methods. *Optimization Methods and Software, 4,* 273–282.

Dennis, J. E., & Moré, J. J. (1974). A characterization of superlinear convergence and its application to quasi-Newton methods. *Mathematics of Computation, 28*(126), 549–560.

Dennis, J. E., & Moré, J. J. (1977). Quasi-Newton methods, motivation and theory. *SIAM Review, 19*(1), 46–89.

Dennis, J. E., & Schnabel, R. B. (1981). A new derivation of symmetric positive definite secant updates. In *Nonlinear programming* (Vol. 4, pp. 167–199). Cambridge, MA: Academic Press.

Dennis, J. E., & Schnabel, R. B. (1983). *Numerical methods for unconstrained optimization and nonlinear equations.* Englewood Cliffs, NJ: Prentice-Hall. Reprinted as *Classics in applied mathematics* (Vol. 16). Philadelphia, USA: SIAM.

Dennis, J. E., & Schnabel, R. B. (1989). A view of unconstrained optimization. In *Optimization. Handbooks in Operations Research and Management* (Vol. 1, pp. 1–72). Amsterdam, The Netherlands: Elsevier Science Publisher.

Dennis, J. E., & Wolkowicz, H. (1993). Sizing and least-change secant methods. *SIAM Journal on Numerical Analysis, 30*(5), 1291–1314.

Deuflhard, P. (1990). Global inexact Newton methods for very large scale nonlinear problems. In *Proceedings of the Cooper Mountain Conference on Iterative Methods*, Cooper Mountain, Colorado, April 1–5.

Dolan, E. D., & Moré, J. J. (2002). Benchmarking optimization software with performance profiles. *Mathematical Programming, 91,* 201–213.

Dollar, H. S., Gould, N. I. M., & Robinson, D. P. (2009). *On solving trust-region and other regularised subproblems in optimization* (Technical Report 09/01). Oxford University Computing Laboratory, Numerical Analysis Group.

Elliott, C. M., & Ockendon, J. R. (1982). *Weak and variational methods for moving boundary problems. Research Notes in Mathematics* (Vol. 50). Pittman.

Epanomeritakis, I., Akçelik, V., Ghattas, O., & Bielak, J. (2008). A Newton-CG method for large-scale three-dimensional elastic full-waveform seismic inversion. *Inverse Problems, 24*(3), 26, Article id. 034015.

Erway, J. B., & Gill, P. E. (2009). A subspace minimization method for the trust-region step. *SIAM Journal on Optimization, 20,* 1439–1461.

Fatemi, M. (2016a). An optimal parameter for Dai-Liao family of conjugate gradient methods. *Journal of Optimization Theory and Applications, 169,* 587–605.

Fatemi, M. (2016b). A new efficient conjugate gradient method for unconstrained optimization. *Journal of Computational and Applied Mathematics, 300,* 207–216.

Fatemi, M. (2017). A scaled conjugate gradient method for nonlinear unconstrained optimization. *Optimization Methods and Software, 32*(5), 1095–1112.

Feder, D. P. (1962). Automatic lens design with a high-speed computer. *Journal of the Optical Society of America, 52,* 177–183.

Fisher, M., Nocedal, J., Trémolet, Y., & Wright, S. J. (2009). Data assimilation in weather forecasting: A case study in PDE-constrained optimization. *Optimization and Engineering, 10* (3), 409–426.

Fletcher, R. (1970). A new approach to variable metric algorithms. *The Computer Journal, 13,* 317–322.

Fletcher, R. (1987). *Practical methods of optimization* (2nd ed.). New York: Wiley.

Fletcher, R. (1991). A new variational result for quasi-Newton formulae. *SIAM Journal on Optimization, 1,* 18–21.

Fletcher, R., & Powell, M. J. D. (1963). A rapidly convergent descent method for minimization. *Computer Journal*, 163–168.

Fletcher, R., & Reeves, C. M. (1964). Function minimization by conjugate gradient. *Computer Journal, 7,* 149–154.

Ford, J. A., & Moghrabi, I. A. (1994). Multi-step quasi-Newton methods for optimization. *Journal of Computational and Applied Mathematics, 50*(1–3), 305–323.

Ford, J. A., & Moghrabi, I. A. (1996a). Minimum curvature multi-step quasi-Newton methods. *Computers & Mathematics with Applications, 31*(4–5), 179–186.

Ford, J. A., & Moghrabi, I. A. (1996b). Using function-values in multi-step quasi-Newton methods. *Journal of Computational and Applied Mathematics, 66*(1–2), 201–211.

Ford, J. A., Narushima, Y., & Yabe, H. (2008). Multi-step nonlinear conjugate gradient methods for unconstrained minimization. *Computational Optimization and Applications, 40*(2), 191–216.

Forsythe, G. E., Hestenes, M. R., & Rosser, J. B. (1951). Iterative methods for solving linear equations. *The Bulletin of the American Mathematical Society, 57,* 480.

Fox, L., Huskey, H. D., & Wilkinson, J. H. (1948). Notes on the solution of algebraic linear simultaneous equations. *The Quarterly Journal of Mechanics and Applied Mathematics, 1,* 149–173.

Ge, R.-P., & Powell, M. J. D. (1983). The convergence of variable metric matrices in unconstrained optimization. *Mathematical Programming, 27,* 123–143.

Gilbert, J. C., & Lemaréchal, C. (1989). Some numerical experiments with variable-storage quasi-Newton algorithms. *Mathematical Programming, Series B, 45,* 407–435.

Gilbert, J. C., & Nocedal, J. (1992). Global convergence properties of conjugate gradient methods for optimization. *SIAM Journal on Optimization, 2,* 21–42.

Gill, P. E., & Leonard, M. W. (2001). Reduced-Hessian quasi Newton methods for unconstrained optimization. *SIAM Journal on Optimization, 12,* 209–237.

Gill, P. E., & Leonard, M. W. (2003). Limited memory reduced-Hessian methods for large-scale unconstrained optimization. *SIAM Journal on Optimization, 14,* 380–401.

Gill, P. E., & Murray, W. (1974). Newton-type methods for unconstrained and linearly constrained optimization. *Mathematical Programming, 7*(1), 311–350.

Gill, P. E., & Murray, W. (1979). *Conjugate gradient methods for large-scale nonlinear optimization* (Technical Report SOL 79-15). Department of Operations Research, Stanford University, Stanford, CA, USA.

Gill, P. E., Murray, W., & Wright, M. H. (1981). *Practical optimization*. New York: Academic Press.

Glowinski, R. (1984). *Numerical methods for nonlinear variational problems*. Berlin: Springer.
Goldfarb, D. (1970). A family of variable metric method derived by variation mean. *Mathematics of Computation, 23,* 23–26.
Goldstein, A. A. (1965). On steepest descent. *SIAM Journal on Control, 3,* 147–151.
Golub, G. H., & O'Leary, D. P. (1989). Some history of the conjugate gradient methods and Lanczos algorithms: 1948–1976. *SIAM Review, 31,* 50–100.
Golub, G. H., & Van Loan, C. G. (1996). *Matrix computation* (3rd ed.). Baltimore, MD, USA: John Hopkins University Press.
Goodman, J., Kohn, R., & Reyna, L. (1986). Numerical study of a relaxed variational problem from optimal design. *Computer Methods in Applied Mechanics and Engineering, 57,* 107–127.
Gould, N. I. M., Orban, D., & Toint, Ph. L. (2003). CUTEr: A constrained and unconstrained testing environment, revisited. *ACM Transactions on Mathematical Software, 29,* 353–372.
Gould, N. I. M., Porcelli, M., & Toint, Ph. L. (2012). Updating the regularization parameter in the adaptive cubic regularization algorithm. *Computational Optimization and Applications, 53,* 1–22.
Gould, N. I. M., Robinson, D. P., & Sue Thorne, H. (2010). On solving trust-region and other regularized subproblems in optimization. *Mathematical Programming Computation, 2*(1), 21–57.
Greenbaum, A. (1997). *Iterative methods for solving linear systems. Frontiers in Applied Mathematics*. SIAM: Philadelphia, PA, USA.
Greenbaum, A., & Strakoš, Z. (1992). Predicting the behavior of finite precision Lanczos and conjugate gradient computations. *SIAM Journal on Matrix Analysis and Applications, 13,* 121–137.
Griewank, A. (1981). *The modification of Newton's method for unconstrained optimization by bounding cubic term* (Technical Report NA/12). Department of Applied Mathematics and Theoretical Physics, University of Cambridge.
Grippo, L., Lampariello, F., & Lucidi, S. (1986). A nonmonotone line search technique for Newton's method. *SIAM Journal on Numerical Analysis, 23,* 707–716.
Grippo, L., & Lucidi, S. (1997). A globally convergent version of the Polak-Ribière conjugate gradient method. *Mathematical Programming, 78,* 375–391.
Grippo, L., & Sciandrone, M. (2002). Nonmonotone globalization techniques for the Barzilai-Borwein gradient method. *Computational Optimization and Applications, 23,* 143–169.
Gu, N. Z., & Mo, J. T. (2008). Incorporating nonmonotone strategies into the trust region method for unconstrained optimization. *Computers and Mathematics with Applications, 55,* 2158–2172.
Guo, Q., Liu, J. G., & Wang, D. H. (2008). A modified BFGS method and its superlinear convergence in nonconvex minimization with general line search rule. *Journal of Applied Mathematics and Computing, 28*(1–2), 435–446.
Hager, W. W. (1989). A derivative-free bracketing scheme for univariate minimization and the conjugate gradient method. *Computers & Mathematics with Applications, 18,* 779–795.
Hager, W. W., & Zhang, H. (2005). A new conjugate gradient method with guaranteed descent and an efficient line search. *SIAM Journal on Optimization, 16,* 170–192.
Hager, W. W., & Zhang, H. (2006a). Algorithm 851: CG-Descent, a conjugate gradient method with guaranteed descent. *ACM Transactions on Mathematical Software, 32*(1), 113–137.
Hager, W. W., & Zhang, H. (2006b). A survey of nonlinear conjugate gradient methods. *Pacific Journal of Optimization, 2*(1), 35–58.
Hager, W. W., & Zhang, H. (2013). The limited memory conjugate gradient method. *SIAM Journal on Optimization, 23,* 2150–2168.
Han, J. Y., Liu, G. H., & Yin, H. X. (1997). Convergence of Perry and Shanno's memoryless quasi-Newton method for nonconvex optimization problems. *OR Transactions, 1,* 22–28.
Han, X., Zhang, J., & Chen, J. (2017). A new hybrid conjugate gradient algorithm for unconstrained optimization. *Bulletin of Iranian Mathematical Society, 43*(6), 2067–2084.

Hestenes, M. R. (1951). Iterative methods for solving linear equations. *Journal on Optimization Theory and Applications, 11,* 323–334.

Hestenes, M. R. (1955). Iterative computational methods. *Communications on Pure and Applied Mathematics, 8,* 85–96.

Hestenes, M. R. (1956a). The conjugate-gradient method for solving linear systems. In *Proceedings of the Sixth Symposium in Applied Mathematics 1953* (pp. 83–102). New York: McGraw-Hill.

Hestenes, M. R. (1956b). Hilbert space methods in variational theory and numerical analysis. In *Proceedings of the International Congress of Mathematicians 1954* (pp. 229–236), North-Holland, Amsterdam.

Hestenes, M. R. (1980). *Conjugate-gradient methods in optimization*. Berlin: Springer.

Hestenes, M. R., & Stiefel, E. (1952). Methods of conjugate gradients for solving linear systems. *Journal of Research of the National Bureau of Standards, 49,* 409–436.

Hsia, Y., Sheu, R. L., & Yuan, Y. X. (2017). Theory and application of p-regularized subproblems for $p > 2$. *Optimization Methods & Software, 32*(5), 1059–1077.

Hu, Y. F., & Storey, C. (1991). Global convergence result for conjugate gradient methods. *Journal of Optimization Theory and Applications, 71,* 399–405.

Huang, S., Wan, Z., & Chen, X. (2014). A new nonmonotone line search technique for unconstrained optimization. *Numerical Algorithms, 68,* 671–689.

Huang, H., Wei, Z., & Yao, S. (2007). The proof of the sufficient descent condition of the Wei-Yao-Liu conjugate gradient method under the strong Wolfe-Powell line search. *Applied Mathematics and Computation, 189,* 1241–1245.

Jiao, B. C., Chen, L. P., & Pan, C. Y. (2007). Convergence properties of a hybrid conjugate gradient method with Goldstein line search. *Mathematica Numerica Sinica, 29*(2), 137–146.

Jian, J., Han, L., & Jiang, X. (2015). A hybrid conjugate gradient method with descent property for unconstrained optimization. *Applied Mathematics and Computation, 39*(3–4), 1281–1290.

Kaporin, I. E. (1994). New convergence results and preconditioning strategies for the conjugate gradient methods. *Numerical Linear Algebra with Applications, 1,* 179–210.

Karimi. S. (2013). *On the relationship between conjugate gradient and optimal first-order methods for convex optimization* (Ph.D. thesis). University of Waterloo. Ontario, Canada.

Karimi, S., & Vavasis, S. (2012). *Conjugate gradient with subspace optimization*. Available from: http://arxiv.org/abs/1202.1479v1.

Kelley, C. T. (1995). *Iterative methods for linear and nonlinear equations. Frontiers in Applied Mathematics*. Philadelphia, PA, USA: SIAM.

Kelley, C. T. (1999). *Iterative methods for optimization. Frontiers in Applied Mathematics*. Philadelphia, PA, USA: SIAM.

Kou, C. X. (2014). An improved nonlinear conjugate gradient method with an optimal property. *Science China—Mathematics, 57*(3), 635–648.

Kou, C. X., & Dai, Y. H. (2015). A modified self-scaling memoryless Broyden-Fletcher-Goldfarb-Shanno method for unconstrained optimization. *Journal of Optimization Theory and Applications, 165,* 209–224.

Kratzer, D., Parter, S. V., & Steuerwalt, M. (1983). Block splittings for the conjugate gradient method. *Computers & Fluids, 11,* 255–279.

Lanczos, C. (1950). An iteration method for the solution of the eigenvalue problem of linear differential and integral operators. *Journal of Research of the National Bureau of Standards, 45,* 252–282.

Lanczos, C. (1952). Solution of systems of linear equations by minimized iterations. *Journal of Research of the National Bureau of Standards, 49,* 33–53.

Laub, A. J. (2005). *Matrix analysis for scientists & engineers*. Philadelphia, PA, USA: SIAM.

Lemaréchal, C. (1981). A view of line search. In A. Auslander, W. Oettli, & J. Stoer (Eds.), *Optimization and optimal control* (pp. 59–78). Berlin: Springer.

Leong, W. J., Farid, M., & Hassan, M. A. (2010). Improved Hessian approximation with modified quasi-Cauchy relation for a gradient-type method. *AMO—Advanced Modeling and Optimization, 12*(1), 37–44.

Leong, W. J., Farid, M., & Hassan, M. A. (2012). Scaling on diagonal quasi-Newton update for large-scale unconstrained optimization. *Bulletin of Malaysian Mathematical Sciences Society, 35*(2), 247–256.

Li, D. H., & Fukushima, M. (2001a). A modified BFGS method and its global convergence in nonconvex minimization. *Journal of Computational and Applied Mathematics, 129*(1–2), 15–35.

Li, D. H., & Fukushima, M. (2001b). On the global convergence of the BFGS method for nonconvex unconstrained optimization problems. *SIAM Journal on Optimization, 11*(4), 1054–1064.

Li, G., Tang, C., & Wei, Z. (2007). New conjugacy condition and related new conjugate gradient methods for unconstrained optimization. *Journal of Computational and Applied Mathematics, 202,* 523–539.

Li, M., Liu, H., & Liu, Z. (2018). A new subspace minimization conjugate gradient algorithm with nonmonotone line search for unconstrained optimization. *Numerical Algorithms, 79*(1), 195–219.

Li, Y., Liu, Z., & Liu, H. (2019). A subspace minimization conjugate gradient method based on conic model for unconstrained optimization. *Computational and Applied Mathematics, 38,* 16. https://doi.org/10.1007/s40314-019-0779-7.

Liao, A. (1997). Modifying BFGS method. *Operations Research Letters, 20,* 171–177.

Lin, Y., & Cryer, C. W. (1985). An alternating direction implicit algorithm for the solution of linear complmentarity problems arising from free boundary problems. *Applied Mathematics & Optimization, 13,* 1–7.

Liu, D. C., & Nocedal, J. (1989). On the limited-memory BFGS method for large optimization. *Mathematical Programming, 45,* 503–528.

Liu, G. H., Han, J. Y., & Yin, H. X. (1995). Global convergence of the Fletcher-Reeves algorithm with an inexact line search. *Applied Mathematics—A Journal of Chinese Universities Series B, 10,* 75–82.

Liu, J. K., & Li, S. J. (2014). New hybrid conjugate gradient method for unconstrained optimization. *Applied Mathematics and Computation, 245,* 36–43.

Liu, H. W., & Liu, Z. X. (2019). An efficient Barzilai-Borwein conjugate gradient method for unconstrained optimization. *Journal of Optimization Theory and Applications, 180*(3), 879–906.

Liu, Y., & Storey, C. (1991). Efficient generalized conjugate gradient algorithms. Part 1: Theory. *Journal of Optimization Theory and Applications, 69,* 129–137.

Livieris, I. E., Karlos, S., Tampakas, V., & Pintelas, P. (2017). A hybrid conjugate gradient method based on self-scaled memoryless BFGS update. In *Proceedings of PCI 2017* (5 p), Larissa, Greece, September 28–30.

Livieris, I. E., & Pintelas, P. (2013). A new class of spectral conjugate gradient methods based on a modified secant equation for unconstrained optimization. *Journal of Computational and Applied Mathematics, 239,* 396–405.

Livieris, I. E., & Pintelas, P. (2016). A limited memory descent Perry conjugate gradient method. *Optimization Letters, 10*(8), 1725–1742.

Livieris, I. E., Tampakas, V., & Pintelas, P. (2018). A descent hybrid conjugate gradient method based on the memoryless BFGS update. *Numerical Algorithms, 79*(4), 1169–1185.

Luenberger, D. G. (1973). *Introduction to linear and nonlinear programming.* Reading: Addison-Wesley Publishing Company.

Luenberger, D. G. (1984). *Introduction to linear and nonlinear programming* (2nd ed.). Reading: Addison-Wesley Publishing Company.

Luenberger, D. G., & Ye, Y. (2016). *Linear and nonlinear programming. International Series in Operations Research & Management Science 228* (4th ed.). New York: Springer.

Lukšan, L. (1992). Computational experience with improved conjugate gradient methods for unconstrained optimization. *Kibernetika, 28*(4), 249–262.

Lukšan, L., Matonoha, C., & Vlcek, J. (2008). *Computational experience with modified conjugate gradient methods for unconstrained optimization* (Technical Report No. 1038). Institute of Computer Science, Academy of Sciences of the Czech Republic, December 2008.

McCormick, P., & Ritter, K. (1974). Alternative proofs of the convergence properties of the conjugate gradient method. *Journal of Optimization Theory and Application, 13*(5), 497–518.

McGuire, M. F., & Wolfe, P. (1973). *Evaluating a restart procedure for conjugate gradients* (Report RC-4382). IBM Research Center, Yorktown Heights.

Meyer, C. D. (2000). *Matrix analysis and applied linear algebra*. Philadelphia, PA, USA: SIAM.

Momeni, M., & Peyghami, M. R. (2019). A new conjugate gradient algorithm with cubic Barzilai-Borwein stepsize for unconstrained optimization. *Optimization Methods and Software, 34*(3), 650–664.

Morales, J. L., & Nocedal, J. (2002). Enriched methods for large-scale unconstrained optimization. *Computational Optimization and Applications, 21,* 143–154.

Moré, J. J. (1983). Recent developments in algorithms and software for trust region methods. In A. Bachen, M. Grötschel, & B. Korte (Eds.), *Mathematical programming: The state of the art* (pp. 258–287). Berlin: Springer.

Moré, J. J., & Sorensen, D. C. (1984). Newton's method. In G. H. Golub (Ed.), *Studies in numerical analysis* (pp. 29–82). Washington, D.C.: Mathematical Association of America.

Moré, J. J., & Thuente, D. J. (1990). *On the line search algorithms with guaranteed sufficient decrease*. Mathematics and Computer Science Division Preprint MCS-P153-0590, Argonne National Laboratory, Argonne.

Moré, J. J., & Thuente, D. J. (1994). Line search algorithms with guaranteed sufficient decrease. *ACM Transaction on Mathematical Software, 20,* 286–307.

Moré, J. J., & Toraldo, G. (1991). On the solution of large quadratic programming problems with bound constraints. *SIAM Journal on Optimization, 1,* 93–113.

Naiman, A. E., Babuska, I. M., & Elman, H. C. (1997). A note on conjugate gradient convergence. *Numerischke Mathematik, 76,* 209–230.

Narushima, Y., Wakamatsu, T., & Yabe, H. (2008). Extended Barzilai-Borwein method for unconstrained optimization problems. *Pacific Journal of Optimization, 6*(3), 591–614.

Narushima, Y., & Yabe, H. (2014). A survey of sufficient descent conjugate gradient methods for unconstrained optimization. *SUT Journal of Mathematics, 50,* 167–203.

Narushima, Y., Yabe, H., & Ford, J. A. (2011). A three-term conjugate gradient method with sufficient descent property for unconstrained optimization. *SIAM Journal on Optimization, 21,* 212–230.

Nash, S. G. (1985). Preconditioning of truncated-Newton methods. *SIAM Journal on Scientific and Statistical Computing, 6,* 599–616.

Nash, S. G., & Nocedal, J. (1991). A numerical study of the limited memory BFGS method and the truncated-Newton method for large-scale optimization. *SIAM Journal on Optimization, 1,* 358–372.

Navon, M. I., & Legler, D. M. (1987). Conjugate gradient methods for large-scale minimization in meteorology. *Monthly Weather Review, 115,* 1479–1502.

Nazareth, J. L. (1975). *A relationship between the BFGS and conjugate gradient algorithms*. Tech. Memo. ANL-AMD 282, Argonne National Laboratory, January 1976. Presented at the SIAM-SIGNUM Fall 1975 Meeting, San Francisco, CA.

Nazareth, J. L. (1977). A conjugate direction algorithm without line search. *Journal of Optimization Theory and Applications, 23,* 373–387.

Nazareth, J. L. (1979). A relationship between the BFGS and conjugate gradient algorithms and its implications for the new algorithms. *SIAM Journal on Numerical Analysis, 16*(5), 794–800.

Nazareth, J. L. (1986). Conjugate gradient methods less dependent on conjugacy. *SIAM Review, 28*(4), 501–511.

Nazareth, J. L. (1995). If quasi-Newton then why not quasi-Cauchy? endif. *SIAG/Opt Views-and-News, 6,* 11–14.

Nazareth, J. L. (1999). Conjugate gradient methods. In C. Floudas & P. Pardalos (Eds.), *Encyclopedia of optimization*. Boston: Kluwer Academic Publishers.

Nazareth, J. L. (2001). Conjugate gradient methods. In C. Floudas & P. Pardalos (Eds.), *Encyclopedia of optimization* (pp. 319–323). Boston: Kluwer Academic Press.

Nemirovsky, A. S., & Yudin, D. B. (1983). *Problem complexity and method efficiency in optimization. Interscience Series in Discrete Mathematics*. New York: Wiley.

Nesterov, Y., & Polyak, B. T. (2006). Cubic regularization of Newton's method and its global performance. *Mathematical Programming, 108,* 177–205.

Nitsche, J. C. C. (1989). *Lectures on minimal surfaces* (Vol. 1). Cambridge, UK: Cambridge University Press.

Nocedal, J. (1980). Updating quasi-Newton matrices with limited storage. *Mathematics of Computation, 35,* 773–782.

Nocedal, J. (1992). Theory of algorithms for unconstrained optimization. *Acta Numerica, 1,* 199–242.

Nocedal, J. (1996). Conjugate gradient methods and nonlinear optimization. In L. Adams & J. L. Nazareth (Eds.), *Linear and nonlinear conjugate gradient-related methods* (pp. 9–23). Philadelphia, PA, USA: SIAM.

Nocedal, J., & Wright, S. J. (2006). *Numerical optimization. Springer Series in Operations Research* (2nd ed.). New York: Springer Science + Business Media.

Nocedal, J., & Yuan, Y. X. (1993). Analysis of self-scaling quasi-Newton method. *Mathematical Programming, 61,* 19–37.

Noether, E. (1918). Invariante variations probleme. *Nachrichten der Könighche Gessellschaft der Wissenschaften zu Göttingen, Mathematisch-Physikalische Klasse*, 235–257. [Noether, E. (1971). Invariant variation problems. *Transport Theory and Statistical Physics, 1*(3), 186–207].

O'Leary, D. P., & Yang, W. H. (1978). Elastoplastic torsion by quadratic programming. *Computer Methods in Applied Mechanics and Engineering, 16,* 361–368.

Oren, S. S. (1972). *Self-scaling variable metric algorithms for unconstrained optimization* (Ph.D. thesis). Department of Engineering-Economic Systems, Stanford University, Stanford.

Oren, S. S. (1974). Self-scaling variable metric algorithm. Part II. *Management Science, 20,* 863–874.

Oren, S. S., & Luenberger, D. G. (1974). Self-scaling variable metric (SSVM) algorithms. Part I: Criteria and sufficient conditions for scaling a class of algorithms. *Management Science, 20,* 845–862.

Oren, S. S., & Spedicato, E. (1976). Optimal conditioning of self-scaling variable metric algorithm. *Mathematical Programming, 10,* 70–90.

Ortega, J. M., & Rheinboldt, W. C. (1970). *Iterative solution of nonlinear equations in several variables*. New York: Academic Press.

Ou, Y., & Liu, Y. (2017). A memory gradient method based on the nonmonotone technique. *Journal of Industrial and Management Optimization, 13*(2), 857–872.

Peressini, A. L., Sullivan, F. E., & Uhl, J. J. (1988). *The mathematics of nonlinear programming*. New York: Springer.

Perry, A. (1976). *A modified conjugate gradient algorithm*. Discussion paper No. 229, Center for Mathematical Studies in Economics and Management Science, Northwestern University.

Perry, A. (1977). *A class of conjugate gradient algorithms with two step variable metric memory*. Discussion paper 269, Center for Mathematical Studies in Economics and Management Science. Northwestern University, Il, USA.

Polak, E., & Ribiére, G. (1969). Note sur la convergence de méthods de direction conjugées. *Revue Francaise d'Informatique et de Recherche Opérationnelle, 16,* 35–43.

Polyak, B. T. (1969). The conjugate gradient method in extremal problems. *USSR Computational Mathematics and Mathematical Physics, 9,* 94–112.

Potra, F. A. (1989). On Q-order and R-order of convergence. *Journal of Optimization Theory and Applications, 63*(3), 415–431.

Potra, F. A., & Shi, Y. (1995). Efficient line search algorithm for unconstrained optimization. *Journal of Optimization Theory and Applications, 85,* 677–704.

Powell, M. J. D. (1970). A new algorithm for unconstrained optimization. In J. B. Rosen, O. L. Mangasarian, & K. Ritter (Eds.), *Nonlinear programming* (pp. 31–66). New York: Academic Press.

Powell, M. J. D. (1975). Convergence properties of a class of minimization algorithms. In O. L. Mangasarian, R. R. Meyer, & S. M. Robinson (Eds.), *Nonlinear programming* (2nd ed., pp. 1–27). New York: Academic Press.

Powell, M. J. D. (1976a). Some global convergence properties of a variable-metric algorithm for minimization without exact line searches. In R. W. Cottle & C. E. Lemke (Eds.), *Nonlinear Programming, SIAM-AMS Proceedings* (Vol. 9, pp. 53–72), Philadelphia, PA, USA.

Powell, M. J. D. (1976b). Some convergence properties of the conjugate gradient method. *Mathematical Programming, 11,* 42–49.

Powell, M. J. D. (1977). Restart procedures of the conjugate gradient method. *Mathematical Programming, 2,* 241–254.

Powell, M. J. D. (1983). *On the rate of convergence of variable metric algorithms for unconstrained optimization* (Report DAMTP 1983/NA7). Department of Applied Mathematics and Theoretical Physics, University of Cambridge, Cambridge, UK.

Powell, M. J. D. (1984a). Nonconvex minimization calculations and the conjugate gradient method. In D. F. Griffiths (Ed.), *Numerical analysis* (Dundee, 1983). *Lecture Notes in Mathematics* (Vol. 1066, pp. 122–141).

Powell, M. J. D. (1984b). On the global convergence of trust-region algorithm for unconstrained optimization. *Mathematical Programming, 29,* 297–303.

Powell, M. J. D. (1986a). How bad are the BFGS and DFP methods when the objective function is quadratic? *Mathematical Programming, 34,* 34–47.

Powell, M. J. D. (1986b). Convergence properties of algorithms for nonlinear optimization. *SIAM Review, 28*(4), 487–500.

Powell, M. J. D. (1987). Updating conjugate directions by the BFGS formula. *Mathematical Programming, 38,* 693–726.

Pytlak, R. (2009). *Conjugate gradient algorithms in nonconvex optimization. Nonconvex Optimization and Its Applications* (Vol. 89). Berlin, Heidelberg: Springer.

Raydan, M. (1997). The Barzilai and Borwein gradient method for the large scale unconstrained minimization problem. *SIAM Journal on Optimization, 7,* 26–33.

Reid, J. K. (1971). On the method of conjugate gradients for the solution of large sparse systems of linear equations. In J. K. Reid (Ed.), *Large sparse sets of linear equations* (pp. 231–254). New York, London: Academic Press.

Ritter, K. (1980). On the rate of superlinear convergence of a class of variable metric methods. *Numerische Mathematik, 35,* 293–313.

Rosser, J. B. (1953). Rapidly converging iterative methods for solving linear equations. In L. J. Paige & O. Taussky (Eds.), *Simultaneous linear equations and the determination of eigenvalues. Applied Mathematics Series 29* (pp. 59–64). Washington, D.C.: National Bureau of Standards, U.S. Government Printing Office.

Saad, Y. (2003). *Iterative methods for sparse linear systems.* Philadelphia, PA, USA: SIAM.

Schlick, T., & Fogelson, A. (1992a). TNPACK—A truncated Newton minimization package for large-scale problems: I Algorithm and usage. *ACM Transactions on Mathematical Software, 18,* 46–70.

Schlick, T., & Fogelson, A. (1992b). TNPACK—A truncated Newton minimization package for large-scale problems: II Implementation examples. *ACM Transactions on Mathematical Software, 18,* 71–111.

Schnabel, R. B., & Eskow, E. (1999). A revised modified Cholesky factorization algorithm. *SIAM Journal on Optimization, 9*(4), 1135–1148.

Schuller, G. (1974). On the order of convergence of certain quasi-Newton methods. *Numerische Mathematik, 23,* 181–192.

Shanno, D. F. (1970). Conditioning of quasi-Newton methods for function minimization. *Mathematics of Computation, 24,* 647–656.

Shanno, D. F. (1978a). Conjugate gradient methods with inexact searches. *Mathematics of Operations Research, 3,* 244–256.

Shanno, D. F. (1978b). On the convergence of a new conjugate gradient algorithm. *SIAM Journal on Numerical Analysis, 15,* 1247–1257.

Shanno, D. F. (1980). Quadratic termination of conjugate gradient algorithms. In A. V. Fiacco & K. O. Kortanek (Eds.), *Extremal methods and systems analysis* (pp. 433–441). New York: Springer.

Shanno, D. F. (1983). *CONMIN—A Fortran subroutine for minimizing an unconstrained nonlinear scalar valued function of a vector variable x either by the BFGS variable metric algorithm or by a Beale restarted conjugate gradient algorithm.* Private communication, October 17, 1983.

Shanno, D. F. (1985). Globally convergent conjugate gradient algorithms. *Mathematical Programming, 33,* 61–67.

Shanno, D. F., & Phua, K. H. (1976). Algorithm 500. Minimization of unconstrained multivariable functions. *ACM Transactions on Mathematical Software, 2,* 87–94.

Shanno, D. F., & Phua, K. H. (1978). Matrix conditioning and nonlinear optimization. *Mathematical Programming, 14,* 149–160.

Shanno, D. F., & Phua, K. H. (1980). Remark on algorithm 500. *ACM Transactions on Mathematical Software, 6,* 618–622.

Shen, W. (2008). *Conjugate gradient methods. Lecture Notes.* Pennsylvania State University, MATH 524 Numerical Analysis II, Spring 2008.

Stiefel, E. L. (1958). Kernel polynomials in linear algebra and their numerical applications. In *Further contributions to the determination of eigenvalues. Applied Mathematical Series* (Vol. 49, pp. 1–22). National Bureau of Standards.

Stoer, J. (1977). On the relation between quadratic termination and convergence properties of minimization algorithms. *Numerische Mathematik, 28,* 343–366.

Stoer, J., & Yuan, Y. X. (1995). A subspace study on conjugate gradient algorithms. *ZAMM—Journal of Applied Mathematics and Mechanics, 75,* 69–77.

Strakoš, Z. (1991). On the real convergence rate of the conjugate gradient method. *Linear Algebra and its Applications, 154–156,* 535–549.

Sun, W., & Yuan, Y. X. (2006). *Optimization theory and methods. Nonlinear Programming.* New York: Springer Science + Business Media.

Sun, J., & Zhang, J. (2001). Global convergence of conjugate gradient methods without line search. *Annals of Operations Research, 163,* 161–173.

Touati-Ahmed, D., & Storey, C. (1990). Efficient hybrid conjugate gradient techniques. *Journal of Optimization Theory and Applications, 64,* 379–397.

Trefethen, L., & Bau, D. (1997). *Numerical linear algebra.* Philadelphia, PA, USA: SIAM.

Trefethen, L., & Schreiber, R. (1990). Average case analysis of Gaussian elimination. *SIAM Journal on Matrix Analysis and Applications, 11,* 335–360.

Van der Vorst, H. A. (1993). *Lecture Notes on Iterative Methods.* Report Mathematical Institute, University of Utrecht.

Wan, Z., Huang, S., & Zheng, X. D. (2012). New cautious BFGS algorithm based on modified Armijo-type line search. *Journal of Inequalities and Applications, 241,* 1–10.

Wan, Z., Teo, K. L., Shen, X. L., & Hu, C. M. (2014). New BFGS method for unconstrained optimization problem based on modified Armijo line search. *Optimization, 63*(2), 285–304.

Wang, H. J., & Yuan, Y. X. (1992). A quadratic convergence method for one-dimensional optimization. *Chinese Journal of Operations Research, 11,* 1–10.

Wang, T., Liu, Z., & Liu, H. (2019). A new subspace minimization conjugate gradient method based on tensor model for unconstrained optimization. *International Journal of Computer Mathematics, 96*(10), 192401942.

Wang, Z. H., & Yuan, Y. X. (2006). A subspace implementation of quasi-Newton trust region methods for unconstrained optimization. *Numerische Mathematik, 104*(2), 241–269.

Watkins, D. S. (2002). *Fundamentals of matrix computation* (2nd ed.). New York: Wiley.

Wei, Z., Li, G., & Qi, L. (2006a). New quasi-Newton methods for unconstrained optimization problems. *Applied Mathematics and Computation, 175*(2), 1156–1188.

Wei, Z., Li, G., & Qi, L. (2006b). New nonlinear conjugate gradient formulas for large-scale unconstrained optimization problems. *Applied Mathematics and Computation, 179,* 407–430.

Wei, Z., & Yang, W. H. (2016). A Riemannian subspace limited-memory SR1 trust-region method. *Optimization Letters, 10,* 1705–1723.

Wei, Z., Yao, S., & Liu, L. (2006). The convergence properties of some new conjugate gradient methods. *Applied Mathematics and Computation, 183,* 1341–1350.

Wei, Z., Yu, G., Yuan, G., & Lian, Z. (2004). The superlinear convergence of a modified BFGS-type method for unconstrained optimization. *Computational Optimization and Applications, 29,* 315–332.

Wilkinson, J. H. (1965). *The algebraic eigenvalue problem.* London: Oxford University Press.

Winfield, D. (1969). *Function and functional optimization by interpolation in data tables* (Ph.D. thesis). Harvard University, Cambridge, USA.

Winther, R. (1980). Some superlinear convergence results for the conjugate gradient method. *SIAM Journal on Numerical Analysis, 17,* 14–17.

Wolfe, P. (1969). Convergence conditions for ascent methods. *SIAM Review, 11,* 226–235.

Wolfe, P. (1971). Convergence conditions for ascent methods. II: Some corrections. *SIAM Review, 13,* 185–188.

Wong, J. C. F., & Protas, B. (2013). Application of scaled nonlinear conjugate-gradient algorithms to the inverse natural convection problem. *Optimization Methods and Software, 28*(1), 159–185.

Wu, G., & Liang, H. (2014). A modified BFGS method and its convergence. *Computer Modelling & New Technologies, 18*(11), 43–47.

Xu, C., & Zhang, J. Z. (2001). A survey of quasi-Newton equations and quasi-Newton methods for optimization. *Annals of Operations Research, 103,* 213–234.

Yabe, H., Martínez, H. J., & Tapia, R. A. (2004). On sizing and shifting the BFGS update within the sized Broyden family of secant updates. *SIAM Journal on Optimization, 15*(1), 139–160.

Yabe, H., Ogasawara, H., & Yoshino, M. (2007). Local and superlinear convergence of quasi-Newton methods based on modified secant conditions. *Journal of Computational and Applied Mathematics, 205*, 717–632.

Yabe, H., & Sakaiwa, N. (2005). A new nonlinear conjugate gradient method for unconstrained optimization. *Journal of the Operations Research Society of Japan, 48*(4), 284–296.

Yabe, H., & Takano, M. (2004). Global convergence properties of nonlinear conjugate gradient methods with modified secant condition. *Computational Optimization and Applications, 28,* 203–225.

Yao, S., Wei, Z., & Huang, H. (2007). A note about WYL's conjugate gradient method and its application. *Applied Mathematics and Computation, 191,* 381–388.

Yang, X., Luo, Z., & Dai, X. (2013). A global convergence of LS-CD hybrid conjugate gradient method. *Advances in Numerical Analysis, 2013*, Article ID 517452. https://doi.org/10.1155/2013/517452.

Yang, Y. T., Chen, Y. T., & Lu, Y. L. (2017). A subspace conjugate gradient algorithm for large-scale unconstrained optimization. *Numerical Algorithms, 76*(3), 813–828.

Yuan, G., Sheng, Z., Wang, B., Hu, W., & Li, C. (2018). The global convergence of a modified BFGS method for nonconvex functions. *Journal of Computational and Applied Mathematics, 327,* 274–294.

Yuan, G., & Wei, Z. (2010). Convergence analysis of a modified BFGS method on convex minimizations. *Computational Optimization and Applications, 47,* 237–255.

Yuan, G., Wei, Z., & Lu, X. (2017). Global convergence of BFGS and PRP methods under a modified weak Wolfe-Powell line search. *Applied Mathematical Modelling, 47,* 811–825.

Yuan, Y. X. (1991). A modified BFGS algorithm for unconstrained optimization. *IMA Journal of Numerical Analysis, 11,* 325–332.

Yuan, Y. X. (1993). Analysis on the conjugate gradient method. *Optimization Methods and Software, 2,* 19–29.

Yuan, Y. X. (1998). *Problems on convergence of unconstrained optimization algorithms* (Report No. ICM-98-028), April 1998, 1–12.

Yuan, Y. X. (2014). A review on subspace methods for nonlinear optimization. In S. Y. Jang, Y. R. Kim, D.-W. Lee, & I. Yie (Eds.), *Proceedings of the International Congress of Mathematics* (pp. 807–827), Seoul 2014.

Yuan, Y. X. (2015). Recent advances in trust region algorithms. *Mathematical Programming, Series B, 151,* 249–281.

Yuan, Y. X., & Byrd, R. (1995). Non-quasi-Newton updates for unconstrained optimization. *Journal of Computational Mathematics, 13*(2), 95–107.

Zhang, L. (2009a). Two modified Dai-Yuan nonlinear conjugate gradient methods. *Numerical Algorithms, 50*(1), 1–16.

Zhang, L. (2009b). New versions of the Hestenes-Stiefel nonlinear conjugate gradient method based on the secant condition for optimization. *Computational & Applied Mathematics, 28*(1), 1–23.

Zhang, H., & Hager, W. W. (2004). A nonmonotone line search technique and its application to unconstrained optimization. *SIAM Journal on Optimization, 14,* 1043–1056.

Zhang, J., Deng, N. Y., & Chen, L. H. (1999). New quasi-Newton equation and related methods for unconstrained optimization. *Journal on Optimization Theory and Applications, 102,* 147–167.

Zhang, J., Xiao, Y., & Wei, Z. (2009). Nonlinear conjugate gradient methods with sufficient descent condition for large-scale unconstrained optimization. *Mathematical Problems in Engineering, 2009,* Article ID 243290. https://doi.org/10.1155/2009/243290.

Zhang, J., & Xu, C. (2001). Properties and numerical performance of quasi-Newton methods with modified quasi-Newton equations. *Journal of Computational and Applied Mathematics, 137,* 269–278.

Zhang, L., & Zhou, W. (2008). Two descent hybrid conjugate gradient methods for optimization. *Journal of Computational and Applied Mathematics, 216,* 251–264.

Zhang, L., Zhou, W., & Li, H. (2006a). A descent modified Polak-Ribière-Polyak conjugate gradient method and its global convergence. *IMA Journal of Numerical Analysis, 26*(4), 629–640.

Zhang, L., Zhou, W., & Li, H. (2006b). Global convergence of a modified Fletcher-Reeves conjugate gradient method with Armijo-type line search. *Numerische Mathematik, 104*(4), 561–572.

Zhang, L., Zhou, W., & Li, H. (2007). Some descent three-term conjugate gradient methods and their global convergence. *Optimization Methods and Software, 22*(4), 697–711.

Zhang, L., & Zhou, Y. (2012). A note on the convergence properties of the original three-term Hestenes-Stiefel method. *AMO—Advanced Modeling and Optimization, 14,* 159–163.

Zhao, T., Liu, H., & Liu, Z. (2019). New subspace minimization conjugate gradient methods based on regularization model for unconstrained optimization. *Numerical Algorithms*, Optimization online, OO Digest: April 2020, http://www.optimization-online.org/DB_HTML/2020/04/7720.html.

Zhou, W., & Zhang, L. (2006). A nonlinear conjugate gradient method based on the MBFGS secant condition. *Optimization Methods and Software, 21*(5), 707–714.

Zhu, M., Nazareth, J. L., & Wolkowicz, H. (1999). The quasi-Cauchy relation and diagonal updating. *SIAM Journal on Optimization, 9*(4), 1192–1204.

Zhu, H., & Wen, S. (2006). A class of generalized quasi-Newton algorithms with superlinear convergence. *International Journal of Nonlinear Science, 2*(3), 140–146.

Zoutendijk, G. (1970). Nonlinear programming, computational methods. In J. Abadie (Ed.), *Integer and nonlinear programming* (pp. 38–86). Amsterdam: North-Holland.

Author Index

N. Andrei, *Nonlinear Conjugate Gradient Methods for Unconstrained Optimization*, Springer Optimization and Its Applications 158,
https://doi.org/10.1007/978-3-030-42950-8

Subject Index

N. Andrei, *Nonlinear Conjugate Gradient Methods for Unconstrained Optimization*,
Springer Optimization and Its Applications 158,
https://doi.org/10.1007/978-3-030-42950-8

The manufacturer's authorised representative in the EU is Springer Nature Customer Service Centre GmbH, Europaplatz 3, 69115 Heidelberg, Germany. If you have any concerns regarding our products, please contact ProductSafety@springernature.com

Printed and bound by CPI Group (UK) Ltd, Croydon, CR0 4YY
05/10/2026
02228126-0014